CZECHOSLOVAK ACADEMY OF SCIENCES

# PROGRESS
# IN SOIL
# ZOOLOGY

CZECHOSLOVAK ACADEMY OF SCIENCES


*Scientific Editor:*
Ing. Jan Vaněk, CSc.


*Scientific Advisers:*
Doc. dr. Miroslav Kunst, CSc.
Dr. Josef Rusek, CSc.


*Graphic Design by*
Miroslav Houska

# PROGRESS IN SOIL ZOOLOGY

PROCEEDINGS OF THE 5th INTERNATIONAL
COLLOQUIUM ON SOIL ZOOLOGY
HELD IN PRAGUE SEPTEMBER 17–22, 1973

*Organized by the Soil Zoology Committee*
*of the International Society of Soil Science*
*and by the Czechoslovak Academy of Sciences —*
*Institute of Landscape Ecology*

## Edited by
## JAN VANĚK

Springer-Science+
Business Media, B.V.

ACADEMIA, Publishing House of the
Czechoslovak Academy of Sciences
PRAGUE

1975

ISBN 978-94-010-1935-4          ISBN 978-94-010-1933-0   (eBook)
DOI 10.1007/978-94-010-1933-0

© ACADEMIA, Publishing House of the Czechoslovak Academy of Sciences,
Prague 1975
Originally published by W. Junk B.V., Publishers in 1975
Softcover reprint of the hardcover 1st edition 1975

# Introduction

L a d i e s    a n d    g e n t l e m e n ,

   I have the pleasure to welcome you here in Prague in the name
of the Czechoslovak Academy of Sciences and to open the Fifth International Colloquium on Soil Zoology. We are very glad that Czechoslovakia was chosen for this important meeting.
   It is clear to all of us that the soil plays and will play a decisive part in providing food for the explosive increase of human population. For this reason we watch with great anxiety the negative
influence of human activities on the environment accompanied also by
the other destructive intervention into the soil ecosystem, its devas-
tation by inefficient management, application of herbicides and pesticides pollution by the waste products of industry and human settlements.
The basis for solving these accumulating and now sometimes latent problems is among others a good knowledge of the role of soil organisms in
the cycles of materials and in the energy flow. Soil zoology as a part
of soil biology is still at the beginning of this trend The lack of information about life in soil is obvious when compared with the results
of a related biological science dealing with the water ecosystem.
In spite of the fact that hydrobiology and soil zoology are almost of
the same age as ecological disciplines if we take into consideration
the publication of Darwin´s famous work on the formating of the
vegetable mould through the action of worms, hydrobiology grew up into
a more advanced stage of causal analysis and modelling of biological
processes. Slower development of the soil biology and especially of
soil zoology was influenced by methodical difficulties coming from the
heterogeneity and daunting complexity of investigated environment and
last but not least by the predominance of the structural, physical and
chemical approach to soil science. But nevertheless we can observe
in the past years a certain qualitative growth of our knowledge of the
edaphon. We have at our disposal data about its composition, num-

bers, biomass, relations  to different soil types,  plant associations,
about internal  and  external forces  working in or on soil animal communities  as well  as data on food chains  and  energy flow.  The first
practical applications of results of soil zoology appear  also in agriculture, forestry and in nature conservation.  We suppose that the positive contribution is given here, among others, by the quicker exchange
of information, the presentation of experiences  from the more advanced
laboratories to the starting ones and other stimulating impulses gained
in symposia on soil zoology.  We hope this jubilee meeting held fifteen
years after the first colloquium  in Herpenden  will also contribute to
the more rapid development  of our investigations, to higher quality of
our approach  to the problems  and  to establishing new productive con-
nections  between single workers or institutions.  It seems to us  that
this last aspect is particularly important. Regarding the rather limited number of soil zoologists I belive  we can achieve a thorough accel-
eration of the research in long-term and far-ranging programs above all
through a close international cooperation.

I wish  this meeting much succes  especially in this point and all
participiants an agreeable stay in our country.

*Miroslav Kunst*

# Contents

SECTION B: INFLUENCE OF ABIOTIC AND BIOTIC
FACTORS ON COMMUNITIES OF SOIL ORGANISMS

SECTION D: MODERN METHODS OF INVESTIGATING SOIL
ORGANISMS AND THEIR INFLUENCE UPON SOIL PROPERTIES

# Die Geschichte der Bodenzoologie und ihre Einbeziehung in die bodenkundliche Forschung

H. FRANZ

**Bodenforschung und Baugeologie,**
**Wien**

Die Tiere waren imstande, einen breiteren Raum der Biosphäre der Erde zu erobern als die höheren Pflanzen, weil sie in Biotopen zu leben vermögen, die das Sonnenlicht nicht erreicht. Sie besiedeln die Meere bis zu den grössten Tiefen, Tiere leben in Höhlen und Spalten der festen Erdrinde und auch in Böden bis zu deren tiefsten Schichten.

Seit den Anfängen der Erforschung der Tierwelt ist bekannt, dass es tierische Organismen gibt, die einen grossen Teil, ja sogar ihr ganzes Leben im Boden zubringen. Allmählich wurde eine sehr grosse Zahl terrikoler Tierarten aus vielen Ordnungen des Tierreiches beschrieben, ohne dass bis zum heutigen Tage eine erschöpfende Kenntnis der gesamten Mannigfaltigkeit der Bodenfauna erreicht worden wäre. Es werden vielmehr alljährlich immer noch viele hunderte bisher unbeschriebene terrikole Arten entdeckt und unsere Kenntnis einzelner Gruppen ist bis heute so lückenhaft, dass eine exakte Erforschung aller an einem bestimmten Standort im Boden lebenden Tierformen unmöglich ist. Es steht ausser Zweifel, dass in keinem anderen Bereich der Zoologie heute auf taxonomischem Gebiete noch so viel Arbeit zu leisten ist, wie auf dem der Bodenfauna. Daraus erklärt sich, dass sich die Bodenzoologie lange Zeit hindurch vorwiegend mit der Beschreibung und taxonomischen Ordnung der Arten befasste und darüber hinaus höchstens die mehr oder weniger extreme Anpassung derselben an ihren Lebensraum eingehender studierte.

Diese Anpassung ist in der Tat in vielen Fällen ausserordentlich weitgehend und in die Augen fallend. Die meisten Bodentiere sind depigmentiert, da sie der Pigmentierung zum Schutz gegen Lichteinwirkung nicht bedürfen. Sie sind mikrophthal oder überhaupt blind, weil sie in den dunklen Hohlräumen des Bodens des Sehvermögens nicht bedürfen und dieses verloren haben. Sie haben mehr oder weniger stark reduzierte oder als Grabwerkzeuge umgebildete Extremitäten und etwa primär vorhandene Flügel haben eine mehr oder weniger weitgehende Reduktion erfahren. Ständig in feuchten Bodenhohlräumen lebende Arten bedürfen des Transpi-

rationsschutzes nicht, sie haben ihn weitgehend verloren und vertrock-
nen rasch, wenn sie aus dem Boden in nicht mit Wasserdampf gesättigte
Luft gebracht werden. Es ist klar, dass derart hochspezialisierte Organismen den Boden nicht verlassen können und demgemäss ein sehr beschränktes Ausbreitungsvermögen besitzen. Eine weitere Folge dieses
Umstandes ist es, dass die hochspezialisierten terrikolen Tiere zumeist
nur eine sehr beschränkte Verbreitung aufweisen und sich die Artenzusammensetzung der Bodenfauna auf geringer Distanz ändert. Auch dies
stellt naturgemäss eine wesentliche Erschwerung für alle bodenzoologischen Untersuchungen dar.

Auf diese Umstände hinzuweisen ist notwendig, um verständlich zu
machen warum es relativ lange gebraucht hat, bis sich die Bodenzoologie
mit den Wechselbeziehungen zwischen dem Bodenleben und dem Boden zu be-
fassen begann. Dazu kam als weiterer Faktor, dass in der Biologie die
Wirkung der Umwelt auf die Organismen auf Grund der in vielen Fällen
sehr auffälligen Anpassungserscheinungen zwar schon früh untersucht
wurde, es jedoch wesentlich länger brauchte, bis man erkannte, dass
auch die Organismen ihrerseits ihren Lebensraum beeinflussen und das
unter Umständen in einem solchen Ausmasse, dass dieser sein Aussehen
und seine Beschaffenheit völlig ändert.

Dieses Erkenntnis bildete aber die Voraussetzung dafür, dass man
die Bedeutung der Bodenorganismen für die Bildung und Entwicklung der
Böden und ihrer Eigenschaften richtig einzuschätzen vermochte, sie bildet die Basis der modernen Bodenbiologie, die sich in den letzten Jahrzehnten zu einem bedeutenden Zweig der Bodenkunde entwickelt hat. Einzelne dem Stande der Forschung ihrer Zeit vorauseilende Forscher haben
allerdings gewisse Leistungen der Bodentiere, vor allem deren Bedeutung
für die mechanische Durchmischung der Böden, früh erkannt. Es sei in
diesem Zusammenhang auf die Untersuchungen von V. Hausen und Ch. Darwin
über die von bestimmten Regenwürmern bewirkte Bodendurchmischung, auf
die Beobachtung russischer Forscher über die Bildung der Krotowinen
in den Steppenböden und nicht zuletzt auf die grundlegenden Forschungen
von H. v. Post (1861-62) und P.E. Müller (1887) über die Bildung der
Humusformen hingewiesen.

Die Gesamtheit der Leistungen der Bodenorganismen wurde in ihrer
Bedeutung für die Bodenbildung erst viel später, zuerst wohl von R.
Francé (1912) erkannt und zugleich durch populär gehaltene Schriften
einem breiten Leserkreis zur Kenntnis gebracht. Trotzdem blieb auch
nachher noch die Bodenzoologie durch Jahrzehnte nahezu ausschliesslich
ein Teilgebiet der zoologischen Forschung und die Anwendung der Bodenbiologie auf die Bodenkunde auf einzelne Teilgebiete beschränkt. Das
war die Folge des Umstandes, dass die Bodenzoologie fast nur von Biolo-

gen, die herkömmliche Bodenkunde aber von biologisch nicht speziell
ausgebildeten Pedologen betrieben wurde. Erst im 2. Weltkrieg und unmittelbar danach erschien eine Reihe von Arbeiten, in denen die Zusammenhänge zwischen dem biologischen Geschehen im Boden und bestimmten
Prozessen der Bodengenese klar herausgearbeitet wurden. Einen wesentlichen Beitrag auf diesem Gebiet stellen die bodenmikromorphologischen
Arbeiten von W. Kubiena (1938) dar, wie die Bodenmikromorphologie auch
für die Bodenzoologie wichtige neue ökologische Erkenntnisse erbrachte.
Auch die Humusmorphologie wurde durch die Arbeiten Kubienas sowie im
Anschluss an diese durch F. Hartmann (1952) wesentlich gefördert. Dass
die Regenwürmer über die mechanische Bodendurchmischung hinaus wichtige
Leistungen im Boden und bei der Zersetzung organischer Reste vollbrin-
gen, haben L. Mayer und A. Fink in Spezialuntersuchungen gezeigt.
Schliesslich konnte ich selbst in einer Reihe von Arbeiten den unmittelbaren Zusammenhang zwischen der Tätigkeit der Bodenorganismen und
der Bodenfruchtbarkeit unter Beweis stellen.

Da zugleich die bodenzoologische Fachliteratur rasch anwuchs, ergab
sich das Bedürfnis, die internationale Zusammenarbeit nicht bloss auf
dem Gebiet der Bodenzoologie selbst, sondern auch zwischen dieser und
den übrigen Zweigen der pedologischen Forschung zu organisieren.

Ich ergriff deshalb die Initiative und bat eine Reihe von Bodenzoologen, für den VI. Internationalen Bodenkundlichen Kongress in Paris
Referate anzumelden, um durch diese breiten Kreise der bodenkundlichen
Fachwelt mit dem derzeitigen Stand der bodenzoologischen Forschung und
deren Bedeutung für die Pedologie bekannt zu machen.

Die am Pariser Kongress anwesenden Bodenzoologen stellten den Antrag auf Gründung eines ständigen bodenzoologischen Komitees innerhalb
der Internationalen Bodenkundlichen Gesellschaft und begründeten ihn
mit dem Hinweis auf die Notwendigkeit eines engen Kontaktes der Bodenzoologischen Forschung mit allen übrigen Teilgebieten der Bodenforschung und mit der Feststellung, dass die Internationale Bodenkundliche
Gesellschaft hiefür das zuständige Gremium sei. Der Antrag fand vor
allem die Unterstützung französischer Kollegen, allen voran J. d´Aquillar
und B. Trouvellot, sogleich aber auch das Interesse des Generalsekre-
tärs der Gesellschaft, Prof. Dr. F.A. van Baren. Seinen Bemühungen war
es zu danken, dass der VI. Internationale Bodenkundliche Kongress die
Gründung eines bodenzoologischen Komitees beschloss und dieses der
Kommission III der Internationalen Bodenkundlichen Gesellschaft ein-
gliederte. Mir wurde die Ehre zuteil, zum ersten Präsidenten dieses Komitees bestellt zu werden.

Auf Grund dieses Beschlusses bildeten die auf dem Kongress anwesenden Bodenzoologen ein Exekutivkomitee bestehend aus den Herren

J. d´Aquilar, Prof. Dr. H.R. Debauche, Prof. Dr. H. Franz, Dr. H. Gisin
und Dr. P.W. Murphy. Das Komitee fasste die folgenden Beschlüsse:

1. Die internationale Zusammenarbeit auf dem Gebiet der Bodenzoologie sollte in Hinkunft durch periodische, in Abständen von 4 Jahren
abzuhaltende Kolloquien gefördert werden, zu denen nicht bloss alle auf
dem Gebiet der Bodenzoologie tätigen Forscher, sondern auch die Boden-
mikrobiologen eingeladen werden sollten. Die Tagungen sollten im Rahmen
der Tätigkeit der Kommission III der Internationalen Bodenkundlichen
Gesellschaft veranstaltet werden.

2. Zur Förderung der gegenseitigen Information aller Bodenbiologen
über eben veröffentlichte einschlägige Arbeiten und im Gange befind-
liche Untersuchungen wurde beschlossen ein Internationales Mitteilungs-
blatt (Bulletin International d´Information) herauszubringen, das zweimal jährlich erscheinen und einen mikrobiologischen und einen bodenzoologischen Teil umfassen sollte. Die Schriftleitung des bodenzoologischen Teils übernahm in dankenswerter Weise J. d´Aquilar. Seiner re-
daktionellen Tätigkeit, die er bis zum Ende des Jahres 1971 beibehielt,
ist es zu danken, dass dieses Blatt in den folgenden Jahren die gegen-
seitige Information der bodenzoologischen Forscher und ihre Zusammenarbeit wesentlich förderte. Die UNESCO hat die Herausgabe diese Blattes
bis zum Jahre 1971 mittels eines Druckkostenbeitrages unterstützt.

3. Als Tagungsort für das erste Kolloquium des bodenzoologischen
Komitees wurde auf Vorschlag von Dr. P.W. Murphy, Rothamsted in England,
ins Auge gefasst und es erging auch bald darauf von der Rothampsted
Experimental Station die offizielle Einladung zur Abhaltung der Tagung
an dieser traditionellen Forschungsstätte.

Das 1. Kolloquium fand vom 10. bis 14. Juli 1958 statt und hatte
die Diskussion der Untersuchungsmethoden der Bodenzoologie in einem
sehr weit gespannten Rahmen zum Gegenstande. Es wurde bewusst darauf
Bedacht genommen, die Verbindung zur Bodenmikrobiologie, aber auch zu
den übrigen Zweigen der pedologischen Forschung herzustellen. Den Teil-
nehmern war Gelegenheit geboten, die Einrichtungen und laufenden Forschungen der Rothamsted Experimental Station in allen Teilgebieten
kennenzulernen, wobei sich reichlich Gelegenheit ergab, die enge Ver-
flechtung der bodenbiologischen Forschung mit allen anderen Fachgebie-
ten der Pedologie und mit der praktischen Bodenwirtschaft zu diskutie-
ren. Von den Vorträgen wurde leider nur ein Teil publiziert, er erschien
in Buchform unter dem Titel Progress in Soil Zoology, London 1962,
Butterworths XVIII u. 398 S. hg. von P.W. Murphy.

Bei dem Kolloquium in Rothamsted wurde Dr. H. Gisin (Genf) für die
folgenden 4 Jahre zum Präsidenten des Internationalen Bodenzoologischen
Komitees gewählt und beschlossen, das nächste Kolloquium dem Thema der

Wechselbeziehungen   zwischen Bodenfauna und Bodenmikroflora   zu widmen.
Als Tagungsort   wurde Oosterbeck bei Arnhem, Niederland, vorgesehen und
Herr Dr. J. van der Drift  gebeten,  die Vorbereitungen  der Tagung  zu
übernehmen.

Das 2. Kolloquium  fand in der Zeit vom 10. bis 16. September 1962
statt, es nahmen daran 112 Forscher  aus 18 Ländern, davon 32 Mikrobiologen und 60 Bodenzoologen teil, die übrigen Teilnehmer  waren in anderen Forschungsbereichen tätige Bodenkundler und interessierte Landwirte.
Die Verhandlungen des Kolloquiums  erschienen in Buchform unter dem Titel: "Soil Organisms,  Amsterdam 1963, North — Holland, Publishing Company, VIII u. 453 S. hg. von J. Docksen und J. van der Drift". Der Band
umfasst  alle  gehaltenen Vorträge  in  die 4 Fachgebiete  gegliedert:
1. Activities of Soil Fauna,  2. Activities of Soil Microflora,  3. Interrelationships between Soil Fauna and Soil Microflora und 4. Biocenological Studies on Soil Fauna and Soil Microflora.

Zum Präsidenten des Komitees  für die folgenden 4 Jahre  wurde Dr.
J. van der Drift gewählt.

Das 3. Internationale Bodenzoologische Kolloquium wurde in der Zeit
vom 5. bis 10. September 1966  auf Einladung  der Forschungsanstalt für
Landwirtschaft, Braunschweig - Volkenrode, abgehalten.  Es nahmen daran
127 Forscher teil. Die 60 dabei gehaltenen Vorträge erschienen in Buchform  unter dem Titel: Progress in Soil Biology, bei Vieweg  und  Sohn
GmbH,  Braunschweig,  und  North-Holland Publishing Company,  Amsterdam
1967, Herausgeber waren Dr. Graff und Dr. J.E. Satchell. Inhaltlich waren  die Referate  wie folgt  gruppiert:  1. Wechselbeziehungen der Bodenlebensgemeinschaft,  2. Äussere  Einflüsse  auf  die  Bodenlebensgemeinschaft,  3. Menschlicher Einfluss  auf die Bodenlebensgemeinschaft,
4. Leistungen der Bodenlebensgemeinschaft, und  5. Weitere Probleme der
Bodenlebensgemeinschaft. Einen Gesamtüberblick über die beim Kolloquium
behandelten Themen  hat am Schluss des Bandes J. van der Drift gegeben.

Zum Präsidenten des Komitees  wurde für weitere 4 Jahre J. van der
Drift gewählt.

Das 4. Internationale Bodenzoologische Kolloquium  fand auf Einladung des Institute National de la Recherche Agronomique in der Zeit vom
14. bis 19. September 1970 in Dijon statt, es nahmen daran 118 Mitglieder teil.  Das Kolloquium war im besonderen dem Einfluss der Bodenorganismen  auf  die Primärproduktion gewidmet,  die Vorträge beschäftigten
sich zum Teil  aber  auch  mit Fragen, die über diesen Rahmen erheblich
hinausgingen.  Die Veröffentlichung der Referate  erfolgte  in Buchform
unter dem Titel "Organismes du sol et production primaire" in den Anna-
les de Zoologie-Ecologie animale,  Numéro hors-série 1971,  590 S., he-
rausgegeben vom Organisationskomitee  bestehend  aus  den Mitgliedern
J. d'Aguilar, C. Athias-Henriot, A. Bessard, M.B. Bouché und M. Pussard.

Der Band  enthält einen historischen Rückblick  auf die. Entwicklung des
Internationalem Bodenzoologischen Komitees, den J. d´Aguilar verfasste,
wofür er  als  eines  der wenigen lebenden Mitglieder,  die dem Komitee
seit seiner Gründung angehören, eine besondere Kompetenz besass. Mit dem
Jahrgang 1971 des Internationalen Mitteilungsblattes legte J. d´Aguilar
die Schriftleitung  des bodenzoologischen Teiles  desselben  zurück, es
trat in dieser Tätigkeit  M.B. Bouché an seine Stelle.  Zum Präsidenten
des Komitees wurde J.E. Satchell gewählt.

Das 5. Internationale Bodenzoologische Kolloquium  wurde  in  Prag
in der Zeit vom 17. - 22. September 1973 abgehalten. Seine Organisation
lag  unter  Leitung  von J. Vaněk  in der Hand  des Instituts für Landschaftsökologie  der Tschechoslowakischen Akademie  der Wissenschaften.
Als Thema  stand  die Abhängigkeit  der Bodenorganismen  von den Bodeneigenschaften  und  von der anthropogenen  Bodenbeeinflussung  zur Diskussion.

An dem Kolloquium nahmen rund 200 Wissenschaftler  aus einer Viel-
zahl  von Ländern teil,  zu dem  daran  anschliessenden Symposium  über
Apterygotenforschung waren 70 Mitglieder gemeldet.  Die grosse Zahl der
Vorträge  machte eine Gliederung  in 4 Sektionen notwendig:  Sektion A:
Stabilität und Verschiedenheit der Bodentiergemeinschaften,  Sektion B:
Einfluss abiotischer  und biotischer Faktoren  auf die Bodentiergemeinschaften, Sektion C: Einfluss menschlicher Aktivitäten,  Sektion D: Mo-
derne Methoden für die Untersuchung der Bodenorganismen  und ihres Ein-
flusses auf die Bodeneigenschaften.

Mit Rücksicht  auf  die stetig wachsende Teilnehmeranzahl  und die
zehlreichen Vortragsanmeldungen  wurde beschlossen,  das nächste Kolloquium schon in 3 Jahren abzuhalten  und eine schwedische Einladung nach
Uppsala anzunehmen.  Es wurde gleichzeitig der Wunsch  zum Ausdruck gebracht,  die Zusammenarbeit mit der Bodenmikrobiologie  und mit den anderen bodenkundlichen Forschungszweigen  weiter  zu intensivieren. J.E.
Satchell  wurde gebeten, die Funktion des Präsidenten  des Komitees bis
zum nächsten Kolloquium beizuhalten.

Mit dem Bericht  über  die bisher abgehaltenen Internationalen Bodenzoologischen Kolloquien ist  die Darstellung der internationalen bodenzoologischen Aktivitäten keineswegs erschöpft. Es ist vielmehr darauf
hinzuweisen, dass auch bei einer Reihe anderer internationaler Kongresse und Symposien bodenzoologische Referate gehalten wurden  und vereinzelt auch bodenzoologische Arbeitsgruppen tätig waren.  Das gilt ebenso
für internationale Kongresse der  gesamte Zoologie, wie auch für solche
der Entomologie und für Spezialtagungen, die einzelnen Bodentiergruppen
gewidmet waren. Auch verschiedene regionale Kongresse waren der Behandlung bodenbiologischer Probleme gewidmet.

Überblickt man die auf den 5 bisher abgehaltenen bodenbiologischen
Kolloquien gehaltenen Vorträge und darüber hinaus die einschlägige Literatur der letzten zwanzig Jahre, so ist man von der gewaltigen Entwicklung beeindruckt, welche die bodenzoologische Forschung in dieser
Zeit genommen hat. Schon J. d´Aguilar konnte in seinem historischen
Rückblick im Jahre 1970 festhalten, dass die Bodenzoologie in dieser
kurzen Zeit zu einer weitverzweigten Wissenschaft geworden ist, die
sich heute nicht bloss auf die Zoologie sondern auch auf die Botanik
und die Bodenkunde stützt.

Seit der Gründung des Internazionalen Komitees für Bodenzoologie
im Rahmen der Internationalen Bodenkundlichen Gesellschaft vor 15 Jahren ist die Bodenzoologie zu einem wesentlichen Zweige der Bodenforschung geworden und aus dieser nicht mehr wegzudenken.Dieser Jubiläums-
band bietet einen willkommenen Anlass dazu, diese heutige Stellung einer Betrachtung zu unterziehen.

Zunächst kann festgestellt werden, dass in den letzten 20 Jahren
die Zahl der Veröffentlichungen auf bodenzoologischem Gebiete ausserordentlich gewachsen ist. Es wurden drei Zeitschriften, die dem Gesamt-
gebiet der Bodenbiologie gewidmet sind, gegründet. Die "Revue d´Eco-
logie et de Biologie du Sol" (Verlag Gautier-Villars, Paris) ab 1964,
die "Pedobiologia" (Verlag Gustav Fischer, Jena) ab 1961 und die Soil
Biology and Biochemistry (Verlag Academic Press, London – New York).
Darüber hinaus gibt es zahlreiche Fachzeitschriften, die laufend bodenzoologische Arbeiten veröffentlichen.

Die terrikolen Tiergruppen gewidmete taxonomische Fachliteratur
ist stark angewachsen. Der Akademie-Verlag in Berlin entschloss sich,
eine Reihe von Bestimmungsbüchern der Bodenfauna Europas herauszubringen, wovon seit 1963 sieben Bände erschienen sind. Für die taxonomische
und ökologische Erforschung einzelner Bodentiergruppen wurden eigene
Fachzeitschriften begründet, ich nenne als Beispiel die Nematologica
und die Acarologia. Für andere Tiergruppen, die zahlreiche terrikole
Arten aufweisen, ich nenne als Beispiel nur die Coleoptera, bestehen
schon seit langem zahlreiche Periodika.

Die Taxonimie der Bodentiere hat in den beiden letzten Jahrzehnten
ausserordentliche Fortschritte gemacht. Diese gehen so weit, dass die
Ergebnisse älterer Arbeiten ohne Revision des ihnen zugrundeliegenden
Untersuchungsmaterials vielfach nicht mehr auswertbar sind. Besonders
grosse Fortschritte wurden in der Taxonomie der schalentragenden Amoe-
ben, der Milben, vor allem der Parasitiformen, und der Collembolen
erzielt.

Diese Fortschritte bilden die Basis für die Erweiterung unserer
Kenntnisse auf den Gebieten der Autökologie und der Synökologie der Bo-

dentiere, die wieder eine Voraussetzung für das Studium der Stellung
der Bodenbiozönosen innerhalb des Ökosystems der Böden bilden.

Je weiter wir im Studium der Bodenbiozönosen vordringen, um so
komplizierter strukturiert erweisen sich diese. Es zeigt sich immer
deutlicher, dass die einzelnen Bodentiergruppen von den verschiedenen
Milieufaktoren sehr verschieden stark beeinflusst werden. So zeigt sich
zum Beispiel, dass die Regenwürmer gegen Bodenfrost sehr empfinglich
sind, während mindestens der Grossteil der frei in der Erde lebenden
Nematoden sehr frostunempfindlich ist. Während gewisse Arten an schwere
oder leichte Böden gebunden sind, reagieren andere auf Texturunter-
schiede wenig. Eine Reihe von Bodentieren, vor allem die gehäusetra-
genden Landschnecken, aber auch Regenwürmer, gewisse Asseln und Tau-
sendfüssler zeigen ein ausgeprägtes Kalkbedürfnis, andere sind kalk-
indifferent. In den meisten Bodentiergruppen gibt es neben euryhygren
und eurythermen streng stenohygre und stenotherme Formen. Hinsichtlich
der Nahrung sind alle Übergänge von monophagen zu polyphagen Arten vorhanden. Dies und die grossen Unterschiede im Lebensraumbedarf bedingen
es, dass jede Bodentiergruppe ein spezifisches Verteilungsmuster innerhalb des Lebensraumes aufweist und zwischen den Synusien der Bodentiere
und den Assoziationen der höheren Pflanzen in dieser Hinsicht eine sehr
geringe Übereinstimmung besteht. Es stellt daher eine die Tatsachen
verfälschende Vereinfachung dar, wenn z.B. ein Autor "die" Bodenbio-
zönose eines Buchenwaldes beschreibt. Eine solche gibt es nicht, sondern
nur ein vielseitiges Mosaik von Bodentiersynusien, das für einen einzigen Buchenwaldbestand aufzunehmen ein sehr mühsames und zeitraubendes
Unterfangen darstellt und das sich nicht so ohne weiteres auf andere
Bestände übertragen lässt. Auch genetische Bodentypen und Bodenbiozönosen lassen sich nicht in einfacher Weise zur Deckung bringen, wenn auch
zwischen den Bodenbiozönosen und Bodentypen eines bestimmten Land-
schaftsbereiches gesetzmässige Zusammenhänge bestehen. Die Erforschung
dieser Zusammenhänge steht wie die der gesetzmässigen Zusammensetzung
der Bodenbiozönosen selbst noch in den Anfängen. Erst wenn sie entspre-
chend vorangetrieben sein wird, können wir erwarten, tiefere Einblicke
in die Rolle zu gewinnen, welche die gesamten Bodenbiozönosen und ihre
Veränderungen durch äussere Einflüsse in der Bodendynamik spielen.

Dies ist eine sehr komplexe Aufgabe, deren Bewältigung zunächst der
subtilen Erforschung der Wechselbeziehungen der Bodentiere untereinan-
der, zur Bodenmikroflora und zu den höheren Pflanzen bedarf. Es müssen
aber auch die Einflüsse des abiotischen Milieus auf die Bodenorganismen,
des Wärme- und Wasserhaushaltes der Böden, ihrer Korngrössenzusammen-
setzung, ihrer Struktur und ihres Porenvolumens, ihres Mineralbestandes
und dessen Verwitterungsgrades sowie der Beschaffenheit der Bodenlösung

untersucht werden. Das sind umfangreiche und weitverzweigte Forschungs-
vorhaben, deren Bewältigung dadurch erschwert wird, dass sich die einzelnen biologischen und abiotischen Faktoren in vielfacher Weise gegen-
seitig beeinflussen. Veränderungen eines Faktors ziehen daher zunächst
Veränderungen in anderen, oft in allen Teilen des Ökosystems nach sich.
Man denke etwa an Veränderungen des Wasserhaushaltes der Böden durch
künstliche Bewässerung, an solche der Bodenstruktur und anderer boden-
physikalischer Eigenschaften durch Bodenbearbeitung, des Bodenklimas,
aber auch des Bodenchemismus durch Waldrodung, um nur einige Beispiele
anthropogener Bodenbeeinflussung zu nennen. Sie führen ausnahmlos, wie
wir heute wissen, unmittelbar zu einer mehr oder weniger weitgehenden
Veränderung der gesamten Bodendynamik und zuzüglich auch zu einer solchen der Bodenbiozönosen. Mit der Veränderung der Bodenlebensgemein-
schaften verändern sich auch deren Einflüsse auf den Boden, woraus
neuerdings Veränderungen in der Bodendynamik resultieren. Dieses Spiel
von Wechselwirkungen setzt sich, wie die biologische Forschung lehrt,
so lange fort, bis sich zwischen allen beteiligten Faktoren ein neues
Gleichgewicht einpendelt, das dann so lange erhalten bleibt, bis es
durch äussere Einwirkungen neuerdings gestört wird.

Die Einwirkungen der Bodentiere auf das Milieu Boden sind mannig-
faltig, man hat sie erst nach und nach kennengelernt und jede für sich
untersucht. Ganz grob kann man physikalische, chemische und biologische
Einflüsse unterscheiden. Unter den ersteren ist die mechanische Durchmischung der Böden durch bestimmte Tiere, allen voran die Regenwürmer,
grabende Kleinsäugetiere und Termiten besonders auffällig. Sie bewirken
nicht bloss eine Homogenisierung des Bodenmateriales, das eine etwa ursprünglich vorhandene Schichtung von Sedimenten zum Verschwinden bringt
und die Grenzen der genetischen Bodenhorizonte verwischt, sondern da-
neben einen gewaltigen aktiven Materialtransport, der die Einmischung
des Auflagehumus in dem Mineralboden, eine teilweise oder nahezu völlige Aufhebung der substanzverlagernden Wirkung der Wasserbewegung im Boden und eine ständige Lockerung des Bodenmaterials bedingt. Die Auswirkungen dieser biologischen Tätigkeit auf die Bodenstruktur, auf die
Bildung von Krotowinen und auf die Horizontfolge innerhalb der Boden-
profile sind dem Feldbodenkundler seit langem bekannt und finden in der
Profilbeschreibung entsprechende Beachtung. Bodentiere sind aber nicht
nur an der Homogenisierung, sondern unter Umständen auch an Entwick-
lungsprozessen im Boden beteiligt, in besonders grossem Umfange die
Erdnester und -galerien bauenden Termiten, die Tonsubstanz akkumulieren
und mehr oder weniger zementieren, wodurch nicht bloss die Bodenstruktur, sondern auch der Bodenwasserhaushalt tiefgehend beeinflusst werden.

Zu den mechanischen Leistungen der Bodentiere gehört auch die Zerkleinerung des pflanzlichen Abfalls, die dessen Humifizierung, bzw. Mineralisierung wesentlich beschleunigt und z.T. in Bahnen lenkt, die von denen wesentlich abweichen, in denen sich die chemischen Umwandlungsprozesse ohne mechanische Aufarbeitung des Materials bewegen würden. Im Gesamtbereich der Verrottung organischer Substanz greifen physikalische und chemische Leistungen der Bodenorganismen eng ineinander und wirken Bodentiere und Mikroorganismen so eng zusammen, dass sich die spezifischen Effekte der Aktivitäten beider Gruppen nur schwer unterscheiden lassen. Vielfach sind klimatische Faktoren dafür entscheidend, welche der beiden Gruppen die Umsetzung der organischen Substanzen vorwiegend beeinflusst und welche Endprodukte entstehen. Vor allem Duchaufour u. seine Schule haben in letzter Zeit wichtige Beiträge zur Erforschung der biologischen Umsetzung der organischen Substanzen im Boden und zu ihrer klimatischen Beeinflussung geliefert, es bleibt aber immer noch viel zu tun, um die äussert komplexen Vorgänge restlos zu klären. Eine wichtige Rolle könnte hiebei das Studium der Umsetzung des pflanzlichen Bestandesabfalles auf ozeanischen Inseln mit einer extrem artarmen Bodenfauna liefern. Beobachtungen, die ich auf den Azoren durchzuführen Gelegenheit hatte, lassen erkennen, dass eine verhältnismässig geringe Zahl von Tierarten ausreicht, wenn sie wenigstens einzelne Vertreter der wichtigsten an der Bestandesabfallzersetzung beteiligten Gruppen umfasst, um einen normalen Ablauf der Rottevorgänge zu ermöglichen. Wesentlich kritischer ist offenbar ein extremer Mangel an Feuchtigkeit und/oder Wärme, da es in diesem Falle zu entscheidenden Veränderungen im gesamten Rottegeschehen kommt.

Besonders komplex sind offenbar die Wirkungen der biologischen Wechselbeziehungen im Boden und ihre Auswirkungen auf die Bodendynamik. Wir sind weit davon entfernt, sie auch nur für den Bereich des gemässigt-humiden Klimas vollständig zu überblicken, kennen aber doch schon eine Fülle von Wechselbeziehungen zwischen höheren Pflanzen, Mikroben und Bodentieren, die von losen und fakultativen Wechselbeziehungen bis zu obligaten Symbiose - und Parasitenverhältnissen reichen. Wir wissen ferner, dass Bodentiere mit Hilfe ihrer Darmflora nicht nur an der mechanischen, sondern auch an der chemischen Zersetzung der organischen Rückstände im Boden wesentlich beteiligt sind und dass sie ferner zur Verbreitung der Mikroorganismen im Boden erheblich beitragen. Auch an der Enzymproduktion haben sie über ihre Darmflora wesentlichen Anteil, wie anderseits mykophage Tiere die Vermehrung von Bodenpilzen in gewissen Grenzen zu halten vermögen.

Wie überall, wo sich in der Biosphäre Lebensgemeinschaften bilden, stehen diese auch im Boden in einem dynamischen Gleichgewicht, innerhalb dessen es nur selten und fast stets nur im Gefolge tiefgreifender Stö-

rungen  zur Massenvermehrung einzelner Arten kommt.  Die Aufrechterhal-
tung dieser biologischen Gleichgewichtszustände, die stets mit ganz bestimmten biologischen Leistungen  für den Boden verknüpft sind,  stellt
sich immer mehr  als vordringlichste Aufgabe  einer nachhaltigen Boden-
pflege und -nutzung dar.  Um eine solche im Zuge der tiefgreifenden Um-
wälzungen,  die sich gegenwärtig  in der Bodenwirtschaft vollziehen, zu
erhalten  oder überhaupt  erst einrichten zu können,  bedarf es umfang-
reicher Forschungen,  die in enger Zusammenarbeit  zwischen  Bodenkunde
und Bodenbiologie geleistet werden müssen.  Es ergibt sich hieraus, wie
aus den  bisher  von der bodenzoologischen Arbeitsgruppe  im Rahmen der
Internationalen Bodenkundlichen Gesellschaft  vollbrachten  Leistungen,
dass die Bodenzoologie in Zukunft noch enger als bisher mit allen übrigen Zweigen der bodenkundlichen Forschung zusammenzuarbeiten haben wird.
Diese Zusammenarbeit  wird sich  auch auf das grosse Gebiet  der Bodenverschmutzung  im Rahmen  des Umweltschutzes  zu beziehen haben,  wobei
diese Aufgabe nur einen Aspekt der Erhaltung einer stabilen und lebens-
freundlichen ökologischen Ordnung in der Natur darstellt.


L i t e r a t u r

H a r t m a n n , F. (1952): Forstökologie. Wien, 461 S.

K u b i e n a , W. (1938): Micropedology. Collegiate Press, Ames, Iowa,
     243 S.

M e y e r , L. (1942): Über die Entstehung  und Bildung der Ton-Humus-
     Komplexe. Forschungsdienst, Sonderh. 17, S. 38 ff.

M ü l l e r , P. E. (1887): Studien über  die natürlichen  Humusformen
     und deren Entwicklung auf Vegetation und Böden.  Berlin, VIII, 324
     S, 6 Taf.

P o s t , H. v. (1888): Nutideus koprogena bildingar: Gythja, Oy,  Torz
     och Mylla. Kong. Svensk. Vatensk. akad. Ngd. 4, 1861-62 (eingehen-
     de Besprechung von E. Raman in Ldw. Jb. 17, 405-420).

# Wandel der Bodenfauna unter dem Einfluss menschlicher Aktivitäten

H. FRANZ
Bodenforschung und Baugeologie,
Wien

Bekanntlich sind viele Bodentiere in hohem Mass an ihren Lebensraum, den Boden, angepasst. Sie haben nicht bloss Körperformen angenommen, die dem Leben in den engen Bodenhohlräumen oder selbst gegrabenen Gängen entsprechen, sondern auch manche Fähigkeiten verloren, die den über der Bodenoberfläche lebenden Organismen in mehr oder weniger hohem Masse zukommen. Dazu gehört der Schutz gegen Lichteinwirkung und Wasserabgabe durch Transpiration, die Fähigkeit rascher Fortbewegung bis zum Verlust des Flugvermögens primär flugfähiger Arten, die Reduktion oder der völlige Verlust des Sehvermögens und die Fähigkeit, sich an einen raschen Wandel der Umweltbedingungen anpassen zu können.

Viele Bodentiere sind aus den angegebenen Gründen ausserordentlich stenotop, sie besitzen ein beschränktes Fortbewegungsvermögen und sehr geringe Verbreitungsmöglichkeiten, um so mehr, als sie zumeist in kurzer Zeit zugrunde gehen, wenn sie den Boden verlassen. Mit diesen Eigenschaften hängt es zusammen, dass die Zusammensetzung der Bodenfauna deutlicher wie irgendeiner anderen Lebensgemeinschaft Veränderungen erkennen lässt, die sich in dem betreffenden Biotop in der Vergangenheit, oftmals bis in weit zurückliegenden geologischen Epochen vollzogen haben. Die Bodenfauna der Erde umfasst eine sehr grosse Zahl von Arten mit sehr beschränkter, oft diskontinuierlicher Reliktverbreiterung, deren erdgeschichtliche Bedeutung der von Fossilien vergleichbar ist, weshalb Jeannel von "fossils vivants" gesprochen hat.

Der Mensch hat im Laufe seiner Existenz auf der Erde deren Bodendecke und damit den Lebensraum der Bodenorganismen in mannigfacher Weise verändert. Ein sehr wirksames Mittel dazu, über das schon die primitivsten Menschen verfügten, war das Feuer und die dadurch hervorgerufene Zerstörung der Vegetation. Je geschlossener und vielschichtiger die Vegetationsdecke ist, um so grösser ist die Wärmeentwicklung und deren Tiefeinwirkung im Boden bei der Zerstörung der Pflanzendecke durch Feuer.

Es liegen zahlreiche Untersuchungen darüber vor, wie weitgehend und nachhaltig auf diesem Wege die Bodenfauna zerstört werden kann. Dabei kommen nicht nur die unmittelbaren Wirkungen des Feuers zum Tragen, sondern auch die Veränderungen des Mikroklimas durch den Mangel der Beschattung der Vegetation nach deren Zerstörung. Der grosse Gegensatz im Bodenmikroklima bewaldeter und waldfreier Standorte bedingt es, dass in der Artenzusammensetzung der Bodenbiozönosen von Natur aus bewaldeter und waldfreier Standorte stets sehr grosse Gegensätze bestehen. Der künstliche Ersatz einer Waldvegetation durch Grasland oder Acker führt aber mangels ausreichender Migrationsfähigkeit nicht zur Einwanderung einer vollwertigen Biozönose offenen Geländes, sondern zu einer Verringerung der Artenmannigfaltigkeit, die um so extremer ist, je intensiver die Zerstörung der Vegetation war.

Seitdem der Mensch gelernt hat, den Boden zu bearbeiten und darauf Nutzpflanzen zu kultivieren, kam zum Feuer ein zweiter Faktor intensiver Beeinflussung des Bodenlebens: die Bodenbearbeitung. Ihre Wirkung war, solange der Mensch mit primitiven Geräten Hackkultur betrieb, noch relativ gering, sie verstärkte sich mit dem Einsatz des Pfluges und erreichte durch die modernen Bodenbearbeitungsgeräte, an ihrer Spitze wohl der Fräse, ihr Maximum. Einmal bearbeiteter Boden bleibt, auch wenn die periodische Bearbeitung wieder eingestellt wird, an Bodentieren verarmt, das gilt für Steppenböden nach meinen Erfahrungen im Osten von Österreich, in Spanien und in Chile ebenso wie für gerodete und bearbeitete Waldböden. Wir wissen heute, dass durch Bearbeitung die Bodenstruktur und deren Stabilität, das Porenvolumen und der Wasserhaushalt des Bodens entscheidend verändert werden und müssen annehmen, dass diese Veränderungen des Milieus sich auf das Bodenleben nicht minder nachteilig auswirken wie die rein mechanischen Schäden, die sie durch den Bearbeitungsvorgang erfährt.

Nicht ganz so einschneidend, aber doch nicht gering einzuschätzen, sind die Wirkungen der Düngung. Wir wissen heute, dass organische Düngung bestimmte Bodentiere fördert, andere zurückdrängt oder ganz zum Verschwinden bringt. Dies gilt z.B. für bestimmte Regenwürmer, aber auch für Nematoden, Milben und Collembolen. Es war seinerzeit für uns sehr überraschend, als wir bei der Untersuchung der Bodenfauna regelmässig gedüngter Wiesen- und Weideböden der Ostalpen eine ganze Reihe von Milben und anderen Bodentieren fanden, die von Island beschrieben worden waren und die sich später in regelmässig organisch gedüngten Grünlandböden des gemässigten Europa als weit verbreitet erwiesen.

Auch mineralische Düngung verändert, wie wir wissen, die Bodenfauna. Kalkung saurer Waldböden kann z.B. eine sehr starke Vermehrung bestimmter Regenwürmer hervorrufen, mineralische Volldüngung bewirkt bei bestimmten Tiergruppen eine Abnahme der Besatzdichte.

Wie angesichts der Bedeutung des natürlichen Wasserhaushaltes für das Bodenleben nicht anders erwartet werden kann, haben Massnahmen der Be- und Entwässerung tiefgreifende Wirkungen auf die Bodenfauna. Stenohygre Arten, die an den betreffenden Standorten heimisch waren, verschwinden im Gefolge der künstlichen Veränderung des Wasserhaushaltes weitgehend. Von verheerender Wirkung ist die Verrieselung von Abwässern städtischer Siedlungen und der Industrien, wobei die Zerstörung der Bodenfauna mit der Rieselmenge zunimmt.

Die Wirkung von Insektiziden und Herbiziden auf die Bodenfauna ist in letzter Zeit viel diskutiert worden, weniger der Einfluss der Bodenverschmutzung durch Abfallstoffe aller Art, nicht zuletzt auch die Bodenverschmutzung von den Verkehrsstrassen her und durch die Luft- und Wasserverunreinigung seitens bestimmter Industrien.

Zu den früher genannten Veränderung des Milieus Boden durch den Menschen kommen im Zeitalter der modernen Technik in zunehmendem Masse Umschichtungen der Bodendecken im grossen Massstab durch Planierungen, Aufschüttungen und künstlichen Abhub. Diese künstlichen Erdbewegungen werden durch die im Gefolge menschlicher Eingriffe verstärkt auftretende Erosion ergänzt, wobei auch hier ein um so vollständigeres Verschwinden der Bodenfauna eintritt, je vollständiger der erosive Abtrag der Bodendecke ist. Fasst man das bisher Gesagte zusammen, so kommt man zu dem Ergebnis, dass sich unter dem Einfluss der menschlichen Tätigkeit seit langem eine fortschreitende Verarmung der Bodenfauna vollzieht, dass deren Ausmasse aber in den letzten Jahrzehnten enorm und zweifellos beängstigend angewachsen sind.

Die Verarmung der landwirtschaftlich genutzten Böden an Bodenorganismen durch unzulängliche Versorgung mit organischem Dünger durch einseitige, das Bodenleben nicht fördernde Fruchtfolgen und biologisch ungünstige Bodenbearbeitungsmassnahmen hat weithin zur Verschlechterung der Bodenstruktur und des Wasserhaushaltes der Böden geführt. Wir wissen, dass in den Ländern mit hoch entwickelter Bodenwirtschaft schon lange nicht mehr die Unterversorgung mit Pflanzennährstoffen, sondern die Verschlechterung der physikalischen Eigenschaften den entscheidenden ertragsbegrenzenden Faktor darstellen. Für die Erstellung günstiger physikalischer Eigenschaften der Böden spielen aber die Bodenorganismen eine massgebende Rolle.

Wir haben bisher nur von der Reduktion der Bodenfauna, z. Teil bis zu ihrer völligen Zerstörung gesprochen, neben·dieser spielt sich aber seit langer Zeit eine mehr oder weniger tiefgreifende Veränderung derselben ab. Durch Handel und Verkehr haben gewisse Pflanzen und Tiere eine nahezu kosmopolitische Verbreitung erlangt. Dies gilt auch für gewisse Bodentiere, so zum Beispiel einige ursprünglich in Europa be-

heimatete Regenwürmer. Sie gelangten bis Australien und Neuseeland, wo
sie nicht bloss die Böden und die darauf wachsende Vegetation veränder-
ten, sondern auch die Bodenfauna stark beeinflussen. So werden durch
sie die heimischen Regenwürmer zurückgedrängt, über ihren Einfluss auf
die Bodenmikrofauna ist noch wenig bekannt. In gleicher Weise haben
sich aber mit der europäischen Humuswirtschaft viele in Anhäufungen
sich rasch zersetzender organischer Substanzen lebende Tiere sehr weit
verbreitet. Von bestimmten terrikolen Pflanzenparasiten ist das seit
langer Zeit bekannt. Dass die Verschleppung von Bodentieren ein ausser-
ordentliches Ausmass erreicht haben muss, wird am deutlichsten auf
ozeanischen Inseln mit einer an sich artenarmen Fauna und dort wieder
besonders in der Umgebung von Hafenplätzen sichtbar. Über die Ausmasse
der Veränderung der Bodenbiozönosen durch die ständige Zufuhr fremder
Elemente besitzen wir aber heute noch eine sehr unzulängliche Information. Auch die zweifellos nicht geringen Rückwirkungen auf die gesamten
Ökosysteme sind uns unzulänglich bekannt.

Welche Bedeutung die Störung ja stellenweise Zerstörung der natürlichen Ökosysteme für die gesamte Naturordnung und vor allem für die
Biosphäre zukommt, hat man in der letzten Zeit zu erkennen begonnen.
Bei den diesbezüglichen Untersuchungen konzentrierte man sich aber vor-
erst vor allem auf die unmittelbaren Auswirkungen auf den Menschen, die
allmöglichen Veränderungen in der belebten Natur werden nach wie vor
wenig beachtet. Dies ist dadurch begründet, dass man bestimmte vordring-
liche Probleme für sich, das heisst weitgehend isoliert betrachtet und
sich nach wie vor nicht der Tatsache bewusst ist, dass in der ökologischen Ordnung der Natur alle Vorgänge miteinander räumlich und zeit-
lich eng verflochten sind.

Ich bin der Meinung, dass es eine der vordinglichsten Aufgaben unserer bodenzoologischen Arbeitsgruppe in der Gegenwart ist, die sich
in immer rascherem Ablauf vollziehenden Veränderungen in den Bodenbiozönosen zu studieren und in ihren Folgen richtig abschätzen zu lernen.
Erst wenn unsere Kenntnisse auf diesem Gebiete genügend vollständig
sein werden, kann es uns gelingen, die breitere Öffentlichkeit für den
Wandel im bodenbiologischen Geschehen zu interessieren und Massnahmen
durchzusetzen, durch die gefährliche Entwicklungen eingedämmt werden
können.

Wenngleich mir bewusst ist, dass dieser praktische Gesichtspunkt
bei dem es letzten Endes um ein Überleben in einer sich immer dichter
besiedelnden und rascher wandelnden Welt geht, heute im Vordergrund
steht, möchte ich meine Ausführungen doch nicht mit ihm beschliessen.

Ich habe selbst ein Leben lang historisch-ökologische und histo-
risch-biogeographische Forschungen betrieben, so dass mir bewusst ist,

wie gross unsere Lücken auch auf diesen Forschungsgebieten sind. Viele
dieser Lücken können nur so lange geschlossen werden, als uns diejenigen
Organismen noch zur Verfügung stehen, die als Fossils vivants die Be-
weise liefern, die zur Schliessung dieser Lücken notwendig sind. Niemand
von uns weiss, wie gross die Zahl der durch den Menschen bereits aus-
gerotteten terrikolen Reliktorganismen ist. Ziehen wir aber die Ausmasse
der Zerstörung der Böden und ganzer Landschaften in Betracht, die wir
vor uns sehen, so müssen wir annehmen, dass es sich um eine sehr grosse
Zahl handelt.

Besonders gefährdet sind naturgemäss eng begrenzte Areale, wie
sie vor allem auf vielen Inseln vorliegen, so dass der Erforschung der
terrikolen Fauna der Inseln heute besondere Dringlichkeit zukommt. Sie
wird überdies auch dadurch gerechtfertigt, dass viele Inselfaunen sehr
lückenhaft sind und ihre Unvollständigkeit wertvolle Aufschlüsse über
die Auswirkung lückenhafter Biozönosen auf die ökologischen Gleichge-
wichte zulässt. Naturgemäss darf neben dieser Aufgabe die der Erforschung der Kontinente nicht vernachlässigt werden und hier besonders
jener Räume, die Ballungszentren menschlicher Aktivitäten sind.

Die bisherige Tätigkeit der bodenzoologischen Arbeitsgruppe hat
vor allem zur Erforschung der Leistungen der Bodentiere für den Boden,
zur Kenntnis ihrer Autökologie und nicht zuletzt auch zur Vervollstän-
digung ihrer Taxonomie wesentliche Beiträge geleistet. Die derzeitige
Situation erfordert, wie mir scheint, in nächster Zukunft die intensive
Behandlung zweier weiterer Gebiete, der Synökologie einschliesslich
ihrer Beeinflussung durch den Menschen und des biogeographischen Aspek-
tes der Bodenfauna. Diese beiden Gebiete scheinen vielleicht auf den
ersten Blick wenig Beziehung zur Bodenkunde im engerem Sinne zu haben.
Diese Beziehungen bestehen aber und zwar einerseits zur gesamten Bodendynamik und andererseits zur Bodengenetik.

# General Trends of Changes in Soil Animal Population of Arable Land

M. S. GHILAROV
Institute of Evolution Ecology and Morphology of Animals,
Moscow

Fifty years ago the Swiss naturalist Burger (1922) compared agricultural soil to a chaotic mass of building material, and natural undisturbed soil to a house built from this material. But probably it would be better to compare undisturbed soil to a building, and agricultural, permanently ploughed soil to debris of this building permanently destroyed with powerful and dreadful weapon.

The founder of our modern soil science, the great Russian naturalist Dokuchaev, regarded the soil as the result of action of many factors upon the earth crust considering the influence of animals as one of the potent factors. Undisturbed soil is a house not only inhabited but also built by soil animals and destruction of this house, of the environment to which soil animals are adapted, is fatal to its dwellers. Numerous soil animal species adjusted to the soil of a given structure and texture cannot exist when their complicate multistratous habitation is destroyed.

Only comparatively few members of the soil animal community can persist the destruction of the soil structure caused by cultivation of arable land.

Soil tillage and agricultural utilization affect soil animals in various ways and are, to a different degree, dangerous to various taxa and ecological groups of soil invertebrates.

When uncultivated virgin soil is tilled the upper layers are destroyed and litter and other plant debris disappear from the surface horizon. Consequently, all animals which are closely related to litter and topsoil do not find suitable conditions after soil cultivation and quickly vanish. Such animal groups abundant in uncultivated soil are scarce or absent in field soil.

An examples are enchytraeids, earthworms such as *Dendrobaena octaedra*, woodlice, many mallipedes, lithobiids, cockroaches, etc. Among the microarthropods, representatives of the so-called hemiedephon are eliminated in the first line. Isotomid spring-tails and morpho-ecologically similar collembolans, oribatid mites such as damaeids, galumnids, etc., deprived of their familiar habitat  cannot survive soil cultivation.

Soil ploughing, harrowing etc. destroy the systems of passage-ways burrowed by larger, active invertebrates like earthworms, ramifying tunnels excavated by minute arthropods  after the decomposition of dead decayed roots and rootlets, interstices between soil-structural  grains and crumps etc. The disappearance of such systems of cavities in plough-ed soil  makes life impossible  for such minute arthropods  whose imme-diate habitat is not the soil as a whole, but cavities, cracks and tun-nels in the soil, where they move as in caves and use tunnels for their permanent up  and down migrations,  influenced by changes in the hydro-thermal regime.  As a result, also  microarthropods  of a bigger size, many euedaphic collembolans  and  relatively big-sized mites  disappear from cultivated soil.  Only very small or elongate microarthropods such as *Oppia* spp. and similarly minute Oribatei, e.g., *Onychiuris* and *Tull-bergia* among  spring-tails, principally such forms  which  are able to migrate deeper  than the depth  of ploughed layer,  are rather numerous in arable soil.  In heavy soils cultivated for a long period, accumula-tion of clay particles  just under the plough sole corks interstices of the deeper layer disturbs their aeration and,  prevents survival in the deep soil horizons, though they have not been disturbed directly.

The destroying  of soil cavities  affects also larger soil animals using existing galleries for their migrations, Cardiophorini-larvae may serve as an example.  Such invertebrates, as well as those constructing their nests in the soil, like ants, beetles of the genus *Lethrus*  etc., encounter an unfavourable situation in ploughed areas. Population densi-ty of larger individuals of soft-skinned soil animals  becomes somewhat reduced during soil cultivation, due to cutting and pressing by plough-share. The change of soil structure is one of principal causes of elim-ination  of various life forms resp. taxa  of soil population in arable land.

Indirect influences of agricultural practice are also significant. Change  and  simplification of floristic composition  of plant cover of field  crops  eliminate  all oligophagous soil insect larvae feeding on underground parts  of plants belonging to other taxa  than those culti-vated, or weeds present  in fields.  Numerous species of weevil maggots are predominating among soil phytophages in virgin steppes and  meadows being absent in fields (*Eusomus* spp.).

Processes of  general impoverishment  of  soil animal populations resulting  from the agricultural utilisation of land proceed are slower in rather humid soil  rich in humus  than in soil poor in humus, and in more arid conditions (Table 1).

T a b l e  1.    Population density of larger soil invertebrates in cultivated and practically virgin soils within one field
(Mikhnevo, Moscow Region)

| Soil type | Humus % in 0-25 cm layer | Soil utilization | All large | Lumbricidae |
|---|---|---|---|---|
| | | | (per 1 sq. m.) | |
| Dark coloured | 5.5 - 3.0 | Uncultivated Cultivated | 249 ± 19  111 ± 11 | 146 ± 14  47 ± 8 |
| Podzolic | 2.2 - 1.6 | Uncultivated Cultivated | 148 ± 25  14 ± 2 | 57 ± 12  5 ± 2 |

This can be explained in the following way.  Among the soil invertebrates, saprophagous species or those capable  of saprophagy predominate being  decomposers  participating  in the matter turnover  in ecosystems.

The conditio sine qua non  of saprophagy  is the presence  of dead
organic matter  (correlating with the humus content) and its availability to saprophagous species depending on necessary soil moistening.

Strictly speaking  many species considered  to be saprophagous are
consumers of moulds, bacteria and other soil-saprophytic microorganisms
having the same demands to organic matter  as saprophagous animals. The
data obtained  by the author in the vicinity of Moscow  may serve as an
illustration (Tables 1 and 2).

T a b l e  2.    Population density of microarthropods in cultivated and
practically virgin soils within one field
(Mikhnevo, Moscow Region)

| Soil type | Humus % in 0-25 cm layer | Soil utilization | Acari | Collembola | all Microarth- ropoda |
|---|---|---|---|---|---|
| | | | (Thousand per 1 sq. m) | | |
| Dark coloured | 5.5 - 3.0 | Uncultivated Cultivated | 24.5  18.1 | 10.0  14.0 | 36.0  32.0 |
| Podzolic | 2.2 - 1.6 | Uncultivated Cultivated | 3.4  0.6 | 1.5  1.1 | 5.5  2.0 |

We have chosen one plain field with two different soil types, both
well expressed, formed on dofferent subsoil. One soil type - chernozem-
like, rich in humus,  dark coloured soil  formed on deep laying calcif-

erous deposit. The other - podzolic soil on moraine clay loam. The texture of both types was rather similar.

Within each soil type, a cultivated field (after small grains) and
an adjacent uncultivated plot, a practically virgin meadow strip, were
investigated. Population density of both larger invertebrates and of
microarthropods was significantly higher in untreated plots of dark
coloured soil. The decrease of soil animal population density following
agricultural usage was significantly higher in podzolic soil than in
dark coloured soil, rich in humus. The same has been observed in many
localities of this country. Table 3 shows comparative population density of oribatid mites in another locality near Moscow (estimations by
M.N. Chugunova). Table 4 refers to soil mites in the Odessa Region
(censuses by O.K. Furman). A lot of other examples could be added, all
being very uniform.

T a b l e  3.  Population density  (thousands per 1 sq. meter) of Oribatei in the soil under various plant cover
(Moscow Region, June)

| *Piceetum oxalidosum* | $70 \pm 17$ |
|---|---|
| *Betuletum herbosum* | $24 \pm 4$ |
| Meadow | $8 \pm ?$ |
| Red clover | $5 \pm 1$ |
| Potato | $1 \pm 0.4$ |

T a b l e  4.  Population density of mites (*Oribatei* - ca 50 %) in uncultivated chernozem soil and under field crops
(Odessa Region)

| Sod land, adjacent a shelter belt | 20 100 | |
|---|---|---|
| Winter wheat | 5 600 | per 1 m$^2$ |
| Sugar beet | 3 200 | |

As to the microarthropods there is a general regularity: when the
humidity of the soil is high enough, the population density of this
animal group is proportionate to the content of organic matter estimated
as humus in standart procedure. Data obtained in humid parts of the
western territory of this country (calculations by V.K. Eglitis) are
demonstrated in Table 5.

T a b l e   5.    Dependence of microarthropod population density   on organic matter of the soil
(average data from numerous fields in the western part of the USSR)

| Organic matter (%) | Microarthropods (indiv./m$^2$) |
|---|---|
| 0 - 2 | 8 700 |
| 2 - 4 | 13 000 |
| 4 - 6 | 15 300 |
| 6 - 12 | 19 300 |
| 12 - 50 | 50 000 |

The decrease  of the soil animal population density  in cultivated
soil indicating a dropping of biological activity  in the soil, results
in a marked decrease  of the humus content.  Of paramount  interest  is
a comparison of data obtained in Moldavia, i.e., data on the humus  content in field chernozem  given by Dokuchaev (1898)  with those obtained
by the author and his colleagues during an expedition in 1960. The humus
content dropped twice in this period,  but in the soil  of small uncultivated sod plots  it was similar  to that recorded by Dokuchaev (estimated by N.A. Prokhina). The population density of larger invertebrates
was significantly lower in fields which did not receive  organic manure
than in soddy soil of uncultivated plots. (Table 6).

T a b l e   6.    Population density of invertebrates extracted  from the
soil by washing method in virgin steppe and in wheat field soil in Orsk
(S. Ural) Region
(individuals per 1 sq. m)

| Land utilization | Microarthropods per 1 sq. m | Invertebrates larger than microarthropods per 1 sq. m |
|---|---|---|
| Virgin steppe | 1 200 | 4 038 |
| Wheat field | 4 900 | 1 300 |

A correlation  between the biological activity  of the soil testified  by censuses  of invertebrates, and of the numus content is rather
evident.
Sometimes, cultivaticr in arid zones  may be favourable for micro-
arthropods since the humidity of ploughed soil  is somewhat higher than
that in virgin steppe, as claimed by T.G. Grigor´eva (Table 7).

T a b l e   7.   Population density of larger invertebrates   and   humus
content in cultivated and neighbouring permanently uncultivated chernozem soil
(Moldavia)

| Type of chernozem (black soil) | Plot | Amount of humus in the layer 0 - 50 cm | Population | Biomass | |
|---|---|---|---|---|---|
| | | | All big inverte- brates | Lumbricidae | |
| leached | Uncultivated | 351 | 171 | 98 | 5.0 |
| | Cultivated | 240 | 71 | 27 | 1.3 |
| common | Uncultivated | 242 | 237 | 123 | 6.0 |
| | Cultivated | 200 | 47 | 23 | 1.0 |

The soil fauna  of fields  is more monotonous  and poor in species
than that of uncultivated territories,  being often rather similar even
in various climatic zones.  Many species of insects developing in field
soil are to be met in field soil, from the Arkhangelsk Region  near the
White Sea to the Rostov Region near the Black Sea, (e.g. *Agriotes lineatus*); the same refers to earthworm species such as *Allolobophora caliginosa*  and  some other so-called  "peregrine" species,  as cultivation
creates changes in the soil regime smoothing differences  in zonal conditions.  Therefore, in some cases, it is possible to increase the biological activity  of the soil  by introducing beneficial soil  invertebrate species  from other territories, sometimes  with a different climate, into fields where such species are wanting.

Formation  of soil  invertebrate fauna,  of communities of arable
land, is the result  of a struggle  for life, survival  of the fittest,
natural selection, i.e., they proceed according to the same laws as the
formation of biocenoses. For invertebrates adjusted  to field conditions
a new rigid selecting factor is included  in the natural selection – the
activity of man.  As to the character of plant cover,  fields under various crops  are somewhat similar to climax formations  in extreme conditions, and change  of vegetation  in the course  of crop rotation  is
somewhat similar to succession. Thus it is possible to study agrocenoses
by using biocenological criteria.

There are some general features enabling the survival of saprophagous animals in arable soil. Among earthworms the survivors are species
inhabiting mineral layers and ingesting mineral particles when feeding,
among *Oribatei* - the small-sized species  are capable to migrate down-
eards  or find ways  between soil particles even in disturbed soil  (as
*Oppia*) etc.

Of interest for both an understanding of tendencies in arable soil
population formation  and for agricultural practical conclusion  is the
study  on general ecological properties  of injurious insects affecting
underground parts of cultivated plants, the so-called soil pests.

Among soil pests there are both polyphagous  and oligophagous species, the last being the so-called specialized ones.

The ability to survive brusque changes  of the environment depends
on the ecological plasticity with respect to the principal vital changing factors.  By changes  of vegetation cover caused  by  crop rotation
insects feeding only on the given kind of cultivated plant  have either
to perish, or to migrate into fields (or other territories) where necessary food is available.

In crop rotation  the same  is to be observed  as in the course of
ecological succession, but more rapidly as schematized.  The ability of
active dispersal  is a property compensating insufficient plasticity as
to the food factor.  In insects it is the adult stage performing active
dispersal, always preceeding propagation.  It is the larval stage which
is feeding  and  destroying roots  and  other underground parts  in the
majority of soil insect pests.

Polyphagous species are more tolerant to the changes of vegetation
cover in crop rotation than oligophagous ones. Oligophagous specialized
species of soil  (and not only soil) pests are more heavily affected by
crop rotation.

Dispersal, as I stressed more than 30 years ago, is the ability to
compensate unsufficient plasticity as to food or any other vital factor,
because any population  of insects must either perish  or emigrate when
environmental changes  exceed  the plasticity  of the given population.
As insects actively disperse  in the adult stage,  the ability  of dispersal compensates the stenotrophy  of larvae,  unable  to migrate  for
long distances. The following comparison allows to understand the major
properties of insects during adjustment to crop rotation  (Table 8). It
is evident  from the data  in Table 8  that  in arable land, in fields,
crop rotation  determines  the natural selection  among  injurious soil
insects  in various directions - the species surviving  and  increasing
in number in field soils are either polyphagous species or those of the
oligophagous species which complete larval development within one vegetation season and in which the adults are capable to disperse.

The above data prove that under the varying conditions of environment, even in the case of one varying factor of importance (plant cover),
the species surviving  under these conditions are selected not only according to the features directly related to the given factor (degree of
stenophagy  and  specific composition of vegetation cover)  but also to

indirectly related ones   (duration of life-cyclus   and   ability of dispersal).

T a b l e  8.   Principal properties  of insects  (in larval stage able
to affect   subterranean  parts  of cultivated plants)   surviving or not
surviving field crop rotation

| Food specialization | Duration of larval stage | Degree of active dispersal of the adults | Do or do not survive crop rotation | Examples |
|---|---|---|---|---|
| Food Polyphagous (not specialized pests) | 1 year or less | high or feeble | yes | Euxoa segetum Opatrum sabulosum |
| "      " | several years | high or feeble | yes | Agriotes sputator Anisoplia austriaca |
| "      " | 1 year but over 2 seasons | high or feeble | yes | Tipula padulosa |
| Oligophagous (specialized pests) | 1 year during 1 season | high | yes | Phyllotreta sp. Sitona spp. Bothynoderes punctiventris |
| "      " | several years | feeble or high | no | Dorcadion spp. Cicadetta sp. |
| "      " | 1 year but over 2 seasons | feeble or high | no | Crambus sp. Zabrus tenebrioides |

The practical importance of such an analysis consists  in the possibility to establish the role of crop-roration in the control of pests
when  recording  the fundamental  biological properties  of the latter.
Crop-rotation  is  of major importance  in the control  of stenophagous
(specialized)  species of low motility  in all stages or with the duration of the larval stage exceeding one vegetation period. The increasing
ability to disperse, or a higher degree of polyphagy interfer  with the
effectiveness  of crop-rotation  as a control measure.  This effect  is

practically nil in the case of polyphagous insects with a high motility and ability of dispersal.

In crop rotations including black fallow (fields kept clean of any vegetation) the majority of survivors among polyphagous soil pests with a low ability of dispersal are species capable to feed on decay in plant remnants, and those capable of facultative saprophagy.

# SECTION A
## STABILITY AND DIVERSITY OF COMMUNITIES OF SOIL ORGANISMS

# On the delimitation of soil microarthropod coenoses in time and space

**W. DUNGER**
Staatsmuseum,
Görlitz

In order to describe mean features of the structure of soil eco-sub-systems many authors deal with species lists of soil microarthropods, especially mites and *Collembola*. Because lists can hardly be compared, abundance and frequence of any single species may be quantified in different ways. By this, the significance of a species within the soil community is not pointed out sufficiently. Much more attention should be paid to the sociological behaviour of soil microarhtropods. The few authors, who had tried to do so, do not agree about the method and the value of this procedure (for example Gisin, 1947; Cassagnau, 1961; Agrell, 1963; Davis, 1963; Dunger, 1968; Lebrun, 1971). This paper does not propose a new method, but compares some methods dealing with some experiments.

As in phytosociology, in the pedozoological literature four main concepts concerning the sociological treatment are discerned:

1. The continuum-concept presumes the co-existence of a undeterminate number of populations in a undeterminate correlation without objective discontinuities. In such a continuum only trends could be stated, and this renders the synbiological viewpoint more or less useless. In the experiments, the continuum-concept is disproved by discontinuities in the occurrence of species or species-groups in time and space, even if they are related.

2. The inventory-concept presumes the existence of a characteristic and homogeneous coenosis of soil microarthropods with the border of any unit of vegetation (phytocoenosis). The enumeration of the species-inventory of mites or *Collembola* is used to describe a soil taxocoenosis. The inventory-concept offers a real heuristic advantage for orientation in the field. However, it is applicable only under favourable circumstances.

3. The discontinuity-concept is trying to delimit coenoses of soil microarthropods by using discontinuities of their species combination. Negatively correlated, i.e., substitutive or vicarious species serve to determinate the limits of a coenosis (differential species).

4. The affinity-concept aims at the contents-determination  of the coenoses. It rests on positive interspecific correlations. If it can be proved that  closely  related species-combinations  (recurrent groups) represent ecologicaly more or less specialized groups of species,  then the affinity-concept enables an analysis of the coenoses  "from the inside".   This  will increase essentially  the indicatory significance of a described coenosis.

As a first step of sociological treatment of soil microarthropods, coenoses may be delimited by using characteristic substitutive species. This paper  will be confined  to methodical considerations  and to some practical results of this procedure.  This may be followed, as a second step, by the definition of the eco-sociological feature of the coenosis using the affinity-concept.

As to find out a "basal coenosis",  the smallest sociological multispecific unit of soil microarthropods,  the synusia,  must be chosen. A synusia is to be defined  as a mero- and/or strato-taxocoenosis. Constructing  any  kind  of hierarchy  of synusiae necessitates  much more knowledge than is available up to date.

In obtaining  and  analysing the primary material the principle of maximal  homogeny  must  be  observed:  homogeny of the area; observing minimal area, mosaic distribution etc.,  homogeny of the structure unit and/or  of the stratum; vegetation- and  humus layer  must be separated from the mineral soil;  in poorly structured soils  at least separation is to be made  between  soil-surface moving epedaphon (caught by a pitfall technic) and soil-inhabiting edaphon  (obtained by core sampling), homogeny  of the time aspect;  analysing either complete year cycles or at least comparing identical aspects;  considering development  of synusiae in time.

Dominance  and  frequence of each species  will be tabulated separately for each homogeneous plot.  Series from different plots are than tested for discontinuities. Negatively correlated, relative or absolute substitutive species will be chosen as differential species, especially if autecological experiments justify it. Owing to such subjective judgement, synusiae delimitated in this way are to be regarded as hypotheses based upon ecological  and  faunistic experiments.  Therefore, synusiae should be delimited  in order  to form working  hypotheses  for further quantitative  ecological  analysis.  The following examples  illustrate both the advantage of synusia-formation using the discontinuity-concept and  the necessity  of exactly separating homogeneous synusiae  in different strata or in point of temporal development.

An interesting  object  for study  is  the spatial  succession  of synusiae within a catena. Figure 1 shows  a schematic profile through

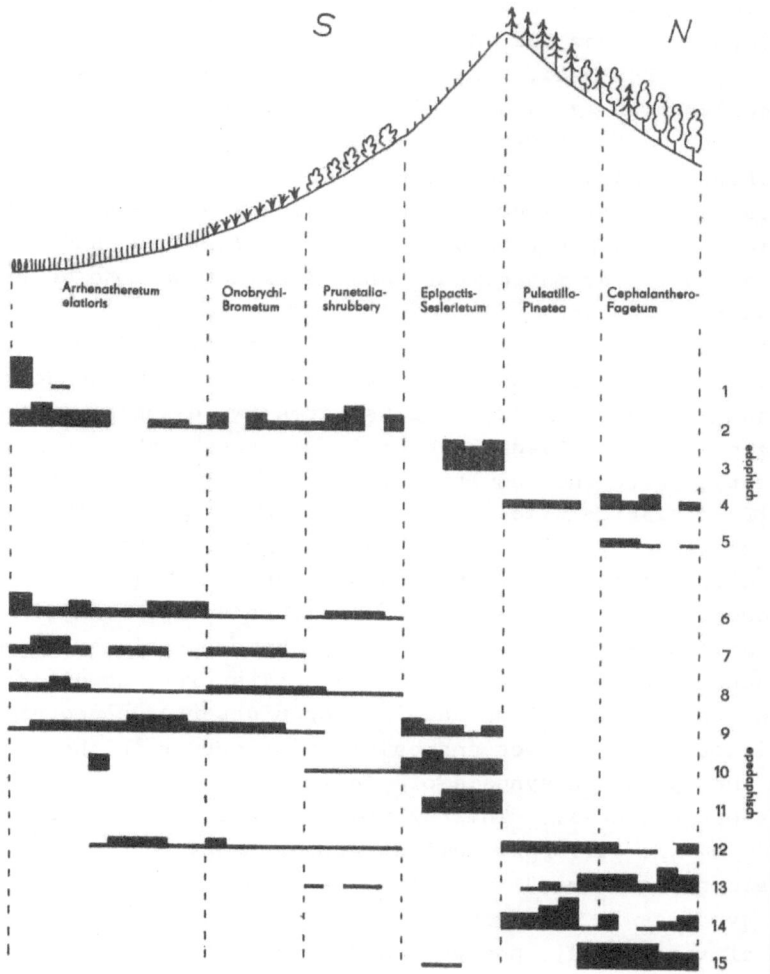

Fig.1. Schematic profile through the northern slope of
a shell lime valley. Above: phytocoenoses; middle: edaphic collembolan synusiae represented by dominance classes
of differential species (1 = *Folsomia quadrioculata*,
2 = *Onychiurus jubilarius*, 3 = *Onychiurus zschokkei*,
4 = *Onychiurus armatus cancellatus*, 5 = *Xenylla boerne-
ri*); below: epedaphic collembolan synusiae (6 = *Isotoma
viridis*, 7 = *Brachystomella parvula*, 8 = *Lepidocyrtus pa-
radoxus*, 9 = *Deuterosminthurus sulphureus*, 10 = *Pseudachorutes angelieri*, 11 = *Lepidocyrtus lignorum*, 12 = *Pseudachorutes parvulus*, 13 = *Entomobrya muscorum*, 14 = *Mi-
crogastrura duodecimoculata*, 15 = *Hypogastrura burkilli*).

the northern slope of a shell lime valley near Jena.   The distribution
of differential species of Collembola demonstrate relative and absolute
discontinuities, but by no means a continuum.  Against the inventory-

-concept the edaphic *Onychiurus jubilarius* synusia extends from the *Arrhenatheretum elatioris* over the *Onobrychi-Brometum* to the *Prunetalia*-shrubbery. On the other hand, the shrubbery can be characterized by an epedaphic collembolan synusia, which is not delimited by differential species, but by the differential combination *Pseudachorutes angelieri - Entomobrya muscorum*. The different behaviour of epedaphic synusiae is obvious. In most cases limits of synusiae correspond to limits of vegetation. The separation between *Arrhenatheretum* and *Onobrychi-Brometum* is caused by management and does not have to be reflected in collembolan synusiae.

This leads to the question of the differentiation of epedaphic and edaphic collembolan synusiae in the same plot under the influence of different human managements. Figure 2 illustrates two examples using the index of homogeneity after Riedl, which in the case mentioned here coincides sufficiently with the results of the sociological analysis. In plot A the differentiation time lasted only 5 to 10 years (Dunger, 1968), in plot B, on the other hand, more than 100 years (Dunger, 1972). In both cases the differentiation of the edaphic synusia will be caused above all by the properties of the substrate (culture soil/clay against raw soil/tertiary acid sand). Contrary to this, the epedaphic synusiae are mainly influenced by the vegetation and soil surface conditions (forest plantation against agriculture against open desert). Consequently, uprooting deciduous forest trees and cultivating spruce in the same plot will alter the epedaphic synusia of *Collembola* far more intensively than the edaphic synusia. This example indicates the necessity of a separate study of the different soil stratocoenoses.

Just the same conclusion results from the study of a succession of microarthropod synusiae in time. In figure 3, a comparison is made of the succession of vegetation, epedaphic and edaphic synusiae of *Collembola* in an afforested brown coal dump (Dunger, 1968). It is obvious, that the epedaphic synusiae, in connection with changes in vegetation, will be altered much more quickly and intensively than in the case of the "conservative" edaphic synusia. The asynchronism of the succession demonstrates once more the necessity to study homogeneous synusiae separately in time and space. Even in this case, one cannot make use of the inventory-concept.

The examples mentioned above confirm the opinion, that the delimitation of microarthropod synusiae in the soil is possible on the basis of the discontinuity-concept. It opens interesting but until now little studied aspects.

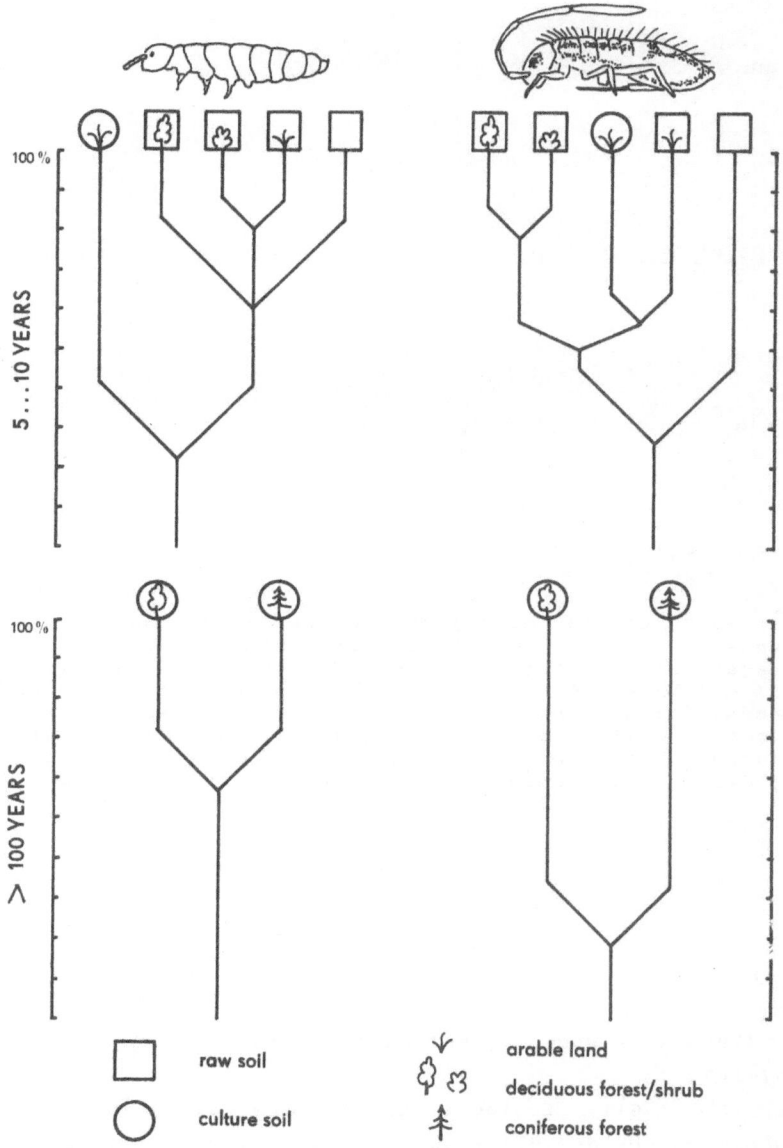

100 %

5...10 YEARS

100 %

> 100 YEARS

raw soil

culture soil

arable land

deciduous forest/shrub

coniferous forest

Fig.2. Differentiation of collembolan synusiae estimated by the index of homogeneity after RIEDL. Plot A (above): short time differentiation of a recultivated brown coal dump; plot B (below): long-term differentiation after spruce cultivation. Left column: edaphic synusiae; right column: epedaphic synusiae.

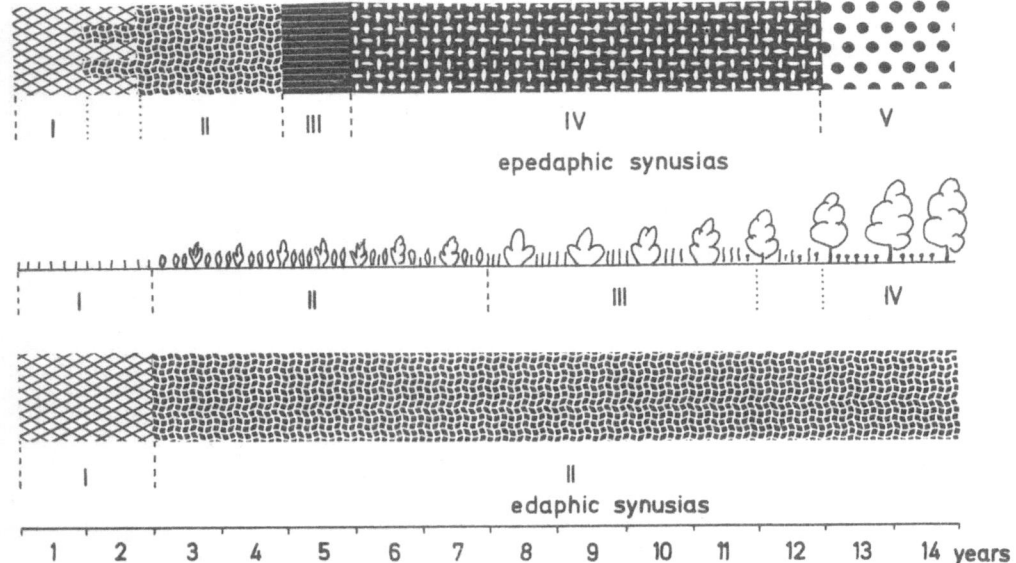

Fig.3. Succession of vegetation, epedaphic and edaphic synusiae of *Collembola* in an afforested brown coal dump. Stages of vegetation: I - *Melilotus*, II - *Artemisia*, III - *Holcus-Achillea*, IV - *Senecio nemorensis*. Synusiae epedaphic: I - *Entomobrya lanuginosa*, II - *Ceratophysella succinea*, III - *Tomocerus vulgaris*, IV - *Lepidocyrtus cyaneus*, V *Orchesella flavescens*; edaphic: I - *Proisotoma minuta*, II - *Isotomodes productus* - *Micranurida pygmaea*.

R e f e r e n c e s

A g r e l l , I. (1963):  A sociological analysis of  soil *Collembola*.
   Oikos 14; 237-247.

C a s s a g n a u , P. (1961):  Ecologie  du sol dans les Pyrénées cen-
   trales.  Les biocénoses des Collemboles. Actualités scient. et in-
   dustr. 1283, Paris, Hermann.

D a v i s , B. N. K. (1963):  A study of microarthropod communities in
   mineral soils near Corby, Northants. J. Anim. Ecol. 32, 49-71.

D u n g e r , W. (1968):  Die Entwicklung  der Bodenfauna auf  rekultivierten Kippen  und  Halden des Braunkohlentagebaues.  Ein Beitrag
   zur  pedozoologischen  Standortsdiagnose. Abh. Ber. Naturkundemus.
   Görlitz 43, 2, 1-256.

D u n g e r , W. (1972): Systematische und ökologische Studien an  der
    Apterygotenfauna des Neissetales bei Ostritz/Oberlausitz. Abh. Ber.
    Naturkundemus. Görlitz 47, 4, 1-42.
G i s i n , H. (1947): Analyses et synthèses biocénotiques. Arch. Sci.
    Phys. Nat. 29, 42-75.
L e b r u n , Ph. (1971): Ecologie et biocénotique de quelques peuple-
    ments d'arthropodes édaphiques. Mém. Inst. Roy. Sci. Nat. Belgique,
    No. 165, Bruxelles.

## D i s c u s s i o n

H. U. T h i e l e : Wenn man eine grössere Anzahl von Tiergruppen
gleichzeitig untersucht, findet man Korrelationen zwischen Phytozönosen
und Zoozönosen. In einem kleineren, klimatisch einheitlichen Gebiet
sind Voraussagen über den Tierbestand einer Pflanzengesellschaft möglich. Umgekehrt kann man aus einer quantitativen Aufsammlung der epi-
gäischen Bodenfauna mit erheblicher Treffsicherheit auf die Pflanzengesellschaft schliessen.

W. D u n g e r : Der praktische Wert einer Orientierung an Phytozönosen bleibt unbestritten. Die Annahme, dass Phytozönosen und z.B. Synusien von Kleinarthropoden übereinstimmen, ist jedoch im Eizefall zu
überprüfen und erweist sich nicht immer als richtig, wie die Beispiele
zeigten.

G. M a r c u z z i : I would like to know Dr. Dunger's opinion about
the reason of negative correlation between 2 or more species.

W. D u n g e r : "Negatively correlated" is used in the common sense
of substitutive or vicarious species.

# The enigma of soil animal
# species diversity

J. M. ANDERSON

Department of Zoology, Animal Ecology Research Group,
Oxford

Temperate  "decomposer"  communities  show their highest levels  of development  in highly organic,  woodland  soils  where up  to a thousand
species  of soil animals,  including several hundred species  of micro-
arthropods, may be present in populations exceeding one to two millions
per square metre.  Investigations  of the trophic interrelationships of
these  communities  has  proved one  of the most intractable ecological
problems.

   Reviews  of the feeding biology of soil animals  in Burges and Raw
(1967)  and by  Wallwork (1970)  show that not only the major groups of
soil animals  but  also individual related  and  unrelated species have
surprisingly  similar feeding habits.  Detailed feeding studies  at the
species  level,  such  as those  of O´Connor (1967)  on *Enchytraeidae*,
Christiansen (1964)  and  Anderson  and  Healey (1972)  on  *Collembola*,
Healey and Russel-Smith (1970)  on *Diptera* larvae  and Luxton (1972) on
*Cryptostigmata*,  show that while generalised feeding groups  can be recognised there is in general less evidence of food differentiation than
in other trophic levels.  Herbivores,  for example,  show  a far higher
degree of food specialisation and monophagy.

   The existence  of these  highly species diverse animal communities
exhibiting  a low degree  of food resource compartmentalisation appears
to be in conflict with current ecological theory.

   There seem  to be three main hypotheses which  could  account  for
this phenomenon:

   1. There is an excess of food available to soil animals.

   2. There are undetected differences in resource utilisation.

   3. There are undetected microhabitat differences  between species.

   1. Slobodkin et al. (1967)  have  argued  on logical grounds  that
there cannot be  an excess  of food available  to decomposers, at least
in ecosystems  in steady state,  as there  is no evidence  of long term
accumulation of organic detritus in terrestrial ecosystems.  It is argu-

able, however, as to whether even mature temperate ecosystems,  such as climax woodlands, attain steady state conditions  but  rather  exist in a highly dynamic state of long term equilibrium.  Witkamp  and  van der Drift (1961),  for example,  have shown for  a mature oak woodland that major annual variations  in the macroclimate can result  in significant year to year differences in the standing crops of leaf litter  and soil organic matter. Most detritivores may therefore exist in non-equilibrium situations,  with food resources  and their own populations fluctuating irregularly from year to year  in varying degrees of relationship  from limiting conditions  to those of excess.  Species which may be actively competing at one time and place may be sharing an excess at others.

A low ratio of biomass to food resources  could also be maintained where soil animal populations are limited by environmental factors such as climate, competition for limiting resources other than food, or pre-dation.  There is no conclusive evidence  that high or low temperatures are a significant mortality factor  in soil animal populations,  though the eggs of many *Cryptostigmata* are killed by exposure to freezing con-ditions  in the laboratory  (Wallwork 1970).  In the field eggs must be deposited  in suitable microhabitats  to avoid  freezing  temperatures, desiccation and predation. Competition for these oviposition sites could be intense and contribute significantly to population regulation as has been shown for co-existing *Psocoptera* species by Broadhead and Wapshere (1966).  In this case  the intensity  of intraspecific competition  for oviposition sites reduced interspecific competition for the shared food resource.

Few soil animals  can survive  long periods  of dry conditions and drought  has been demonstrated  as a major mortality factor  in popula-tions of  *Enchytraeidae* (Nielsen 1955)  and *Tipulidae* (Cragg 1961). The extent to which microarthropod populations  are affected  by dry condi-tions, other than by indirect effects  such as the reduced availability of food resources, remains uncertain.

Predators may have  a fundamental role  in maintaining the species diversity of animal communities  by holding down their prey populations to levels where competition between the prey species is reduced.  Paine (1966) has demonstrated this effect experimentally in marine intertidal communities  where the top carnivore, a starfish, prevented the exploi-tation of space  by bivalve molluscs and barnacles.  The removal of the starfish resulted in a 15 species system  being reduced to an 8 species system.  Similar effects of herbivores on plant species diversity  have been shown by Harper (1969) and Janzen (1970).

In soil communities most groups  of animals contain predatory spe-cies and what little information  is available on their activities sug-

gests   that their effects on other trophic levels   may be considerable.
Spiders have been implicated as major arthropod predators by Clarke and
Grant (1968), Moulder and Reichle (1972).   Clarke and Grant removed the
spiders from an area of woodland  and found that the populations of the
major prey species, *Collembola* and centipedes, significantly increased.

The gamasid mites  *(Mesostigmata)*  feed extensively  on *Collembola*
and *Cryptostigmata*  (Karg 1961, Lebrun 1970).  Martin (1969) has shown
experimentally that the gamasids  are extremely efficient predators under the confined conditions of soil microhabitats.  The effects of mite
predation  on *Collembola* populations  has been demonstrated  by  Sheals
(1966) who found that *Collembola* increased following the application of
DDT insecticide to soils. This effect was attributed to the sensitivity
of the gamasids to DDT while the *Collembola*  were relatively unaffected
and increased in response to the reduced predation pressures.

2. In most studies  of the feeding biology  of soil animals a des-
criptive, trophic classification (such as that proposed by Luxton 1972)
has  been used  to compare   species.  A qualitative comparison  of this
type is insensitive  at the level  of competitive species interactions.
A sensitive quantitative method  of gut content analysis,  by  counting
food particles precipitated on to grid-marked filters,  has been developed by Anderson and Healey (1970).  This method was used in a study of
the feeding relationships of 3 species of surface living  *Entomobryidae*
*(Collembola)*.  Significant differences were found in the proportions of
fungal  and  higher plant material selected by two genera  but no major
feeding differences  were detected  for apparently sympatric congeners.
There are,  however,  a number of strategies which can be used  by soil
animals to subdivide shared food resources which would be difficult, or
impossible, to detect by this type of analysis.

It is important  to distinguish  between what is ingested  by soil
animals  and  what is actually assimilated as food.  For example,  soil
animals apparently feeding on a mixture of fungal  and  plant material,
and  regarded  as non-specialised feeders,  may be  utilising plant material, fungal material, bacteria or other fine particles in suspension
or materials in solution.

Zinkler (1970) and Luxton (1972) have investigated the enzyme com-
plements of a number of *Collembola* and *Cryptostigmata* species  and conclude  that enzymes capable of digesting structural polysaccharides are
more widespread  than was formerly thought.  Most of the animals investigated  which ate higher plant materials  could apparently produce xy-
lenase, pectinase and cellulase  which would enable them to digest cell
wall components.  While some animals may vary  in the range of activity

of their own digestive enzymes, others may have intestinal symbionts which serve the same function.

Von Törne (1967a,b) has shown, for example, that some species of *Collembola* vary in the ability of their specific gut microflora to digest particular components of their food. This situation, plus coprophagy, would enable species to eat a common range of food materials whilst subdividing the resources they contained.

Szabó and Buti (1969) have shown a different form of soil animal/micro-organism interaction in bibionid larvae *(Bibionidae; Diptera)*. Instead of these diptera larvae maintaining a specific gut flora there is selection for two or three of the bacterial species ingested with the food which proliferate rapidly in the favourable gut environment. These bacterial populations may assist in the breakdown of gut content material and/or form a food resource for the larvae.

The soluble, exocatabolic products of micro-organisms could form a food resource for soil animals ingesting decomposing plant materials. Walsh and Harley(1962), for example, have shown that the fungus *Chaetomium globosum* is capable of breaking down cellulose to monosaccharides. *Chaetomium* referentially absorbs glucose so that where hydrolysis exceeds uptake an excess of fructose over glucose is set free and is available to other organisms.

The majority of soil animals are apparently obligate or facultative microphytophages (Anderson in prep., Dash and Cragg 1972). This is not surprising in view of the high nutritive value of microbial protoplasm compared with the substrates on which they are growing. Fungal protoplasm frequently has carbon/nitrogen ratios lower than 10/1 in contrast with the C/N ratios of dead plant material which are usually over 50/1 and may exceed 100/1 (Wittich in Wallwork 1970). Some soil animals may not digest fungal material passing through their guts. Healey (pers. comm.) showed that some *Collembola* may digest materials present on the surface of fungal hyphae and spores without significantly reducing the viability of the fungal material.

The simplest strategy which could be adopted by soil animals for sub-dividing microbial food resources would be by selecting bacterial or fungal species. Selection of individual fungal species has been shown in the laboratory by Hartenstein (1962), Luxton (1966) and Mignolet (1970) but Dash and Cragg (1972) have demonstrated, by placing fungal baits in the field, that little selectivity is shown by natural populations of a wide variety of soil animals. Petersen (1971) has reviewed the literature on the feeding biology of *Collembola* and suggests that preference is related to the availability of items in microhabitats occupied by the *Collembola* at a particular instand in time.

3. Microhabitat differences  between species  are difficult to establish because of the complexity of the soil micro-structure. Anderson
(1971) has used soil sections to make detailed observations  on the microdistribution of *Cryptostigmata*. Marked preferences were detected for
particular micro-horizons,  often separated  by only a few millimetres,
which  appeared to be closely related  to the feeding habits and to the
physiological tolerances of these mites. These micro-distributions were
maintained  with little change throughout the year and the desiccation-
-resistant surface-living species only showed a downward redistribution
during a drought when the dry leaf litter  became  physically unstable.
The implication of these well defined  and maintained microdistribution
patterns is that their spatial separation reduces competition  and·ver-
tical  redistribution  may bring species  into  competitive interaction
which are normally spatially separated.  If this hypothesis  is correct
species undergoing vertical migration  should cease feeding  when  they
move into more 'species packed' horizons.

In addition to vertical micro-distribution patterns, the distribution of soil animals  may be affected  by the horizontal composition of
leaf litter layers.  Anderson (1973) has shown how the litter layers in
a mixed species woodland  could form a mosaic,  unpredictable  in space
and time,  under  the influence of wind,  frost and season  on the leaf
fall of individual tree species.  Under these conditions a single layer
of the profile  could represent  a large number  of 'compartments' each
containing  a range  of different resources, and between which movement
is so restricted  that they function effectively  as a mosaic of microhabitats each favouring different species of soil animals.

It is not clear  which of these hypotheses can account for the low
trophic specialisation  and  high species diversity of soil animal communities. It is likely that all interact to reduce the pressure of competition  between  the large number  of species utilising superficially
rather uniform resources in most soil habitats.


R e f e r e n c e s

A n d e r s ● n ,  J. M. (1971):  Observations  on the vertical distri-
     bution of *Oribatei (Acarina)* in two woodland soils. IV. Coll. Pedo-
     biologiae.  C.R. 4eme  Coll. Int. Zool. Sol. Ed.  I.N.R.A.  Paris,
     257-272.

A n d e r s o n ,  J. M. (1973):  Stand structure  and  litter  fall of
     a coppiced beech *(Fagus sylvatica)*  and  sweet chestnut  *(Castanea
     sativa)* woodland, Oikos 24.

A n d e r s o n , J. M. and H e a l e y , I. N. (1972): Seasonal and interspecific variations in major components of the gut contents of some woodland *Collembola*. J. Anim. Ecol. 41, 359-368.

B r o a d h e a d , R. and W a p s h e r e , A. J. (1966): Mesopsocus populations on larch in England - the distribution and dynamics of two closely-related co-existing species of *Psocoptera* sharing the same food resource. Ecol. Monogr. 36, 327-388.

B u r g e s , A. and R a w , F. (Eds.) (1967): Soil Biology. Academic Press, London and New York.

C h r i s t i a n s e n , K. (1964): Bionomics of *Collembola*. Ann. Rev. Entomol. 2, 147-178.

C l a r k e , R. D. and G r a n t , P. R. (1968): An experimental study of the role of spiders as predators in a forest litter community. Ecology 49, 1152-1154.

C r a g g , J. B. (1961): Some aspects of the ecology of moorland animals. J. Anim. Ecol. 30, 205-234.

D a s h , M. C. and C r a g g , J. B. (1972): Selection of microfungi by *Enchytraeidae (Oligochaeta)* and other members of the soil fauna. Pedobiologia 12, 282-286.

H a r p e r , J. L. (1969): The role of predators in vegetational diversity. In: Diversity and stability in ecological systems. Brookhaven Symp, Biol. 22, 48-62.

H a r t e n s t e i n , R. (1962): Soil Oribatei. I. Feeding specificity among forest soil Oribatei. Ann. Entomol. Soc. Am. 55, 202-206.

H e a l e y , I. N. and R u s s e l - S m i t h , A. (1971): Abundance and feeding preferences of fly larvae in two woodland soils. IV. Coll. Pedobiologiae. C.R. 4eme Coll. Int. Zool. Ed. I.N.R.A. Paris, 177-191.

J a n z e n , D. H. (1970): Herbivores and the number of tree species in tropical forests. Am. Naturalist 104, 501-528.

K a r g , W. (1961): Ökologische Untersuchungen von edaphischen Gamasiden *(Acarina, Parasitiformes)*. Pedobiologia 1, 77-98.

L e b r u n , P. (1970): Ecologie et biologie de *Nothrus palustris* (C.L.Koch) Acarien - Oribate Acarologie 12, 827-848.

L u x t o n , M. (1966): Laboratory studies on the feeding habits of saltmarch *Acarina,* with notes on their behaviour. Acarologia 8, 163-175.

L u x t o n , M. (1972): Studies on the oribatid mites of a Danish beech wood. I. Nutritional Biology. Pedobiologia 12, 434-463.

M a r t i n , F. J. (1969): Searching success of predators in artificial litter. Am. Mid. Nat. 81, 218-227.

M i g n o l e t , R.(1971):Etude des relations entre la flore fongique et quelgues espēces d'oribates *(Acari)*. IV. Coll. Pedobiologiae. C.R. 4ème Coll. Int. Zool. Sol. Ed. I.N.R.A. Paris, 155-164.

M o u l d e r , B. C. and R e i c h l e , D. E. (1972): Significance of spider predation in the energy dynamics of forest floor arthropod communities. Ecol. Monogr. 42, 473-498.

N i e l s e n , C. O. (1955): Studies on the Enchytraeidae. 2. Field studies. Natura jutl. 4, 1-58.

O' C o n n o r , F. B. (1967): The *Enchytraeidae*. In: Soil Biology. (Ed. A.Burges and F.Raw). Academic Press, London and New York, 213-257.

P a i n e , T. (1966): Food web complexity and species diversity. Am. Naturalist 100, 65-75.

P e t e r s e n , H.(1971): Collembolernes ernaeringsbiologi og dennes økologiske betydning. Entomol. Meddl. 39, 97-118.

S h e a l s , J. G. (1955): The effects of DDT and BHC on soil *Collembola* and *Acarina*. In: Soil Zoology. (Ed, D.K. McE. Kevan). Butterworths, 241-252.

S l o b o d k i n , L. B., S m i t h , F. E. and H a i r s t o n , N. C. (1967): Regulation in terrestrial ecosystems, and the implied balance of nature. Am. Naturalist 101, 109-124.

S z a b ó , I., M a r t o n , M. and B u t i , I. (1969): Intestinal microflora of the larvae of St.Mark's Fly. IV. Studies on the intestinal bacterial flora of a larval population. Acta microbiol. Acad. Sci. Hung. 16, 381-397.

T ö r n e , E. von. (1967): Beispiele für mikrobiologene Einflüsse auf den Massenwechsel von Bodentieren. Pedobiologia 7, 296-305.

W a l l w o r k , J. A. (1970): Ecology of Soil Animals. McGraw-Hill. 283 pp.

W a l s h , J. H. and H a r l e y , J. L. (1962): Sugar absorption of *Chaetomium globosum*. New Phytol. 61, 299-313.

W i t k a m p , M. and v a n  d e r  D r i f t , J. (1961): Breakdown of forest litter in relation to environmental factors. Plant and Soil 15, 295-311.

Z i n k l e r , D. (1971): Carbohydrasen streubewohnender Collembolen und Oribatiden. IV. Coll. Pedobiologiae. C.R. 4ème Coll. Int. Zool. Sol. Ed. I.N.R.A, Paris, 329-334.

D i s c u s i o n

G. M a r c u z z i : 1.The utilization of plant materials versus microorganisms depends mainly on animal size = macrofauna utilizes mainly

plant materials, microfauna utilizes often microorganisms.  2. The non-
equilibrum has an important role particulary in non-climax communities.

J. M. A n d e r s o n : 1. I agree that there are these differences but
they are generalisations and our methods of analysis are insensitive at
the species level. 2. Yes.

C. A t h i a s : I think, the 3 hypotheses you propose are true at least
partly.  I think also there  is a 4th one- specialisation  in time: for
instance soil gamasids  may be separated into 2 groups owing to the way
they exploit  the medium  in time:  1st group species exploit the medium
throughout the year,  other ones  do it only during spring,  or summer,
etc. Regarding your 1st hypothesis,  I think  my colleague Bouché could
explain to us some ideas on the consequences of glaciations  on temperate ecosystems.

J. M. A n d e r s o n : Yes, in addition to animals  such as Collembola
which show anamorphic development  have a non-feeding period before and
after moulds so that at any instant in time a significant proportion of
the population are not utilising food resources.

H. U. T h i e l e : In my opinion competition  can only be convincingly
demonstrated  by laboratory  experiments.  But one cannot realize field
conditions  in such experiments,  so that it is impossible  to conclude
from a competition effect found in the laboratory that competition does
take place also in the field.

J. M. A n d e r s o n : I agree. In many laboratory experiments the in-
teracting species may not be presented  with optimal environmental conditions corresponding to their niches.  Under these conditions competitive exclusion in the laboratory  might be in the opposite direction to
that in the field. Most species are probably not competing actively for
resources  because  of the strategies  I have described  for the subdivision of food and space.

M. B o u c h é : 1. Le 3 hypothèses sont en fait complementaires et résultant de nos démarches  de recherches. 2. Pour les differentes para-
metres il y a de grandes problèmes d'échelles "geologique".

J. M. A n d e r s o n : Any consideration of space and time must be related to the organism's relationships  with  these parameters.  We must
also recognise  how slow  some ecological phenomena can be.  A competi-
tively superior species might take tens  of years  to exclude a weaker
species and if we sampled the community  during this time we might  assume that the two species were co-existing on a stable basis.

# Ecological characteristics of xerothermic rendzinas

JARMILA KUBÍKOVÁ AND JOSEF RUSEK

Entomological Institute of the Czechoslovak Academy of Sciences, Prague

The aim of this work was to get more details about a developmental series of rendsina soil types. The problem was approached from a botanical, microbiological, zoological, ecological, and pedological viewpoint.

SITE AND LOCALITIES

The model site was situated on the hill Doutnáč in the Czech Karst, 40 km SW from Prague 428 m above sea level. The parent rock are limestones of silurian and devonian origin. The climate of the Czech Karst is defined as moderately warm, moderately dry, mostly with moderate winter. From the biogeographical point of view, the locality lies in the region of thermophilic flora, the so-called Pannonicum. In this region the favourably situated southern slopes of hills host remarkable xerophilous and thermophilic communities. They are very well developed on the southern slope of the chosen hill Doutnáč, where a developmental series from a rocky steppe, and xerophilous grassland to thermophilic oak wood can be differentiated. A scheme of their distribution on a profile through the studied site is shown in Fig. 2.

The soils of the Czech Karst were studied previously. The limestone slopes are covered by various developmental stages of xerothermic rendsinas.

BRIEF DESCRIPTION OF THE LOCALITIES

L o c a l i t y 1: 410 m above sea level, $15^u$ inclination, parent rock silurian limestones, closed xerothermic grassland, ass. *Carex humilis- -Festuca sulcata* Klika 1932. Soil type: Mullartige Rendsina (Kubiěna

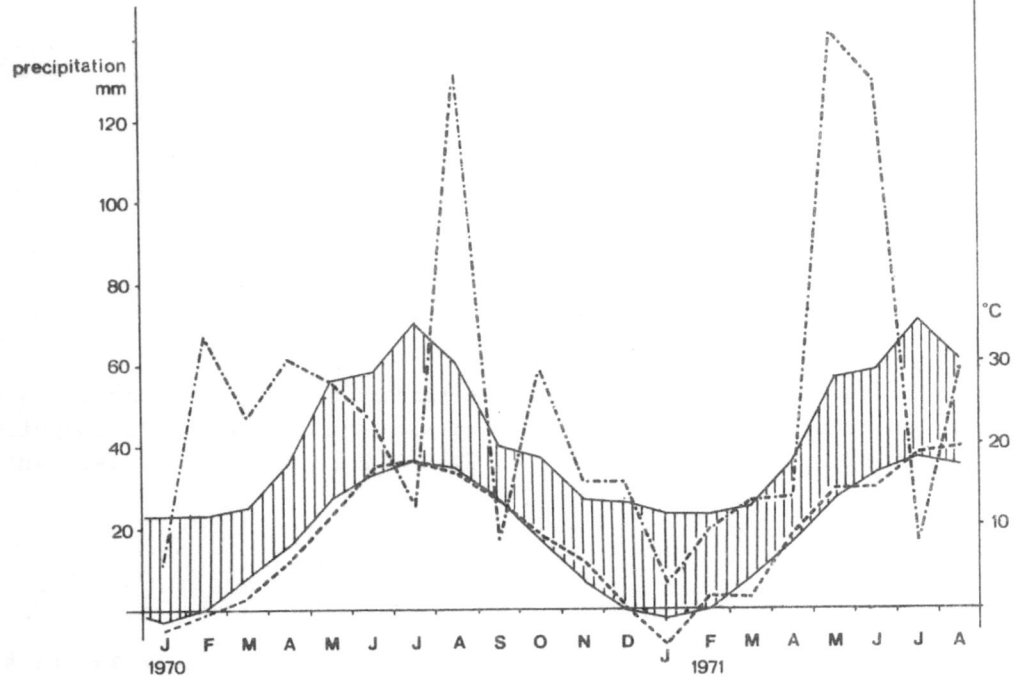

Fig.1.  Monthly  averages of temperature and monthly sum
of  precipitation  measured  in the station  Králův Dvůr
during the studied period  are figured on the grounds of
"Klimadiagramm" for this station.

1953), soil profile 20 cm deep, A-C, A horizon differentiated in a thin
subhorizon $A_{oo}$ (0 - 1 cm)  formed by not decomposed died off mosses and
plant remnants, and in $A_1$ subhorizon of dark brown colour, densely grown
through by roots of perennial grasses,  with a high portion of skeleton
of different categories from fine sand to gravel, the fine earth forming
crumbs of two diameters (3 mm and 0.5 mm).

L o c a l i t y  2: 395 m above sea level, 20° inclination, parent rock
silurian limestones,  stunted  thermophilic oak wood  with  dense shrub
layer, ass. *Lathyro (versicoloris) - Quercetum pubescentis* (Klika 1938)
Jakucs 1960.  Soil type:  Moderrendsina, soil profile 25 to 30 cm, A-C,
$A_{oo}$ (0 - 2 cm) - not decomposed plant leaves  and  branchlets, $A_{o1}$ (2 -
- 10 cm) - Arthropoden Moder of dark brown colour, $A_1$ (10 - 25 cm) light
**brown** Mull  with earth worm pellets.  Great portion  of skeleton mainly
in the layer of 5 to 8 cm.

L o c a l i t y  3: 380 m above sea level, 30° inclination, parent rock
silurian limestones, opened rocky steppe, ass. *Festuca duriuscula-Seseli
osseum*  Klika 1933.  Soil type:  Protorendsina only on small horizontal

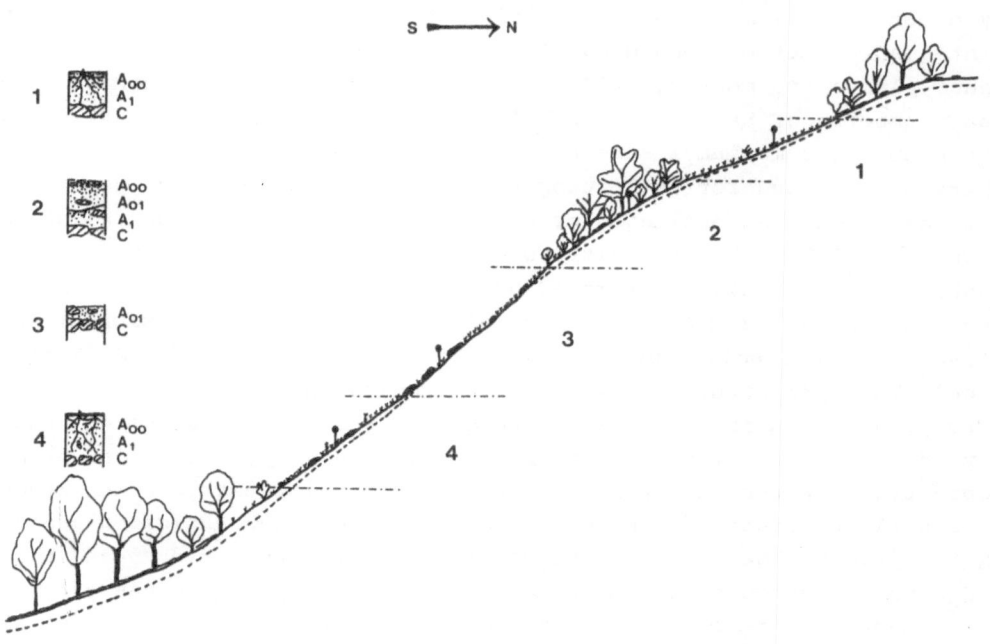

Fig.2. Transect across the southern slope of Doutnáč
showing the localities of studied ecotopes: 1-xerothermic
grassland (ass. *Carex humilis-Festuca sulcata*), 2-thermo-
philic oak wood (ass. *Lathyro (versicoloris) - Quercetum
pubescentis*), 3-rocky steppe (ass. *Festuca duriuscula-
Seseli osseum*), 4-xerothermic grassland (ass. *Festuca
valesiaca-Erysimum crepidifolium*).

terraces, soil profile 2 to 5 cm, composed of coprogenic humus and
skeleton. The microclima of this locality is most extreme and highest
values of air and soil temperatures and also evaporation were estimated
here (Pachnerová 1960, Chocholová 1960/.

L o c a l i t y  4: 375 m above sea level, $25°$ inclination, closed xerothermic grassland, ass. *Festuca valesiaca - Erysimum crepidifolium* Klika
1933. Soil type and profile the same as Loc. 1.

The samples were taken from the centre of a well differentiated
phytocoenosis, repeatedly from a previously delimitated quadrat.


METHODS


To obtain the completest possible picture about biotic and abiotic
factors in the studied habitats, these various methods were used:

phytocoenological relévés were estimated according to Braun-Blanquet, point-quadrat method was used for estimating relative frequency of higher plants and mosses, dilution method was used for estimating the total numbers of bacteria (on MPA), and total numbers of fungi (on Martin nutrient medium); the extraction of soil samples in Tullgren's apparatus was used for estimating the numbers of Tracheata; the biomass of mesoedaphon was estimated by computation using the index according to Dunger (1968). The activity of soil organisms was followed by $CO_2$ uptake from the soil, estimated by the incubation of soil samples in vitro and the titration of eliminated $CO_2$; the disentegration of cellulose in intact soil was followed according to Tesařová and Úlehlová (1968). The microstructure of soil was studied on soil slides according to Kubiena's methods; soil water relations were estimated using momentary soil humidity, maximal capillary water capacity, number of hygroscopicity. For the estimation of CaO and $K_2O$ 1 % citric acid extract and its flame photometry was used, total nitrogen was estimated according to Kjeldahl-Foerster, total carbon according to Springer and Klee, pH $H_2O$ was measured electrometrically. The soil temperature was estimated by inversion of sucrose according to the method of Pallmann et al.

The sampling was repeated several times during April 1970 to June 1971 to follow the seasonal dynamics of above-mentioned factors.

## DYNAMICS OF ENVIRONMENTAL AND BIOTIC FACTORS

Environmental factors, soil and biota of studied ecosystems were estimated in approximately three month-intervals. The main interest was centered on the soil and on the processes in it. The results are figured and the ranges of the values for individual ecotopes are given in Table 1.

## BIOTIC FACTORS

In the studied ecotopes, the phytocoenosis of higher plants and communities of soil Arthropoda were studied in detail, together with the changes in their population structure during one year. The diversity of studied ecotopes is given in Figs. 3 and 4.

T a b l e  1.   Characteristics of studied ecotypes

| Ecotope and No. of locality | Thermophilic oak-wood No.2 | Xerothermic grassland No.1, No.4 | Rocky steppe, No.3 |
|---|---|---|---|
| Plant community | Lathyro (versicoloris)-Quercetum pubescentis (Klika) Jakucs 1960 | Carex humilis-Festuca sulcata Klika 1932 (1) Festuca valesiaca-Erysimum crepidifolium Klika 1933 (4) | Festuca duriuscula-Seseli osseum Klika 1933 |
| Characteristic plant species | Quercus pubescens, Cornus mas, Lathyrus versicolor, Dictamnus albus, Primula veris | Festuca sulcata, Carex humilis, Thalictrum minus, Festuca valesiaca, Filipendula hexapetala | Festuca glauca, Seseli devenyense, Potentilla arenaria, Thuidium abietinum |
| Characteristic species of soil Arthropoda | Folsomia quadrioculata, Cryptopygus bipunctatus, Isotomiella minor, Metaphorura bipartita | Isotomodes productus, Sphaeridia sp., Cryptopygus bipunctatus, Doutnacia xerophila (1) Isotomodes sexsetosus, I. armatus, I. productus (4) Sphaeridia sp.,Leptophyllum nanum (1,4) | Xenylla boerneri Entomobrya cf. handschini |
| Characteristic species of soil fungi (according to Fassatiová 1966) | Absidia heterospora, Penicillium brevicompactum, P. diversum, Fusarium javanicum | Aspergillus ochraceus, A. ustus, Penicillium casei, P. frequentans, Fusarium oxysporum, Myrothecium verrucaria (1) | |

Table 1 continued

| Ecotope and No. of locality | Rocky steppe, No.3 | Xerothermic grassland No.1, No.4 | Thermophilic oak-wood No.2 |
|---|---|---|---|
| Total numbers of soil bacteria (in 1 g soil dry matter) | $1,65 - 4.94 \cdot 10^6$ | $2,64 - 10.66 \cdot 10^6$ (1) <br> $3.74 - 6.83 \cdot 10^6$ (4) | $3.45 - 6.90 \cdot 10^6$ |
| Total numbers of soil fungi (in 1g soil dry matter) | $0.55 - 1.83 \cdot 10^5$ | $1,11 - 2.33 \cdot 10^5$ (1) <br> $0.84 - 2.15 \cdot 10^5$ (4) | $1.04 - 3.55 \cdot 10^5$ |
| Total numbers of soil Arthropoda (on 100 $cm^2$) | *Acarina* 280 - 420 <br> *Collembola* 20 - 110 | 320 - 1340 <br> 40 - 360 | 470 - 1130 <br> 60 - 260 |
| Biomass of soil Arthropoda (mg/$m^2$) | *Acarina* 120 - 510 <br> *Collembola* 20 - 250 | 160 - 1620 <br> 100 - 440 | 840 - 1540 <br> 160 - 780 |
| Disintegration of cellulose (mg/$dm^2$. day) | 0.08 - 9.75 | 0.61 - 8.97 (1) <br> 0.57 - 9.47 (4) | 0.19 - 6.89 |
| Uptake of $CO_2$ (mg $CO_2$/day. 100 g soil dry matter) | 24.29 - 64.61 | 33.3 - 55.5 (1) <br> 29.5 - 61.1 (4) | 33.9 - 62.7 |
| Soil type (according to Kubiena) | Protorendsina | Mullartige Rendsina | Moder Rendsina |
| Soil profile | $A_{Ol}$ 0 - 5 cm <br> C | $A_{OO}$ 0 - 1 cm <br> $A_1$ 1 - 15 cm <br> C | $A_{OO}$ 0 - 1 cm <br> $A_{Ol}$ 1 - 10 cm <br> $A_1$ 10 - 25 cm <br> C |
| Structure | Without structure | Crumby structure | uper horizon without structure, lower horizon crumby str. |
| Minerals | limestone 0.025 mm <br> quartz 0.2 - 0.002 mm | limestone 0.025 mm <br> quartz 0 2 - 0.002 mm | limestone 0.025 mm <br> quartz 0.2 - 0.002 mm |

Table 1 continued

| Ecotope and No. of locality | Rocky steppe, No.3 | Xerothermic grassland No.1, No.4 | Thermophilic oak-wood No.2 |
|---|---|---|---|
| Arthropoda pellets | Oribatidae, Collembola 30-50 m | Diplopoda 500 m<br>Oribatidae 30-50 m<br>Collembola<br>Enchytraeidae 120-200 m<br>Lumbricidae Ø 5 mm | Diplopoda 500 m<br>Isopoda 350 m<br>Diptera larv. 1400 m<br>Oribatidae 35 m<br>Collembola 30-75 m<br>Enchytaeidae 120 m<br>Lumbricidae Ø 5 mm |
| Maximal capillary water capacity (g/100 g soil dry matter) | 80.2 | 149.5 (1)<br>143.1 (4) | 115.1 |
| Fluctuations of soil humidity in the studied period (in percent of max. cap. water capacity) | 9.7 - 139,3 | 17.5 - 59.8 (1)<br>14.1 - 59.7 (4) | 22.9 - 81.4 |
| pH | 7.29 - 7.8 | 7.09 - 7.6 (1)<br>7.18 - 7.65 (4) | 7.12 - 7.7 |
| CaO (% of dry matter) | 0.35 - 1.08 | 0.24 - 0.81 (1)<br>0.29 - 0.78 (4) | 0.26 - 0.93 |
| C/N | 8.9 - 16.8 | 10.4 - 21.4 (1)<br>11.0 - 15.4 (4) | 13.8 - 20.6 |
| Humus content (%) | 22.9 - 28.9 | 25.3 - 30.2 (1)<br>21.9 - 28.6 (4) | 33.5 - 34.5 |
| $K_2O$ (mg %) | 9.5 - 20.0 | 9.5 - 20.0 (1)<br>10.0 - 18.0 (4) | 17.5 - 35.0 |

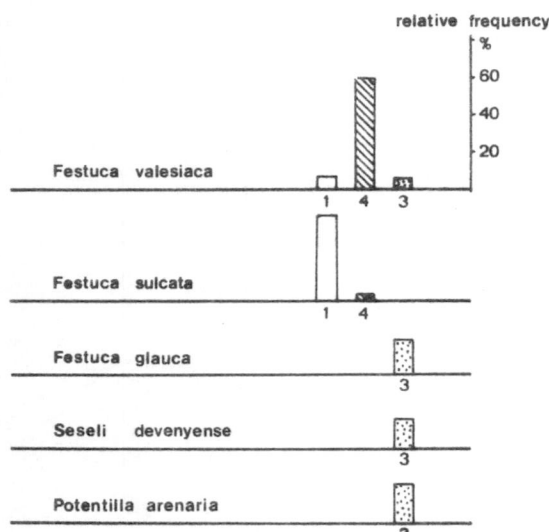

Fig.3. Relative frequency of some characteristic plant species on grassland ecotopes, showing the differences in plant cover.

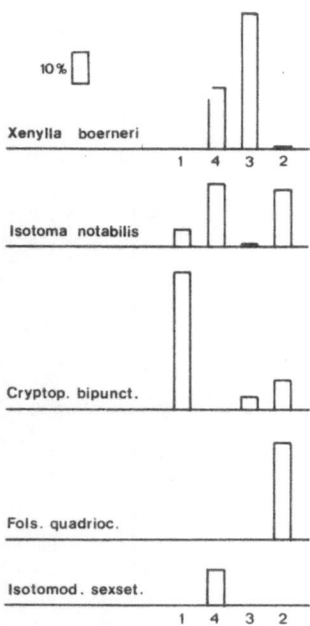

Fig.4. Dominancy of characteristic species of soil *Arthropoda* on studied ecotopes, showing the differences of animal communities.

THE DYNAMICS OF SOIL MICROORGANISMS

The results for soil bacteria are shown in Fig. 5. The total numbers of
bacteria correspond well with similar results of other authors. Loub
(1960), who studied microbiological conditions of different soil types,

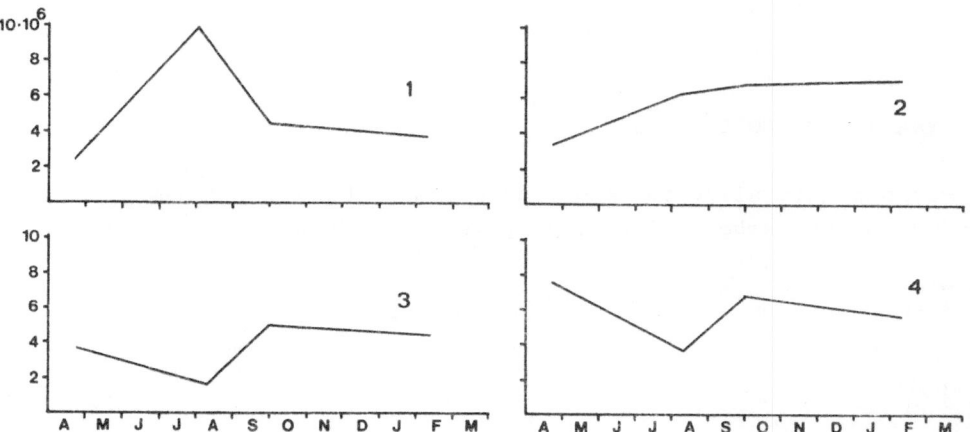

Fig.5. Dynamics of total numbers of bacteria in 1 g soil
dry matter on the ecotopes studied (see Fig.2).

gives for rendsinas values of total numbers of bacteria 3 to $4.10^6$ in
1 g soil dry matter; their seasonal fluctuations form a two-peak curve
with maximum in spring and autumn and minimum in summer and winter. Our
results revealed summer minima in rocky steppe and grassland, but no
decrease in the oak wood. Similar results were obtained by Seifert
(1960) for an analogical phytocoenosis in the Czech Karst. Lowest numbers of bacteria were found in the rocky steppe, highest in the oak wood.

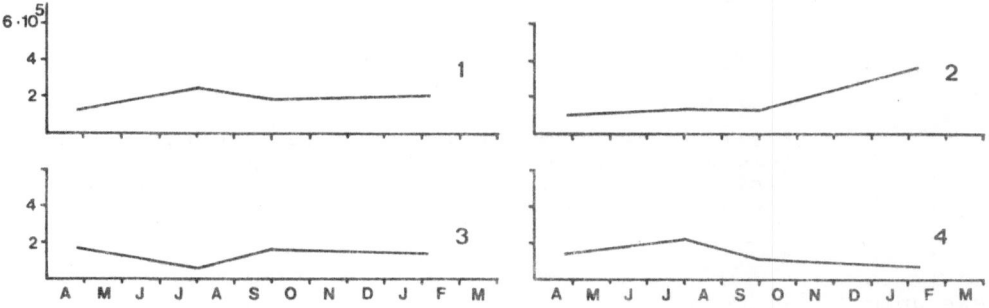

Fig.6. Dynamics of total numbers of fungi in 1 g soil
dry matter on the ecotopes studied (see Fig.2).

The results for soil fungi are in Fig. 6. There was no significant decline in soil fungi numbers in the summer, with the exception of the most extreme rocky steppe. The qualitative changes in the populations of soil fungi during the year shown by Loub (1960) for rendsinas, were confirmed for Doutnáč by Fassatiová (1966). The lowest numbers of fungi were found in the rocky steppe.

## THE DYNAMICS OF SOIL FAUNA

The methods used allow to make conclusions about the dynamics of soil *Acarina* and *Collembola*. Fig. 7 gives the dynamics of abundance of

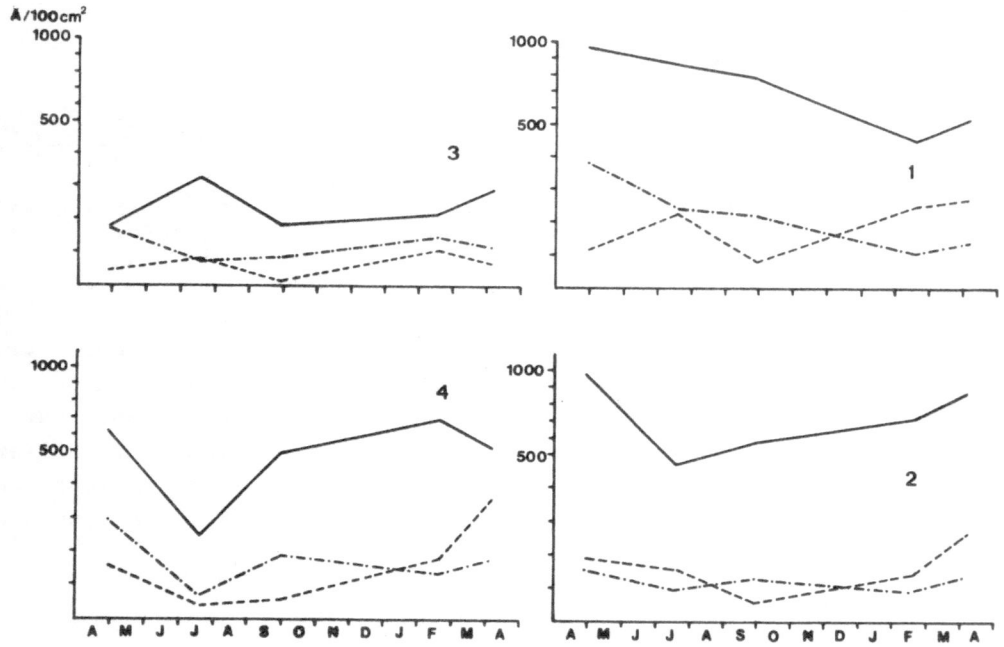

Fig.7. Dynamics of abundancy of soil *Acarina* and *Collembola* on 100 cm². on the ecotopes studied (see Fig.2) full line - Oribatidae, dashed line - Apterygota, dot-and--dashed line - Acarina excl. Oribatidae.

these animals on 100 cm². The highest values of abundance achieve the *Oribatidae* in all habitats, other *Acarina* and *Collembola* are much lower and approximately similar. The total abundance is lowest in the rocky steppe, highest in the grassland. The curve has a typical two-peak form, with maximum in spring and autumn.  A remarkable exception are *Oribatidae* in rocky steppe, where the maximum is achieved in summer.   Somewhat different   are the relations   in the dynamics of biomass,   which is affected by the age categories of the populations (see Fig. 8). The highest biomass was found in *Acarina* in the grassland ecotope (162o mg$/$km$^2$),

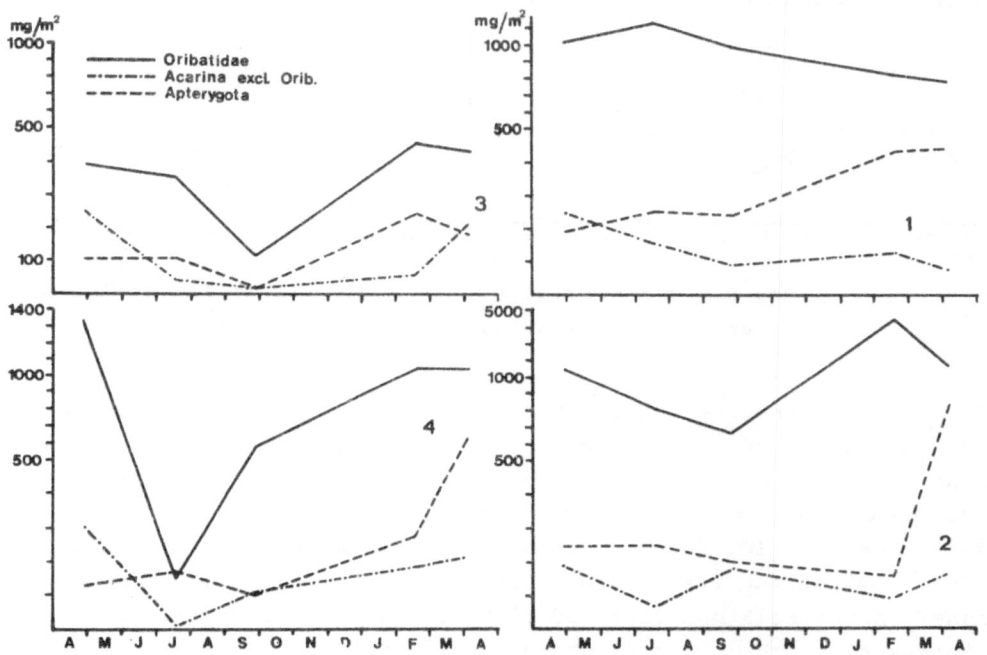

Fig.8. Dynamics   of biomass   of soil *Acarina* and *Collembola* (in mg/m$^2$) on the ecotopes studied (see Fig.2).

*Collembola*   on   the contrary   had   the highest biomass   in the oak wood (780 mg/m$^2$) This may be caused by the occurrence of big species in this ecotope, which are absent from the grasslands.


THE ACTIVITY OF SOIL ORGANISMS

The activity of soil organisms   was measured by the $CO_2$ production from the soil (see Fig. 9). Most authors study this activity on soil samples wetted on 60 % maximal capillary water capacity. In our experiments, we

mg $CO_2$/day·100g soil dry matter

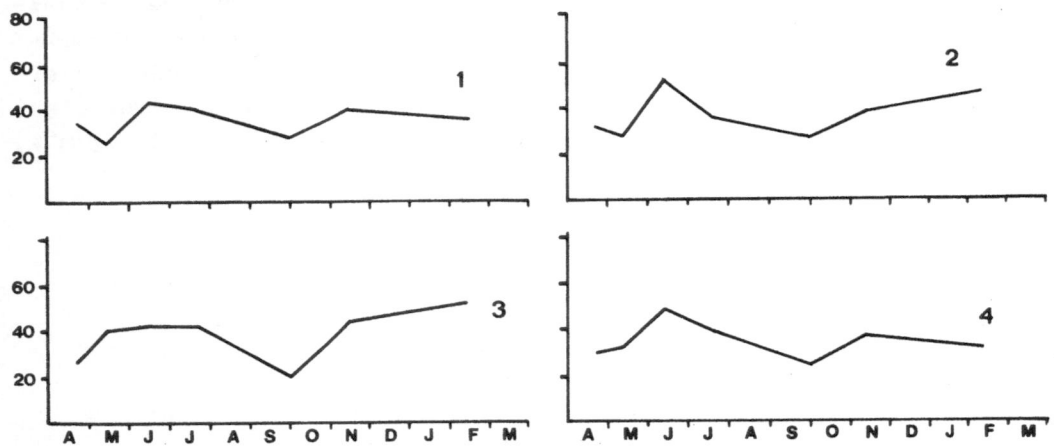

Fig.9.   Dynamics of $CO_2$ production from the soil samples
taken from the studied ecotopes (see Fig.2).

did not change the soil humidity of soil samples,  we wanted to get information  about the actual activity of soil organisms  under momentary
humidity conditions. The results have shown that under constant temperature there is direct linear regression of $CO_2$ production on the soil humidity, see Fig. 10. The significance of the regression coefficient was
proved.  Out of the regression equation it may be computed that the increase  in the soil humidity  of 10 % maximal capillary capacity causes
an increase  of the $CO_2$ production  of 4 mg/day.100 g  soil dry matter.
This is true in the range between 20 % to 100 % maximal capillary water
capacity. In very wet soils where also the non-capillary pores are full
of water, no further increase in the $CO_2$ production was found. Also very
dry soils enable very retarded activity of soil organisms  and the values do not fall in the above pattern.  A similar dependence of $CO_2$ production on soil humidity  was found by Seifert (1960) for rendsina soil
in the Czech Karst.  The differences in $CO_2$ production  between studied
ecotopes  are  thus  manifested  through the humidity conditions of the
habitat.


THE DECOMPOSITION OF CELLULOSE

The dynamics  of cellulose decomposition  in studied ecotopes  is given
in Fig. 11. There are remarkable differences in cellulose decomposition

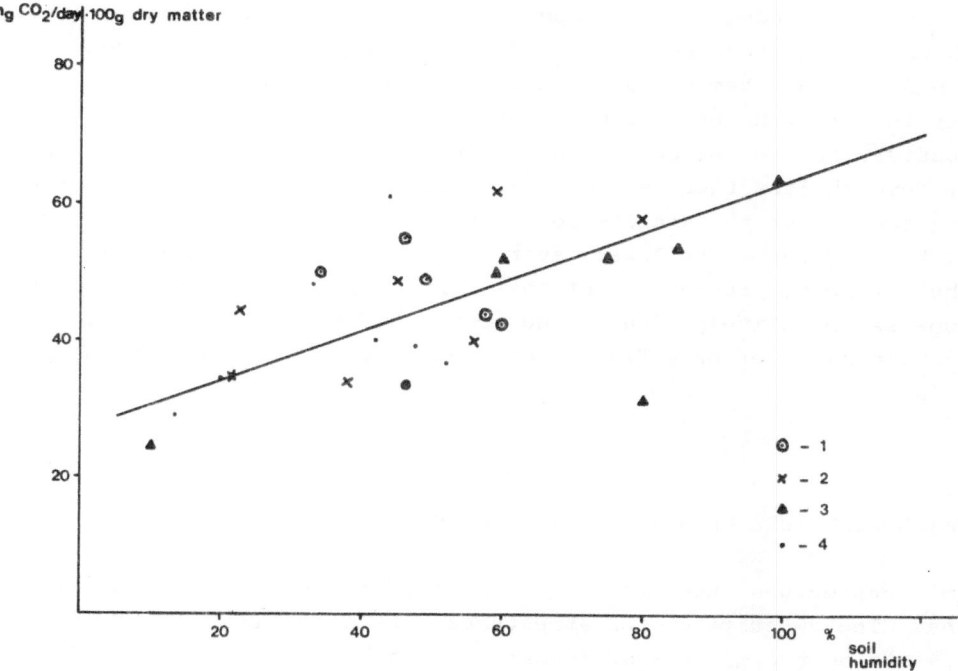

Fig.10. Regression of $CO_2$ production on soil humidity (in percent of maximal capillary water capacity). The significance of the regression coefficient was proved.

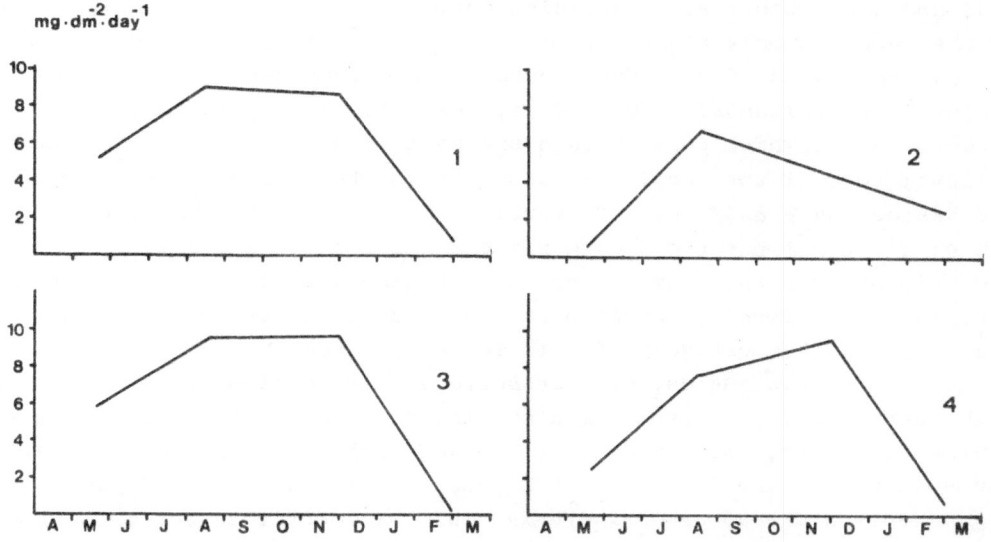

Fig.11. Dynamics of · cellulose disintegration on the ecotopes studied (see Fig.2).

found in grasslands, rocky steppe  and   oak wood.  The seasonal average
was highest  in rocky steppe,  an ecotope with the greates fluctuations
in humidity  and  temperature conditions.  Many authors have shown that
mainly the changing water conditions have a stimulatory affect on various activities  in the soil;  the cellulolytic activity  may be  one of
them. Nevertheless, there was no cellulose decomposition in rocky steppe
during the winter, the shallow soil profile of this open community being
completely frozen.  The lowest average for the period studied was found
in the oak wood;  regardless of this, the cellulolytic activity in this
ecotope was relatively high in the winter.  This is in correlation with
the development  of specific winter populations  of soil fungi and *Collembola*.

THE MICROSTRUCTURE OF SOIL PROFILE OF THE STUDIED ECOTOPES

The microstructure  was studied  on soil slides  embedded in artificial
resins.  The description  of structures  was done  according to Kubiěna
(1967) and on the grounds of Zachariae´s works (1965). T h e  r o c k y
s t e p p e  -  l o c .  3. The soil is without aggregate structure(according to Beckmann et Geyger in Kubiěna 1967), the humus form is the so
called Arthropoden Moder. The minerals are limestone, which forms gravel,
coarse and fine sand, and quartz  in the categories of coarse  and fine
sand and silt. Other soil particles are: pellets of soil animals - some
of them contain only plant remnants, e.g. pellets of *Oribatidae* Ø 50 µm,
of  *Collembola*  Ø 30 µm,  some contain also  some quartz  and limestone
grains, e.g. *Collembola*  30 - 50 µm,  exceptionally *Diplopoda*  500 µm;
fungous hyphae, which grow through the soil pores and connect the animal
pellets; some of them could be identified on the slides as Endogone sp.
and *Cenococcum graniforme*.  No vertical zonation of the profile exists.
T h e  x e r o t h e r m i c  g r a s s l a n d  -  l o c .  1  and 4.
The soil has typical crumby structure (Krümmelstruktur).  The minerals
are: limestone forming coarse particles with smallest ones about 25 µm,
and quartz in the category of silt and clay.  Pellets are of two kinds:
without  minerals inside, e.g. *Oribatidae*  and  *Collembola* Ø 50 µm, and
with quartz  and  limestone  grains  inside, e.g. *Collembola*  Ø 100 µm,
*Diplopoda* 500 µm, *Enchytraeidae* Ø 200 µm, *Lumbricidae* Ø 1000 - 2000 µm.
The soil pores  are inhabited by hyphae of soil fungi, out of which *Endogone* sp. and *Cenococcum graniforme* could be identified on soil slides.
The roots  of herbs  are very dense. · Vertical zonation  on the slides:
0 to 2 cm: *Oribatidae, Collembola* and individual *Diplopoda;*  2 to 5 cm:

pellets of *Lumbricidae*, disintegrated pellets of *Oribatidae* and *Collembola*, pellets of *Enchytraeidae*, individual *Diplopoda*.

T h e   t h e r m o p h i l i c   o a k   w o o d   -   l o c . 2 .   The microstructure differs  in different horizons, the upper $A_{O1}$ horizon is without structure, composed of loose Arthropoden Moder, the lower horizons  have a typical crumbly structure. Minerals are the same as in the above described localities. Pellets are of two kinds: without minerals, e.g. *Oribatidae* Ø 35 µm; with minerals, e.g. *Collembola* Ø 75 µm to 90 µm, *Enchytraeidae* 120 µm,  *Isopoda* 340 x 160 µm, *Diplopoda* 500 µm, *Diptera-*-larv. 1450 µm, *Lumbricidae* up to 5 mm. There are conspicuous hyphae of mycorrhizal fungi,  and some spores  of  *Endogone* sp. Vertical zonation on slides: 0 to 5 cm a layer  of free lying pellets  of *Diplopoda, Isopoda, Collembola* (the width of this layer  is subdued to seasonal fluctuations);  5 to 15 cm - pellets  of *Lumbricidae,  Diptera,* plant roots and plant remnants, disintegrated pellets of *Collembola,* much limestone skeleton.


CONCLUSIONS

The results of this study are summarized in Table 1.

The studied ecotopes form  a successive developmental series.  The primary stage  is represented here  by the rocky steppe  with  its open one-layer plant community, with an initial community of soil Arthropoda, poor in species, and by undeveloped structure of Protorendsina soil type. The whole profile of this soil is formed by small pellets of *Oribatidae* and *Collembola* mixed in limestone skeleton.  The conditions of humidity and  temperature  are most extreme here  and  the ranges  of most other values studied are very wide.

The middle stage  is formed  by the xerothermic grassland with its high plant cover,  differentiated into moss  and  herbaceous layer, inhabited by a rich community of soil animals. The soil may be identified as  a Mullartige Rendsina soil type,  with  more complicated structure, whose  A horizon  is densely rooted through higher plants,  and divided in subhorizons.  The upper part of the horizon  is formed by pellets of *Diplopoda, Oribatidae, Collembola* and *Enchytraeidae,*  the lower part by pellets of *Lumbricidae*. The ecological conditions are less extreme here; that enables the growing complexity of space  and population structure.

The most developed stage in the southern slope of the hill Doutnáč is the thermophilic oak wood,  divided in a tree,  shrub and herbaceous layer, with a very diversified community of soil animals. The soil is of the forest Moderrendsina soil type with very complex structure. A horizon is divided in subhorizons. The upper subhorizons are formed by a mixture of pellets of *Diplopoda, Isopoda, Diptera* larv., *Oribatidae, Collembola, Enchytraeidae,* by decaying leaves, roots and limestone skeleton; the lower subhorizons contain mainly pellets of *Lumbricidae* and limestone skeleton.

These results suggest a direct dependence on the succession of biocoenosis and soil development. With the growing complexity of the communities of living organisms the heterogeneity of soil structure grows and the factors effecting its development are getting more and more complicated. In the course of the succession of the ecosystem, a specific inner environment is formed, which transforms the influence of the outer environment (insolation, temperature, precipitation). Thus Kubiěna's idea expressing the soil development as a growth of heterogeneity was confirmed in detail.

R e f e r e n c e s

D u n g e r , W. (1968): Die Entwicklung der Bodenfauna auf rekultivierten Kippen und Halden des Braunkohletagebaues. - Abh. u. Ber. Naturkunde Mus. Görlitz, 43/2, 1-256.

F a s s a t i o v á , O. (1966): Bodenmikromyceten am Hügel Doutnáč im Böhmischen Karst. Preslia 38, 1-14.

K l i k a , J. (1939): Die Gesellschaften des *Festucion valesiacae*-Verbandes in Mitteleuropa. Studia Bot. Čech. 2, 117-157, Praha.

K u b i ě n a , W. L. (1953): Bestimmungsbuch und Systematik der Böden Europas. Stuttgart

K u b i e n a , W. L. (1955): Animal activity in soils decisive factor in establishment of humus forms. In: K e v a n Mc E. D. K.: Soil Zoology, 73-82, London.

K u b i ě n a , W. L., ed. (1967): Die mikromorphometrische Analyse. Stuttgart.

L o u b , W. (1960): Die mikrobiologische Charakterisierung von Bodentypen. Die Bodenkultur 11, 38-68, Wien und München.

S e i f e r t , J. (1960): Biogennost, nitrifikace a biologická aktivita půd rostlinných společenstev na Kódě u Srbska. Rozpravy ČSAV, řada MPV, 70/5, 1-35.

T e s a ř o v á , M., Ú l e h l o v á , B. (1968): Abbau der Zellulose unter einigen Wiesengesellschaften. In: Mineralisation der Zellulose, Tagungsber. 68, Deut. Akad. d. Landwirtsch., Berlin, 277-287.

Z a c h a r i a e , G. (1965): Spuren der tierischen Tätigkeit im Bo-
    den des Buchenwaldes. Hamburg und Berlin.

## D i s c u s s i o n

U. B a b e l : Wird der Zersetzung durch Collembolen an diesen Standorten eine so grosse Rolle zugeschrieben, wie man aus der Häufigkeit von
Collembolenlosungen schliessen musste? Sind es wirklich immer Collembolenlosungen? Das wäre im Widerspruch zu Ergebnissen von Zachariae.

J. R u s e k : Ja, die Rolle von Collembolen bei der Streuzersetzung
ist hier sehr gross. Zachariae hat andersartige Standorte untersucht.

L. S. K o z l o v s k a j a : Welche Arten von Wurzeln wurden von den
Mikroarthropoden befallen? Enthalten Collembolenexkremente auch Mineralpartikeln, oder nicht?

J. R u s e k : Alle abgestorbenen Wurzeln der Pflanzen wurden ange-
griffen. Collembolenexkremente, vor allem die grösseren, enthalten auch
die Mineralpartikeln und davon kann man auf die Wichtigkeit der Collembolen für die Bodenentwicklung schliessen.

H. J a n e t s c h e k : Wie wurde der $CO_2$ Gehalt des Bodens gemessen?

J. K u b í k o v á : Nach Steubing L. (1965).

# Phytogenic microstructure
# of Collembola associations
# in steppes and forests of Siberia

S. K. STEBAYEVA

Biological Institute,
Novosibirsk

It is a well-known fact that springtails react readily to the slightest changes in their environment. Population density and the pattern of their spatial distribution are frequently defined by microconditions in neighbouring plant samples. Environmental influences of the plants are particularly marked under conditions of the continental climate of Siberia, where soils are frozen to a considerable depth in winter and desiccate in summer (Stebayeva 1963). In the present paper, associations of *Collembola* with the individual plants are discussed.

*Collembola* of dead litter in a cultivated pine forest were studied around the individual trees (*Pinus silvestris* L.), and around large bushes (*Caragana bungei* Ldb.) in semidesert steppes with sparse grassland vegetation.

A schematic illustration of sampling around four pine trees is shown in Fig.1. Closest to the trees, the density of *Collembola* populations was 15 - 20 times higher than that at sites more removed from the trees (40 - 75 cm). There, the number of *Collembola* appeared to be in correlation to the height of the humus layer (Fig.2). This phenomenon was not observed near the trunk where litter thickness (not less than 1 cm) indicated a certain "critical" height. In all samples taken from sites closest to the trees, the number of species was 9 - 11, while the number of species taken from maximally distant sites was 3 - 4 species (exceptionally 6 - 7) per sample unit. This indicated that *Collembola* aggregations were densest near the trunk of the trees.

The microdistribution of *Collembola* was studied in detail on 55 litter samples (10 x 10 cm unit/sample) taken from the environment of a pine tree in mid-August.

The pattern of *Collembola* distribution was arranged into three principal types in accord with their distance from the base of the tree (fig.3). Type one was typical of *Tomoceridae* (particularly *Tomocerus minutus*) with the highest density at the base of the trunk, which re-

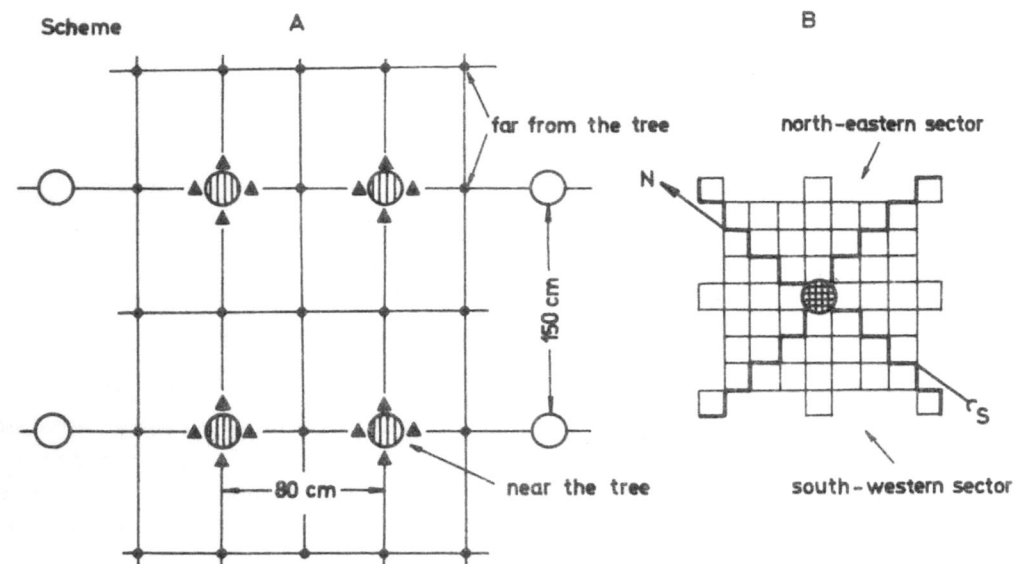

Fig.1. Two schemes of the distribution of sample units
around the trees (*Pinus silvestris* L.).
A. Near the trees (▲) and at a maximum distance from
them (●).
B. Around one tree. Investigated area shaded. Diameter
of trees 12 - 13 cm, age 18 - 20 years. Sectors of exposure marked with bold lines. Size of sample unit 10 x 10
cm.

duced significantly towards the periphery of the area of distribution,
mainly at a distance of more than 30 - 40 cm. Also *Sminthuridae* were
abundant at the tree base. The second type of *Collembola* distribution
occurred at a distance of 15 - 17 cm from the tree base (approximately
in the centre of the sampling area). *Collembola* numbers decreased in
direction towards the periphery (*Lepidocyrtidae, Schoettella ununguiculata* Tullb., *Isotoma notabilis* Schäff., etc.). Type three was characteristic of species which increased in number at a distance of 21-31 cm
from the tree, and decreased again in direction towards the periphery
(fig.3). This type of distribution pattern was typical of species of
the genus *Onychiurus* and *Tullbergia*. A variant of this "peripheral"
distribution pattern were the species *Drepanura quadrilineata* Steb. and,
to some extent, *Isotoma fennica* Reut., which were present at 17 - 49 cm
from the tree, but practically absent near the trunk. Our observations
indicated that atmobiotic and hemiedaphic forms with large, chiefly
pigmented, bodies and of high density concentrated mainly in the litter
layer near the tree trunk. Most of these species feed on plant remnants.
In the peripheral zone we found a few species belonging to euedaphic

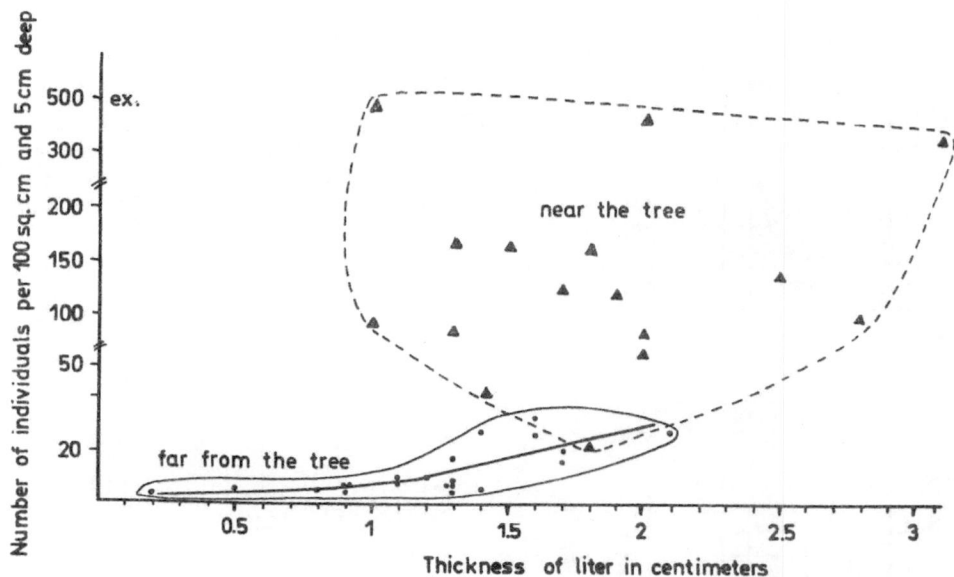

Fig.2. Dependence of *Collembola* numbers on the height of
the litter layer in plots close to (▲) and at a maximum
distance from the trees (●). Total number of *Collembola*
in litter and soil up to a depth of 5 cm in 10 x 10 cm
plots.

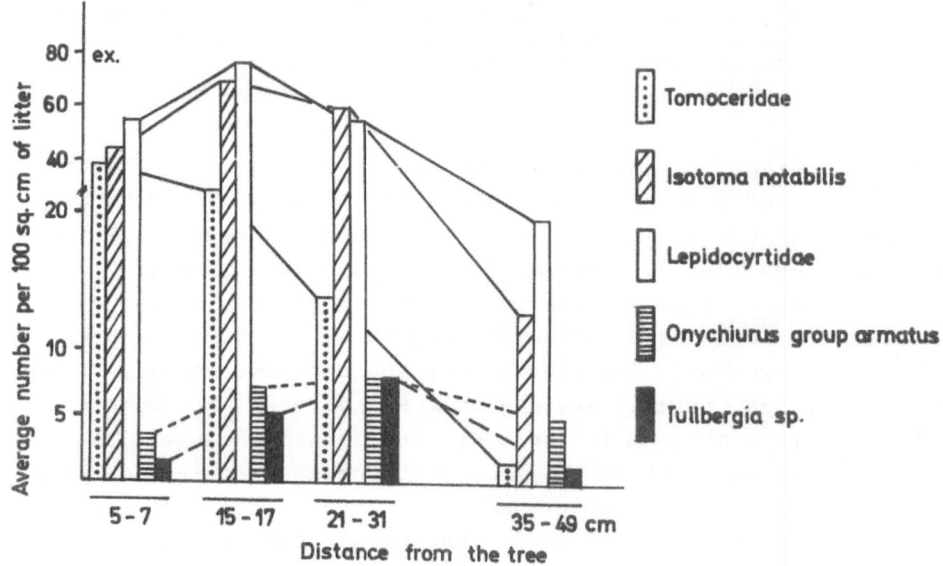

Fig.3. Number of *Collembola* at different distances from
the tree.

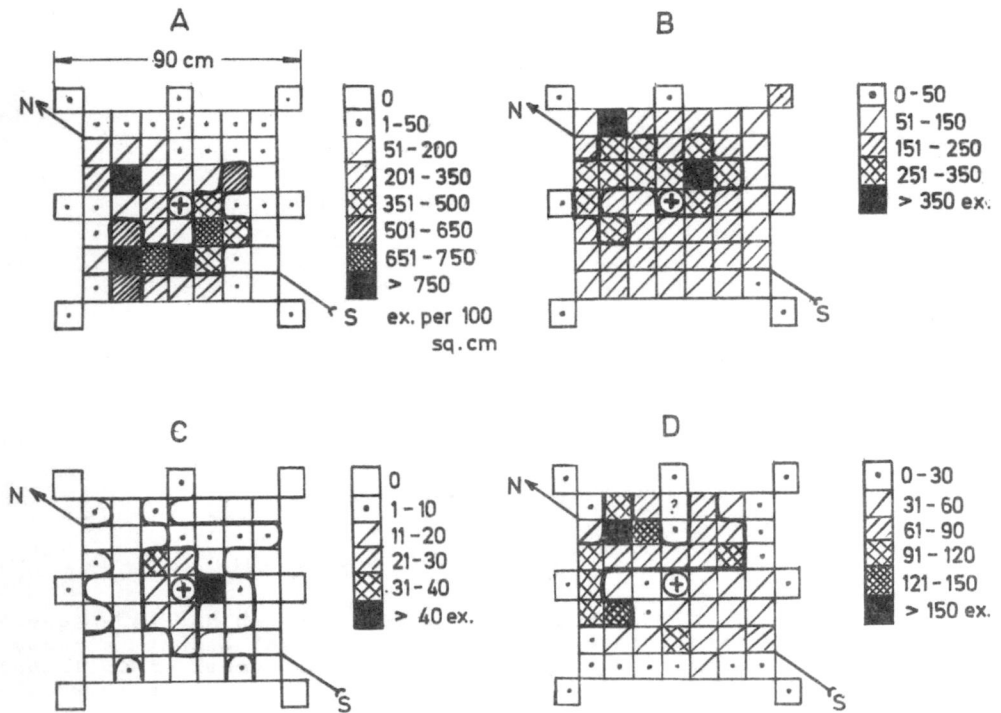

Fig.4. Quantitative distribution of *Collembola* around the trunk of *Pinus silvestris*. A – *Schoettella ununguiculata*; B – Remaining *Collembola*; C – Adults of *Sch. ununguiculata*; D – *Isotoma notabilis*.

forms and feeding on hyphae of microscopic soil fungi and detritus. Also the number of several upper-hemiedaphic forms appeared to be low.

In general, the situation resembled that of successional changes studied by Kubiëna (1943), Stebayev (1963), Chernova (1966).

For abundant species we observed their dependence on the cardinal points faced by the sample unit with regard to the tree. This was reflected in an asymmetric distribution of the zones of density of these species. E.g., *Sch. ununguiculata* was most abundant in sample units facing West, with a maximum in the S.W. Considering, however, the total number of *Collembola* without this abundant species, their zone of high density was the N. and N.E. (fig.4 B). The zone of highest density of I.notabilis (fig.4 D) was similar to that of all *Collembola* and, therefore, it may be regarded as an antipode of *Sch. ununguiculata*. Of the remaining species, *L. lanuginosa* and *Tullbergia* sp. displayed trends of concentrating in the North, *Entomobrya* sp. in the N.N.-E., species of

the genus *Onychiurus* ("armatus" group)  in the S.,  *Isotoma fennica*  in
the S.-E.,  *Sch. ununguiculata* and *Lepidocyrtus cyaneus* Tullb. in the W.

Interactions of factors such as sites of preference, distance from
the tree and height of the litter layer  are demonstrated for *Sch. ununguiculata* in fig.5. Around the tree trunk,up to a distance of 5-7 cm,
the density of this species  was high  with a maximum in the S.-W. Differences in abundance increased from the zone of 15- 17 cm up to 31 cm,
with  a preference  of exposure to the W.  and  the lowest incidence in
sites exposed  to the N.-E.  which are coldest.  At distances  of 35 cm
from the tree and above, the effect of exposure  to any of the cardinal
points on species density ceased; this was everywhere a 100 times lower
than density at the base of the tree.

Changes in the height of the litter layer are shown in fig.5. This
layer was high and medium high at the base of the tree, medium high and
low at 15 – 31 cm, extremely low (2.5 cm and less) at a distance of more
than 35 cm. West of the tree, at 21 – 31 cm, species density was high in
spite  of the low litter layer, which indicated  that exposure  to this
·ardinal point  was more important than the height of the litter layer.

The results  suggest that, in addition  to a spatial discontinuity
in direction  from the tree,  there is also a discontinuity  in the incidence of species at the trunk base with regard to the cardinal points.
In our material the maximum density  of two abundant species  did never
occur at the same point. We observed even a discontinuity in the spatial
distribution of adult and larval forms of *Sch. ununguiculata*: the highest density  of adults occurred  near the trunk,  that of the larvae at
some distance from it (fig.E A,C).

The effect  of exposure  was  even more important  in  semideserts
where the soil  is dry  in the summer.  The arrangement  of 50 sampling
plots (15 x 15 cm) to the South and North of a big *Caragana bungei* bush
is shown  in fig.6.  From the centre  of each plot,  a sample measuring
125 cm$^2$  was taken  from  the layer 0-5 cm and 5 - 10 cm.  South of the
bush, large numbers of *Collembola* were found only near the bush (7.5 cm),
where  the litter layer (Fig.6 A)  and  soil moisture (Fig.6 D-E)  were
highest.  This was followed by a narrow zone of low density (up to 22.5
cm).  At a distance  of more than 30 cm S. of the bush, where  the soil
was very warm, *Collembola* were practically absent  in the upper layer.
The influence of sparse grassland vegetation,  which did  not grow near
the bush, was observed in a considerable distance from it in the deeper
soil layers  (5 - 10 cm).  Soil moisture  was increased in  these plots
(Fig.6 E).  On the northern side, the maximum density was recorded near
the bush where soil moisture was higher.  The zone of lower density was
wider especially in the layer 5 - 10 cm. Apparently, the number of *Col-*

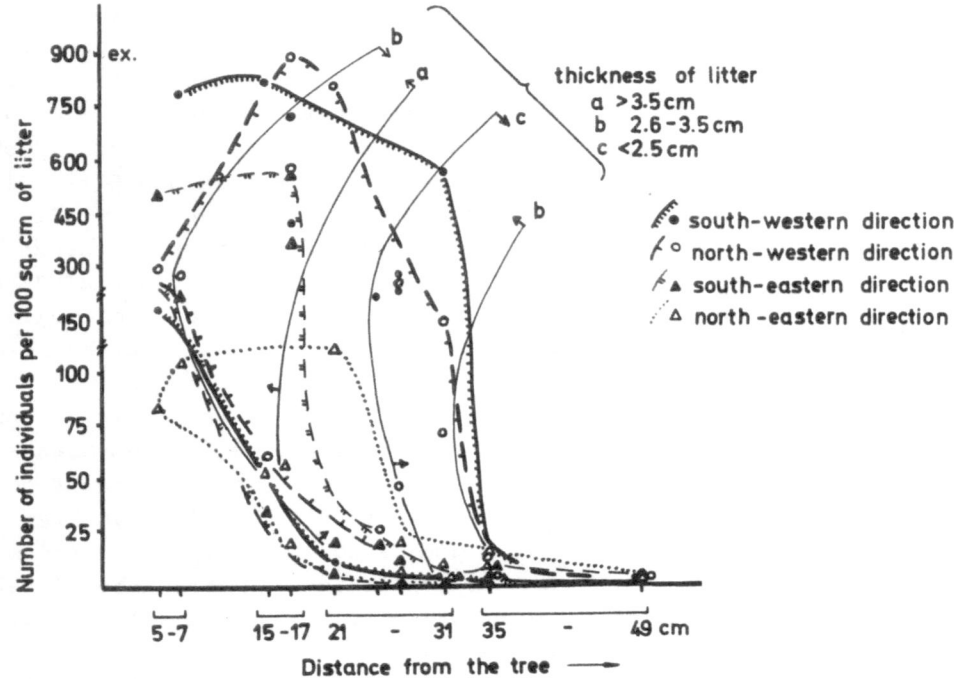

Fig.5. Scatter diagram showing the dependence of *Schoet-tella unungüiculata* on the joint influence of three factors: exposure to one of the cardinal points, distance from the tree, height of litter layer.

*lembola* in the lower soil layers was not influenced by the height of the litter layer. In these steppes, abundant species close to the bushes were members of the genus *Xenylla*; they appeared to be a kind of xerophilous vicariant of *Sch. unungüiculata* from the pine forests. The sites of maximum density of *Xenylla* species were situated S. of the bush, those of *Paruzelia furcata* Grinb. and *Onychiuridae* towards the North and in the deeper soil layers.

The relationship between *Collembola* and plants was more marked in the steppes, where conditions are extreme, than in the forests, and also the influence of the factor of exposure surpassed that of other factors. As a result, we observed for almost all species a marked asymmetry in the distribution of species density in the zones close to the bush; simultaneously, changes in the concentration of species in the more distant plots were indistinct. The concentrations of *Collembola* in the vicinity of the plants bear the character of true microstructures in the biogeocoenosis similar to those shown by Rapoport (1966) in his

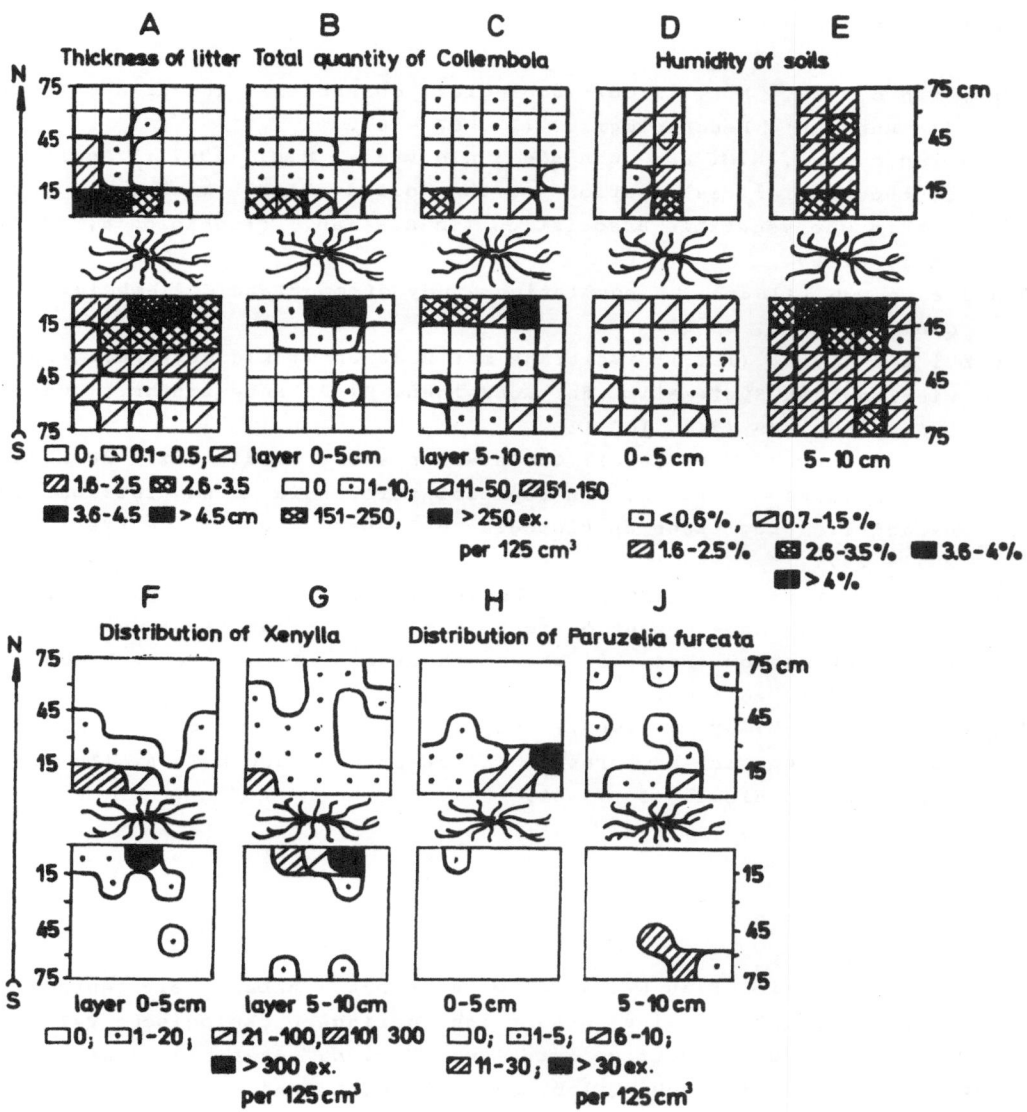

Fig.6. Number of *Collembola* and environmental conditions in sites situated to the North and South of the bush *Caragana bungei* in semidesert steppes.

micromap. Data on the microdistributional pattern of various species under different plants were given by Hale (1966), Chernova and Chugunova (1967) and others.

R e f e r e n c e s

C h e r n o v a ,   N.M. (1966): Zoological characteristics of a compost.
    (In Russian). Pedobiologia 2 (1), 67 - 87.
C h e r n o v a , N.M. and  C h u g u n o v a ,   M.N. (1967):  Analysis
    of the spatial distribution  of the soil-dwelling  microarthropods
    within one vegetable association.(In Russian.) Pedobiologia, Bd.7,
    Hf.1; 67-87.
H a l e ,  W.G. (1966):  A population study of moorland *Collembola*. Pe-
    dobiologia, 6, 65-99.
K u b i ë n a , W. (1943):  Beiträge zur Bodenentwicklungslehre.   Ent-
    wicklung und Systematik  der Rendsinen. Z. Pflanzenernähr., Düng.,
    Bodenkunde, 29 (74), 108-119.
R a p p o p o r t , E.H.(1966): Comentarios sobre la diantaxis de algu-
    nos animales des suelo, con especial referencia  a su distribucion
    espacial. Progressos en Biologia  del Suelo.  Unesco, Montevideo,
    283 - 297.
S t e b a y e v , I.V. (1963):  Changes  in  soil  animal  populations
    during their development on rock  and loose products of weathering
    in forest-meadow areas  of the South Ural.  (In Russian). Pedobio-
    logia, 2(4), 265-309.
S t e b a y e v a , S.K. (1963):  Ecological  distribution  of spring-
    tails (Collembola) in forests  and steppes of South Tuva. (In Rus-
    sian). Pedobiologia, 3, 75-85.

D i s c u s s i o n

J. R u s e k : Wie  Frau Stebaeva  gezeigt hat,  haben viele "euröke"
Arten  in Wirklichkeit  eine  wesentlich  geringere ökologische Valenz.
Dies gilt auch  für so häufige Arten  wie *Isotoma notabilis*. Unsere An-
sicht  über die ökologische Valenz der meisten Arten  muss daher  revi-
diert werden.

R. C o v a r r u b i a s : Did you also sample the trunks of the pines
for those species typically distributed around  (and near)  the trunks?
Did you follow the distribution patterns through the time?

S. K. S t e b a e v a : The species of *Collembola* occurring  on the
trunks are different  from those found  around the trunks.  The distribution patterns change with the time, but the data given in this report
are relative to August only.

# Distribution of dropping fabrics in central European humus forms

U. BABEL

University Hohenheim,
Stuttgart

In humic soil horizons, aggregates of uniform size - about 60 µm - which are edged and isometric occur frequently. Where they are concentrated, they form dropping fabrics. (This term was used in Babel, 1968). It was proved directly in loess-para brown earth that these aggregates are enchytraeid droppings; this has been confirmed by a number of authors, e..g. Bal, (1973, fig. 45).

These fabrics occur in smaller or larger domains from the Fm-horizon (fig. 1) to the Ahu-horizon, not seldom also in lower horizons (nomenclature of horizons see Babel, 1971). Thus they were found in the Bv-horizon of acid brown earths in depths of 30 cm and once in the C-horizon (with traces of Bt) of a para brown earth of loess even at a depth of 120 cm, still well developed. In the literature their occurrence in the Bh-horizon of podzols is pointed out frequently where they constitute part of the agglomeratic fabric (e.g. Righi, 1969). Dropping fabrics exist within sand, loam and clay. The primary aggregates in sand consist of pure organic substance whereas in loamy soil and in clay they are of the same composition aś the enclosing material. Very clear developments are found in loess. In clay these dropping fabrics often occur as "character aggregates" ("Schriftzeichenaggregate", Kubiëna, Beckmann, Geyger, 1962, fig. 2).

Micromorphological papers often show photographs of these fabrics, not always pointing out their origin as excrements (e.g. Babel, 1972; Ghitulescu and Stoops, 1970 (in aggrotubules); Kowalinski, 1969/70).

There are also less characteristic developments, obviously resulting from deformations of the typical developmental form (fig. 2). Finally there are also clearly different developments, believed to be partially of different origin.

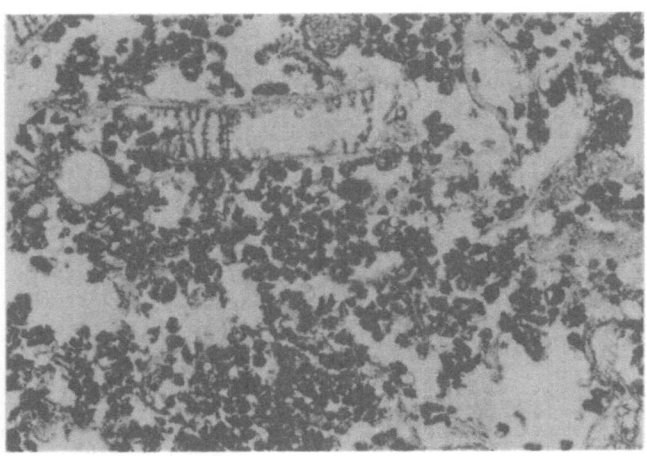

Fig.1. Well developed, loose dropping fabric in the
Fm-horizon of a wet moder under spruce. (Distinctness 5,
area share 5: class a). The dark aggregates of 40-60 µm
which compose the fabric are enchytraeid excrements.
Some needle residues. Thin section (18 µm), transmitted
light, natural height of the photo 1.4 mm.

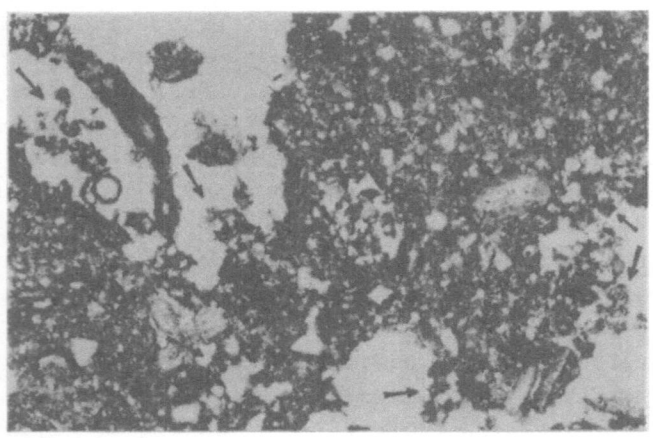

Fig.2. Small areas of badly developed dropping fabrics
(indicated by arrows). (Distinctness 2, area share 3:
class d). Ahh-horizon of a mull under beech and oak. Thin
section (18 µm), transmitted light, natural height of
photo 1.4 mm.

## DISCUSSION OF THE PHENOMENOM

O r i g i n :   As said before,  the typical formation  the primary aggregates contain enchytraeid droppings. They cannot be mistaken for the
egg-shaped,  well-rounded droppings of oribatids;  sometimes,  however,
they may be mistaken  for droppings of smaller diptera larvae. -  Other
aggregates  of probably always larger size  can originate  from  animal
movements in loose soil structure ("Rollaggregate",  Zachariae,  1962).
Sometimes they may be mistaken for round aggregates which occur in soilss
under strong frost conditions  (e.g. the silt granules of Romans et al.
(1966), which, however, mostly appear to be larger).

Occurrence  of such fabrics,  well developed,  in deeper  horizons
at first seems  surprising.  But enchytraeids can, although seldom, de-
finitely be found in depths of 30 cm and more (Nielsen, 1955, once found
29 % of the total population at a depth of 25 - 29 cm). Since the amount
of dropping fabrics is not only a function of the formation process the
following formula is valid:

$$\text{Amount of droppings} = f \text{ (formation processes)} - \frac{f \text{ (destruction processes)}}{f \text{ (stabilization)}}$$

Concentration of droppings  must be expected when destruction processes are missing. This is believed to explain generally the existance
of well-developed dropping fabrics in lower horizons.

D e s t r u c t i o n   p r o c e s s e s   are  caused  mainly  by the
activity of earthworms which again consume the enchytraeid droppings as
food, also by swelling and shrinkage. Finally, the fabric is being  destroyed by mechanical decay of the aggregates  which occurs if, for instance, the organic binding agent is dissolved or decomposed (Bal,1970).

## SPECIAL INVESTIGATION ABOUT THE DISTRIBUTION
## OF DROPPING FABRICS

M a t e r i a l   a n d   M e t h o d :

Subject of the investigation  were humus profiles  from different sites
in Central Europe which had extremely different types of climate, vegetation and soil.  Vegetation consisted of forests  on 2/3 of the sites,
on 1/3 of extensively utilized or natural grass land  and pioneer vegetation.

Humus forms were determined, if not differently stated, by Kubiëna (1953), sand mull by Hoover and Lunt (1952). The term mull-moder describes a macro-morphological connecting link between mull and moder under forest. Typical raw humus profiles, characterized by a mighty Lv and Fr, strongly interspersed with fungal hyphae after Kubiena, were not examined. As wet moder humus, profiles with greasy H-horizons developed under moist conditions were described.

For each depth of each profile one to several Vestopal thin sections, size 20 x 30 mm, thickness 15 - 30 μm were examined under a stereomicroscope at magnification 40 x. (By stereomicroscopic observation dense dropping fabrics can be recognized better than under a microscope). For each section it was determined separately:

D i s t i n c t n e s s of development of the dropping fabric (estimation scale 1 - 5): as clearly developed are considered: isometrical and edged aggregates of a diameter of 40 - 80 μm which are of the same size within a given section and can be single or welded; as not clearly developed we consider: larger (up to 130 μm) or smaller (to 20 μm) primary aggregates, other shapes, great density; characteristically formed oribatid droppings were not included in the estimation).
A r e a   S h a r e   in the section (0 -5),
W e l d i n g ,
S i z e   of the single fabric domains.
Reflections concerning the origin of the aggregates did not affect the estimation values.

Based upon the afore mentioned characteristics the thin sections were classified into 6 groups (a - f) whereby class "a" contains the best development of dropping fabric, "f" the poorest or none at all (table 1). The number of sections of a given class was expressed in

T a b l e   1.   Classes of thin sections according to distinction and area share of dropping fabrics
(previously determined in an estimation scale 0 - 5 each)

| Class | a | b | c | d | e | f |
|-------|-----|-----|-------|-------|-----|-----|
| Distinctness | 5;4 | 5;4 | 5;4 | 3;2 | 3;2 | 1;0 |
| Area share | 5;4 | 3;2 | 1;1/2 | 5;4;3 | 2;1 | any |

percentage of all examined sections of the same humus form. When several sections of the same depth class were examined an average rate for

distinctness and area share was figured and in this way the corresponding group  for the depth class  was found  (depth classes: humus cover,
0 - 10 cm, lower than 10 cm).


R e s u l t s    a n d    D i s c u s s i o n

The results  are demonstrated in table 2 and 3 and in fig. 3.  Approxi-

T a b l e  2.     Occurrence of dropping fabrics  and  similar fabrics in
humus profiles from Central Europe
(percent of investigated depth classes).

| Depth | n | a | b | c | d | e | f | a + b + d |
|-------|---|---|---|---|---|---|---|-----------|
| Hs cover | 82 | 16 | 27 | 5 | 13 | 27 | 12 | 56 |
| 0-10 cm | 123 | 20 | 19 | 7 | 11 | 20 | 24 | 50 |
| 10 - cm | 50 | 10 | 30 | 8 | 8 | 12 | 32 | 48 |
| Total | 255 | 17 | 24 | 6 | 11 | 20 | 22 | 51 |

mately 1/6  of the examined sections  shows good  development  and high
percentage  of area shares (class a).  About half  of the sections show
good  to fair  development  (a+b+d);  this does not only pertain  to the
humus cover  and  to the upper 10 cm of the mineral soil  but  also  to
greater depths. Lower than 10 cm, however, sections with low area share
are more frequent (class b instead of class a).
     Fig. 3  shows  that the amount  and  the change according to depth
of the dropping fabrics are  characteristic of most humus forms.

S a n d m u l l  shows a high rate of dropping fabric  (class a as well
as b and d) and thus differs from the other mull profiles.  This can be
seen  as follows:  enchytraeid droppings  are not or only partially destroyed since clay  as binding agent  is missing  which normally  would
result in welding under biological  (earthworms)  or abiological (moisture, shrinkage)  influence.  Also the embedded sand grains  render the
origination of welding more difficult.  But in any case there are often
deformations which  explain  the high rate of class d sections.

M u l l   only seldom shows well developed  numerous  dropping  fabrics
(a), but more often sections  with smaller quantities (b).  Interpreta-

T a b l e   3.   Occurrence of dropping fabrics   and   similar fabrics in
humus forms from Central Europe
(percent of investigated thin sections)

| | | n | a | b | c | d | e | f | *)<br>a + b + d |
|---|---|---|---|---|---|---|---|---|---|
| Sandmull | hs cover | 0 | - | - | - | - | - | - | - |
| | 0-10 | 10 | 30 | 20 | 0 | 30 | 20 | 0 | 80 |
| | 10- | 7 | 29 | 29 | 0 | 29 | 14 | 0 | 86 |
| Mull | hs cover | 1 | 0 | 0 | 0 | 0 | 0 | 100 | - |
| | 0-10 | 28 | 14 | 32 | 7 | 11 | 14 | 21 | 57 |
| | 10- | 15 | 0 | 27 | 13 | 0 | 0 | 60 | 27 |
| Mull-Moder | hs cover | 4 | 50 | 25 | 0 | 0 | 25 | 0 | - |
| | 0-10 | 10 | 50 | 20 | 10 | 0 | 10 | 10 | 70 |
| | 10- | 7 | 14 | 43 | 14 | 0 | 29 | 0 | 57 |
| Moder | hs cover | 44 | 18 | 23 | 2 | 23 | 18 | 16 | 64 |
| | 0-10 | 33 | 30 | 18 | 12 | 9 | 12 | 18 | 58 |
| | 10- | 12 | 17 | 25 | 8 | 17 | 0 | 33 | 58 |
| Mull-like Moder<br>(Kubiëna 1953) | hs cover | 1 | 0 | 100 | 0 | 0 | 0 | 0 | - |
| | 0-10 | 10 | 20 | 10 | 0 | 20 | 20 | 30 | 50 |
| | 10- | 4 | 0 | 50 | 0 | 0 | 50 | 0 | - |
| Wet Moder | hs cover | 6 | 33 | 17 | 0 | 0 | 33 | 17 | (50) |
| | 0-10 | 2 | 0 | 50 | 0 | 50 | 0 | 0 | - |
| | 10- | 0 | - | - | - | - | - | - | - |
| Pitch Moder | hs cover | 7 | 0 | 57 | 14 | 14 | 0 | 14 | 71 |
| | 0-10 | 3 | 0 | 0 | 0 | 33 | 33 | 33 | - |
| | 10- | 0 | - | - | - | - | - | - | - |
| Rendzina Moder | hs cover | 5 | 20 | 40 | 0 | 0 | 40 | 0 | (60) |
| | 0-10 | 5 | 0 | 0· | 0 | 0 | 40 | 60 | ( 0) |
| | 10- | 0 | - | - | - | - | - | - | - |
| Tangel | hs cover | 14 | 0 | 21 | 14 | 0 | 64 | 0 | 21 |
| | 0-10 | 9 | 0 | 22 | 11 | 0 | 67 | 0 | 22 |
| | 10- | 1 | 0 | 0 | 0 | 0 | 100 | 0 | - |
| Anmoor | hs cover | 0 | - | - | - | - | - | - | - |
| | 0-10 | 13 | 8 | 0 | 0 | 0 | 15 | 77 | 8 |
| | 10- | 4 | 0 | 25 | 0 | 0 | 0 | 75 | - |

*)  Rates with standard deviation   ca 29 % ($n$  4) not listed.
    Rates with standard deviation   ca 21-25 % ($n \approx$ 5,6) in parenthesis.

tion: enchytraied droppings  are destroyed here   (contrary to sandmull)
by earthworm activity Earthworm activity decreases slower with increasing depth than that of enchytraeids: therefore, below the depth of 10 cm
even lower values for a + b + d  appear.

M o d e r   shows medium to fairly high values   in all three depth classes which shows   the great influence   of the enchytraeids   on this most
important   humus form   in the forests   of Central Europe   (compare   for
example Babel, 1972).

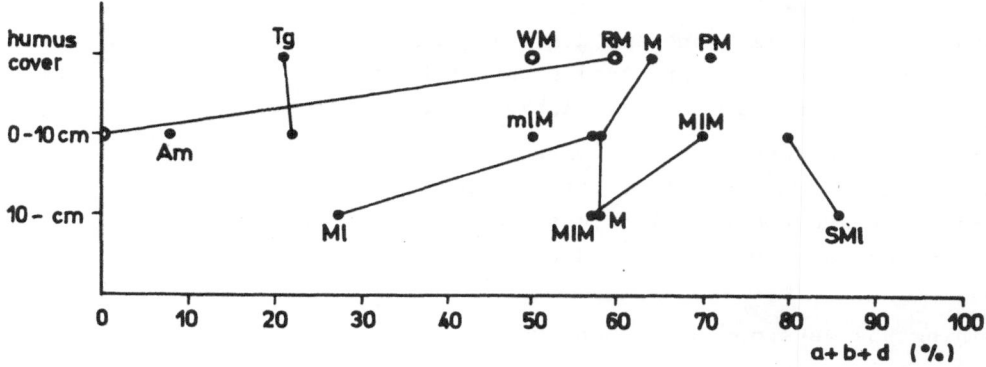

Fig.3. Comparison of the examined humus forms as to percentage   of sections   with good   to fair developments of
dropping fabrics (classes a+b+d).   Circles: n=5,6; dots:
n=7...44, Am=anmoor - Tg-tangel - Ml=mull - WM=wet moder
- mlM = mull-like moder - RM = Rendzina moder - MlM= mull-
moder - PM≠pitch moder - SMl=sand mull.

M u l l - M o d e r,   macromorphologically   placed   between   mull   and
moder, shows   at O - 10 cm, probably also   in the cover,   higher values
than mull and than moder as well.   Interpretation: formation processes,
i.e. activity of enchytraeids, are stronger   than in moder; destruction
processes, i.e. earthworm activity, is slower than in mull.   (This conclusion, alike important for biology   and   classification of this humus
form, requires confirmation by further examination   of additional mull-
-moder profiles.   The difference   between   the a-values (O - 10 cm)   for
mull-moder and moder is not statistically significant).

M u l l - l i k e   m o d e r   (Kubiĕna, 1953)   shows conditions of medium type, similar to those of mull.

The few available profiles of   w e t   m o d e r   are not enough   for
a final interpretation.

P i t c h   M o d e r   of which also   only few profiles   were examined
shows   neither extremely   well developed   nor extremely frequent occurrence   of dropping   fabrics   (none of the 7 profiles   was classified a,
only one d). Thus the statement of Kubiĕna (1953) that dropping fabrics
are characteristic of pitch moder   cannot be supported.   The fact that

very well developed dropping fabrics are rare  in pitch moder, however, corresponds with the generally low biological activity there.

The few examinations of R e n d z i n a - M o d e r s   only show the biological shallowness of these soil profiles.

T a n g e l  shows little dropping fabric in the humus cover and in the mineral soil as well,  thereby differing clearly  from all types previously discussed. An interpretation cannot be given since further examinations  of Tangel  are not available  in micromorphology  and biology.

A n m o o r  shows nearly none or none of dropping fabric  (class f) in .10 of 13 sections; this fact  is interpreted  with the rareness of enchytraeids there and with little stability of the soil fabric which results in fast disappearance of droppings, also of those from other animals.

Examining the development of the dropping fabrics within one humus form but  d i f f e r e n t  t y p e s  o f  s o i l  one finds also differences.  For example, moder seems to have higher rates  of dropping fabrics in brown earth than in podzol brown earth  [within humus cover and O  10 cm in brown earth 4 of 10 sections (40 %) were class a, in podzol brown earth it was 6 of 27 (22 %)]. These differences must be considered minor differences  within one humus form  and  may assist in further classification of humus forms.

Close relationships  between  the development  of dropping fabrics on the one side and vegetation  as well  as smaller differences  of the site conditions on the other side  were reported earlier  (Benecke  and Babel, 1969).


CONCLUSIONS


Examination  of dropping fabrics supplies information  about biological and abiological processes  in humus profiles.  Since humus profiles can be distinguished clearly  by the state of development  of dropping fabrics these can furnish  a feature for the classification  and sub-classification of humus forms. As an interesting example mull-moder is mentioned once more, the high rate values on dropping fabric of which show that it is not only an intergrade  between mull and moder.  The results show that it is possible to utilize dropping fabrics  as one of several quantitive features  to create  similarity groups  of humus profiles by the use of statistical means and, thereby, develop a new approach to the classification of humus profiles  (compare discussion on Howard, 1969).

R e f e r e n c e s


B a b e l ,   U. (1968): Enchytraeen-Losungsgefüge  in Löss. Geoderma 2,
    57-63.

B a b e l ,   U. (1971):  Gliederung und Beschreibung  des  Humusprofils
    in mitteleuropäischen Wäldern. Geoderma 5, 297-324.

B a b e l ,   U. (1972): Moderprofile  in Wäldern - Morphologie  und Um-
    setzungsprozesse. Hohenheimer Arbeiten 60. 120 p., Ulmer, Stuttgart.

B a l ,  L. (1970): Morphological  investigation  in two moderhumus  pro-
    files and the role of the soil fauna in their genesis. Geoderma 4,
    5-36.

B a l ,   L. (1973) : Micromorphological  analysis  of soils. (Lower levels
    in the organization  of organic soil  materials.)  174 p., Thesis,
    University of Utrecht, The Netherlands.

B e n e c k e ,  P. und  B a b e l ·,  U.  (1969):  Untersuchungen  über
    die Auswirkungen  des Fichtenreinanbaues  auf  Parabraunerden  und
    Pseudogleye des Neckarlandes. III. Humusmorphologie  und  Bodenge-
    füge. Mitt. Ver. Forstl. Standortskunde 19, 81-89.

G h i t u l e s c u ,  N. and  S t o o p s ,  G.  (1970):  Etude micro-
    morphologique de l'activité biologique  dans  quelques sols  de la
    Dobroudja (Roumanie). Pédologie 20, 339-356.

H o o v e r ,  M. D. and  L u n t ,  H. A. (1952):  A key for the clas-
    sification  of  forest  humus  types. Proc. Soil Sci. Soc. Am. 16,
    368-37o.

H o w a r d ,  P. J. A.  (1969):  The classification  of humus types in
    relation  to soil ecosystems. p. 41-54  in Sheals, J. G. (editor):
    The soil ecosystem,  E. W. Classey, Ltd., Hampton,  Great Britain.

K o w a l i n s k i ,S.(1969/70): Interdependence  between micromorpho-
    logical and chemical properties in some zonal soils  of the Karko-
    nosze Mountains (Poland). Geoderma 3, 89-115.

K u b i ë n a ,  W. L. (1953): Bestimmungsbuch und Systematik der Böden
    Europas. 392 p., Enke, Stuttgart.

K u b i ë n a , W.,  B e c k m a n n , W. and  G e y g e r , E. (1962):
    Zur Untersuchung  der Feinstruktur  von Bodenaggregaten  mit Hilfe
    von Strukturphotogrammen. Zeiss-Mitt. 2, 256-273.

N i e l s e n ,  C. O. (1955):  Studies on  Enchytraeidae. 2. Field stu-
    dies. 58 p., Natura jutlandica 5, Aarhus, Danmark.

R o m a n s ,  J. C. C.,  S t e v e n s , J. H. and  R o b e r t s o n ,
    L. (1966):Alpine soils in northeast Scotland J. Soil Sci. 17, 184-
    199.

R i g h i ,  D. (1969):  Aspects morphologiques et physico-chimiques de
    la podzolisation en Fôret de Rambouillet. 116 p., Thèse, E.N.S.A.,
    Grignon,  France.

94      Babel, U.

Z a c h a r i a e ,   G. (1962):   Zur Methodik bei Geländeuntersuchungen
    in der Bodenzoologie. Z. Pfl. Ern. Düng. Bodenk. 97, 224-233.

# Energy flow through a mountainous pasture ecosystem

A. KAJAK

Institute of ecology, P.A.S.,
Dziekanow Leszny near Warszawa

The studies were conducted by a team of 11 workers for 3 years, 1969 to 1971. The objective of these studies was to examine the paths of energy flow through an ecosystem, i.e., to determine the amount of energy stored and transferred between trophic levels, and to give an analysis of animal contribution to some key processes involved in energy flow. The study area was a mountain pasture grazed by sheep. It was located in the Pieniny mountains, southern Poland, 700 m above see level on the area of former fields.

The soils of this area are formed of weathered flysch rocks on globotruncana marles. These are brown pseudogley soils with low Ca and P content, of pH value 5.7.

This pasture belongs to the *Lolio Cynosuretum cristati* association

The study was carried out in the plots fertilized at different rates with sheep manure. The main study plot was a so-called "sheep tract" i.e., a part of the pasture that sheep passed along 2-4 times a day, what resulted in the moderate fertilizer input of 360 g manure dw/m$^2$/season.

The other plots studied were sheep folds, where sheep were kept for several successive nights and left large quantities of manure, about 500 g dw/m$^2$ in a short period of time.

It is difficult to describe in details all the methods used during investigations, because a large number of them was applied, so only those concerned with the experiments will be described.

The experiments were concerned with three basic processes:

1. consumption by sheep
2. manure decomposition rate
3. contribution of earthworm to decomposition of soil detritus.

The obtained results are demonstrated in a diagram showing carbon transfers in the ecosystem. All the transfers are expressed in g C/m$^2$/season, all the compartments in g C/m$^2$ (fig. 1).

Above-ground plant production equalled to 239 g C per growing season. A large part of it, up to 86 % was grazed by sheep (Andrzejewska in press).

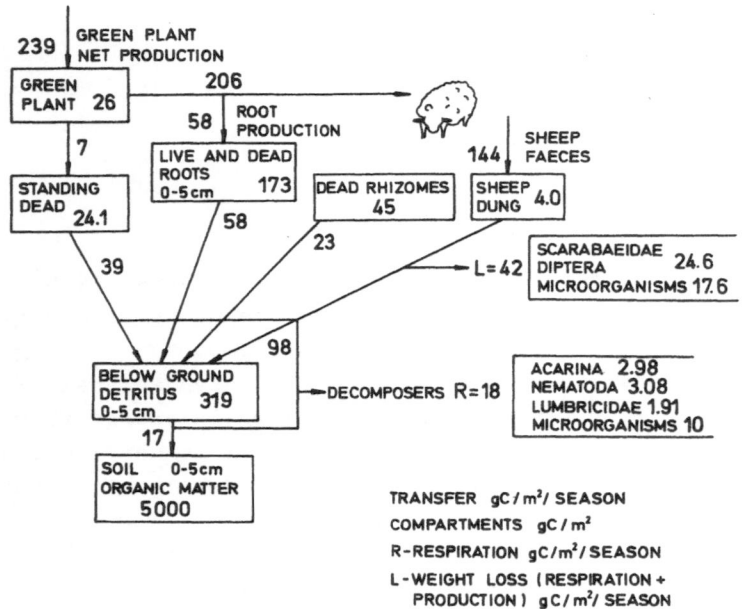

TRANSFER  gC/m²/ SEASON
COMPARTMENTS  gC/m²
R-RESPIRATION  gC/m²/SEASON
L-WEIGHT LOSS (RESPIRATION +
        PRODUCTION) gC/m²/SEASON

Fig.1. Transfer of carbon in pasture ecosystem.

The consumption by sheep was estimated by comparing plant biomass on small areas (0.5 m²) protected from grazing by means of wire cages, and on the surrounding grazed areas. 15 such cages were used, moved on to new places every 5 weeks.

Very short sward only remains on the pasture, 26 g C/m² as an average. This value is very stable over a year cycle.

Input of sheep faeces into the study site was 144 g C/m²/season. This very high proportion of faeces as compared to consumption does not characterize digestibility of food, but rather the way of distribution of manure in the pasture which was unregular, a relatively large amount was left in the analysed "sheep tract".

The decomposition rate of manure was studied in field experiments. The portions of manure (150) of known weight were placed on the pasture on plastic bags filled with sand or lignin, which were partly dug into the ground. A part of them was isolated from meso- and macrofauna by a dense net, 0.3 mm mesh size, hence the others were exposed for invertebrate action, protected from trampling by sheep only. The weight loss of manure was recorded for a month period as well as the composition and biomass of animals  and microorganisms living in sheep faeces. Such experiments  were repeated five times in a growing season (Olecho-wicz 1974).

The sheep faeces  are decomposed  rather  rapidly.  In experiments 32 % of the input  was lost  due to respiratory processes  of organisms and production  of their bodies in a month time   (Olechowicz    1974). Mainly  Diptera larvae, scarabaeid beetles  and  microorganisms are re-sponsible for decomposition processes in the manure.

The weight losses  were much higher  in the series  of experiments exposed to the action of invertebrates. The total losses of C amounted to 42 g C in a season.  The losses  in the series of experiments  where animals were excluded equalled to 17.6 g C only,  98 g C were transfer-red to the soil.  The total losses of manure weight, amount transferred into the soil as well as animal impact in those processes were estimat-ed basing on the results of 5 experiments running during a season, each per a month time.  So, it was assumed  that  decomposition processes of each manure portion lasted above the ground in the pasture  for a month period.  In fact  in a longer run microorganisms might  become more im-portant, as far as animal biomass  was the highest  in the fresh manure (Olechowicz 1974, Breymeyer 1974).

The biomass of roots  and  below-ground detritus  was estimated by sieving and washing soil cores,  and then by separating them  into live and dead components.

The compartment  of below-ground detritus comprises organic debris of the size ranging  from 0.25 to 2 mm.  C content  in that compartment was estimated by combustion.  This fraction could include parts of liv-ing roots  which failed to be separated.  It was very difficult  to es-timate  root  production.  It was assumed to be the difference  between maximum and minimum yearly root biomass.  The results  are in agreement with those obtained when a three-year-period of root life cycle  is as-sumed. This method probably under-estimated real root production.

The matter input  into  the detritus compartment  originated  from dead roots and rhizomes, dead above-ground material (very scarce in the pasture) and from manure.

Basing  on the known values  of biomass  of the dominant  organism groups  and  on literature data  concerning metabolic rate of organisms (*Acarina*, earthworms, Bacteria),  or using our own estimates  on metab-olism  *(Nematoda)*  (Wasilewska in press,  Nowak in press)  it has been calculated  that respiratory losses  for these dominant groups  of soil organisms  amount  to 18 g C  per season.  The values  for soil animals obtained  that way are reasonable,  but values  for microorganisms  are probably  underestimated*.

For the next transfer, from soil detritus to amorphous soil organic matter, a value of 17 g C per season was assessed on the basis of carbon content in worm casts  pushed up on the soil surface  during the season (Czerwiński, Jakubczyk, Nowak in press).  It was found that chemical composition of worm casts (C/N ratio, nutrient content)  is the same as in the soil humus.  Therefore, earthworm activity accounts  for humus formation.  In an experiment  that  consisted  in collecting worm casts from marked plots every four days, the number of worm casts  pushed  up on the soil surface  throughout a season  was estimated.  From the difference in the carbon content between original material, i.e. earthworm food, and worm casts, it was possible to calculate the amount of carbon transferred to humus  due to the activity of earthworms.

In the experiment, only the input  of worm casts  deposited on the surface could be determined;  basing on its results, however, and using Barley's (1959)  data  on the earthworm consumption  total  worm  casts production in the habitat was calculated.

The transfer  of carbon  to humus due to earthworm activity is not the only way of transformation, but as far as it contributes to approximately as much  as 5 % of primary production it seems to be  an important way of transfer.

The data  just discussed prove  that  invertebrates intensify considerably decomposition processes.  It is clear at least  in manure decomposition  during the first month  and  in earthworm contribution towards humus formation.

The decomposition  proceeded  more  rapidly  not only due  to the consumption  and  physical transformation of material, but also  to the stimulating effect of animals on microorganisms.

Several experiments were carried  out  to analyse interactions between invertebrates and bacteria (Breymeyer, Jakubczyk, Olechowicz in press; Czerwiński, Jakubczyk, Nowak in press).  It was found in laboratory experiments that the number of ammonifying bacteria depends on the number  of Diptera larvae and the number of scarabaeid beetles  present in the cultures.

In both experiments – with *Diptera* and with scarabaeids – the number of fungi decreased.  The number of ammonifying bacteria in first 6 days of experiment increased considerably, with increasing number of Diptera larvae, decreased when cultured with A. fimetarius Scarabaeidae (fig.2).

The rapid growth of bacteria in worm casts  is well known.  In our experiments the number of ammonifying bacteria was significantly larger

---

* The dilution plate method  was used  to estimate number  of bacteria, and respiratory rates are based on Jensen's (1936) data.

in cylinders with the soil containing normal density of earthworm - *Al-lolobophora caliginosa*, than in cylinders with the same soil from which earthworms were removed (fig. 2).

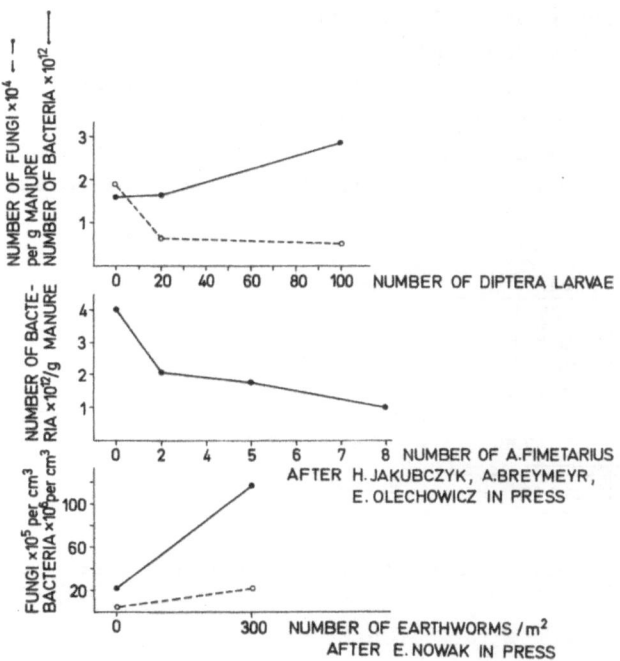

Fig.2. Influence  of  invertebrates  on the number of mi-croorganisms (after 6 days of culture).

The biomass  of animals  per gram of manure,  and per gram  of the surface layer of soil (0-0.5 cm) was compared (table 1). The biomass of

| | SOIL SURFACE | MANURE |
|---|---|---|
| LUMBRICIDAE | 0.01 | 0.43 |
| ACARINA | 0.06 | 0.05 |
| NEMATODA | 0.03 | |
| DIPTERA | | 0.80 |
| COLEOPTERA | | 6.60 |
| TOTAL | 0.10 | 7.88 |

T a b .  1.  Biomass of soil invertebrates mg d.w/lg

invertebrates  in manure  is by  two orders  of magnitude higher than in the soil. In the soil *Acarina, Nematoda* and *Lumbricidae* predominate and in manure *Diptera* larvae and *Scarabaeidae*.

This intensity of life  certainly determines decomposition rate of
the substrate.  Dead plant material  lying  on  the groung reaches  the
belowground detritus compartment  in a year´s time, while manure during
several weeks.

Also the effect of manuring on soil invertebrate biomass  was compared.  The pasture discussed so far, which was fertilized at a rate of
360 g d.w. of manure per season, was compared  with a sheep fold intensively fertilized (500 g of manure within a short time period).

Fertilizing  is followed by a significant increase  in soil herbivore  and  saprovore biomass independently  of their taxonomic position
(table 2).

| | | PASTURE | SHEEP PEN |
|---|---|---|---|
| HERBIVORES | ELATERIDAE | 219 | 850 |
| | NEMATODE | 58.2 | 66.8 |
| DECOMPOSERS | EARTHWORMS | $4.5 \cdot 10^3$ | $20.6 \cdot 10^3$ |
| PREDATORS | NEMATODA | 76.2 | 21.0 |
| | ANTS | 52.3 | 3.5 |
| | SPIDERS | 4.2 | 2.6 |
| ELIMINATION OF EARTHWORMS (IN PERCENT) | | 22 | 13 |

T a b .  2. Biomass of soil invertebrates  after manuring (mg dw/m$^2$)

Only the predator biomass decreases  after  fertilizing,  which is
accompanied  by  a decrease  in the earthworm elimination rate  in more
fertile habitats.

An increase in trophic resources is followed by a decline in elimination rate,  and then  by  a rise  in numbers  of animals using these
resources.

The pasture  is an ecosyste,·where the amount  of plant production
removed is very high, almost as high as in arable lands.  But a significant part of that production is rapidly recycled through manure input.
Manure, in turn, determines a number of soil processes;  it contributes
to an increase  in saprophage numbers,  mainly earthworms,  which again
intensify further decomposition mainly humus formation.

In all experiments carried out the impact of invertebrates appears
to be important, but it is really very difficult to distinguish between
the action of animal and microorganisms — but presumably studying them
together gives more reliable results than studying them separately.

R e f e r e n c e s

A n d r z e j e w s k a ,  L. (1974): Analysis of the pasture ecosystem
    in Pieniny Mountains. VI. Herbivores and their influence on primary
    production. Ekol. Pol. 23.
B a r l e y ,  K. P. (1959):  The influence of earthworms  on soil fer-
    tility.  Consumption of soil and organic matter  by  the earthworm
    *Allolobophora caliginosa* (Savigny). Auste. J. Agric. Res. 10, 171-
    185.
B r e y m e y e r ,  A. (1974):  Analysis  of the pasture ecosystem in
    Pieniny Mountains. XII.  The role  of coprophagous beetles *(Coleo-
    ptera, Scarabaeidae)*  in sheep manure utilisation.  Ekol. Pol. 23.
B r e y m e y e r , A.,  J a k u b c z y k , H.,  O l e c h o w i c z ,
    E. (1974): Analysis of the pasture ecosystem in Pieniny Mountains.
    XII. The influence  of coprophages  on microorganisms in sheep ma-
    nure. Ekol. Pol. 23.
C z e r w i ń s k i ,  Z.,   J a k u b c z y k ,  H.,   N o w a k ,  E.
    (1974): Analysis  of the pasture ecosystem  in  Pieniny Mountains.
    XIV. The influence of earthworms on soil. Ekol. Pol. 23.
J e n s e n ,  H. L. (1936):  Contributions to the microbiology of Aus-
    tralian soils. IV.  The activity  of microorganisms  in the decom-
    position of organic matter. Proc. Linn. Soc. 56: 27-55.
N o w a k ,  E. (1974):  Analysis  of the pasture ecosystem  in Pieniny
    Mountains. XII. Population dynamics  and some parameters of earth-
    worm production. Ekol. Pol. 23.
P l e w c z y ń s k a - K u r a ś ,  U. (1974): Analysis of the pasture
    ecosystem  in  Pieniny Mountains.  IV. Biomass  of aboveground and
    belowground plant organs and organic detritus. Ekol. Pol. 23.
W a s i l e w s k a ,  L. (1974):  Analysis  of  the pasture  ecosystem
    in Pieniny Mountains. XV. Quantitative distribution and respiratory
    metabolism  with suggestion  on production of Nematoda on mountain
    pastures. Ekol. Pol. 23.

C o n t r i b u t o r s   t o   t h e   s t u d i e s   (Department of
Grassland Ecosystem, Institute of Ecology): dr. L. Andrzejewska - sheep
and invertebrate herbivore consumption, dr. A. Breymeyer - coprophagous
beetles,  dr. J. Jakubczyk - microorganisms,  mgr. E. Olechowicz - coprophagous *Diptera*,  mgr. U. Plewczyńska-Kuraś - belowground plant material,  dr. A. Tatur - geology,  hab. dr. T. Traczyk - primary production,  dr. L. Wasilewska - *Nematoda*,  dr. H. Zyromska-Rudzka - *Acarina;*

Academy of Agriculture,  Warsaw:  dr. Z. Czerwiński -  soil properties;
Department of Zoology, University of Warsaw: mgr. E. Nowak - earthworms.

## D i s c u s s i o n

**L.  V l y m** :  Sheep manure produces  patchiness  in the field.  Is  it
possible to discuss the carbon cycle  of these two "subsystems" and how
they are interrelated?

**A.  K a j a k ,**  The carbon cycle could be analysed in both subsystems,
and  in fact  it was analysed to some extent.  The "subsystem" with low
manuring treatment  can be considered  as a general plot from which migration towards treated plots occurs. For example in spider population:
the species composition, structure of dominant species  and  population
dynamics  is identical  in both subsystems,  excluding  the first month
after manuring  when vegetation  was completely destroyed. In dominant
saprophagous group  earthworms, there is  also  the same species composition in both subsystems, only dominance structure differs considerably.

# Comparaison des communautés d'Oribates de litières de chênaies

G. WAUTHY, Ph. LEBRUN
Laboratoire d'Écologie générale et expérimentale,
Louvain

## INTRODUCTION

*a) Intérêt de l'étude*

Cette étude a pour objet l'analyse des communautés d'Oribates hémiéda-phiques de 10 réserves naturelles du sud de la Belgique. Ces réserves présentent un intérêt certain du point de vue botanique et ornitholo-gique principalement (Noirfalise, Huble et Delvingt, 1970) mais on n'y avait jamais prospecté la faune du sol.

Les réserves naturelles ont pour vocation de protéger certains sites particuliers de l'action de l'homme. Dès lors, en se basant sur une prospection exhaustive des Oribates, il nous sera possible de suivre l'évolution spatiale et temporelle des communautés afin d'y déceler, à l'avenir, d'éventuelles modifications ou perturbations.

*b) Buts*

Par l'étude de divers facteurs phytosociologiques, physico-chimiques et pédologiques on peut établir un premier classement des 10 chênaies que nous avons étudiées. En analysant les 10 communautés d'Oribates, on en réalise ensuite un classement. Enfin, nous avons essayé de com-parer et d'interpréter ces deux classifications.

## METHODES

Chaque station est représentée par un carré de 20 m de côté situé dans le sous-bois. On y a prélevé 20 carottes dans les couches holorganiques, réparties sur toute la surface de la parcelle, à l'écart des arbres,

des plantes herbacées  et des débris ligneux,  du moins ceux  de grande
taille. L'unité de prélèvement avait un volume de 90 cc (16 cm$^2$ de surface).   Cent trente deux espèces ou formes on été recensées; elles sont
représentées  par 32.218 adultes  et immatures  d'Oribates (Wauthy en
préparation).

Nous avons réalisé une autre série  de prélèvements inspirée de la
méthode  des "petites faunes" de Grandjean (1947)  dont on trouvera des
applications  dans Aoki (1967), Lebrun (1971) et Lions (1972).  Six habitats ont plus particulièrement été explorés:  la litière, l'humus, le
milieu endogé,  les mousses à terre  et  les milieux arboricoles  (épi-
phytes et écorces).  Ces prélèvements ont été effectués pour trois raisons principales: d'abord pour compléter les relevés spécifiques de nos
prélèvements quantitatifs,  ensuite pour préciser  la répartition  verticale  des espèces  et enfin  pour établir  les correspondances  entre
stases adultes et immatures (voir Trave, 1964).

LES MILIEUX
a) *Critères de classification*

1$^o$ L a   s a t u r a t i o n   e n   b a s e s   d u   c o m p l e x e
a b s o r b a n t   désigne le rapport entre la quantité de cations mé-
talliques absorbés  et la quantité maximale  de ces cations  que le sol
est capable de fixer. N'ayant pas eu l'occasion  de faire  des analyses
précises, nous nous en tiendrons à une échelle relative  (de 0 à +++++)
qui désigne des niveaux de saturation observés  dans des types de végétation analogues à celles que nous avons prospectées.

2$^o$ L e s   3   t y p e s   d' h u m u s   classiquement décrits ont
été observés: les mor, moder et mull.  Les différentes variantes seront
décrites ci-après.

3$^o$ D u   p o i n t   d e   v u e   p h y t o s o c i o l o g i q u e,
les chênaies étudiées  relèvent  de 2 classes différentes: celle des
*Quercerea robori - petraeae* Br.- Bl. et Tx. 1943  qui regroupe les chênaies (R$_1$ et R$_2$) silicicoles et oligotrophes  et celle  des *Querco-Fa-
getea* Br. - Bl. et Vlieg. 1937 qui rassemble  les forêts  à humus doux.
Dans  cette  deuxième classe,  les chênaies R$_4$, R$_8$  et R$_9$ appartiennent
à l'alliance édaphique  sous le climat de Belgique  du *Carpinion* Oberd.
1953 et les chênaies restantes  font partie de l'alliance climacique du
*Fagion* Tx. et Diem. 1936.

## b) *Description des stations*

Pour chaque station, nous nous en tiendrons à signaler la région naturelle, le nom de la forêt et de la localité ainsi qu'un aperçu des
principales espèces végétales.

$R_1$ : Plateau des Hautes Fagnes (Brachkopf, Eupen). alt. 545[1] H: hydro-
mor. S.B.: pratiquement nulle ( 0). C'est un petit bois de chênes
et de bouleaux installé sur un sol podzoloque.

$R_2$ : Ardenne condruzienne (Colonster, Angleur). alt. 165. H: dysmoder.
S.B: très faible (+). Vieux taillis clairsemé de chêne, de hêtre
et de bouleaux provenant de l'exploitation de la hêtraie silicicole.

$R_3$ : Basse Ardenne (Bois du Pays, Amonines). alt. 340, H:mmoder. S.B:
très faible (+). C'est un facies de chênaie silicicole issue de la
hêtraie acidophile à luzule et myrtille.

$R_4$ : Ardenne condruzienne (Grand Pré, Courrière). alt. 270. H: hydro-
moder. S.B: faible (++). Cette futaie relève de l'aile la plus acidophile du *Carpinion*.

$R_5$ : Ardenne septentrionnale (Hertogenwald, Eupen). alt. 420. H: moder.
S.B.: faible (++). Futaie mélangée de chêne sessile et de hêtre.

$R_6$ : Ardenne méridionale (vallée de la Semois). alt. 240. H: mull-moder.
S.B.: faible à moyenne [++ (+)]. C'est un taillis de chênes, de
bouleau et de charme.

$R_7$ : cf. $R_6$. alt. 340. H: mull acide. S.B.: moyenne (+++). Taillis de
charme, de coudrier et d'érable sycomore sous une futaie claire
de chênes, de hêtre et d'érable.

$R_8$ : Basse Ardenne (Bois du Pays, Erezée). alt. 375. H: mull acide.
S.B.: assez élevée [+++(+)]. C'est une chênaie alluviale submon-
tagnarde assez particulière (voir Sougnez, 1973).

$R_9$ : Lorraine belge (Plainsart, Musson). alt. 275. H: "mull neutre".
S.B.: élévée (++++). C'est une chênaie-charmaie édaphique méso-eu-
trophe.

$R_{10}$: Famenne calcaire (Dourbes) alt. 165. H: mull calcique. S.B: saturé
(+++++). C'est une chênaie-charmaie calcicole à orchidées et éllé-
bore fétide.

[1] Alt. désigne l'altitude en mètres. H. le type d'humus et S.B.
l'état de saturation du complexe absorbant.

*c) Classification phytosociologique, physico-chimique et pédologique*

La figure 1 tente d'intégrer les observations reprises ci-dessus.  Nous avons envisagé 5 facteurs de classification:

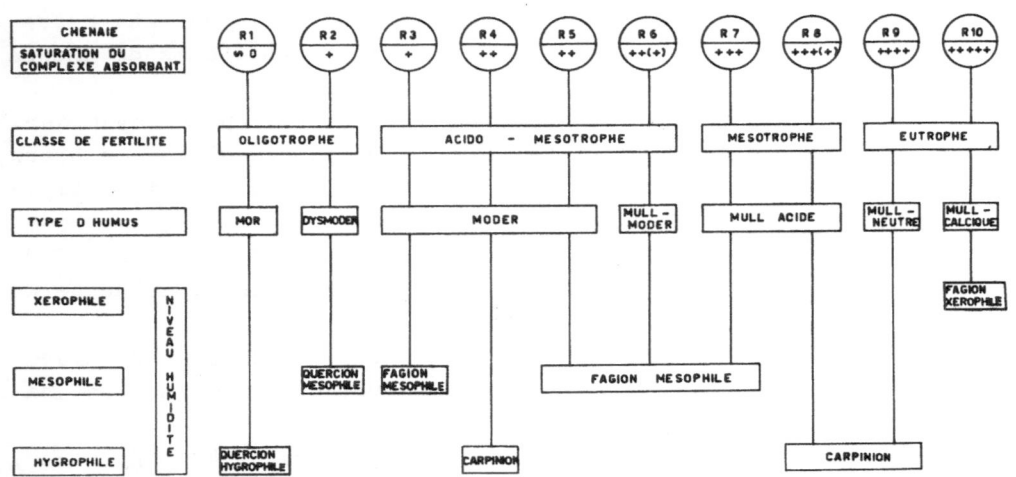

Fig.1.  Classification  phytosociologique,  physico-chimique des dix chenaies.

    $1^{\circ}$  L' é t a t   d e   s a t u r a t i o n   d u   c o m p l e x e
a b s o r b a n t    est la base  de notre classification.  On a ordonné les 10 biotopes selon un gradient croissant  de saturation: $R_1$ possede la valeur la plus faible, $R_{10}$ la plus élevée.
    $2^{\circ}$  L a   f e r t i l i t é   exprime la saturation en "bases" et en sels biogènes d'un sol; c'est donc une mesure de sa productivité poten-tielle. Nous avons observé 4 niveaux: oligotrophe ($R_1$, $R_2$), acido-méso-trophe ($R_{3,4,5,6}$), mésotrophe ($R_7$, $R_8$) et eutrophe ($R_9$, $R_{10}$).
    $3^{\circ}$  L e s   7   f o r m e s   d' h u m u s .   On remarquera que $R_3$, $R_4$ et $R_5$  sont des moder tapiques, que $R_7$ et $R_8$ sont des mull acides et que $R_9$ et $R_{10}$ des mull neutro-basiques. Nous avions donc à notre dispo-sition un large éventail des différents types d'humus des milieux aérés (voir Duchaufour 1965).
    $4^{\circ}$  A u   d é p a r t   d e   l a   s y s t é m a t i q u e   p h y -
t o s o c i o l o g i q u e ,   on répartit  les dix biocénoses étudiées en fonction de deux groupes de facteurs:
- suivant le niveau  de fertilité, les chênaies  du *Quercion* ($R_1$ et $R_2$) occupent l'aile la plus acide tandis que les chênaies du *Fagion* et du *Carpinion* sont cantonnées sur des sols mésotrophes.

- suivant les niveaux d'humidité du sol, les chênaies mésophiles ($R_{3,5,6}$) et xérophiles ($R_{10}$) sont celles du *Fagion*. Celles du *Carpinion* sont hygrophiles tandis que celles du *Quercion* peuvent être méso- ou hygrophiles.

ANALYSE BIOCENOTIQUE
a) *Indice d'affinité entre les communautés d'Oribates*

Dans le but d'établir une classification écologique des communautés d'Oribates, on a utilisé la fonction T (Kullback, 1959). Cette fonction compare globalement le degré de concordance entre le spectre d'abondance relative des espèces de deux communautés et mesure, en fait, l'homogénéité des proportions des espèces ou, si l'on préfère, l'homogénéité de la structure numérique des deux communautés. Il va de soi, par conséquent, que cet indice tient implicitement compte du nombre total d'individus, du nombre d'espèces et de la répartition des individus entre les espèces.

A partir de la fonction T, on définit un nouvel indice A d'affinité qui est une fonction décroissante de T.

b) *Méthode de classement*

Au départ de la matrice d'affinité des communautés prises 2 à 2, on choisit la valeur de l'indice la plus élevée ($R_5$ U $R_6$ = 0,80) et on considère ces deux communautés comme une seule entité. On construit une nouvelle matrice d'où l'on sélectionne la valeur la plus élevée. La méthode se poursuit jusqu'à épuisement des valeurs. La figure 2 montre le processus suivi.

La figure 3 tente de clarifier les processus de regroupements biocénotiques. Se regroupent d'abord les communautés $R_5$ et $R_6$ puis $R_1$ et $R_3$ auxquelles s'ajoute $R_4$. Ensuite, on rassemble ces deux premiers groupes. Enfin, les communautés $R_9$ et $R_{10}$ se réunissent toutes deux, en occupant une position marginale.

Trois communautés occupent une position intermédiaire en ce sens qu'elles présentent des affinités avec le regroupement $R_1$,$R_3$,$R_4$,$R_5$ et $R_6$, affinités qui vont en décroissant dans le sens $R_2$,$R_7$,$R_8$.

|  |  |  |  |  |  |  |  |  |
|---|---|---|---|---|---|---|---|---|
| R5 | R5 | R5 | R5 | R5 | R5 | R5 | R5 | R5 |
|  | U 0,80 | U 0,73 | U 0,81 | U | U | U | U | U |
| R6 | R6 | R6 | R6 | R6 | R6 | R6 | R6 | R6 |
|  |  |  |  | U | U | U | U | U |
| R1 | R1 0,73 | R1 | R1 | R1 0,81 | R1 | R1 | R1 | R1 |
|  |  | U 0,79 |  | U | U 0,85 | U | U | U |
| R3 | R3 0,72 | R3 | R3 0,83 | R3 | R3 | R3 0,81 | R3 | R3 |
|  |  |  | U | U | U | U | U 0,76 | U 0,69 |
| R4 | R4 0,78 | R4 0,83 | R4 | R4 | R4 | R4 | R4 | R4 |
|  |  |  |  | U | U | U | U | U |
| R2 | R2 0,73 | R2 0,80 | R2 0,69 | R2 0,85 | R2 | R2 | R2 | R2 |
|  |  |  |  |  | U | U | U | U |
| R7 | R7 0,77 | R7 0,69 | R7 0,69 | R7 0,79 | R7 0,81 | R7 | R7 | R7 |
|  |  |  |  |  |  | U | U | U |
| R8 | R8 0,60 | R8 0,74 | R8 0,71 | R8 0,79 | R8 0,76 | R8 0,76 | R8 | R8 |
|  |  |  |  |  |  |  | U | U |
| R9 | R9 0,60 | R9 0,61 | R9 0,70 | R9 0,51 | R9 0,67 | R9 0,70 | R9 0,58 | R9 |
|  |  |  |  |  |  |  |  | U 0,63 |
| R10 | R10 0,42 | R10 0,37 | R10 0,47 | R10 0,44 | R10 0,41 | R10 0,44 | R10 0,57 | R10 |

Fig.2. Matrice de regroupement des communautes d'Oribates selon l'indice A par selection de la valeur le plus élevée.

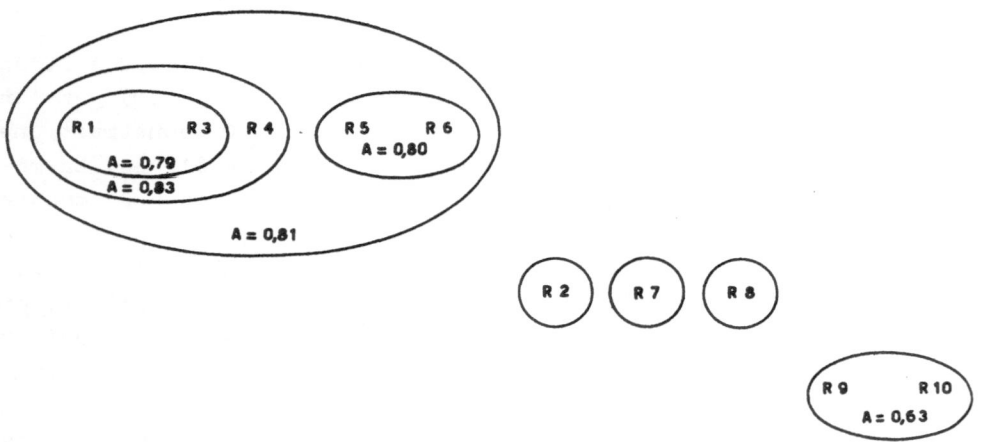

Fig.3. Processus de regroupement des communautes d'Oribates.

*c) Interprétation du classement biocénotique*

1° Le groupe $R_9$ - $R_{10}$

Les communautés $R_9$ et $R_{10}$ appartiennent à des chênaies dont le niveau de fertilité et la saturation du complexe absorbant sont élevés.

2° Les groupes $R_1$ - $R_3$ - $R_4$ et $R_5$ - $R_6$

Les communautés $R_1$ et $R_4$ proviennent de chênaies hygrophiles. Le sol de la chênaie $R_3$ se range dans le groupe mésophile mais la saturation du complexe absorbant est faible comme dans $R_1$ et $R_4$. Ce sont toutes trois des chênaies pauvres et acidophiles.

Les communautés $R_5$ et $R_6$ montrent l'affinité la plus élevée: ces chênaies sont issues toutes deux da la hêtraie climacique. On notera cependant le caractère "mull-moder" de l'humus du biotope $R_6$.

La juxtaposition de ces deux groupes peut s'expliquer par le caractère acido-mésotrophe du sol.

3° Les biocénoses intermédiaires

Une fertilité du type oligotrophe et l'appartenance au Quercion de la chênaie $R_2$ la rapprochent de la chênaie $R_1$. Son type d'humus et la saturation faible du complexe absorbant l'apparente aux chênaies $R_3$, $R_4$, $R_5$ et $R_6$.

Quant à $R_7$ et $R_8$, l'humus est du type mull ce qui les rapproche de $R_9$ - $R_{10}$. De plus, par leur caractère mésotrophe, on doit les apparenter aux autres chênaies.

Ces observations peuvent expliquer la position intermédiaire des communautés $R_2$, $R_7$ et $R_8$ dans le classement biocénotique.


SYNTHESE

La figure 4 harmonise la classification phytosociologique, physico-chimique et pédologique des dix chênaies reposant principalement sur lŝétude de la saturation du complexe absorbant et la classification de ces mêmes chênaies basées sur l'étude de leurs communautés d'Oribates. Il y a parfait parallélisme entre les deux types de classement sauf en ce qui concerne la chênaie $R_2$ puisque l'analyse biocénotique la range dans le groupe intermédiaire.

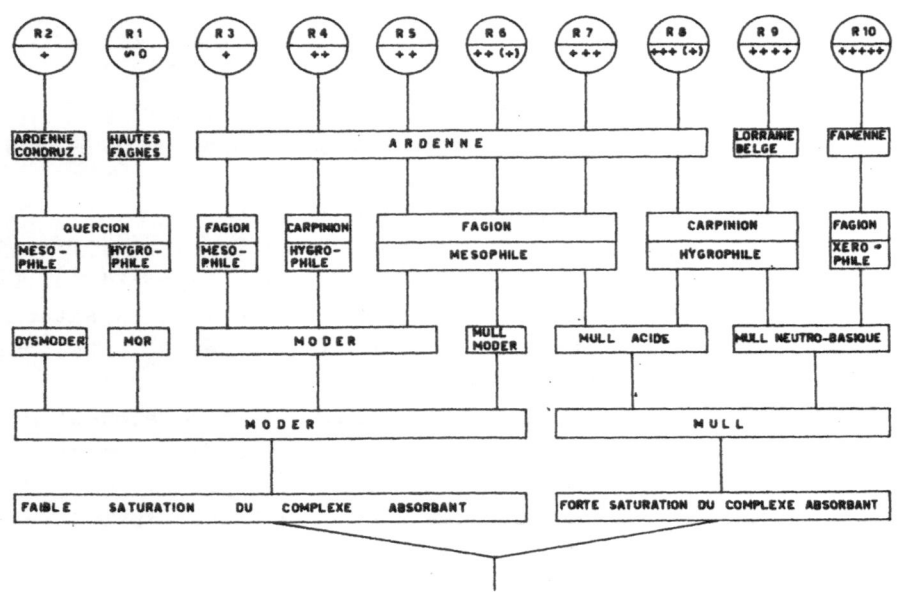

Fig.4. Integration de la classification phytosociologique, physico-chimique et pédologique et de la classification biocénotique des communautés d'Oribates.

La saturation en bases du complexe absorbant semble, des lors,
être un facteur important de différenciation ou d'association des communautés d'Oribates hémiédaphiques de chênaies. Bien plus, et en co-
rollaire, on remarquera la subdivision "mull-moder" (le terme moder
étant pris au sens large puisqu' on regroupe mor, dysmoder, mull-moder
et moder typiques). Ensuite, on notera la concordance entre la systématique phytosociologique et notre étude biocénotique. Enfin, on remarquera le regroupement des chênaies ardennaises.


RÉSUMÉ

Dans ce travail, nous avons réalisé deux types de classification des
chênaies du sud de la Belgique: la première repose sur des critères
phytosociologiques, physico-chimiques et pédologiques; la seconde, sur
l'analyse biocénotique des Oribates hémiédaphiques de ces chênaies,
à partir de la fonction T.

Nous avons montré  que ces  deux classifications  se juxtaposaient
et que divers facteurs conditionnaient  le regroupement ou la différenciation des communautés  et plus particuliérement  l'état de saturation
en bases des colloides du sol.

Remerciements

Nous exprimons toute notre reconnaissance au Dr. J.C. Lions  qui a bien
voulu nous aider et nous conseiller dans les déterminations spécifiques.

References

A o k i , J. I. (1967): Microhabitats  of Oribatia Mites  on a Forest
    Floor. Bull. Nat. Sci. Mus., Tokyo. 1ø/2/, 133-138 + I - II pl.

D u c h a u f o u r , Ph. (1965): Précis  de Pédologie, Masson,  ed.,
    Paris, 481 p.

G r a n d j e a n , F. (1947): Etude  sur les Smarisidae  et quelques
    autres  Erythroides  (Acariens).Arch.Zool.Exptl.Gen. 85/1/; 1-126.

K u l l b a c k , S. (1959): Information theory and statistics.Willey,
    ed., New-York, 395 p.

L e b r u n , Ph. (1971): Ecologie et Biocénotique de quelques peuple-
    ments  d'Arthropodes  édaphiques. Mem. Inst. Roy, Sci. Nat. Belg.
    165, 203 p.

L i o n s , J. C. (1972): Ecologie  des  Oribates  (Acariens)  de  la
    Sainte-Baume  (Var).  Thèse  Doct. Sci. Nat.,  Marseille. C.N.R.S.
    n°A.0.7248, 549 p.

N o i r f a l i s e , A.,  H u b l e , J. and D e l v i n g t , W.
    (1970): Les réserves naturelles de la Belgique. Ministère de l'Ag-
    riculture  (Administration  des eaux  et forêts),  ed., Bruxelles,
    143 p.

S o u g n e z , N. (1973): Le chênaie mélangée à bistorte de l'Ardenne
    (Polygono bistortae-Quercetum roboris). Bull. Jard. Bot. Nat. Belg.
    43, 37-81.

T r a v e , J. (1964): Importance  des stases immatures  des Oribates
    en systématique et en écologie, Acarologia, Fasc.h.s. 1964, 47-54.

# The role of soil fauna of arid shrubland in South Australia

P. GREENSLADE
AUSTRALIAN MUSEUM,
Adelaide

## INTRODUCTION

Arid shrublands of southern Australia are of interest to the soil zo-ologist as an example of a plant association whose stability seems to depend on the cycling of plant nutrients at a rate that appears to be surprisingly rapid for a severe, arid environment (Charley and Cowling, 1968). It is possible that the soil fauna plays a significant part in this system through its effects on decomposition of organic matter. General accounts of arid shrublands in Australia are given by Perry (1970) and Specht (1972). Their principal use is for sheep grazing. All work described here is concerned with the salt bush (*Atriplex vesicaria* Chenopodiaceae) association of the low shrubland formation (Specht, 1972), and is being carried out in South Australia at Koonamore, 340 km north-northeast of Adelaide, rainfall 190 mm per annum. The climate, soils and vegetation are described by Carrodus, Specht and Jackson (1965).

## THE SALT BUSH ASSOCIATION AND SOIL FERTILITY

Low shrublands of *Atriplex* and *Kochia* spp., occur in southern Australia south of the zone in which summer rainfall predominates and where mean annual precipation is less than 350 mm. These shrublands are most ex-tensive in the strictly arid zone with average rainfall of less than 250 mm on calcareous, gypseous and saline soils. In this area, rainfall is fairly evenly distributed throughout the year, but that falling in winter is relatively predictable and tends to occur as light showers, while most summer rain falls as very occasional, heavy thunderstorms. Falls of rain are classes as effective for plant growth (sufficient to

penetrate  the soil  to root level)  if they exceed   12 mm according to
Cowling  (cited by Charley, 1972)  or 6 mm  according to Specht (1972).
Ineffective small showers  do not wet more  than the surface or the top
few mm of the soil profile. The precise dividing line between effective
and ineffective rain  is  obviously  arbitrary, and will vary  with the
season,  previous weather,  penetrability of the soil etc.  On average,
in arid South Australia,  rain occurs on 30 days per annum,  on nine of
which  it is effective (Specht, 1972).  At Broken Hill  with a similar
climate to that  of Koonamore,  88 % of falls  of rain  are regarded by
Charley (1972)  as ineffective.  Generally, in the arid zone in summer,
day time temperatures  and evaporation rates are high, but relative humidity exceeds 80 % for six hours each night.  In winter, days are mild
but frosts at night are frequent.

The typical salt bush association of shallow soils  consists of an
incomplete shrub cover (10-30 %) the density varying with minor changes
in soil  and  topography, over *Bassia* spp. *(Chenopodiaceae)*, annual and
perennial  grasses, and lichens, liverworts  and algae.  The associated
plant species  may cover the ground  between salt bushes but are liable
to be eliminated  by overstocking  with sheep;  soil erosion may ensue,
resulting  in a ʽmonocultureʼ of spaced  salt bushes with bare soil exposed between them.

The nutrient status of the salt bush association has been described by Charley and Cowling (1968), and by Charley (1972)  who emphasised
nitrogen.  Salt bush soils have a low organic content and nutrients are
concentrated  near  the surface  so that the nutrition  of the plant is
tied to their availability  in the top 5-10 cm  of the soil profile. If
soil moisture  is continually available,  the growth  of salt bush continues until so much labile soil nitrogen has accumulated in the standing crop that further growth  is limited  by the rate of release of nitrogen from decomposing organic matter.  Annual litter fall  may exceed
the weight of leaf  on the plant  when in good condition, and, together
with a negligible accumulation  of litter, this indicates rapid  breakdown. Disappearance of an input of litter probably  takes about a year.
Decomposition is limited by moisture. It is well established that bursts
of microbial activity follow the moistening of soil in wetting and drying cycles and that the intensity of microbial activity  after moistening tends to vary directly with the length of the preceeding dry period.
According to Charleyʼs functional model  of the relation between water,
nitrogen and plant performance, falls of rain which are ineffective for
plant growth  do stimulate microbial activity  leading to breakdown of
litter, release of nutrients  and  their buildup in soil.  But, on re-
sumption  of plant  growth after  an effective rain,  the rate of with-

drawal of nutrients from the soil is likely to exceed the rate of accumulation  from organic matter breakdown.  This  ultimately reduces  the
growth rate  and  also the efficiency of water  use by the plant. It is
clear then, that the decomposition of litter  is important if the plant
community  is to persist  and  is closely related to moisture.  At this
point  we can enquire whether  the soil fauna  plays a significant part
in the decomposition process,  either by directly attacking  dead plant
material, communiting it  and  thereby improving the habitat for microorganisms,  or through locomotor  and  feeding activities,  maintaining
the complexity of spatial pattern  and the diversity of the microflora.


THE SOIL FAUNA

Preliminary  observations  on the fauna  of salt bush litter  have been
made on a number of visits to Koonamore  covering  various combinations
of high  and  low temperatures,  when dry,  or after  and during effective and ineffective rain.  The fauna has been sampled and collected by
Tullgren  funnel extraction of litter and soil, before and after moistening if dry,  flotation of soil samples,  pitfall trapping,  and  hand
collections including beating  and sweeping vegetation.  What appear to
be the major components of the fauna judged by abundance and frequency.,
are shown in Fig.1, arranged in a tentative food web. A notable feature
is the lack  of earthworms, diplopods  and  termites, of which at least
the last group is tolerant of arid climates.  This can be accounted for
by the rapid decomposition of litter  and the small quantity present at
equilibrium,  between input and disappearance.  These three groups tend
toward large population biomass,  and individual mobility is low;  they
depend therefore  on the long term presence of accumulations of organic
matter.  They seem to be replaced in salt bush by larvae of *Lepidoptera*
and *Diptera*  whose adults have superior migratory ability  and are thus
adapted to exploit transient resources. Seed feeders, such as *Lygaeidae*
and ants of the genera *Melophorus, Chelaner, Monomorium*  and *Meranoplus*
are well represented  in the salt bush fauna.  The numerically most important groups are ants, of which 59 species  are known  from salt bush
and other low shrublands at Koonamore, and *Collembola*  with 17 species.
Ants are likely to be  of functional significance  as they are the dominant predators, while *Collembola*  are the most abundant 'decomposers'.
Densities of an Isotomid species equivalent to 30,000 per m$^2$  have been
recorded under salt bush. Among the mites, *Mesostigmata* and *Cryptostig-*
*mata* are relatively scarce;  *Prostigmata* are represented by large pred-

ators (e.g. caeculids, bdellids, erythraeids) and by very small species
(e.g. tydeids, erenytids, rhaphignathids) which seem too small to be
predacious on other arthropods.

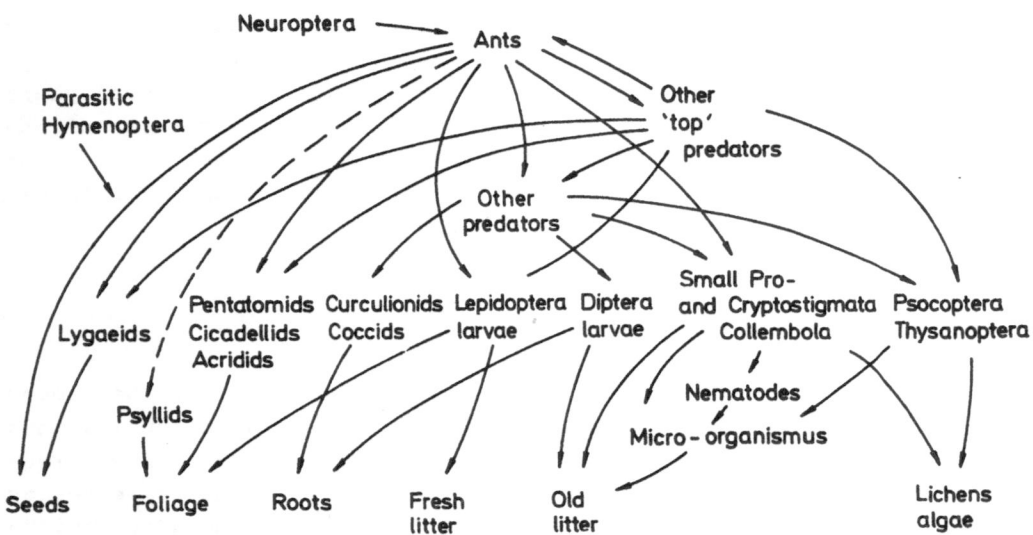

Fig.1. Tentative food web of salt bush fauna. 'Other top
predators' are Carabidae, spiders and scorpions; 'other
predators' include Staphylinidae, Pselaphidae, small
spiders, pseudoscorpions and Prostigmata.

    Figure 1 conceals the complicating factor of spatial pattern. Ver-
tically there is segregation of species associated with soil, foliage
etc., although this is blurred since a large proportion of the total
fauna can be found in the litter from time to time, many above ground
species moving down and sheltering there under hot, dry conditions
while other species move up into the litter from the soil under moist
conditions. In the horizontal plane, pattern is represented by the
bush - interbush alternation, some species being restricted to salt-
bushes, others to intervening grasses and other ground cover. However
the bush cannot be considered in isolation since there is much lateral
movement of species so that processes in and under it are likely to be
influenced by the interbush fauna.
    Evidence of variation in the activity of the fauna in relation to
rainfall comes from pitfall traps, very many species being active on
the surface, extraction of soil samples by flotation, which allows
estimates of total populations, and Tullgren funnel extraction of dry
and moistened samples, as described for the Collembola of a semi-arid

site (Greenslade and Greenslade, 1973).   As yet it is difficult to dis-
sociate rainfall  and seasonal effects, but, taking the extreme case of
hot, summer conditions, it is possible to divide the soil  and  surface
fauna and its activities into two parts.  This is shown in Fig. 2 which

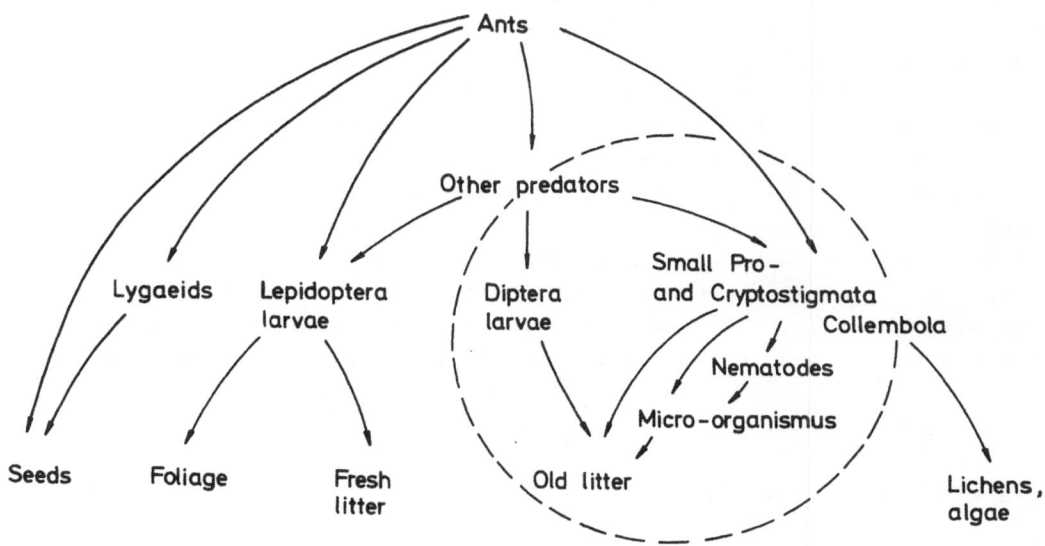

Fig.2.  Simplified food  web  showing  major groups. The
broken line encloses that part  which  seems to function
only under moist conditions.

consists of the food web  in Fig. 1  greatly simplified  and reduced to
what seems to be the key elements. ·That part outside the circle represents groups  that seem  to be relatively  independant  of rainfall and
whose activities for the most part  take place above  the soil surface.
For example  even at midday  in  very hot weather  workers  of numerous
species of the ant genus *Melophorus* are active on the soil surface. The
activity of many other groups  is restricted to cooler parts of the day
and the night  when humidity is higher  and  entomobryid *Collembola* and
a wide range of ants, other insects  and  arachnids appear  on the soil
surface.  The encircled groups in Figure 2, representing what is generally regarded as the true soil fauna, includes the ´decomposer´ complex
of mites and Collembola  whose exact trophic relationships  have yet to
be established.  This part of the food web can be subdivided into those
species active  after ineffective rains  and  those requiring effective
rain.  In the first category,  in the *Collembola*  for example,  are two
species of *Folsomides,* and two *Brachystomellinae*.  They survive hot dry

conditions as adults in an inactive state  and are rapidly reactiviated
when moistened.  On the other hand *Sphaeridia* species  and  apparently,
a *Proisotoma* species  are only active  after heavy effective rain, when
larval Diptera  and  Coleoptera also become frequent-  This can be confirmed for  *Sphaeridia*  in the laboratory.  They are not obtained  when
samples which are dry on collection  are extracted in Tullgren funnels.
But if these samples  are thoroughly moistened,  subsequent  extraction
yields initially immatures of *Sphaeridia* and finally adults after three
to five days.  In this case eggs  appear to be desiccation - resistent
and the complete life history  is extremely short.  In groups  in which
the adults are active after ineffective rain, effective rain  is accompanied by reproduction. The main features of the soil and surface fauna
after effective rains are the presence and activity of immature stages.
Activity by animals resembles that shown by microorganisms  in increasing with rainfall, but it differs  in that some activity  occurs  under
hot dry conditions, and in being greatly intensified by effective rain.

A c k n o w l e g m e n t s

I should like to thank the Department of Botany, University of Adelaide
for permission  to use their facilities at Koonamore, and Mr. D. Lee of
South Australian Museum, for identification of Acari.

R e f e r e n c e s

C a r r o d u s ,  B. B.,  S p e c h t ,  R. L.  and  J a c k m a n ,
    M. E. (1965): The vegetation of Koonamore station, South Australia.
    Trans. Roy. Soc. Sci. Aust. 89, 40-57.
C h a r l e y ,  J. L. (1971): The role of shrubs  in nutrient cycling.
    In: Wildland Shrubs their biology  and  utilisation.  International
    Symposium Utah State University  July 1971, USPA  Forest  Service
    Technical Report Int.-I.
C h a r l e y ,  J. L.  and  C o w l i n g ,  S. W. (1968):  Changes in
    soil nutrient status  resulting  from overgrazing  and  their con-
    sequences  in  plant communities  of  semi-arid areas. Proc. Ecol.
    Soc. Aust., Vol. 3, 28-38.
G r e e n s l a d e ,  P. J. M.  and  G r e e n s l a d e ,  P. (1973):
    Epigaeic *Collembola*  and their activity in a semi-arid locality in
    Southern Australia during summer. Pedobiologia 13, 227-235.

P e r r y ,   R. A. (1970): Arid Shrublands and Grasslands in Australian
        Grasslands. Ed. Moore, R.M. (1970): A.N.U. Press Canberra, 246-259.
S p e c h t ,   R. L. (1972): The vegetation of South Australia, Govern-
        ment Printer, Adelaide, South Australia.

D i s c u s s i o n

J. A.   W a l l w o r k :     In my Californian desert work   I carried out
experiments similar to yours   on litter wetting   but could recover only
newly-hatched specimens of Folsomides.   This suggested that perhaps the
life cycle   is related   to seasonal changes   with   the egg   as the main
drought-resistant stage.   This evidently contrasts   with   your observations since apparently you recovered adults from wetted litter.

P.   G r e e n s l a d e :   It is possible   that   there are two kinds of
egg one of which is drought-resistant   and   whose occurence varies with
the seasons. In our work with dry litter collected in summer and wetted,
initially   we obtained adults   and immatures since all stages appear to
be able to survive hot dry conditions in an "anhydric state".   However,
during   successive wetting   and   drying cycles   the population   of Folsomides sp. obtained   becomes   progressively   more adult   until finally
only adult specimens   were obtained.   The litter   was only damp for 3-5
days during these cycles. It was not until the litter was kept damp for
10 days   or more   that immatures   were obtained and these results   have
been confirmed in culture.   It does appear   in our samples that the egg
is susceptible to desiccation.

# Testacida in two types of soil in Leningrad region

G. A. KORGANOVA, J. G. GELTZER

Laboratory of Soil Zoology,
Institute of Evolutionary Morphology and Ecology
of Animals,USSR Acad. Sci.,
Moscow

*Testacida* are of special interest ᴄo the biological soil diagnostics, since some species show certain requirements for the main environmental factors and therefore can occur in certain types of soils (Ghilarov, 1955; Chardez, 1967).

The abundance and faunal composition of *Testacida* in the USSR soils is but slightly touched. Their presence in some soils is briefly reported by Yakimoff and Zerem (1924), Dogel (1926), Beljaeva (1929), Brodsky (1935), Bozhko (1936, 1940), Geltzer (1964, 1967, 1970, 1972). Only 56 *Testacida* species belonging to 15 genera were listed by these authors, while Chardez (1965) points about 200 species in the soils of various geographical regions. The following genera are the most numerous in the USSR soils: *Centropyxis* (9 species), *Plagiopyxis* (8 species), *Difflugia* (7 species), *Euglypha* (6 species), *Arcella* (5 species).

The authors have studies *Testacida* found in dern-feebly-podzolic and dern-calcareous (rendzina) soils in the south-west part of Leningrad region (Izhory Hills). One plot under study with dern-feebly-podzolic loamy soil was situated on the top of a rather low moraine ridge under *Piceetum-oxalidosum*. The other plot with dern-calcareous leached loamy soil (rendzina) was located on calcareous moraine, 300 m northeast from the first one, near the foot of the north-east slope under *Piceetum-alneto-aegopodiosum*. The compared soils differed greatly in their chemical properties: in the upper layer humus content in dern-feebly-podzolic soil - 2.3 %, in dern-calcareous one - 7.7 %; pH - 5.2 and 7.6 respectively; free carbonates in the first soil type - from the depth 230 cm, in the second one - already at the surface.

Soil samples were taken from both plots in $A_1$ layer using methods accepted in soil microbiological analyses (Krasilnikov, 1958). Two methods were used for *Testacida* isolation from the soil: the direct method and flotation. Using the direct method tests have been taken out from soil suspension (10 samples, 100 mg each); at the flotation they are extracted by $CO_2$ bubbling (Bonnet et Thomas, 1958).

From these  two  soil types 27 *Testacida*  species  were  isolated:
17 species  from  dern-feebly-podzolic  and  23 - from  dern-calcareous
(rendzina).  13 species were  common  to  both: *Centropyxis aerophila*,
*Centropyxis aerophila* v. *sphagnicola*, *Centropyxis* sp., *Cyclopyxis kahei*,
*Cyclopyxis* sp., *Plagiopyxis callida*, *P. declivis*, *P. intermedia*, *P. ob-
longa*, *P. penardi*, *Plagiopyxis* sp., *Phryganella acropodia*, *Trigonopyxis
arcula*[+].[1]

In dern-feebly podzoloc soil  *Centropyxis aerophila* v. *sphagnicola*
proved to be the most abundant.  This species  prefers acid conditions,
often occurring in aerophilic mosses and  in well aerated feebly decomposed moor type litter  as well as in brown  and  acid forest soils; in
dern-calcareous (rendzina) soil  the most familiar species  was *Arcella
arenaria* -  a widly distributed species  mostly  inhabiting mosses  and
lichens and rough calcareous soils as well.

The following species have been isolated from dern-feebly-podzolic
soil only: *Arcella excavata*, *Cyclopyxis ambigua*,  *Trinema complanatum*,
*Trinema* sp.  They occur  in sapropel,  can tolerate the lack of oxygen
and prefer slightly acid media.  The species  *Arcella arenaria*, *A. arenaria* v.*sphagnicola*, *Centropyxis halophila*, *C. plagiostoma*[+], *C. plagiostoma* v.*terricola*,  *C. sylvatica* v.*minor*, *Cyclopyxis eurystoma*, *Plagiopyxis angularis*[+],  *P. minuta*, *Schwabia terricola*  were isolated exclusively from dern-calcareous soil.  *Arcella arenaria*,  *Centropyxis halophila*, *C. plagiostoma* - are aerophilic forms, preferring alkaline conditions (till pH 8.5-9.0).  They are rather characteristics of calcareous
soils (Bonnet et Thomas, 1960; Schönborn, 1962, 1967).

Though the above mentioned species  are probably not the only *Testacida* species,  inhabiting these soils,  the data obtained demonstrate
certain  patterns  of  *Testacida* distribution  in the soil types  under
study.

R e f e r e n c e s

B e l a j e v a ,  K. V. (1928):  K voprosu o protistofaune počv Ak-Ka-
    vakskoj opytno-orosit. stanzii  (On the problem  of soil protisto-
    fauna of the Ak-Kavakskaja experim. station). Vestnik irrigazii, 1.
B o n n e t ,  L. et  T h o m a s ,  R. (1958):  Une technique d'isole-
    ment des Tchécamoebiens *(Rhizopoda,  Testacea)*  du sol et ses re-
    sultats. C. R. Acad. Sci., 247, 21.

[1] Species marked  with [+] were found by T. Ratkova.

B o n n e t ,  L.  et  T h o m a s ,  R. (1960):  Thécamoebiens du sol.
  In: Fauna terrestre  et d'eau douce des Pyrénées-Orientales, fasc.
  5.

B o ž k o ,  M. P. (1936): O pitanii protofauny rasličnych počv Ukrainy
  (On the problem  of Protozoa  feeding  in  different soils  of the
  Ukraine). Učenye zapiski Char'kov. Gosud. Univ., 6-7.

B o ž k o ,  M. P. (1940): Reakzija protistofauny počv na normy udobre-
  nija sveklovičnych polej  (The reaction  of soil protistofauna  on
  field fertilization). Trudy Char'kov. Gosud. Univ., 8-9.

B r o d s k i j ,  A. L. (1935): Protozoa počvy i ich rol' v počvennych
  processach (Soil Protozoa and their role in soil processes). Bull.
  Sredneas. Gos. Univ., 3.

C h a r d e z ·,  D. (1965):  Ecologie générale des Thécamoebiens. Bull.
  Inst. Agr. et Stat. Rech. Gembloux, 33, 3.

C h a r d e z ,  D. (1967):  Histoire  naturelle  des protozoaires Thécamoebiens. Les naturalistes belges 48, 10.

D o g i e l ,  V. (1926):  Sovremennoe sostojanie voprosa  o počvennych
  prostejšich (The modern approach to the problem of soil Protozoa).
  Isv. Gos. Inst. opytn. agron., 4,3.

G e l' t z e r ,  Ju. G. (1964): Protosoinaja fauna pojmennych i derno-
  vo-podzolostych počv i eje svjaz' s rizosferoj  nekototych selsko-
  cozjajstvennych rastenij (Protozoan fauna of flood-lands and dern-
  podzolic soils and its relation  to the rhizosphere  of some agricultural plants). Avtoreferat, Moskva.

G e l' t z e r ,  Ju. G. (1970):  Sravnitel'naja charakteristika proto-
  zojnoj fauny rizosfery  nekotorych  sel'skochosjajstvennych rastenij na dernovo-podsolistoj počve (The comparison of protozoan fauna in the rhizosphere of some agricultural plants on dern-podzolic
  soil).  Sbornik "Povyšenie plodorodija počv necernozemnoj polosy",
  Izd. Moskovskij Univers.

G e l' t z e r ,  Ju. G. (1970):  Prostejšie počv  pojmy  reki Kljaz'my
  i metody  ich  identifikacii i količestvennogo uceta  (Protozoa of
  their counts and identification) Sborn. "Mikroorganizmy v sel'skom
  chozjajstve", Izd. Moskovskij. Univers.

G e l' t z e r ,  Ju. G. (1972):  Počvennye rakovinnye kornenozki (Soil
  Testadida). Sbornik "Problemy počvennoj zoologii", Baku.

G h i l a r o v ,  M. S. (1955):  Počvennye rakovinnye ameby (Testacea)
  i ich značenie slja diagnostiki bolotnych počv  (Soil Testacea and
  their role in bog soils diagnostics). Počvovedenie 10.

K r a s i l' n i k o v ,  N. A. (1958):  Mikroorganizmy počvy  i vysšie
  rastenija  (Soil microorganisms  and higher plants). Moskva,  Izd.
  Acad. Nauk SSSR.

S c h ö n b o r n ,   W. (1962):   Zur Ökologie der sphagnikolen, bryoko-
     len und terrikolen Testacean. Limnologia 1.
S c h ö n b o r n ,   W. (1967):   Taxonomik   der beschalten   Süsswasser-
     Rhizopoden. Limnologia 5, 2.
J a k i m o f f ,   M. L., Z e r e m ,   S. (1924): Contribution a l'etu-
     de des protozoaires   du sol   de Petrograde. Z. Bact., Abt. 11, Bd.
     63.

Discussion

Y. D. P a n d e : Can you tell me the economic importance of *Testacida?*
What is the effect  of too much dryness   and   moisture   in soil   on the
density of population of *Testacida?*

G. A. K o r g a n o v a : Biomass of soil protozoan organisms, including Testacida, being about 300 kg per 1 ha of arable layer of the soil,
increases soil fertility.

     More humid soils contain a richer and more numerous *Testacida* fauna.

M. M. C o u t e a u x : Vos resultats rejoignent ceux  de Bonnet  dans
les Pyrenées orientales concernant la difference entre sols podzoliques
et rendzines Avez vous trouve les espèces carasterictiques de rendzines
telleque *Geopyxella sylvicola?*

     Avez vous trouve une difference entre le rapport individus vivants
theques vides dans les 2 types de sol?

G. A.  K o r g a n o v a : I have not found *Geopyxella sylvicola* in
the soils investigated.  May be, it can be explained  by the difference
between the same soils in different geographical regions.

     I have not differentiated between dead  and live organisms.  It is
the subject of my future work.

# Ecologie des Thecamoebiens de quelques humus bruts

M. M. COÛTEAUX
Station de Recherches Cytopathologiques,
Saint-Christol-Les-Ales

## INTRODUCTION

L'analyse quantitative de trois peuplements thécamoebiens a été pour-suivie pendant une année et a permis le recensement de 56 à 71 espèces selon les biotopes dont 39 à 48 étaient représentés par des individus vivants (Coûteaux, 1973).

## TECHNIQUES ET METHODES

La fréquence des échantillonnages est hebdomadaire. Vingt échantillons sont mélangés par dix, fixés et colorés et mis en suspension dans de l'eau. Une fraction aliquote de la suspension est analysée sur membrane filtrante Millipore (Coûteaux, 1967).

Pour la mise en évidence des périodicités, nous avons utilisé une méthode d'autocorrélation (Kendall, 1945). Celle-ci consiste à établir une série de corrélations avec les données de densités hebdomadaires et les mêmes données décalées d'une semaine.

Par l'analyse factorielle (Dagnelie, 1960), il a été possible d'expliquer des corrélations entre relevés hebdomadaires par la mise en évidence de phénomènes fondamentaux sous-jacents justifiant en tout ou en partie les valeurs de corrélations.

## DESCRIPTION DES BIOTOPES
### Situation géographique

Deux stations sont situées en Moyenne-Belgique, dans la Forêt de Meer-dael, la troisième dans la région parisienne en Forêt de Sénart.

## Composition floristique dominante

- F o r ê t   d e   M e e r d a e l   -   C h ê n a i e   à   l u z u -
l e  *Fagus silvatica, Quercus petraea, Deschampsia flexuosa, Luzula pilosa, campestris* et *luzuloides.*

- F o r ê t   d e   M e e r d a e l   -   P e s s i è r e  *Picea abies*

- F o r ê t   d e   S é n a r t   -   C h ê n a i e   à   m o l i n i e
*Quercus petraea, Deschampsia flexuosa, Molinia coerulea*


## Caractéristiques physico-chimiques

Le pH  des trois stations  est très bas (3,7 à 4,3) et le C/N est élevé
(18 à 25).


## Climat

Les trois stations sont situées  dans une région où le climat peut être
qualifié d'océanique avec une légère tendance continentale.  L'humidité
du sol est toujours élevée.  Il est rare qu'elle descende en dessous du
point de flétrissement permanent.


## RESULTATS
## Description de la faune

Le peuplement  des trois stations offre des similitudes manifestes. Les
corrélations entre les densités moyennes annuelles sont comprises entre
$0,714^{xxx}$ et $0,936^{xxx}$.  Quatre espèces  sont dominantes partout: *Phryganella acropodia, Euglyphidion enigmaticum, Trinema complanatum*  et *Plagiopyxis declives*.  C'est  au niveau  de la cinquieme espèce par ordre
d'importance quantitative que l'on peut distinguer  les trois stations:
*Hyalosphenia subflava* caractérise la chênaie à luzule, *Trigonopyxis arcula* la pessière  et  *Centropyxis aerophila* var. *sphagnicola* la chênaie
à molinie. Toutes ces espèces ont une densité relative supérieure à 4 %.
C'est dans la chênaie à molinie que l'on rencontre le plus d'espè-

ces, 14 ont une densité relative comprise entre 1 et 4 % alors que dans la chênaie à luzule, il n'y en a que 5 et dans la pessière, 2.


## Dynamique des peuplements et des populations

La méthode d'autocorrélation citée plus haut nous a permis de mettre en évidence des cycles de variation de densité chez les Thécamoebiens. Ces cycles ont une durée de 6 à 8 semaines, 3 mois, 6 mois et 40 semaines.

Dans le chênaie à luzule et dans la pessière, le cycle des peuplements est de 6 mois. Dans la première station c'est à *Phryganella acropodia* et à *Plagiopyxis declivis* qu'il faut attribuer la responsabilité de cette périodicité, dans la deuxième, seule *Phryganella acropodia* est responsable. C'est également à *Phryganella acropodia* et *Plagiopyxis declivis* qu'il faut attribuer le cycle de 40 semaines du peuplement de la chênaie à molinie.

Des cycles de 6 à 8 semaines se rencontrent chez les individus enkystés dand la chênaie à luzule. *Trinema complanatum* et *Trachyleuglypha acolla* ont un cycle de 3 mois dans cette même station.


## Relations avec quelques facteurs climatiques

Parmi les facteurs climatiques étudiés (température, humidité actuelle, pluviométrie, niveau de la nappe phréatique, évaporation) c'est avec le niveau de la nappe phréatique que la corrélation avec la densité des Thécamoebiens est le plus souvent significative. Cette corrélation est négative c'est-à-dire que plus la nappe est haute, plus la densité des Thécamoebiens est élevée. Il ne faut cependant pas voir là de relation directe car, au plus haut niveau, la nappe n'est jamais qu'à - 50 cm de la surface, aucun rapport donc avec les Thécamoebiens qui se trouvent dans les premiers cm. Le niveau de la nappe phréatique traduit, en quelque sorte, l'état général du complexe hydrique du sol.


## Synécologie dynamique

Dans la chênaie à luzule, l'analyse factorielle en composantes prin-

cipales des relevés hebdomadaires permet de souligner une affinité
entre les relevés où *Trigonopyxis arcula* est toujours présent et où *Euglypha cuspidata* est toujours absent et une autre affinité entre les
relevés où *Euglypha cuspidata* et *Plagiopyxis declivis* sont présents
alors que *Trigonopyxis arcula* est absent. Deux autres affinités ont pu
être mises en évidence: celle entre les relevés où *Centropyxis aerophila* var. *sphagnicola* était présent et *Phryganella paradoxa* absent et
celle entre les relevés où *Phryganella paradoxa* était présent et *Centropyxis aerophila* var. *sphagnicola* absent. Ces résultats sont représentés graphiquement sur les figures 1 et 2 où les coordonnées sont les
saturations en divers facteurs principaux extraits par l'analyse.


DISCUSSION

S'il a été possible de mettre en évidence des cycles, leur signification n'est pas toujours manifeste. L'influence de l'humidité explique
assez bien les cycles de 6 mois. En effect, le printemps et l'automne
sont deux périodes favorables, l'été et l'hiver sont deux périodes de
sécheresse (en hiver, le gel fréquent du sol rend l'eau inutilisable
par les Thécamoebiens). Le cycle de 40 semaines dans la chênaie à molinie s'explique aisément aussi par une diminution considérable de l'humidité actuelle en automne. Les cycles de 9 à 10 semaines ou de trois
mois sont plus difficiles à interpréter. Peut-être pourrait'on trouver
une réponse grâce à des travaux sur les relations entre Protozoaires
et Bactéries, Cryptogames et Phanérogames tels que ceux présentés par
Bamforth, Darbyshire, Geltzer et Korganova, Nikolyuk et Tapinskaya au
IVè Congrès International de Protistologie.
    On a vu la symétrie de forme des trois peuplements étudiés au ni-
veau des 5 espèces dominantes. Au delà de ces espèces, la divergence
apparaît et grâce à l'étude des variations saisonnières il est possible
de mesurer l'évolution de la diversité. Dans la pessière ($A_1$), (fig. 3)
la diversité est la plus faible, elle va croissant: il s'agit d'un peuplement en évolution. Dans la chênaie à lazule (FM), la diversité est
moyenne et assez stable à l'exeption d'un maximum isolé au mois d'avril.
Ce maximum témoigne de l'expansion printanière. Dans la chênaie à molinie (FS) où le diversité est la plus élevée, le peuplement est le
plus stable.
    L'étude d'un peuplement dans sa dynamique par l'analyse factorielle
permet de voir dans les affinités entre relevés une tendance antago-
niste entre certaines espèces: entre *Trigonopyxis arcula* et *Plagiopyxis*

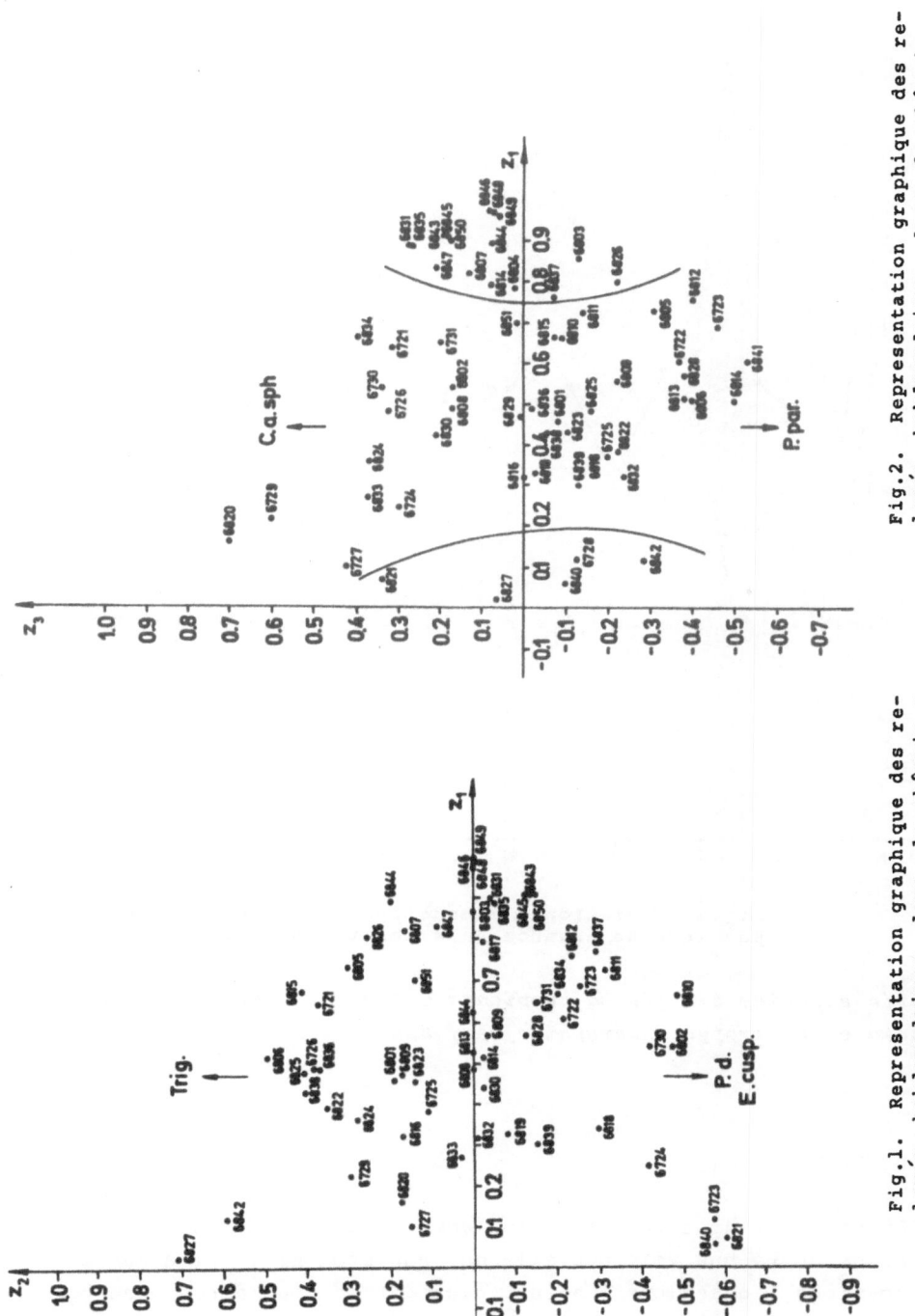

Fig.2. Representation graphique des relevés hebdomadaires dans la chênaie a luzule en fonction de leur saturation en premier et troisième facteur.

Fig.1. Representation graphique des relevés hebdomadaires dans le chênaie a luzule en fonction de leur saturation en premier et deuxième facteur.

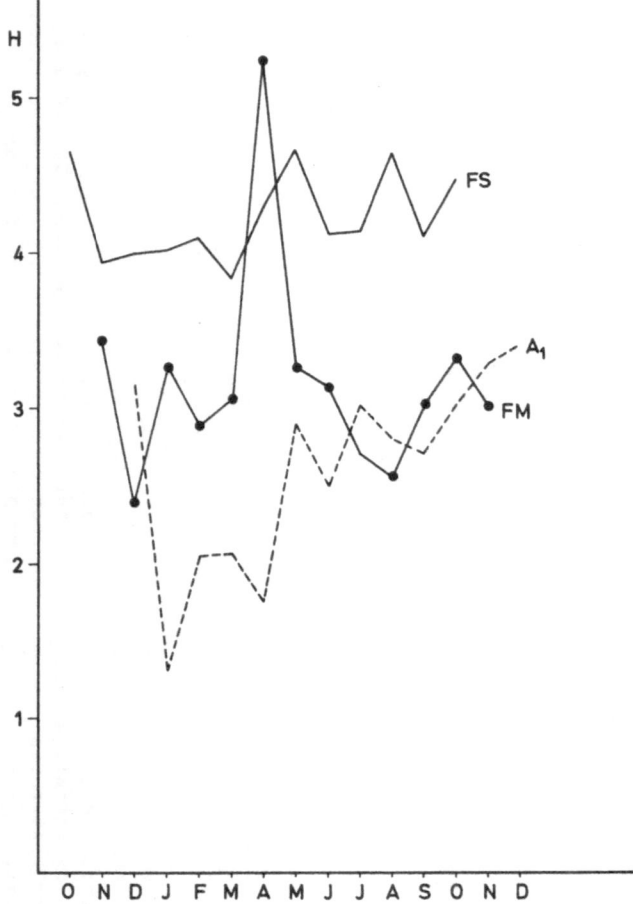

Fig.3. Variations mensuelles de la diversite exprimée
par le H de shannon dans les trois stations.

*declivis* associée à *Euglypha cuspidata* d'une part  et entre *Phryganella*
*paradoxa* et *Centropyxis aerophila* var. *sphagnicola* d'autre part.

CONCLUSIONS

Si l'étude de la dynamique des peuplements  et des populations  de Thécamoebiens ouvre une voie nouvelle en Protistologie, elle suscite bien
des questions,  questions d'autant plus difficiles  à résoudre  que nos
connaissances sur la biologie et l'éthologie de nombreuses espèces nous
est  totalement inconnue.  On a vu, en effet, que les facteurs écologi-

ques  ne sont pas les seuls en cause  dans  les variations saisonnières
mais que d´autres facteurs propres à l´espèce ou aux facteurs biotiques
du milieu devraient être pris en considération tels les autres Protozoaires (Bamforth, 1971) et la microflore.

R e f e r e n c e s

B a m f o r t h , S. S. (1971): The numbers  and  the proportions  of
        Testacea and ciliates in litters and soils. J.Protozool. 18, 24-28.
C o u t e a u x , M. M. (1967): Une technique d!observation  des Thé-
        camoebiens du sol pour l'estimation  de leur densité absolue. Rev.
        Ecol. Biol. Sol. 4, 593-596.
C o u t e a u x , M. M. (1973): Ecologie des Thécamoebiens de quelques
        humus bruts forestiers: L'espèce dans la dynamique de l'équilibre.
        Thèse doct. Etat, Paris VI. 311 p.
D a g n e l i e , P. (1960): Contribution à l'étude  des communautés
        végetales par l'analyse factorielle. Bull. Serv. Carte phytogéogr.,
        sér. B, 5, 2, 93-195.
K e n d a l l , M. G. (1945):  On the Analysis of Oscillatory Time Se-
        ries? J. R. Stat. Soc. 106, 91.

D i s c u s s i o n

P. L e b r u n :  1. Quelle est la signification biologique du rapport
Cilies (Rhizopodes utilisé par certains auteurs?) 2. Les cycles de 6 semaines ou 3 mois serraient-ils dus a un rythme de reproduction? 3. L'en-
kystement est-il seulement provoqué par les facteurs climatiques?

M. M. C o u t e a u x :  1. L'indice Cilies (Rhizopodes  est un indice
de fertilité utilisé surtout dans des sols de cultures.)   2. Oui, quand
j'ai parle d'innéité, c'est aussi aux problèmes  de la reproduction que
je pensait. Notons que cette innéité ne peut être dissociée du contexte
du peuplement.  Une même espèce peut ainsi presenter  un cycle  dans un
biotope alors que dans un autre, elle n'en presente pas.  3. L'enkyste-
ment de resistance  est provoqué  par des facteurs climatiques et principalement l'humidité. Les kystes de sexualité sont déterminé par d'autres facteurs plus complexes qu'il reste a definir.

R. C o v a r r u b i a s :  Concernant les cycles  de 6 semaines  et 3 mois, est-ce que l'on pourrait chercher une interprétation parmi les interrelations predateur/proie? Aussi quels sont les relations trophiques des Thecamoebes?

M. M. C o u t e a u x : Nous ne disposons que de très peu de données en ce sujet. Les Thécamoebiens se nourriraient essentiellement de Bactéries. Ceci n'est cependant pas prouvé parmi les Thécamoebiens des sols. Quant aux prédateurs, aucun travail precis n'a été fait sur ce sujet.

M. B o u c h é : Les apparentes "oppositions" que vous avez mis en évidence dans votre analyse factorielle de correspondence concerne-t-elle les espèces appartenant au même groupe de morphologie (morphologie reliée à la distribution spatiale)?

M. M. C o u t e a u x : Non, elles n'appartiennent pas au même groupe morphologique.

# Quantitative distribution and respiratory metabolisms with suggestion on production of nematodes on mountain pastures

L. WASILEWSKA

Institute of Ecology,
Warszawa

The community of soil nematodes was studied in a pasture fertilized at different rates with sheep manure. The pasture was situated in the Low Pieniny Mountains, near Jaworki village, southern Poland. According to the way of its utilization, 3 plots were selected for study:

1. An ordinary pasture, grazed by sheep over the growing season, thus regularly fertilized with sheep manure.

2. Unutilized pasture being a fenced off part of the pasture mentioned above, neither grazed by sheep nor mown.

3. Pasture used as a sheep fold, where sheep stayed for 2 - 3 successive nights and during milking. It was used as a grazing land several months later. Consequently, the manure input was limited to a very short period, but at a very high rate (about 500 g dry manure/$m^2$). This way of utilization resulted first in the destruction of vegetation, which was followed by its luxurious growth. As a result of putting sheep in a fold, there was an increase in the content of available mineral compounds in the soil, a rise in plant production and in the number of bacteria.

Simplifying the situation it is possible to arrange the three study plots according to the increasing rate of fertilization with sheep manure:

unutilized pasture                    pasture                    sheep fold

Nematodes were sampled mainly once a month. The results presented here are average annual values. Numbers, biomass and respiratory rate of nematode trophic groups were estimated. To estimate the respiratory rate, a regression formula was used as reported in a paper by Klekowski, Wasilewska, Paplińska (1972). This regression concerns the relation of nematode body weight to respiratory rate ($R = 1.40 w W^{0.72}$ at 20°C, where R is the amount of oxygen consumed by an individual per time unit, and

W is the body weight). Cumulative respiratory metabolism was calculated
for the particular nematode trophic groups   at the actual soil temperatures and developmental stages as reported by Wasilewska (in press, b).
The following trophic groups  of nematodes  were distinguished:  microbivorous, fungivorous, parasites of higher plants, omnivorous and predators.

In a joint analysis of all nematode trophic groups, their numbers,
biomass and cumulative respiratory metabolism attained t'-e highest values in the unutilized pasture, and the lowest  in the sheep fold, where
sheep manure input was the highest (Tab. 1).

T a b l e  1.   Numbers, biomass  and  cumulative respiratory metabolism
of the total community of nematodes

| Type of pasture | Numbers | Wet biomass | Respiratory metabolism |
|---|---|---|---|
| | $10^3/m^2$ | $g/m^2$ | $kcal/m^2/year$ |
| Unutilized pasture | 3 695 | 3.0 | 65 |
| Pasture | 3 500 | 2.2 | 50 |
| Sheep fold | 3 185 | 1.4 | 41 |

When particular trophic groups are analysed one by one, then:
1. Numbers, biomass  and  cumulative respiratory metabolism of the
microbivorous group are higher in the pasture and sheep fold. The input
of manure  is followed  by a rise  in bacterial activity, which in turn
influences this group (Fig. 1).
2. Cumulative  respiratory  metabolism  of the  fungivorous group,
which was not numerous, reveales the same trend (Fig. 2).
3. No significant differences  were observed among the three plots
for the most abundant group  of parasites of higher plants.  A slightly
increasing trend was recorded in the sheep fold (Fig. 3).
4. Considerable differences  were found in the omnivorous group of
the three plots.  In the  grazed pasture plot  and the sheep fold plot,
numbers, biomass  and  cumulative respiratory metabolism of this omnivorous group  were lower  than  in the unutilized pasture plot.  For the
ordinary grazed pasture intermediate values  were  found  (Fig. 4). The
values  of the parameters  in question significantly decreases  with an
increasing fertilizing rate. They were highest in the habitat not modified by sheep, i.e., in the unutilized pasture. Intensive fertilization
by sheep had, as a result, a threefold decrease in numbers of the omni-

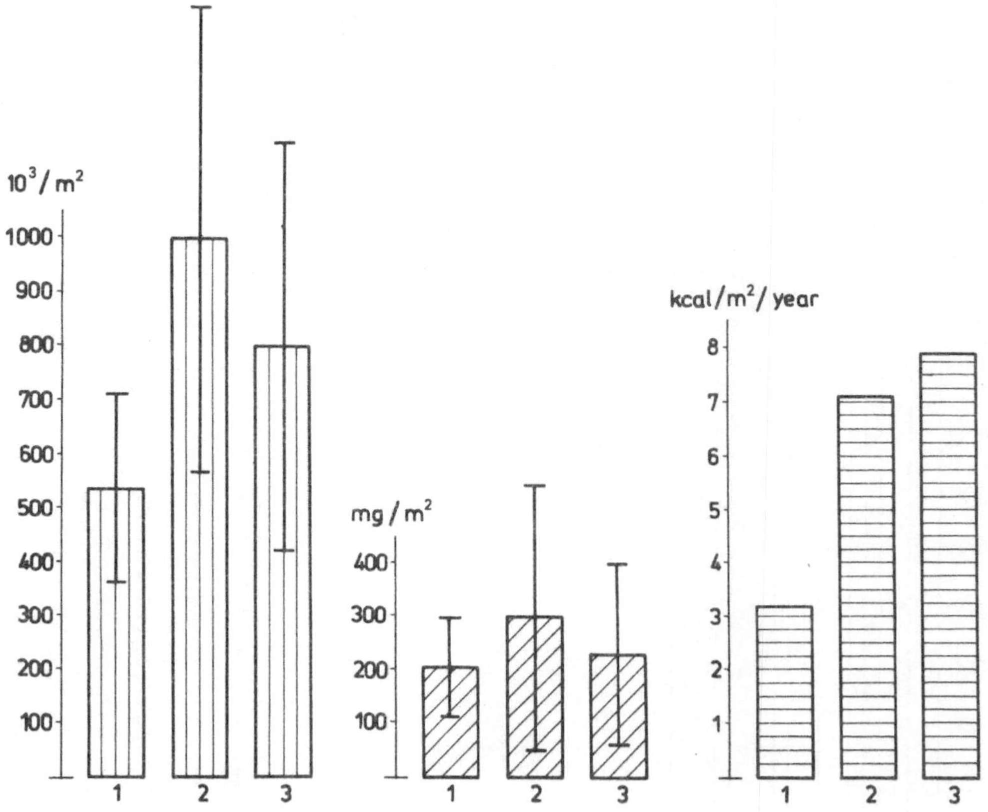

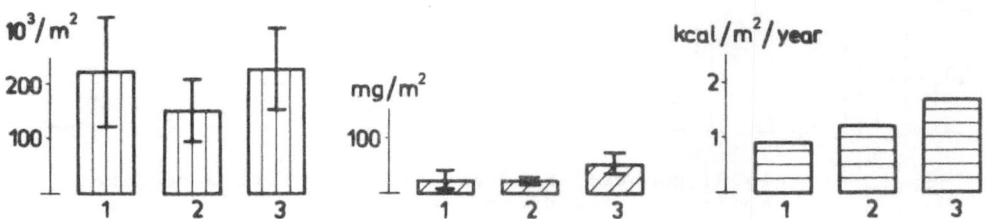

Fig.1. Numbers, biomass   and   cumulative respiratory metabolism of the microbivorous group in examined pasture.
1 - Unutilized pasture   2 - Pasture   3 - Sheep fold.

Fig.2. Cumulative   respiratory metabolism   of the fungivorous group.

vorous group. It seems that the nematodes in this omnivorous group prefer more natural, undisturbed habitats. It is significantly reduced, for
instance, in agrocoenoses, where its number is low (Sandner, Wasilewska

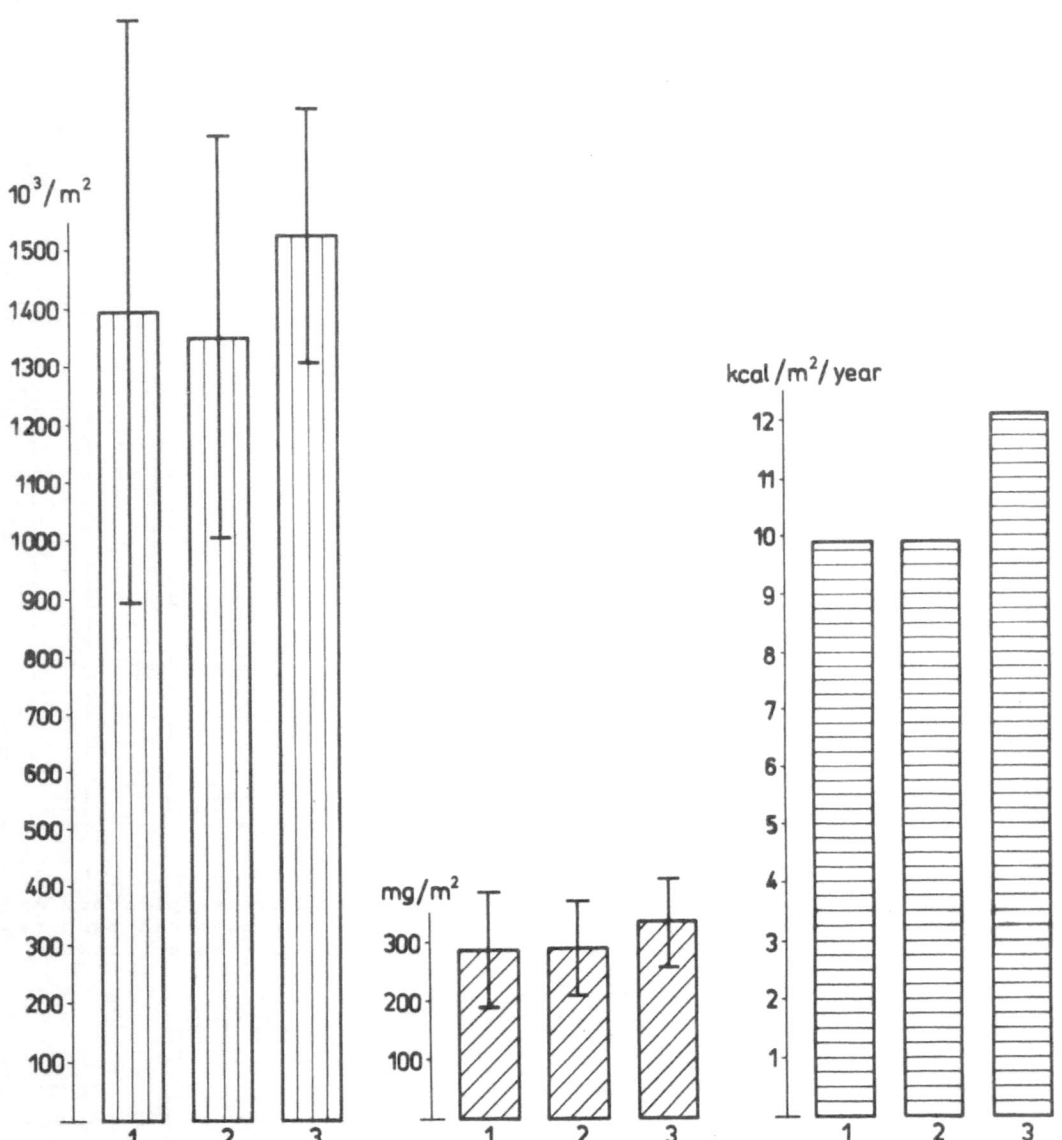

**Fig.3. Numbers, biomass and cumulative respiratory me-**
tabolism of the group of parasites of in examined pasture. 1 - Unutilized pasture  2 - Pasture  3 - Sheep fold.

1970, Wasilewska in press a). Percentage contribution  of this group to
the total nematode community for several habitats  is presented in figures 5 and 6.

     If one assumes  that  the proportions  among  the various  trophic
groups of nematodes  under different habitat conditions  are one of the

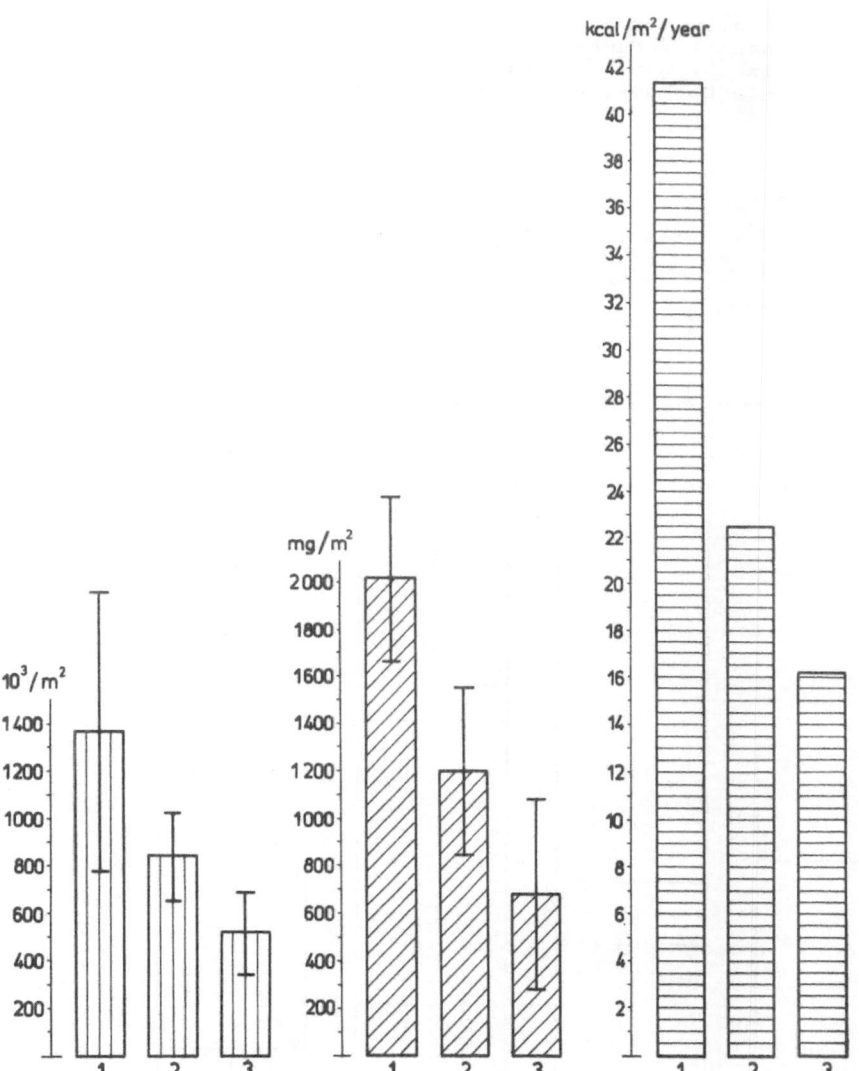

Fig.4. Numbers, biomass and cumulative respiratory metabolism of the omnivorous group in examined pasture. 1 - Unutilized pasture 2 - Pasture 3 - Sheep fold.

possible responses to an outside impact on the natural environment, the omnivorous group can be considered as the most sensitive in this respect.

In the group of predators, a similar trend was observed as in the omnivorous group (Fig. 7).

Fig.5. Percentage of the omnivorous group in different habitats. Numbers of individuals from all nematode groups in the given habitat (from Sandner and Wasilewska 1970). **M - Meadows**, F - Forest, D - Dunes, PF - Potato fields.

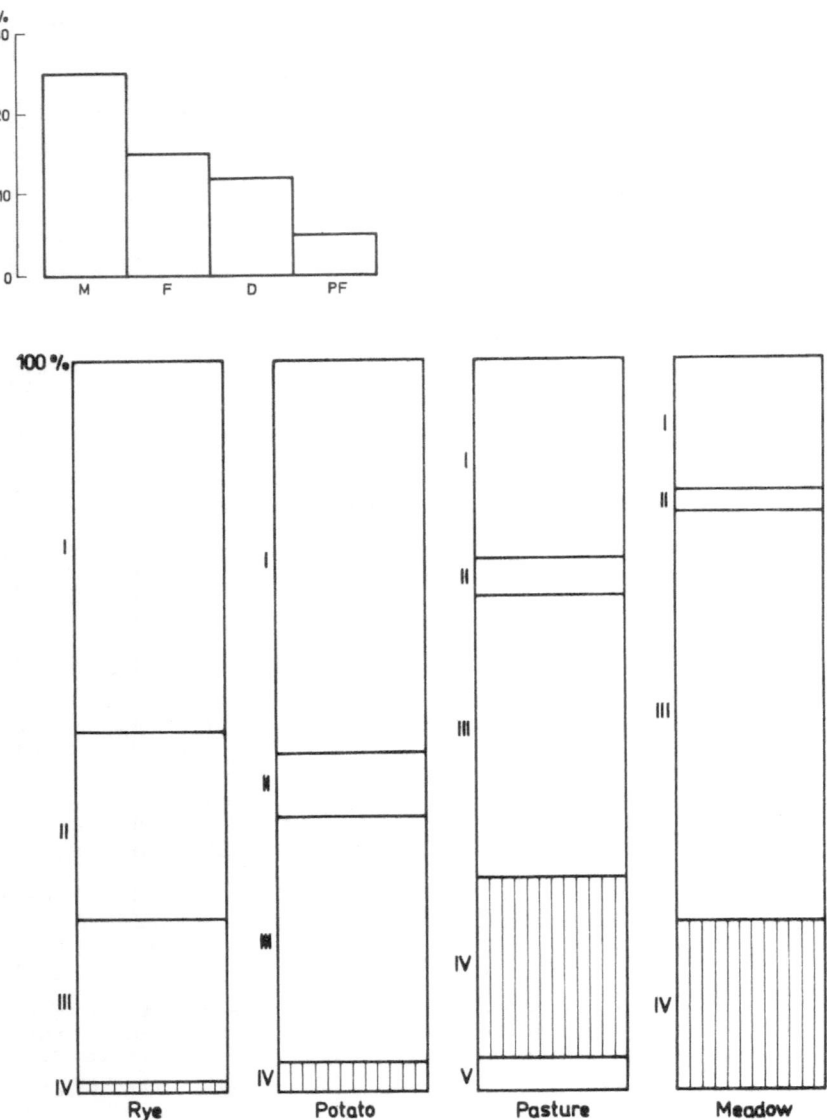

Fig.6. Omnivorous group in crops and grasslands in Poland (from Wasilewska in press a,b). I - Microbivorous, II - Fungivorous, III - Parasites of higher plants, IV - Omnivorous, V - Predators.

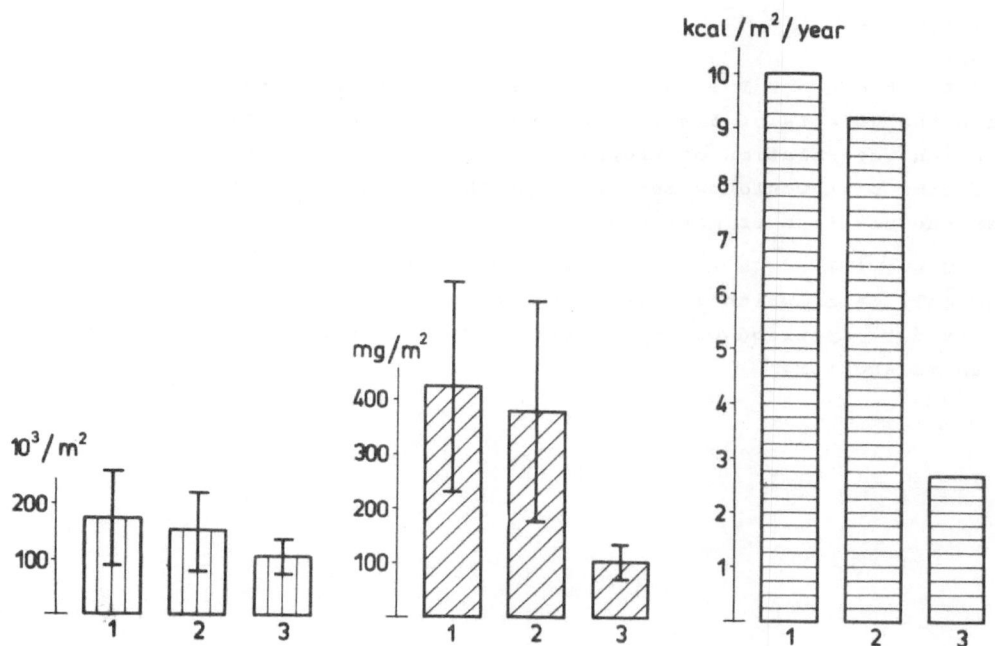

Fig.7. Numbers, biomass  and  cumulative respiratory metabolism of the predator group  in examined pasture. 1 -
Unutilized pasture  2 - Pasture  3 - Sheep fold.

References

K l e k o w s k i , R. Z.,  W a s i l e w s k a , L.  and  P a p -
    l i ń s k a , E. (1972): Oxygen  consumption  by soil-inhabiting
    nematodes. Nematologica 18, 391-403.
W a s i l e w s k a , L. (in press):  Numbers, biomass  and metabolic
    activity  of nematodes  in two crops  in Turew. Zesz. Probl. Post.
    Nauk Roln.
W a s i l e w s k a , L. (in press):  Quantitative  distribution  and
    respiratory metabolism of nematodes on mountain pastures. Ekologia
    Polska.
S.a n d n e r , H.,  W a s i l e w s k a , L. (1970): The role of the
    habitat  in  forming communities  of  soil nematodes. Zesz. Probl.
    Post. Nauk Roln. 92, 391-408.

# Discussion

J. Satchell : Dr. Wasilewska concluded from her experiments that the effects of disturbance of natural ecosystems might be changes between the relation of trophic groups. Do the results give any idea of the time that would be required for these relations to be restored under the new form of management?

L. Wasil'ewska : I have not studied the rate of this process. The only data which I have, are based on my observations which were limited to a period of one year. I suppose nevertheless that the time span is shorter.

# Die Verbreitung von Regenwürmern in verschiedenen Waldtypen der Mischwaldzone des europäischen Teiles der UdSSR

T. S. PEREL

**Laboratorium für Forstkunde,**
**Akademie der Wissenschaften der UdSSR,**
**Moskau**

Die Zone der Mischwälder (Laubholzarten mit Fichte) stellt einen eigenartigen Landschaftsstreifen Europas dar, der sich zwischen der Taiga-Zone und der Nemoralwald-Zone befindet. Manche charakteristische Züge beider letztgenannter Zonen sind auch der Zone der Mischwälder eigen, doch besitzt diese Zone auch eine Reihe von Eigenarten. Die Haupteigenart der Zone der Mischwälder besteht darin, daß hier über Jahrhunderte andauernde zyklische Beziehungen zwischen Schwarznadelhölzern und Laubhölzern in den Phytozönosen existieren. Diese sehr enge Zone ist von den Alpen bis zum Ural-Gebirge ausgedehnt. Ungeachtet mannigfaltiger edaphischen Bedingungen sind die Mischwälder dieser Zone ziemlich einheitlich in Bezug auf die Kombination der Leittypen der Gras- und Krautschicht. So ist diese Zone durch eine begrenzte Zahl von Waldtypen-Reihen und Waldtypen-Zyklen charakterisiert. In der russischen forstkundlichen Literatur bezeichnet man als Waldtypen-Reihe die Gesamtheit von Grundtypen und von sekundären, im Laufe der Sukzession entstehenden Typen, wie zum Beispiel: *Piceetum oxalidosum* und *Betuletum oxalidosum*. Unter dem Waldtypenzyklus versteht man eine solche Gesamtheit von verschiedenen Waldtypenreihen (z.B. *Piceeta, Querceta, Fageta, Betuleta* - alle *oxalidosa*), die in der unteren Vegetationsschicht einen ähnlichen Artenbestand von Dominanten und auch ähnliche Struktur haben. Dieser Einheitlichkeit der Zone wegen war es von Interesse, die Resultate bodenzoologischer Forschungen als ein Kriterium bei der Einschätzung des ökologischen Haushaltes in verschiedenen Waldtypen innerhalb derselben Waldtypenzyklen zu überprüfen (Perel und Utkin, 1972).

Im Rahmen des europäischen Teiles der UdSSR kann man 3 Sektoren der Mischwald-Zone unterscheiden: den zentralen, den westlichen und den östlichen. Die zonale Stellung der Mischwälder ist im zentralen Sektor am besten ausgeprägt. Wo in Flußscheiden-Geländen oberflächlich Lehmböden verbreitet sind, sind die Mischwälder mit *Carex pilosa* als ein Urtypus der Vegetation zu beurteilen. Nach den Resultaten der Untersuchung der Bodenfauna (einzelne Quadratproben zu 50 x 50 cm, Auslese der Bodentiere durch Handsortierung) sind der Artenbestand und das Verhältnis einzelner Gruppen von Bodenwirbellosen denen in Eichenwäldern in der Zone der sommergrünen Laubwälder, z.B. im Tulskije Sasseki Forstgebiet (Tula-Gebiet) sehr ähnlich (Ghilarov, Perel, 1970).

Etwa die Hälfte aller gesammelten Wirbellosen stellen die Regenwürmer dar, unter welchen die nemoralen Arten überwiegen: *Allolobophora caliginosa* (Sav.) und *Lumbricus*-Arten. Fast überall in solchen Wäldern ist *Lumbricus terrestris* L. anzutreffen, ein typisches Element der Fauna der Laubwälder Europas. Diese Art wurde in 14 von 16 untersuchten Waldmassiven gefunden, die zum *Carex pilosa*-Zyklus gehören. In Eichenwäldern der Laubwald-Zone ist der Artenbestand und das Verhältnis der Populationsdichte verschiedener ökologischer Gruppen der Regenwürmer solchen in Mischwäldern des *Carex pilosa*-Zyklus sehr ähnlich. In den letztgenannten Wäldern sind außer den auch in der Taiga verbreiteten kleinen streubewohnenden Arten *(Dendrobaena octaedra* [Sav.] und *D. rubida f tenuis* [Eisen]) auch große Arten vorhanden, die tiefe beständige senkrechte Gänge im Boden anlegen und noch nicht zersetzten Abfall fressen, ebenso wie solche Arten, die die Mineralschicht des Bodens bewohnen und sich von humifizierten Pflanzenresten ernähren.

Die Vertreter dieser letzten Gruppe sind in Wäldern des *Carex pilosa*-Zyklus besonders zahlreich (Abb. 1).

Zu dieser Gruppe gehört die dominierende Art *Allolobophora caliginosa* (Tafel 1), die für gut dränierten Boden besonders charakteristisch ist (Perel, 1964). Es wurden keine besonderen Unterschiede des Bodentier-Bestandes zwischen den Waldtypen, die zu verschiedenen Reihen innerhalb des Zyklus *Carex pilosa* gehören, bemerkt.

Der Gruppenbestand der Wirbellosen bleibt beim Sukcessionswechsel von Nadelholzarten zu Laubholzarten fast unverändert (Abb. 1, 3, 10, 12, 13). In Laubholzbeständen nimmt die Populationsdichte der Mehrzahl von Regenwurmarten zu, doch bleibt der Artenbestand und das Verhältnis der Individuen-Zahlen von Vertretern verschiedener ökologischer Gruppen fast unverändert, auch bei der Sukzession der Baumarten.

Unter typischen zonalen Bedingungen sind kleinere Flächen mit Wäldern vom *Aegopodium podagraria*-Zyklus besetzt. Sie sind an solche Standorte gebunden, die eine besondere Befeuchtung durch Grundwasser haben. Solche Wälder, von denen Bestände der Reihe *Carex pilosa - Aegopodium podagraria* untersucht worden sind, unterscheiden sich von Wäldern des "Carex pilosa"-Zyklus durch den Charakter des Tierbesatzes des Bodens.

In Wäldern des *Carex pilosa*-Zyklus dominiert die Art *A. caliginosa* nicht nur unter den Regenwürmern, sondern auch unter allen aufgenommenen Wirbellosen (etwa die Hälfte aller gesammelten Individuen), während

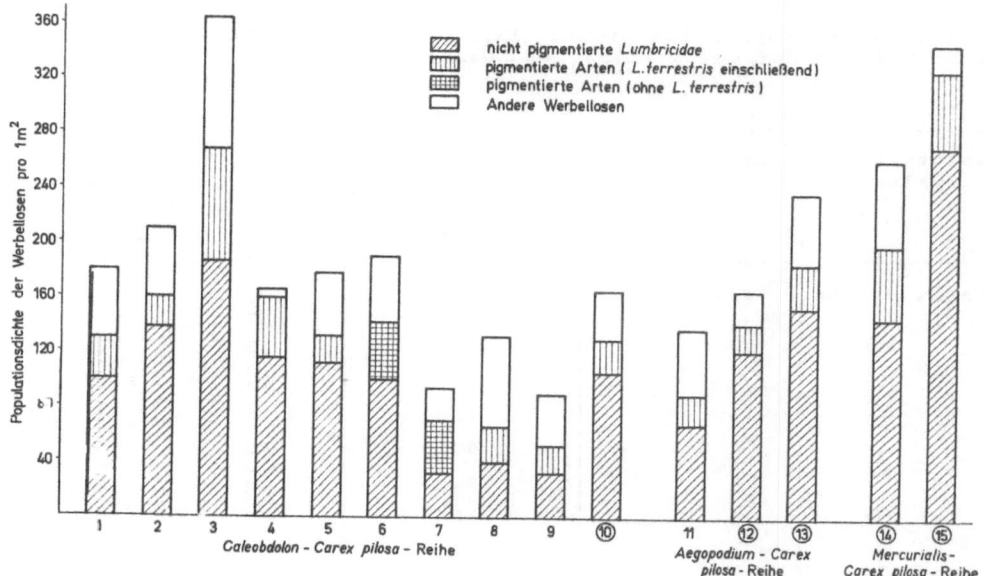

Abb.1. Populationsdichte von Regenwürmern und anderen bodenbewohnenden Wirbellosen in Wäldern *(Piceeta mixta)* des *Carex pilosa*-Zyklus.

in *Aegopodium podagraria*-Waldtypen die Populationsdichte dieser Art viel niedriger ist; selten überstieg hier die Zahl von *A. caliginosa* 10 Exemplare pro 1 $m^2$ (Tafel 2).

In Wäldern der *Carex pilosa - Aegopodium podagraria*-Reihe überwiegen in den Mineralschichten des Bodens solche Regenwurm-Arten, die eine Vernässung des Bodens ertragen. wie *Allolobophora rosea* Sav. und *Octolasium lacteum* Oerley. Die letzte Art ist in Wäldern des *Carex pilosa*-Zyklus nur selten zu treffen. In den untersuchten Primär-Beständen der *C. pilosa - Ae. podagraria*-Reihe wurde *O. lacteum* nur einmal nicht gefunden; in zwei Fällen war diese Art dominierend (Tafel 2).

*O. lacteum* ist kalziophil und besiedelt entweder solche Böden, die sich auf kalkhaltigem Gestein entwickeln, oder Böden mit reichlichem Zufluß von hartem Wasser, einschließlich der Niedermoore. Umgekehrt wurde *Lumbricus terrestris* in keiner Probe aus diesen Wäldern gefunden; eine Art, die schwach dränierte vernässte Böden, schlecht verträgt. Ganz deutlich überwiegen die Bewohner der Streu- und Humusschicht in dieser Waldtypen-Reihe (Abb.2), was eine Eigenart feuchterer Biotope ist. Die Wälder, die in der Reihe *C. pilosa - Ae. podagraria* vereinigt sind, sind von den Wäldern der Reihe *Ae. podagraria - C. pilosa* des Zyklus *Carex pilosa* schwer zu unterscheiden, da verhältnismäßig

T a f e l  1.  Populationsdichte (Ex/m$^2$) der unpigmentierten *Lumbrici-dae* in Wäldern der *Carex pilosa* und *Oxalis-Corylus* Reihen

| Waldtypen-Zyklus | Waldtypen-Reihe | Regenwurmarten | | |
|---|---|---|---|---|
| | | *Allolobophora caliginosa* | *A. rosea* | *Octolasium lacteum* |
| *Carex pilosa* | *Galeobdolon-C. pilosa* | 100 | - | - |
| | | 103 | 33 | 1 |
| | | 121 | 64 | 1 |
| | | 111 | 6 | - |
| | | 96 | 16 | - |
| | | 96 | 4 | - |
| | | 31 | - | - |
| | | 33 | - | 5 |
| | | 22 | 9 | 1 |
| | | 96 | 8 | 2 |
| | | 174 | 201 | - |
| | *Aegopodium-C. pilosa* | 62 | 4 | 2 |
| | | 104 | 14 | 3 |
| | | 140 | 14 | - |
| | *Mercurialis-C. pilosa* | 137 | 9 | - |
| | | 110 | 162 | - |
| *Oxalis nemorosa* | *Oxalis-Corylus* | 49 | 37 | 1 |
| | | 84 | 178 | 1 |
| | | 126 | - | - |
| | | 114 | - | - |

hygrophile Arten der Krautdecke relativ schwach vertreten und haupt-sächlich an sekundäre biozönotische Parzellen gebunden sind.

Auf der Grundlage der Methode der bodenzoologischen Diagnostik ist es möglich, nach dem Charakter des Bodentierbesatzes die Nomenklatur dieser Gruppe von Waldtypen zu präzisieren. Den phytozönotischen Kriterien nach ist diese Gruppe mit der *Ae. podagraria - C. pilosa*-Reihe verwandt, doch gehört sie nach den bodenzoologischen Angaben zum *Ae. podagraria*-Zyklus.

T a f e l   2.    Populationsdichte (Ex/m$^2$) der Regenwürmer in *Aegopodium podagraria*  und  *Oxalis nemorosa* -Zyklus = Wäldern im Zentralsektor der Mischwald-Zone

| Waldtypen-Zyklus | Waldtypen-Reihe | Regenwurmgruppen und Arten | | | | |
|---|---|---|---|---|---|---|
| | | Unpigmentierte Mineralboden-Bewohner | | | Streu-bewohner | |
| | | *Allolobophora caliginosa* | *A.rosea* | *Octolasium lacteum* | *Dendrobaena octaedra* | *Lumbricus rubellus* |
| *Aegopodium podagraria* | *Carex pilosa-Aegopodium* | 13 | 38 | 2 | 9 | 9 |
| | | – | 33 | 5 | 15 | – |
| | | 30 | 60 | 1 | 4 | 54 |
| | | 8 | 2 | – | 6 | 6 |
| | | 13 | 15 | 49 | 6 | 11 |
| | | 1 | 10 | 7 | 23 | 9 |
| | | 25 | – | 30 | 10 | 7 |
| | *Galeobdolon-Aegopodium* | 3 | 1 | 3 | 36 | 26 |
| | *Mercurialis-Aegopodium* | – | – | – | 13 | – |
| *Oxalis nemorosa* | *Oxalis-Galeobdolon* | 1 | 10 | 6 | 24 | 22 |
| | *Oxalis-Carex pilosa* | – | 1 | 2 | 18 | 2 |
| | | 3 | 8 | – | 12 | – |
| | *Oxalis-Aegopodium* | 9 | 1 | – | 3 | 14 |
| | *Oxalis-Mercurialis* | – | – | – | 34 | – |
| | *Oxalis-Acer* | – | – | – | 12 | – |

Die Wälder,  die zum *Oxalis nemorosum*-Zyklus  gehören, besitzen in der Zone der Mischwälder azonale Standorte,  und zwar an Kontaktstellen von Moränen-Hügeln  mit  Sander-Ebenen,  wo  eine dünne Sand-Lehmboden-Schicht  die Moräne  bedeckt.  In solchen Standorten,  wo eine nur kurz dauernde Vernässung  (durch Oberflächenwasser)  auftritt,  sind die Bo-

denwirbellosen-Gemeinschaften solchen der  der Fichtenwälder   Süd-Taiga
ähnlich (vergl. Perel, 1965).

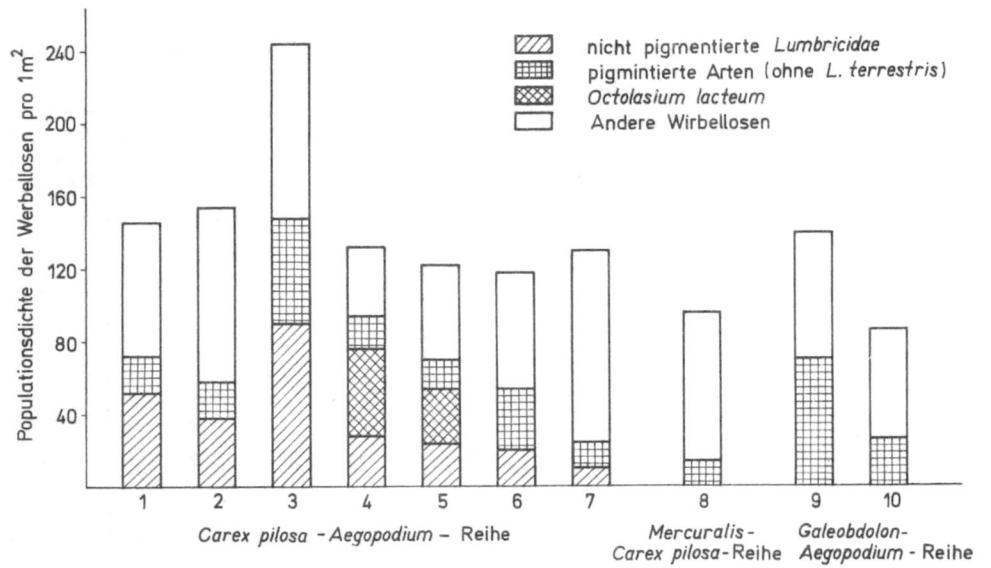

**Abb.2. Populationsdichte von Regenwürmern und anderen bodenbewohnenden Wirbellosen in Mischwäldern vom *Aegopodium podagraria*-Zyklus.**

Regenwürmer  sind hier zahlenmäßig gering  und  fast nur durch die
boreale Art  *D. octaedra*  (ein Streubewohner) vertreten. Von  nemoralen
Arten ist hier nur *Lumbricus rubellus* Hoffm. anzutreffen, eine Art, die
auch  in die Taiga-Zone  eindringt,  wo sie  in sekundären Zönosen  der
Birken und Espenwälder vorkommt.

Es gibt keine wichtigen Unterschiede  zwischen den Wirbellosen-Ge-
meinschaften in Böden der Wälder vom *Ae. podagraria*-Zyklus und vom *Oxalis nemorosum*-Zyklus,  die auf zweigliedrigem Unterboden wachsen  (Abb.
2 und 3).

Eine ähnliche Bodenbevölkerung ist auch anderen Wäldern des *Oxalis
nemorosum*-Zyklus eigen,  die auf Territorien mit  schwach gegliedertem
Relief und mit schwachem Wasserausfluß stocken, d.h. an Orten  mit fla-
chen Schluchten  und  mit temporärer Vernässung,  wie an der Vergleyung
des unteren Teiles des Bodenprofils zu sehen ist (vergl. Abb. 2). Hier-
von sehr verschieden sind  die Bodenwirbellosen-Komplexe in Wäldern der
*Oxalis nemorosum - Corylus avellana*-Reihe  des *Oxalis nemorosum*-Zyklus,
die auf abschüssigen Abhängen auf dünnen, oberflächlich liegenden Lehm-

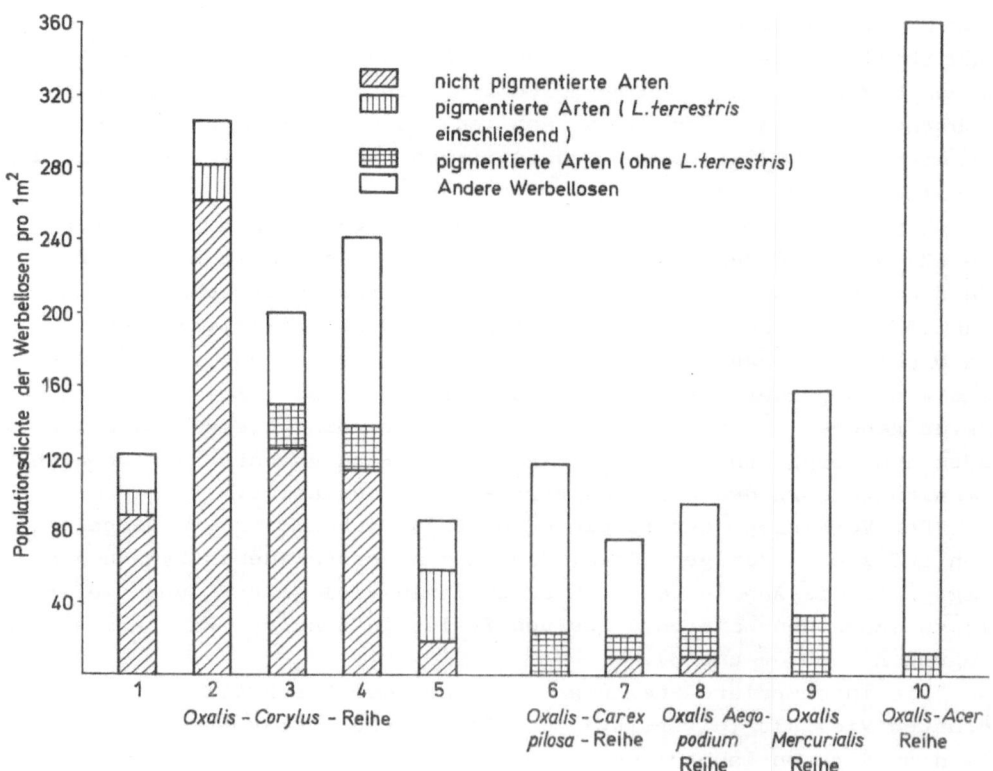

**Abb.3. Populationsdichte von Regenwürmern und anderen bodenbewohnenden Wirbellosen in Mischwäldern vom *Oxalis nemorosa*-Zyklus.**

böden wachsen. Im ganzen sind die Wälder der *Oxalis-Corylus*-Reihe nach dem Artenbestand der Bodenwirbellosen den Wäldern des *Carex pilosa*-Zyklus ziemlich ähnlich (Abb.1 und 2). Charakteristisch ist die oft beobachtete Dominanz von *Allolobophora rosea* (und nicht von *A. caliginosa!*) unter den Lumbriciden (vergl. Tafel 1).

Die angeführten Angaben zeigen, daß die Waldtypen der Zone der Mischwälder durch eine hohe Populationsdichte von größeren Bodenwirbellosen zu charakterisieren sind, unter denen die Regenwürmer als dominierende Gruppe zu bezeichnen ist.

Die Lumbriciden besiedeln hier hauptsächlich die Mineralschichten des Bodens. Die Populationsdichte solcher Arten ist hoch, über hundert Exemplare pro Quadratmeter. Die Komplexe der Bodenwirbellosen in solchen Wäldern sind solchen in Waldgrauböden sehr ähnlich (vergl.Ghilarov, Perel, 1970).

Mit Verringerung der Dränage nimmt die Populationsdichte der Bodenwirbellosen im engeren Sinne ab, während die Streubewohner zu dominieren beginnen. In extremsten Fällen ist der Artenbestand solchen in Fichtenwäldern der Süd-Taiga ähnlich, wo sich die Wirbellosen in der obersten Schicht des Bodens konzentrieren. Die Lumbriciden sind durch *D. octaedra* vertreten.

Der angeführte Vergleich von Bodenwirbellosen-Populationen läßt die Wälder des *Carex pilosa*-Zyklus und *Ae. podagraria*-Zyklus präzisieren und voneinander unterscheiden. Die starken Unterschiede in der Bodenbevölkerung der Mischwälder (*Piceeta, Querceto-Tilieta* und andere) vom *Oxalis nemorosum*-Zyklus zeigen, wie verschieden der ökologische Haushalt in Wäldern ist, die nach phytozönotischen Kriterien zu einem Zyklus gehören und in dieser Weise falsch vereinigt sind. Dasselbe wird durch den Vergleich der Bodenwirbellosen-Populationen von zonalen und azonalen Wäldern des *Ae. podagraria*-Zyklus bestätigt.

Der Vergleich der Komplexe von Bodenwirbellosen in azonalen Wäldern auf zweigliedrigen Böden, die zu den verschiedenen Zyklen gehören, zeigt, daß die Ähnlichkeit zwischen ihnen viel größer und tiefer ist, als zwischen den Wäldern desselben Zyklus in zonalen und azonalen Bedingungen (Abb. 2 und 3).

Die intermediäre Stellung der Zone der Mischwälder kommt in der höchsten Vielfalt (diversity) der Bodenwirbellosen-Bevölkerung in gemischten Fichten-Laubholz-Beständen zum Ausdruck.

Diese Unterschiede widerspiegeln die Mannigfaltigkeit des ökologischen Haushaltes in Wäldern dieser Zone in der ersten Linie der Bodenfeuchtigkeit, da sie ein Leitfaktor der Verteilung der Bodenwirbellosen darstellt.

Die unterschiedliche Zusammensetzung der Bodenwirbellosen ist auch einer der Grundfaktoren, die für die Unterschiede in der Zersetzung des Bestandesabfalles und in der Geschwindigkeit der Eingliederung der Pflanzennährstoffe in den biologischen Stoffkreislauf innerhalb der Biogeozönose verantwortlich sind.


L i t e r a t u r

G h i l a r o v , M. S., P e r e l , T. S. (1970): Ratio of population density of various groups of saprophagous invertebrates as an indicator of differences between brown forest soils and gray forest soils. Dokl. Akad. Nauk SSSR 192, 2, 296-299 (in Russian).
P e r e l , T. S. (1964): Die Verbreitung der Regenwürmer (*Lumbricidae*)

in Wäldern des ebenen Landes des europäischen Teiles der UdSSR.
Pedobiologia 4, 92-110 (russisch, deutsche Zusammenfassung).

P e r e l , T. S. (1965): Die Bodenbewohner der natürlichen Fichten-
bestände in der südlichen Taiga und ihre Veränderung im Zusammenhang mit Kahlschlag und Holzartwechsel. Pedobiologia 5, 102-121.

D i s c u s s i o n

A. Z i c z i : In Ungarn ist *Lumbricus terrestris* nicht verbreitet,
nur in Parkanlagen und unweit von Menschen-Wohnorten. Wie ist es in
UdSSR?

T. S. P e r e l : In der UdSSR ist *L.terrestris* in der Zone der Mischwälder in natürlichen Wäldern verbreitet. In der Wald-Steppe Zone ist
diese Art, wie in Ungarn, nur in anthropogenen Wäldern zu treffen und
an Süden dieser Zone und in Wäldern der Steppe-Zone ist diese Art nicht
zu finden.

J. S a t c h e l l : The data obtained in NW Europe on earthworm distribution in forests support your data.

T. S. P e r e l : It is very gratifying.

# Die Bodenfauna
# der Waldsteppen-Eichenwälder

I. V. KUDRJASHEVA
Laboratorium für Forstkunde der Akademie
der Wissenschaften der UdSSR

Weit bekannt  ist die Bedeutung  der im Boden lebenden Wirbellosen beim
Funktionieren der Waldbiogeozönosen verschiedener Naturzonen. Besonders
gross ist die Rolle dieser Tiere in den breitblättrigen Wäldern, wo sie
besonders zahlreich  und verschiedenartig sind, insbesondere in den Eichenwäldern der Waldsteppenzone, wo man  die Besonderheiten ganzer Komplexe  dieser  Bodeninvertebraten  in  den  Wäldern  verschiedenen  Typus
untersucht hatte.  Diese Untersuchungen  wurden auf dem Territorium der
Försterei  Tellerman  (Süd-Ost   des Gebiets Voronež)  zustandegebracht,
welche  im Ostsektor  der europöischen Waldsteppe liegt.  Im Sommer 1967
unternahm man  eine Abschätzung grosser Invertebraten  in überständigen
220jährigen Eichenwäldern  von sechs Typen,  wo *Quercetum aegopodiosum*,
*Fraxineto-Quercetum aegopodioso-caricosum, Tilieto-Quercetum  caricosum*
auf dunkelgrauen Waldböden,  *Quercetum acerosum campestris*  auf dunkelgrauen  schwach podsolierten solonezartigen Waldböden,  *Quercetum evony-
mosum* und *Quercetum salinum*  auf Böden mit verschiedenartiger Bodenver-
salzung wachsen.

Die untersuchten Forsttypen unterscheiden sich durch Wachstumsver-
hältnisse (Moltschanov,  1953, 1963, 1967;  Jelagin, 1963).  Besonders
wesentlich  ist der Unterschied in den hydrologischen Bodenregimen, wodurch in erster Linie die Produktivität dieser Anpflanzungen  zu erklä-
ren ist; ihre Bonität gestattet uns, darüber zu urteilen (Tafel 1).

Bei der Kontrolle  der Wirbellosenbesatzdichte  wurde  die Methode
der Ausgrabungen angewandt,  wobei  jede Bodenschicht  mit der Hand ge-
trennt untersucht wurde.  In jedem Waldtypus wurden je 16 Proben (50x50
$cm^2$, bis 40 cm tief) entnommen.

Überwiegend  sind  in den untersuchten Waldtypen  unter den Hauptgruppen der Bodenfauna  die Saprophagen,  vertreten durch *Lumbricidae,
Enchytraeidae, Oniscoidea, Julidae,* Raubmyriapoden *Chilopoda,* unter de-
·nen *Geophilomorpha* dominieren, auch verschiedene *Coleoptera*  und Larven
von *Cicadidae* (Tafel 2).

T a f e l   1.   Änderung der Regenwurmbesatzdichte im Zusammenhang mit
einigen Faktoren, welche die Waldverhältnisse beeinflussen

| Waldtypus / Faktoren | 220jährige Eichenwälder | | | | | |
|---|---|---|---|---|---|---|
| | 1[*] | 2 | 3 | 4 | 5 | 6 |
| Bonität | I | I-II | II-III | III | IV | Va |
| Bodentypus | dunkelgrauer Wald-boden | | | dunkel-grauer schwach podsolier-ter Boden | Boden mit verschiedenar-tiger Boden-versalzung | |
| Mächtigkeit des Humushorizonts (cm) | 44 | 35 | 28 | 23 | 7 | 2 |
| Regenwurmbesatz-dichte (Ex/m$^2$) | 52 | 36 | 44 | 24 | 16 | 5 |

[*]1-Quercetum aegopodiosum, 2-Fraxineto-Quercetum aegopodioso - caricosum, 3-Tilieto-Quercetum cariosum, 4-Quercetum acerosum campestris,
5-Quercetum evonymosum, 6-Quercetum salinum.

In Eichenwäldern  auf dunkelgrauem Waldboden  sind geringe Schwankungen der Invertebratenbesatzdichte  zu erwähnen,  in Quercetum evony-
mosum ist die Besatzdichte maximal,  in Quercetum salinum - minimal. In
allen Eichenwäldern, ausser in denen mit Salzboden, dominieren die Lumbricidae, Geophilomorpha und Elateridae.  In Quercetum salinum sind neben Lumbricidae und Geophilomorpha auch Enchytraeidae zahlreich.

Wie bekannt, sind in plakorischen Waldsteppeneichenwäldern  höchst
verschiedenartige grosse Saprophagen zu beobachten (Alejnikova, Utrobina,
1967;  Tchernow, 1967).  Bei der Verarbeitung organischer Reste  nehmen
die Regenwürmer einen bedeutenden Anteil. In verschiedenen Entwicklungs-
stadien der Biogeozönosen des Waldes  ändert sich bedeutend die Besatzdichte der grossen Saprophagen.  Das wurde bei der Untersuchung der Bodenfauna  in Fraxineto-Quercetum  aegopodioso-caricosum  verschiedenen
Alters beobachtet (Kudrjaschewa, 1970).  In überständigen Eichenwäldern
ist  ihre Besatzdichte minimal.  Doch  sogar bei einer geringer Besatzdichte grosser Saprophagen  überständiger Eichenwälder  ist  ihre Rolle
bei der Verarbeitung des Abfallaubs  sehr bedeutend.  Darüber  kann man
urteilenn, indem man das Resultat der Versuche bei dem Abbau des Abfallaubs verschiedener Baumartem unter Anteilnahme grosser Saprophagen und
bei deren Isolation beobachtet  (Das Abfallaub  wurde in Säcken aus Ka-
pronnetz isoliert. Tafel 3).

T a f e l   2 .    Besatzdichte der Hauptgruppen der Wirbellosen  in den Eichenwäldern verschiedener Typen

| Tiergruppen | Eichenwaldtypen | | | | | |
|---|---|---|---|---|---|---|
| | i* | 2 | 3 | 4 | 5 | 6 |
| Mittelbesatzdichte der Wirbellosen pro 1 m$^2$ | 148,0 | 114,8 | 156,0 | 158,8 | 215,2 | 108,4 |
| Unter ihnen | | | | | | |
| *Lumbricidae* | 21,7 | 39,5 | 17,0 | 13,2 | 35,7 | 14,2 |
| *Enchytraeidae* | 3,2 | 0 | 0 | 2,0 | 15,5 | 28,0 |
| *Oniscoidea* | 3,0 | 1,2 | 2,5 | 1,2 | 9,5 | 6,0 |
| *Julidae* | 4,2 | 1,7 | 0,5 | 8,0 | 4,2 | 2,2 |
| *Geophilomorpha* | 61,7 | 29,5 | 42,0 | 43,7 | 58,2 | 15,7 |
| *Lithobiidae* | 8,2 | 6,0 | 7,7 | 14,5 | 15,0 | 10,2 |
| *Carabidae* | 3,7 | 4,5 | 3,0 | 1,2 | 1,0 | 1,7 |
| *Elateridae* | 12,5 | 10,2 | 22,0 | 30,7 | 25,0 | 9,5 |
| *Staphylinidae* | 1,7 | 0,5 | 0,5 | 2,7 | 5,5 | 1,0 |
| *Cicadidae (Cicadetta montana* Scop.*)* | 5,0 | 10,5 | 38,4 | 2,0 | 9,7 | 0 |

*Siehe Tafel 1.

In den untersuchten Eichenwäldern ändert sich die Besatzdichte der
*Lumbricidae* (Mittelbesatzdichte nach Malewitsch, Perel, 1958; Moltscha-
now, 1954, 1961;  und dem Material  des Verfassers),  die vom Waldtypus
und den Bodenverhältnissen abhängt  (Tafel 1).  Der Vergleich der Mächtigkeit  des Humushorizonts  des Bodens  verschiedenartiger Wälder  und
auch der Besatzdichte der Regenwürmer  zeigen die direkte Korrelations-
abhängigkeit  mit dem Korrelationsquotienten 0,78.  In den untersuchten
Eichenwäldern  stehen  somit die Exponente  der Besatzdichte der Regenwürmer mit der Mächtigkeit der Humushorizonte und der Bonität des Baum-
bestandes in Verbindung.

In überständigen Eichenwäldern  sind  die *Lumbricidae*  durch  fünf
Arten vertreten:  *Eisenia nordenskioldi* (Eisen),  *E. uralensis* Mal., *E.
rosea* (Sav.), *Dendrobaena octaedra* (Sav.), *Octolasium lacteum* (Oerley).
Die Änderung  des Artbestandes  nach Waldtypen  bestimmt eine Reihe von
Faktoren: hydrologische Bodenregime, Bodenvergrasung, Tiefe der Lagerung
der Karbonathorizonte a. a. m.  In allen Eichenwäldern herrscht *E. nordenskioldi*.  In trockenen Eichenwäldern *(Quercetum evonymosum* und *Quercetum salinum)* fehlt *E. uralensis,* welche in Eichenwäldern anderen Typus

T a f e l  3.    Die Verwesung des Abfallaubs im Eichenwald  *(Fraxineto-Quercetum aegopodioso-caricosum)*   im Alter   von 220 J. unter Mitwirkung von grossen Saprophagen und auch bei ihrer Isolierung (Versuche ab IX.1967 bis IX.1968)

| Verschiedene Arten des Abfallaubs | Die Menge des verwesenen Abfallaubs in % | |
| --- | --- | --- |
| | Unter Mitwirkung der Invertebraten | Bei ihrer Isolierung |
| Abfallaub der Eiche | 34,0 ± 1,2 | 25,5 ± 4,7 |
| Abfallaub der Linde | 70,0 ± 5,0 | 46,0 ± 5,0 |
| Abfallaub der Esche | 91,4 ± 1,0 | 69,3 ± 2,5 |

mit besseren Feuchtigkeitsverhältnissen weit verbreitet ist. In Eichen-wäldern  mit vorwiegenden Halmpflanzen  hauptsächlich  der *Carex pilosa* (Scop.)  in der Grasnarbe,  erscheint eine Art,  die in der Steppe weit verbreitet ist - *E. rosea. O. lacteum* ist gewöhnlich in *Quercetum evony-mosum* zu finden, wo die geringste Tiefe  des Karbonathorizontes fixiert wird.

In allen Waldtypen wurden unter anderen Saprophagen *Oniscoidea* ge-funden,  die ziemlich zahlreich  im Salzboden sind,  und auch die Viel-füssler *Julidae,* deren maximale Besatzdichte in *Quercetum acerosum cam-pestris* fixiert ist. *Enchytraeidae,* die in *Fraxineto-Quercetum aego-podioso-caricosum* und  *Tilieto-Quercetum caricosum*  fehlen, findet man in grossen Menge in *Quercetum salinum.*


L i t e r a t u r

A l e j n i k o w a ,  M. M.,  U t r o b i n a ,  N. M. (1967):  Struk-tura  životnogo  naselenija  potschw  Srednego  Powolzja.  Sbornik "Struktura  i  funkcionalno-biogeocenototscheskaja  rol  ziwotnogo naselenija suschi", Moskwa, 35-36.

J e l a g i n ,  I. N. (1963): Tipi lesa nagornoj tschasti Tellermanow-skogo  opitnogo  lesnitschestwa  i  ich hozjajstwennoe znatschenie. "Biogeocenotitscheskie issledowanija  w dubrawach lesostepnoj  zo-ni". Moskva, Izd. AN SSSR, 52-99.

K u d r j a s c h e v a ,  I. V. (1970):  Izmenenie  potschwennoj fauni w processe razwitija jasenewo-osokowo-snitewoj dubrawi  i w swjazi s rubkoj.  Sbornik "Wzaimootnoschenija  komponentow  biogeocenoza w listwennich molodnjakach", Moskwa, Nauka, 163-177.

M o l t s c h a n o w , A. A. (1953): Izmenenie biologitscheskich, ökologitscheskich i gireologitscheskich faktorow w razlitschnich tipach dubowogo lesa. Soobsschenija In-ta lesa AN SSSR, 2. Moskwa, 107-157, 1963.

Öksperimentalnoje kompleksnoje (biogeocenotitscheskoe) izutschenie schirokolistwennich lesow kak nautschnaja osnowa lesochozjajstwennich meroprijatij. Sbornik "Biogeocenoticheskie issledowanija w dubrawach lesostepnoj zoni". Moskwa, Izd. AN SSSR, 20-50, 1967.

Itogi biogeocenotitscheskogo izutschenija lesostepnich dubraw i ich hozjajstwennoe znatschenie. Lesowedenie, 2, 11-26, Moskwa, Nauka.

M a l e w i t s c h , I. I., P e r e l , T. S. (1958): Dozdewie tscherwi Tellermanowskogo lesnitschestwa i ich rasıredelenie w nagornoj dubrawe i lesach pojmi. Utschen. zapiski MGPI im. Potemkina, 84, 254-268.

T s c h e r n o w , J. I. (      ): Nekotorie osobennosti struktury ziwotnogo naselenija ewropejskoj lesostepi na primere bespozwonotschnich. Sbornik "Struktura i funkcionalno-biogeocenotitscheskaja rol ziwotnogo naselenija suschi", Moskva, 24-26.

Diskusion

A. S z u j e c k i : What layers considered the ratio of C/N in various humus?

I. W. K u d r j a s c h e w a : In the paper under discussion these data are not used, but numerous data on the chemical, physical and microbiological characteristics of the forest under study are published in Proceed. of Laboratory of Forest Sci., Acad. Sci. USSR.

# The weight structure and oxygen consumption of oribatid communities

P. BERTHET

Institute of Zoology
Louvain

In the study of animal communities, it is common to introduce the concept of relative abundance of the different species. If $n_i$ indicates the number of individuals* belonging to the $i^{th}$ species and $N$ the total number of individuals of the S species, than the ratio $n_i/N$ will express the numerical abundance of the $i^{th}$. species. This concept is of use for the analysis of community but does not describe the functional importance of the species in its community.

On the other hand if $w_i$ is the mean weight of an individual of the $i^{th}$ species, the biomass *(B)* of the community* will be $B = \sum_{i=1}^{S} n_i\, w_i$, and the relative biomass of the $i^{th}$ species *(b_i)* will be expressed by $b_i = \dfrac{n_i\, w_i}{B}$ .

However, it is known that the metabolism of a poikilotherm is not related to the weight by a linear function but by an exponential function, in those temperatures encountered in natural conditions.

Metabolic studies of adult *Oribatei* and *Collembola* have shown that their oxygen consumption (O) is related to the weight by: $O = o(T) \cdot w^{.7}$, where $o(T)$ is a function of temperature (Engelmann, 1961; Berthet, 1964; Zinkler, 1966).

With these conditions it seems reasonable to define a new expression for the biomass of the *Oribatei* and *Collembola* community, which I propose to call ´active biomass´*, and is estimated by $AB = \sum n_i\, w_i^{0.7}$; for the $i^{th}$ species the expression $ab_i = \dfrac{n_i\, w_i^{0.7}}{AB}$ will show the contribution of the $i^{th}$ species to the active biomass of this community.

By studying the characteristics of an oribatid community of an oakwood forest, the relative constancy in time of the ratio $R = AB/B =$

---

\* $n_i$, B and AB are expressed per unit area.

$\pm \dfrac{\Sigma \; n_i \; w_i^{0.7}}{\Sigma \; n_i \; w_i}$ became apparent. This work is an attempt to verify if this ratio remains constant in other environments  and also if it is a characteristic of the system.

SITES AND SAMPLING

The analysis  is based  upon the data  collected  by Lebrun (1971) in 3 ecosystems.  Monthly samples were collected in the litter (L) and humus (F + H) sub-horizons.  Sampling  was carried out  during a period of 20 months  in an oak-hornbeam forest,  (Querceto Carpinetum convallarieto-sum), 13 months  in a poplar plantation (Populus canadensis) and  14 months in a mowed grassland.  Besides the above important set  of data, information collected  by Elsen (1965)  was also included;  he analysed the oribatid communities of 3 adjacent oakwoods, one on a plateau (moor-type humus), another  in the middle of a valley (moder-type humus), and the third one in the bottom  of the valley  (mull-moder type humus). In this study 3,6 and 4 samples  have been included from the first, second and third site respectively.

In the above studies,  on each sampling occasion,  25 sample units were collected  from each  of the two organic layers.  Thus a faunistic list of 12 biotopes  studied at different periods  of the year  was obtained.

RESULTS AND DISCUSSION

The results  of analysis  of variance  performed on the values of R obtained  from 6 sites,  are summarized  in Table 1.  It appears that the time effect is not significant while layer effect  is generally significant at 1% level. It appears that variations between biotopes are considerably larger  than variations  in time  within a biotope  and thus, R could be considered as a biotope characteristic.  It is also observed that the oak-hornbeam forest  and the oakwood of the bottom of the valley present similar values of R when respective layers are compared. It is interesting  to note  that these two sites  are extremely similar in their composition: presence of hornbeam and mull-moder type humus.

The constancy in time of the value of R appears to be fairly  surprizing.  In the upper layer of oakwood-hornbeam, for example, from one

sampling  to another, that is to say, a few months apart, the number of
species recorded varied from 20 to 60,  the total number of individuals
collected varied    in the ratio of 1: 42, biomass in the ratio of 1: 34
and the active biomass  in the ratio  of 1: 29.  More detailed analysis
showed  that these variations  induce modifications  in the $H$ "index of
diversity" (Margaleff, 1957; Pielou, 1969) but that the index $J$  $H/H$max
which  Pielou calls   "evenness"  (which can be translated  in French by
"indice de parité interspécifique") remains remarkably constant as well

T a b l e  1. :  Mean values of the ratio $R = \dfrac{\Sigma\ n_i\ w_i^{.7}}{\Sigma\ n_i\ w_i}$ and mean squares
of the analysis of variance

| | Mean of $R$ | | Mean square | | |
|---|---|---|---|---|---|
| | Upper layer | Lower layer | Layer effect | Time effect | Residual |
| Oak-hornbeam forest | .2334 | .3003 | .04469* | .00059 | .00106 |
| Poplar plantation | .2951 | .3097 | .00138 | .00051 | .00062 |
| Grassland | .4130 | .4398 | .00503* | .00036 | .00028 |
| Oakwood on plateau | .4048 | .4543 | .00383* | .00044 | .00010 |
| Oakwood in the middle of valley | .2951 | .3532 | .01012* | .00054 | .00058 |
| Oakwood in the bottom of valley | .2445 | .2988 | .0590* | .00005 | .00027 |

   * 1% level  of significance, $n_i$  is the number  of individuals  of
species $i$, $w_i$ is the mean weight (in µg) of an individual of species $i$.

when it is calculated  from abundance,  biomass  or the active biomass.
This indicates that the partition of individuals in different abundance
categories or among the different weight classes  follows the same pattern, which is  a characteristic  of the community  in a given biotope.
In other words the low variability of evenness and of the ratio $R$ seems
to indicate  that the phytosociological  and/or pedological environment
imposes a defined structure on the community and this structure  is not
affected by the annual climatic variations.
   A second consequence  of the low variability of $R$  is of practical
use. As has been reported earlier (Berthet, 1971), a single measurement
of the value of $R$ in a given biotope allows the possibility of estimating the annual $O_2$-consumption of oribatid community by determining only
variations in temperature and total biomass.

Daily oxygen consumption (O) of the community can be calculated by

$$O = o(T) \cdot \Sigma \, n_i \, w_i^{0.7}$$

$$= o(T) \cdot R \cdot \Sigma \, n_i \, w_i = o(T) \cdot R \cdot B \, .$$

where $o(T)$ is the relationship between daily oxygen consumption and temperature. For example, if we have monthly sampling, then the annual $O_2$-consumption will be $O* = R \sum_{t=1}^{12} o(T_t) \, J_t \, B_t$ , where $T(t)$ is the mean temperature of the $t^{th}$ month with $J_t$ days and $B_t$ the total biomass of the community at the $t^{th}$ month.

It would then suffice to take a careful census of the community at a given time and determine the values for gross biomass, active biomass and their relationship $(R)$. Subsequently, the ecologist, knowing the annual temperature regime of a site and the effect of temperature variations on metabolic activity, could confine himself to determining gravimetrically the gross biomass of the community. He would no longer need to take into account the composition of species or distribution of weight classes and his work would be considerably simplified.

## SUMMARY

The term active biomass is proposed for $AB = \sum_{i=1}^{S} n_i \, w_i^{.7}$ where $n_i$ is the number of individuals per unit area belonging to the $i^{th}$ species, $w_i$ is the mean weight of the $i^{th}$ species and $S$ is the total number of species. It is observed that the ratio between the active biomass and the gross biomass is of low variability and is a characteristic of the oribatid community in its biotope.

## Acknowledgements

This work was carried out with the assistance of Miss M.P. Maisin to whom I offer my warm thanks. The discussion which I had with Dr. G. Gerard on this subject has been extremely valuable. Mr. Y. Pande helped in translating this paper from the original French. I am happy to acknowledge my debt to them.

References

B e r t h e t ,   P. (1964): L'activite des Oribatides d'une chênaie.
      Mémoire n° 152, Inst. Roy. Sci. Nat. Belgique, 152 pp.
B e r t h e t ,   P. (1971):   Une nouvelle méthode pour l'estimation du
      bilan annuel de consommation d'oxygène des communautés édaphiques.
      In Productivité des Ecosystèmes forestiers. Unesco, 479-482.
E n g e l m a n n ,   M. D. (1961):   The role of soil arthropods  in the
      energetics  of an old field community.  Ecol. Monogr. 31: 221.238.
M a r g a l e f f ,   D. R. (1957): La teoria de la Information  en Eco-
      logia. Mem. Acad. Ciencias y Artes Barcelony. n° 661, Vol. 33, 13,
      79 pp.
P i e l o u ,   E. C. (1969):   An introduction  to mathematical ecology.
      J. Wiley, 286 pp.
Z i n k l e r ,   D. (1966):   Vergleichende Untersuchungen  zur Atmungs-
      -physiologie von Collembolen  und anderen Bodenkleinenarthropoden.
      Z. vergl. Physiol. 52, 99-144.

Discussion

V a n   d e r   D r i f t :   What do you mean in this context by community and are the different feeding habits  no objection  to take them together?

P. B e r t h e t :   In this study  the community concept  does not correspond to a given  trophic level, but is only based  upon a systematic
classification: it includes all the adult *Oribatei*, xylophagous, fungivorous as well as non-specialized feeders.

R. C o v a r r u b i a s :   Oribatid adults relative  to the whole soil
fauna (includind xerobionts  and hydrobionts) are not so heterogeneous,
mainly including only herbivores.  This fact  could be affected  by the
remarkable constancy of the two indexes  (evenness and active biomass).
This probably would not be so if the studied subcommunity  would include predators, such as *gamasids or prostigmatid* mites.

M. S. G h i l a r o w :   Is the formula proposed by Krivolutsky  to establish the population density of Oribatoid mites  according to data on
the thickness of litter in a forest  and hydrothermal conditions of the
locality applicable in your case?

P. B e r t h e t :   In the present study, unfortunately  all the information which is essential for making use of Dr Krivolutsky's method was
not collected.

C. G r e g o i r e - W i b o :  Vous considérez un unique poids (moyen) pour chaque espèce. Est-ce suffisant étant d né que la composition ju-vénile-adulte de la population peut varier temporellement? Dans ce cas, la méthode ne se complique-t-elle pas considérablement?

P. B e r t h e t :  Effectivement, mais jusqu'à présent nous ne dis-posons que de très peu d'informations concernant les besoins métaboli-ques des larves ou des nymphes. L'étude a porté uniquement sur les adultes pour lesquels on peut admettre que le poid ne varie pas sensi-blement durant l' année.

K. L a v e l l e :  Quelle forme a la distribution des biomasses (Log normale? Log linéaire?) et ceci a-t-il un rapport avec la constance de ce coefficient R.

P. B e r t h e t :  Je n'ai pas tente d'ajuster la distribution des bio-masses à une distribution théorique, n'ayant pas d'hypothèses préalables relatives a un modèle déterminé.

# Distribution patterns of lithobiomorpha on the territory of the USSR

N. T. ZALESKAJA

Institute of Animal Evolutionary Morphology and Ecology,
Moscow

*Lithobiomorpha* are typical dwellers of the forest litter and soil, they play an essential role in the regulation of the densities of soil invertebrates. Studying the ecology of this important group was restrained due to lack of knowledge of the species somposition. The first stage of studying of the *Lithobiomorpha* fauna in the USSR can now be summed up with the ecological and morphological characteristics of the centipedes described in detail for a number of zones and regions of the country.

At the present time 131 species and subspecies of *Lithobiomorpha* belonging to 12 genera and 3 families are known over the territory of the USSR. They inhabit all the landscape zones except desert zone. *Lithobiomorpha* sometimes form an important part of the complexes of soil invertebrates in different habitats. They are the only representatives of myriapodes in tundra (Chernov, 1964, 1972) and at the nivale belt in Alpines (Valiachmedov, 1972) as well.

Though the territory of the USSR is not yet studied sufficiently, the increase of *Lithobiomorpha* diversity is however found in the north-south direction. For example, in the typical tundra subzone (the Taimyr Peninsula) the only species *Monotarsobius chernovi* sp. n. (fam. *Lithobiidae)* is present in rather large numbers, up to 120 spec. per m$^2$. They however inhabit only grass-herb soil of southern slopes (Chernov, 1972), being absent in other sites. The same species has been observed in the pine and mixed forests of the Irkutsk region and Buriat ASSR from where it must have migrated to the North.

The coniferous ·north- and middle-taiga forests of the European part are also inhabited by one species only, *Monotarsobius curtipes* C. K. reaching 42 species/m$^2$ in the coniferous forests of the Komi ASSR (Krylova, 1970). This euritopic European species is the most numerous in the country, particularly in its European part. In the East it approaches Altai, it is found in the Caucasus in the South, it penetrates the Polar zone in the North and thus covers three main zones: taiga, broad-leaved forests and steppes.

Besides the representatives of the genus *Monotarsobius*, in the mixed forests of Chuvashia, Tataria and Ulyanov Region centipedes belonging to the genera of *Lamyctes* (fam. *Henicopidae)* and *Lithobius* (fam. *Lithobiidae*) are recorded.

The fauna of the mixed and leaf forests of the mountain areas of the South of the country is the most various. Up to 10 species of *Lithobiidae* are distributed in the subtropical broad-leaved forests of Lenkoran (see Table 1).

T a b l e   1.    The fauna of the mixed and leaf forests of the mountain areas

| Regions \ Biotope | Lenkoran | Sakhalin | Kursk Region | Chuvash ASSR Ulyanov Region | Tatar ASSR |
|---|---|---|---|---|---|
| Subtropical broad-leaved forest | IO | | | | |
| Bamboo | | 5 | | | |
| Oak forest | | | 4 | | |
| Mixed forest | | | | 3 | |
| Mixed forest with coniferous type prevailing | | | | | 4 |
| Meadows and arable lands | I | I | I | I | 2 |

The fauna of meadows and arable soils is mentioned to be very poor. In the European part of the country *M. curtipes* is prevailing in meadow soil. while *Lam. fulvicornis* Mein. prevails in arable soil. In the fields of Middle Asia the ecological niche of *Lam. fulvicornis* is occupied by *M. ferganensis* Trotz. The poor fauna of the above biotops

should probably be attributed to lack of developed layer of forest litter and the effect of cultivation on the soil.

Thus, most of the species of this group are connected with forest
litter which is also testified by the morphological form of *Lithobiomorpha*. The effect of certain conditions of existence resulted in
the formation of some adaptive morphological peculiarities of centipedes. A decrease in the diameter and number of coxal pores along with
a dense cover of small setae growing on the legs of the last pairs and
back sternites can be observed with the species of the genus *Hessebius*
inhabiting mostly dry mountain habitats of Middle Asia and the Caucasus
and with a number of species of the genus *Lithobius* existing under similar conditions. Their antennae are also covered with smaller and denser
setae as compared with the inhabitants of humid areas. A remarkably
wider diameter of coxal pores is observed in most of the lithobiid
species inhabiting areas of high humidity (Pacific islands, Primorye,
the Kolkhida Depression). So, decrease of humidity first of all results
in morphological changes in the structure of the front and back parts
of the body which contact the air of the open surface and contain or-
gans sensible to the temperature and humidity of the environment.

The peculiarities of the structure of the claws of female gonopodes
used to bury eggs in the soil can also be connected with humidity of
the habitat and substrate structure. In the plain and mountain forests
with a thick layer of loose soil, there prevail species with three-teeth
claws of female gonopodes which seem more suitable for digging and
loosening soft and humid soil. On the other hand in dry and stony habitats like those in Middle Asia, *Lithobiidae* with ordinary strong claus
of gonopodes prevail.

*Lithobiomorpha* fauna of Middle Asia seems to be the most various,
including 21 species. *M. turkestanicus* att. alone can be found beyond
the region (namely in the Caucasus). Only here in the USSR there are
representatives of the subfamily *Zygethobiina* (*Esastigmatobius kirgi-
sicus* Zal.) known from Japan, Korea and Taiwan. Besides, it was in this
very region where *Ghilaroviella* gen. n. was found, being the first representative of *Anopsobiidae* in the fauna of our country. Endemic percentage is also rather high (51 %) in the Caucasus. Of the *Lithobiomorpha*
fauna of the Far East, 47 % consist of the forms inhabiting Japan.
*Dakrobius* gen. n., morphologically peculiar, appeared to be the only
high-ranking endemic. There are no endemics in the fauna of the European
part of the USSR., 88 % of which is formed of European elements. As to
the *Lithobiomorpha* fauna of the remaining regions of the USSR, it has
not as yet been studied well enough.

Zaleskaja, N.T.


References

V a l i a c h m e d o v , B. V. (1972): Invertebrate fauna in mountain
    soils of the East Pamir and its change under the influence of cul-
    tivation.  Dokl. IV. Vsesoyuznogo sovestchan. probl. pochven. zoo-
    log., Nauka, M., 26-27.

K r y l o v a , L. P. (1970): Soil fauna  of  coniferous  habitats in
    Kochchoy-Jaga  (Komi ASSR).  Uchenye zapiski MGPI, N 394, 217-223.

C h e r n o v , Yu. I. (1964):  Relation  between  the soil fauna  and
    turf vegetation  in certain types  of tundra.  Problemy Severa, 8,
    254-267.

C h e r n o v , Yu. I. (1972): Animal population structure in the sub-
    zone of typical tundras  of the Western Taimyr., In: Tundra Biome.
    Proc. IV. Int. Meeting  on  the Biol. Productivity  of the Tundra,
    Leningrad USSR, October 1971; 63-74, Stockholm.

# Dispersion patterns of diplopods and their activity in the litter decomposition in the Carpathian foothills

B. R. STRIGANOVA

Institute of Evolutionary Morphology, Acad. Sci.,
Moscow

Diplopods are known to reach high densities in broadleaved and mixed forests of the temperate zone. Their role in litter decomposition was discussed repeatedly and their importance as primary decomposers was shown to be the most significant in the brown forest soils. Diplopods belong to the surface forms. During periods of active feeding they concentrate in forest litter, decaying wood, under stones etc. The pattern of the spatial dispersion within a habitat seems to be the important factor determining the activity of diplopod populations. Dispersion patterns in different habitats can modify widely depending on both specific peculiarities of animals and habitat structure.

The present report provides data on the surface dispersion and feeding activity of diplopods in foothills of the East Carpathians. This group was found to be the most numerous in the forest brown soil communities. More than 50 species were recorded there (Lang, 1954); Fasulati, 1959; Kurcheva, 1972). Our investigations were carried out in the Transcarpathian province of the USSR. A model broadleaved forest *(Querceto-Fagetum)*, 10 ha in area, was selected on the northern slope at an altitude of 130 m.

3 types of forest plots different in stand and litter composition were distinguished in the forest under investigation: 1. Beech plots with scarce hornbeam underwood. 2. Dry oak plots with an admixture of beech and pine and hornbeam undergrowth on elevations. 3. Moist hornbeam-oak plots in hollows and near a stream. The oak plots are included in the beech stand on the slope. Their distribution is determined by the relief and soil moisture.

64 litter and soil samples (size 25 x 25 cm) were taken in the forest. The density of invertebrates averaged 204,8 spec./m$^2$, that of diplopods - 56,0 spec./m$^2$. The least numbers of diplopods were found under beeches - 30,4 spec./m$^2$. The population density in the dry oak plots increased up to 46,4 spec./m$^2$, while the total abundance of in-

vertebrates was lower than that in beech plots. The maximum density of
invertebrates as well of diplopods was found in the moist hornbeam-oak
plots - 87,2 spec./m$^2$ (Table 1).

T a b l e  1.  Dispersion of diplopods in different forest plots

| Species | Density of animals spec./m$^2$ | | |
|---|---|---|---|
| | beech plots | dry oak plots | moist hornbeam-oak plots |
| *Chromatoiulus projec-tus kochi* Verhoeff | 5.6 | 24.3 | 29.0 |
| *Cylindroiulus bursen-landicus* Verhoeff | 13.6 | 3.8 | 17.6 |
| *Leptoiulus triloba-tus* (Verhoeff) | 7.2 | 9.1 | 8.8 |
| *Unciger foetidus* (C.L.Koch) | 2.4 | 3.1 | 2.2 |
| *Polyzonium germani-cum* Brandt | - | - | 29.8 |
| *Heteroporatia bos-niense* Verhoeff | - | 3.8 | - |
| *Polydesmus compla-natus* L. | 0.8 | 2.2 | - |
| Diplopods | 30.4 | 46.4 | 87.2 |
| Invertebrates | 193.6 | 168.0 | 248.0 |

7 species of dipopods belonging to 4 families were recorded in
soil samples. They all are widely distributed in Europe and were found
in the Carpathians in all altitude belts. The detail analysis showed
the definite regularities in their surface dispersion on the territory
of the forest under investigation.

*Chr. projectus* was the predominating species, its frequency averag-
ing 51 %. It occurred in all three types of forest plots, but its abundance in beech stands was much lower than that in oak ones where it
reached 2o-25 spec./m$^2$. The dispersion and density of *Chr. projectus*
seemed to be determined by the composition of the leaf litter. The soil
moisture was not significant, since they were abundant in the dry oak
plots as well in the moist ones. The freguency of *C. bursenlandicus*
averaged 42 %. It was numerous (13-17 spec./m$^2$) in the moist hornbeam-
oak and beech forest plots and scarce in dry places on elevations (3,8

spec./m$^2$). *C. bursenlandicus* appeared   susceptible to     the soil moisture   and   avoided the dry sites.   *L. trilobatus*   and   *U. foetidus* were
found to be evenly distributed in all plots of the forest.   The density
of *L. trilobatus*   averaged 7-9 spec./m$^2$, and that   of *U. foetidus* - 2-3
spec./m$^2$.   *H. bosniense* was recorded in the dry oak plots only, it was
found there in 5 samples. The distribution of *Polyzonium germanicum* was
limited   by   the moist plots   where   it was found   in   significant numbers - 29,8 spec./m$^2$.   This species   is distributed   in the forest zone
and occurs in northern coniferous forests.   *P. germanicum*   inhabits the
swamp habitats near the streams   and   in hollows in the model forest as
well in other regions of the Carpathians.   *P. complanatus*   was found in
the beech   and   dry oak plots where it was scarce.   Thus the pattern of
diplopod dispersion depends on the plant cover and in the first turn on
the leaf litter composition and the soil moisture as well. The majority
of diplopods   was found to be concentrated   in sites with oak and hornbeam leaf fall.

   Diplopods showed distinct selection with respect to different leaf
litters from that forest. They were fed   with a mixture   of hibernated
beech, oak and hornbeam litters   and ate presumably the latter one. The
traces of the feeding were observed   on the oak leaves and beech litter
was   disliked   by all diplopods species   sampled   in that forest.   Such
a tendency in the litter selection   is characteristic of many diplopods
and other *Saprophaga* (Lyford, 1943; Dunger, 1958; Striganova, 1971).

   A comparative study   of the feeding activity   was carried out in 4
species of diplopods which were the most abundant in soil samples: *Chr.
projectus, C. bursenlandicus, L. trilobatus* and *U. foetidus*. The feeding
rate   and   assimilability of food   were determined by the weight method
(Striganova 1969).   Adult specimens were used in the experiments.   They
were fed   with   hornbeam litter.   The results are shown in Table 2. The
food consumption in *C. bursenlandicus*   was found to be half that of the
other 3 species.   The low rate   of food consumption   was compensated by
a high assimilability - 43.9 %, while *Chr. projectus,* and *L. trilobatus*
assimilated 22-25 %   and   *U. foetidus* - 13.8 % of the food consumed. As
a result the amount   of the food   assimilated   in *C. bursenlandicus* appeared to be not less than in other species   and   reached in all diplopods 4-5 mg (dry weight)/day   as well the increase of the body weight -
2 mg/spec./day.

   The weight of the hornbeam litter decomposed by diplopods in different plots was calculated   by taking. into account the feeding activity of separate species and their population density (Table 3).   Diplopods destroy 522.6 mg of litter/day in the beech plots.   The decomposition activity   was distributes evenly among *Chr. projectus, C. bursenlan-*

T a b l e   2.   Feeding activity of diplopods

| Species | Mean weight of animals | Food consumed mg dry w./ spec./day | Assimilation % | Amount of food assimilated mg dry w./spec./ day |
|---|---|---|---|---|
| *Chromatoiulus projectus kochi* Verhoeff | 388.0 | 24.8 ± 8.0 | 21.6 ± 7.0 | 5.4 |
| *Leptoiulus trilobatus* (Verhoeff) | 171.5 | 25.8 ± 2.5 | 22.0 ± 4.6 | 5.6 |
| *Unciger foetidus* (C.L.Koch) | 132.6 | 26.9 ± 6.4 | 13.8 ± 3.9 | 3.7 |
| *Cylindroiulus bursenlandicus* Verhoeff | 120.3 | 9.8 ± 2.2 | 43.5 ± 8.6 | 5.4 |

T a b l e   3.   Rate of litter decomposition by Diplopods in different forest plots

| Species | Amount of litter decomposed mg dry weight/m$^2$/day | | |
|---|---|---|---|
| | beech plots | dry oak plots | moist hornbeamoak plots |
| *Chromatoiulus projectus kochi* Verhoeff | 138.9 | 602.6 | 719.2 |
| *Leptoiulus trilobatus* (Verhoeff) | 185.8 | 234.8 | 227.0 |
| *Unciger foetidus* (C.L.Koch) | 64.6 | 83.4 | 59.2 |
| *Cylindroiulus bursenlandicus* Verhoeff | 133.3 | 37.2 | 170.5 |

*dicus* and *L. trilobatus*. 958-1176 mg of litter/day was decomposed in oak plots with the hornbeam admixture and underwood. *Chr. projectus* contributed there to the greatest extent due to its high population density. They decomposed more than 60 % of litter. *L. trilobatus* decomposed there 20-24 % in spite of low numbers The activity of *C. bursenlandicus* was significant in the moist hornbeam-oak plots - 60 %. *P. germanicum* does not participate in the mechanical litter decomposition, they have the mouth apparatus adapted for sucking and can consume liquid food only. So the rate of litter decomposition in different plots of the forest depends not only on a population density of diplopods but also on their species composition and feeding peculiarities of separate species.

The activity level of diplopods in the Carpathian foothills seems to be very high. Field and laboratory observations have revealed that hornbeam litter is the predominant source of food in the forest soil. The litter of beech the main tree species in mountain forests on brown soils of the Carpathians is not virtually decomposed by these invertebrates. The density, dispersion patterns and activity of diplopods in broad-leaved forests seem to be determined by the tree species, accompanying the beech.

SUMMARY

The surface dispersion of diplopods in s broad-leaved forest is described. The surface dispersion was found to correlate with relief peculiarities, soil moisture and litter composition. The density of diplopods averaged in different plots 30-87 spec./m$^2$. Diplopods consume 0.5-1.2 g of leaf litter/m$^2$/day. Intensity of the litter decomposition depends on the population density, species composition and specific peculiarities of the feeding activity.

R e f e r e n c e s

D u n g e r , W. (1958): Über die Zersetzung der Laubstreu durch die Boden-Makrofauna im Auenwald. Zool. Jahrb. Abt. Syst. 86, I, 139-180.

F a s u l a t i , K. K. (1959): About the fauna of terrestrial invertebrates of the East Carpathians. In: Fauna and animal world of the Soviet Carpathians, Uzhgorod, 40, 120-141.

K u r c h e v a , G. F. (1972): Soil invertebrates in forests of the Transcarpathians. Pedobiologia 12, 5: 381-400.

L a n g , J. (1954): Mnohonožky-*Diplopoda*. Fauna ČSR, svazek 2, 1-183, Praha.

L y f o r d ,  W. H. (1943):  The palatability of freshly fallen forest
    tree leaves to millipedes. Ecology 24, 2, 252-261.
S t r i g a n o v a ,  B. R. (1969):  The evaluation  of assimilability
    of different .leaf-litter  by millipedes (Diplopoda). Zool. Zhurnal
    48, 6, 821-826.
S t r i g a n o v a ,  B. R. (1971):  Significance of diplopods' activ-
    ity  in  the leaf litter decomposition.  In:  Organismes du sol et
    production primaire, I.N.R.A., Publ. 71-7. 409-415, Paris.

D i s c u s s i o n

J. V a n  d e r  D r i f t :  What was the activity period  of the millipedes in your samples  and is it possible to calculate  from your results the annual consumption?

B. R. S t r i g a n o v a :  The measurements  of the milipedes feeding
activity  were carried out in May.  Millipedes in the foothills  of the
Carpathians  were observed to feed actively in April - June and September - October.  The rate of food consumption  can widely vary in millipedes along the periods of activity. Thus the results of the calculation
of the litter consumption, made for a 1-day-period, concern the peak of
the spring feeding activity only.

A. S z e p t y c k i :  What do you understand under the term "Invertebrate" in the Table 1?

B. R. S t r i g a n o v a :  We have recorded  the large-sized invertebrates belonging to the mesofauna only: millipedes, isopodes, earthworms,
insect larvae etc. Microarthropods were not collected.

A. S z u j e c k i :  I have  one question  to  the first part  of this
very interesting paper  presented  by Dr Striganova:  Can you inform us
about soil conditions occurring  under the beech stands.  In the Polish
part  of the E. Carpathian Mountains  these  conditions vary on various
locations of beech stands from dry to very moist  and from high acidity
to a low one.

B. R. S t r i g a n o v a :  The site under consideration  was selected
on the brown forest soils  slightly podzolized, which are characteristic
of the broad-leaved forest region in the foothill belt.

R. C o v a r r u b i a s :  1. Because of your comparisons between densities  of "Diplopoda"  and  "Invertebrata" - was  the sampling method
a simultaneous one? 2. Anyway what was that method?

B. R. S t r i g a n o v a :   1-2. The invertebrates belonging   to   the "mezofauna" which were recorded in soil samples were picked out by hand from the soil.   For the Microarthropods there are used other methods of extraction but those little invertebrates  were not recorded in our in-vestigations.

R. C o v a r r u b i a s :   -then the group  of   invertebrates does not include either *Acarina* (mites) or *Collembola*.

P. B e r t h e t :   1. What do you call   "amount of litter decomposed"? 2. Did you observe veriations  in assimilation rates   according  to the quantity of food available?

B. R. S t r i g a n o v a :   The mechanical destruction  of leaf litter was considered only.   The amount of food ingested was recorded.   Amount of food available under the natural condition is usually much more then millipedes can consume.   During the laboratory experiments   the animals received the amount of food  exceeding sometimes   the food requirements of millipedes.

W. D u n g e r :   I shall come back to the stated differences   in assi-milation rate. Have you studied the pellets or the gut contents to look at a higher or lower degree of breakdown?

B. R. S t r i g a n o v a :   We have not carried out   chemical analyses of millipedes'pellets.   The degree of the breakdown   of leaf litter can differ significantly in different species.

# Evolution spatio-temporelle d'une population de Folsomia quadrioculata (Tullberg) (Insecta: Collembola)

C. G. WIBO

Laboratoire d'Ecologie Générale et Expérimentale,
Louvain

Le présent travail a pour but l'analyse des propriétés spatio-tempo-
relles d'une population de *Folsomia quadrioculata* en prairie. Des unités
de sol (quadrats) ont été prélevées par paire (24,25 ou 26 paires) au
cours de 16 échantillonnages effectués de juin à octobre 1970 et de 2
échantillonnages à des dates ultérieures.

Une sonde de 3.6 cm de diamètre et l'appereil d'extraction type
Macfadyen on été utilisés (Wibo, 1971).

## RÉSULTATS

1. Les distributions de recensement. Le
nombre moyen d'individus recensés par paire de quadrats (20.35 cm$^2$)
varie de 6.2 à 65.1 individus (Fig.1). Les intercalles de confiance déterminés approximativement en supposant une distribution normale des
individus par paire de quadrats sont relativement grands.
2. Structure d'âge. Pour certaines populations, de.....
tels pics de densité pourraient résulter d'une reproduction massive
à certaines époques. Celle-ci peut être causée par différents facteurs:
adaptation du cycle de vie au régime climatique, cannibalisme de certains stades, etc... Dans le cas de la population étudiée les fluctuations ne semblent pas être liées à un cycle de reproduction comme le
montre la composition juvénile - adulte de la population (Fig.2).

La distinction entre individus matures et immatures est hasée
d'une part sur la longueur de la griffe de la troisième paire de pattes,
organe fort chitinisé, mesurée au microscope (640x) et d'autre part sur
le développement de la plaque génitale. Le graphique indique en échelle
semi-logarithmique les nombres moyens d'individus adultes femelles et
juvéniles obtenus au cours des 14 semaines successives. Les proportions

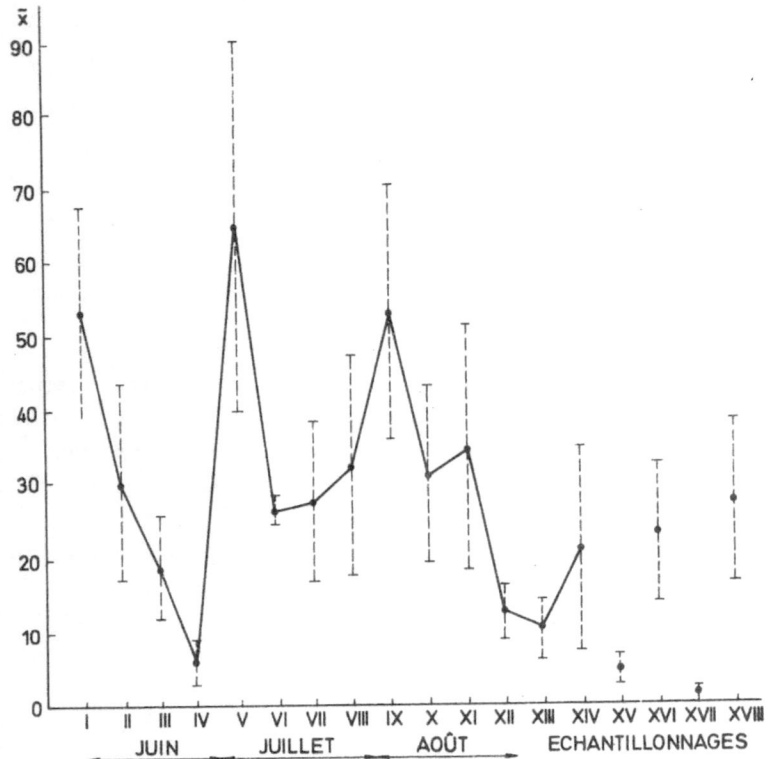

Fig.1. Nombre moyen d'individus recensés $(\bar{x})$ et intervalle de confiance au sours de 18 échantillonnage's.

d´individus juvéniles  et adultes restent quasi constantes: les pics de
densité ne reflètent des lors pas  dès vagues d´apperition de juvéniles
et d´adultes au sein de la population.

    3. R é p a r t i t i o n   s p a t i a l e .  La répartition spa-
tiale des populations de Microarthropodes édaphiques  dépend d´une part
de la diversité  des conditions  de vie  dans l´habitat et d´autre part
des associations  entre  les individus.  Ces deux facteurs  peuvent en-
gendrer la formation  de groupes d´individus  ou agrégats,  répartis de
façon aléatoire ou non dans l´habitat.

    Etant donné l´influence particulière  de l´un  ou l´autre facteur,
Gerard (1970)  a établi quatre modèles  de répartition spatiale  sucep-
tibles  de s´appliquer aux populations de Microarthropodes.

    Dans ce travail  la discrimination entre les modèles est basée sur
l´analyse de paires de quadrats contigus.

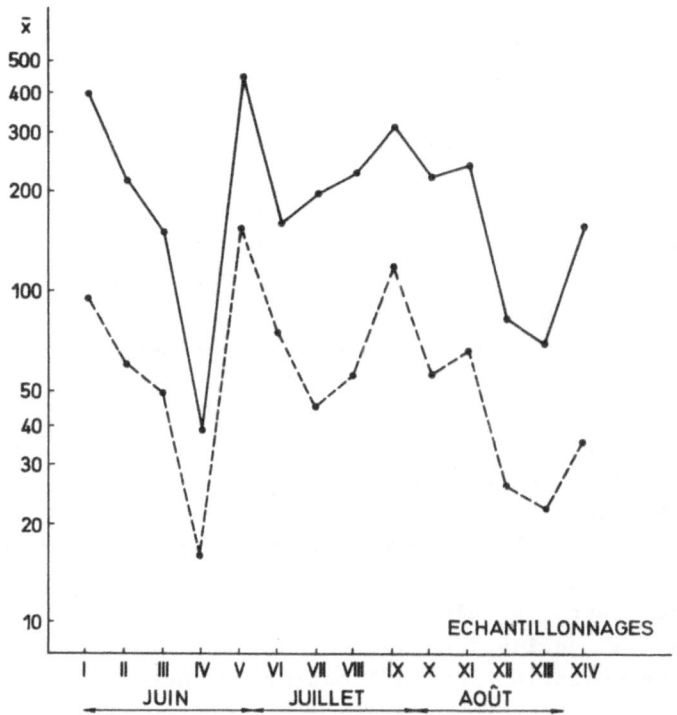

Fig.2. Evolution du nombre moyen d'individus femelles adultes (-----) et d'individus juvéniles (————) dans les différents échantillonnages (x̄).

L'analyse de la covariance entre quadrats contigus permet d'éprou-
ver l'hypothèse d'une répartition homogène ou aléatoire des individus
ou des agrégats contre l'hypothèse d'une répartition hétérogène. En
effet, la covariance est nulle dans le cas d'une répartition homogène.
Les valeurs de covariances obtenues sont toutes positives. Un test de
corrélation, qui cependant n'est strictement applicable qu'à une distribution normale, donne de nombreux coéfficients supérieurs aux valeurs critiques au seuil 0,05, en particulier lorsque les moyennes observées dépassent 20 individus, à l'exception des distributions I,IX
et X. Dès lors, la population de *F. quadrioculata* est caractérisée par
une répartition hétérogène au cours de son évolution (échantillonnages
II, V, VI, VII, VIII, XI, XIV, XVI et XVIII). Les distributions III,
IV, XII, XIII, XV et XVII ont une moyenne faible: il n'est pas possible
de caractériser la répartition spatiale par l'étude de la corrélation
entre quadrats contigus.

D'autre part, un test d'homogénéité de proportions basé sur le quotient de vraisemblance (Sokal et Rohlf, 1969) permet de déterminer si la répartition des individus présente des agrégats ou non, en supposant toutefois une répartition homogène pour des quadrats de petite surface (Gerard, 1970,a). Pour les différentes distributions l'hypothèse d'homogénéité est à rejeter au seuil 0,01 à l'exception des distributions XII et XVII.

Sur la base de ces deux tests et lorsque la moyenne dépasse 20 individus, il est vraisemblable de supposer que la répartition spatiale de F. quadrioculata résulte à la fois d'une forte diversité des conditions de vie dans l'habitat et d'associations entre les individus.

Sous l'hypothèse d'une répartition hétérogège en agrégats, l'état spatial de la population peut être caractérisé par trois paramètres donnant (Gerard, 1970,b): 1. le niveau moyen de densité dans l'habitat; 2. l'intensité des associations intraspécifiques, mesurée par l'indice A et 3. le degré d'hétérogénéité spatiale mesurée par l'indice W.

La figure 3 met en évidence les valeurs de A et W pour les différentes distributions. Elles semblent bien être independantes de la densité.

En tenant compte des tests précédents apparaissent différents groupes de distributions: 1. celles dont la moyenne est faible (III, IV, XII, XIII, XV, XVII) ces distributions à l'exception des distributions XII et XVII, présentent néanmoins des valeurs de A différentes de zéro. 2. celles qui présentent une hétérogénéité spatiale faible (I, IX, X) et des valeurs de A différentes de zéro. 3. celles qui présentent à la fois une hétérogénéité spatiale et des valeurs de A différentes de zéro (II, V, VI, VII, VIII, XI, XIV, XVI, XVIII).

La même étude a été faite pour les juvéniles et les adultes. Elle met en évidence une plus grande hétérogénéité spatiale et des valeurs d'association plus élevées pour les juvéniles que pour les adultes.


DISCUSSION ET CONCLUSIONS

L'étude met en évidence des variations temporelles importantes de l'abondance de Folsomia quadrioculata en prairie. Des échantillonnages hebdomadaires montrent des pics de densité séparés de un mois (I, V, IX) durant l'été. Comme l'évolution des juvéniles et des adultes est semblable, ces variations ne peuvent s'expliquer par un cycle de reproduction. Bödvarsson (1973) observe également de telles fluctuations en forêt suédoise. En forêt de Meerdael (Belgique) une population de F. quadrioculata recensée durant une année toutes les trois semaines

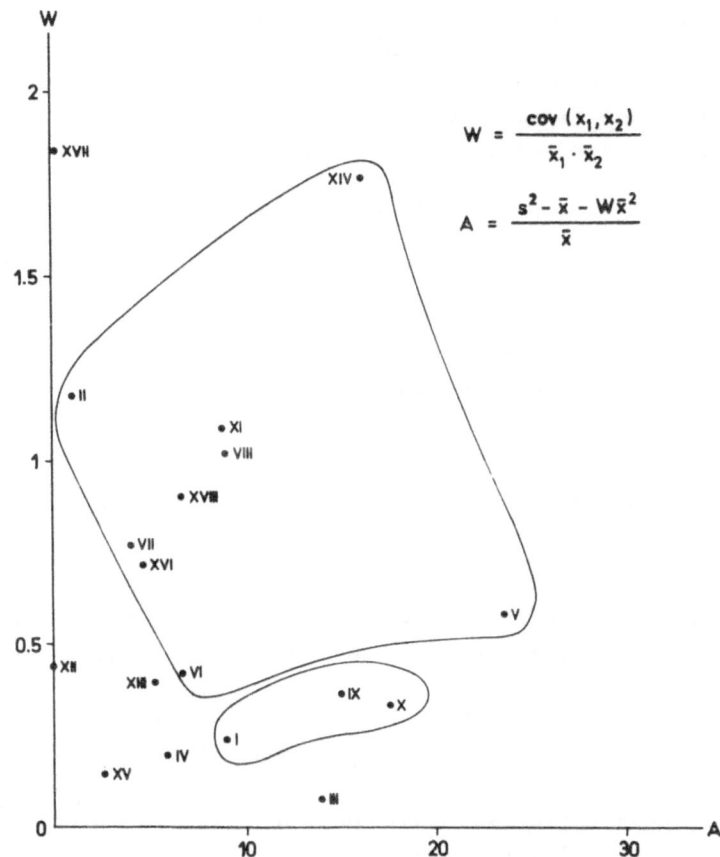

$$W = \frac{cov\ (x_1, x_2)}{\bar{x}_1 \cdot \bar{x}_2}$$

$$A = \frac{s^2 - \bar{x} - W\bar{x}^2}{\bar{x}}$$

Fig.3. Valeurs de l'indice d'association intraspécifi-
que (A) et de l'indice d'hétérogénéité spatiale (W). $\bar{x}_1$,
$x_2$ et cov($x_1$, $x_2$) sont les moyennes et la covariance
pour les différents couples de quadrats contigus. $\bar{x}$ et
$s^2$ sont la moyenne et la variance pour la somme des
quadrats contigus.

montre durant l'été un important pic de densité; la composition juvénile adulte étant par ailleurs constante (Priemen, Mignolet, Gregoire-
Wibo, en préparation). Dans le nord de l'Angleterre, d'après Hale (1966),
F. quadrioculata présente deux maxima d'abondance et d'après Usher
(1970) un seul caractérisé par la présence de 90 % de juvéniles.

L'analyse de paires de quadrats contigus montre que la répartition
spatiale de F. quadrioculata est fortement agrégative caractérisée par
une hétérogénéité spatiale importante et des associations entre les individus.

D´après des expériences d´élevage   (Gregoire-Wibo, 1973) celles-ci
peuvent résulter:   1. du mode de reproduction:   en élevage les femelles
déposent leurs oeufs par paquet   et 2. du phénomène de mue:   en élevage
les individus se groupent pour la mue.   Ces deux mécanismes constituent
un facteur de survie dans un environnement hétérogène.


R e m e r c i e m e n t s

Les Drs Ph. Lebrun et G. Gerard   ainsi que Mrs J. Feron et E. Jal   nous
ont aimablement aidé au cours de l'élaboration   de ce travail.   Qu'ils
en soient remerciés.

B i b l i o g r a p h i e

B e r t h e t ,   P. (1960): La mesure écologique   de la température par
    détermination   de la vitesse d'inversion du saccharose.   Vegetatio
    9, 197-204.
B ö d v a r s s o n , H. (1973): Contribution to the knowledge of Swed-
    ish forest *Collembola*. Institute of Forest Zoology. Research Notes
    13, 1-43.
G e r a r d , G. (1970): Répartition spatiale de populations d'Oribates
    édaphiques. Thèse Inédit. Université de   Louvain. 1-171.
    - : Modèles  de répartition spatiale  en  écologie animale. Biomé-
    trie-Praximétrie 11: 124-190.
G r e g o i r e - W i b o , C. (1973): Bioécologie de *Folsomia qua-
    drioculata*  (Tullberg) V MCPZ Prague 1973.
H a l e , W. G. (1966): A population study  of  moorland *Collembola*.
    Pedobiologia 6, 65-99.
M i g n o l e t , R.,   P r i e m e n , J. P.,   G r e g o i r e  -
    W i b o , C. (en préparation): Colonisation  par les Collemboles
    de litières en décomposition.
S o k a l , R. R. et R o h l f , F. J. (1969):  Biometry.  The prin-
    ciples and practice of statistics in biological research. Freeman,
    San Francisco, 776 pp.
U s h e r , M. B. (1969): Some properties  of the aggregations of soil
    arthropods: *Collembola*. J. Anim. Ecol. 607-622.
W i b o , C. G. (1971): Etude  de la régulation numérique  de quelques
    populations d'Oribates et de Collemboles.(Note préliminaire). Rev.
    Ecol. Biol. Sol  8, 103-109.

Discussion

C. A t h i a s : Avez-vous une opinion sur le mécanisme des variations
d'abondances numériques dans le temps? M. Bouché (comm. pers.) a observé qu'a certaines périodes de l'année, les galeries de vers de terre
contiennent de grandes quantités de collemboles. On peut penser qu'à
certaines saisons une partie des populations des collemboles que vous
avez étudiés, se trouve dans des habitats "de survie" (p. ex. galeries
de lombrics) et échappe ainsi à votre observation.

C. G r e g o i r e - W i b o : Mon attention n'a pas été attirée
par de tels phénomènes dans la prairie étudiée. S'ils existaient, ils
constitueraient évidemment une cause d'hétérogéneité spatiale.

P. B e r t h e t : L'examen de coupes de sol indique que les Oribates
sont, proportionellement à leur taille, répartis à grande distances les
uns des autres. Les associations inter-individuelles ne doivent donc
pas être considérables. Croyez-vous que, chez *Folsomia*, la situation
soit fort différente?

C. G r e g o i r e - W i b o : Chez *F. quadrioculata* nous avons ré-
colté jusqu'à 266 individus pour une unité de sol de 10 $cm^2$. En particulier là où il y a une concentration de radicelles près des touffes
d'herbes l'abondance est nettement supérieure à celle avoisinante. Ces
endroits semblent favorables.

F. A t h i a s : Vous avez noté que les échantillons au niveau des
touffes d'herbes étaient plus riches en Collemboles que le sol environ-
nant. Vous avez sans doute remarqué que les autres Microarthropodes
aussi se groupent sous les touffes. Ce congroupement est plus facile
à observer dans les savanes où j'ai travaillé, car les touffes de graminées sont plus dispersées et mieux individualisées que dans les prairies européenes. Il apparaît donc que la base des touffes d'herbe et
leur rhizosphère est un milieu favorable, et principalement à cause de
nombreux facteurs d'ordre trophique. L'agrégation des Microarthropodes
dans ces microbiotopes est due à des causes trophiques principalement,
plutôt qu'à des causes d'origine uniquement comportementales.

C. G r e g o i r e - W i b o : Ces microhabitats présentent également
une structure du sol plus aérée et dès lors plus favorable à l'agrégation.

Z. M a s s o u d : Est-ce que vous avez observé le phénomène de ponte
collective chez *F. quadrioculata*?

C. G r e g o i r e - W i b o : Pour *Folsomia quadrioculata* en ele-
vage, le nombre d'oeufs pondus par paquet semble indépendant du nombre
de femelles présentes (à l'inverse d'autres *Isotomidae*).

G. J o s s e n s :   Est-ce que les groupements  de *F. quadrioculata* que vous avez observés pourraient  être justifi.  par une défence vis-à-vis des prédateurs;  le regroupement des individus diminuant la probabilité de rencontre avec les prédateurs?

C. G r e g o i r e  -  W i b o :  Vous avez raison.  Les Gamasides con- stituent les principaux  prédateurs  s'attaquant  aux *Isotomidae*.  Nous avons pu observer en élevage des "poursuites" entre Gamide  et individu isolé de *F. quadrioculata*.

# Dominante Coleoptera auf der Bodenoberfläche der Wiesen- und Waldökotone

R. K. CYKOWSKI

A. R. – Zoologie,
Szczecin

Die Käfer sind eine der zahlreichsten Faunagruppen auf der Bodenoberfläche der Wiesen- und Waldökotone. Dem Vorkommen der *Coleoptera* (vorwiegend ernste Pflanzenschädlinge) auf dem Boden der Wiesen- und Waldökotone hat man bisher verhältnismässig wenig Beachtung geschenkt. Mit
diesen Problemen befasst sich zum Teil die Arbeit von Honczarenko J.
(1973). Nach Ghilarov M. S. (1954) sind den Faunauntersuchungen Domi-
nanzanalysen der Familien und Arten zugrundezulegen, was auch in der
vorliegenden Arbeit berücksichtigt wird.

METHODIK UND MATERIAL

Die Untersuchungen über *Coleoptera* wurden 1969, 1970 und 1971 auf zwei
Wiesen- und Waldökotonen innerhalb und in Nachbarschaft grosser Wald-
gesellschaften bei Szczecin durchgeführt. Einer der Ökotone befand sich
am Rand der Bukowa Puszcza und umfasste die Wiesengesellschaft *Festuca
rubra* L. und die Waldassoziation *Fageto-Quercetum*, das zweite erstreck-
te sich am Rand der Goleniowska Puszcza und bestand aus der Wiesengesellschaft *Poa pratensis* L. und der Waldassoziation *Pineto-Vaccinietum*.
In diesen Ökotonen wurden aufgrund ökologischer Analysen je vier Standorte bestimmt: einer im Wald (50 m vom Waldrand im Waldinneren) und
drei auf der Wiese (5 m, 50 m und 100 m vom Wald entfernt).

     In allen Untersuchungsjahren, von Mai bis Oktober, wurden in jedem
Monat (Monatsende) je 4 Bodenproben von jedem Standort entnommen. Die
Einheitsgrösse aller Bodenproben betrug: 50 x 50 x 30 cm, unter Einbe-
ziehung von drei Bodentiefen: 0 - 10, 10 - 20, 20 - 30 cm. Im Verlaufe
der drei Untersuchungsjahre. wurden 480 Bodenproben entnommen.

     Das gesammelte Faunamaterial wurde im Labor nach Arten determi-
niert und wird gegenwärtig in den Sammlungen der Bodenentomofauna auf-
bewahrt.

ERGEBNISSE

In den drei Untersuchungsjahren  wurden 6109 Individuen  von *Coleoptera*
gefangen und bestimmt, die 20 Familien angehören.

 Die Abbildung 1  stellt die mittlere dreijährige Populationsdichte
(Col./1 m$^2$)  der Käfer  auf  den einzelnen Standorten  der untersuchten
Ökotone dar. Daraus ergibt sich, dass die grösste Populationsdichte auf

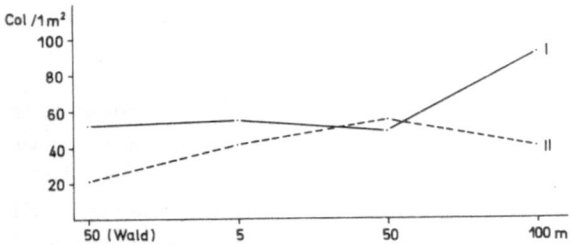

Abb.1. Populationsdichte der *Coléoptera* (1969-1971): I -
Laubwald- und Wiesenökoton, II - Nadelwald- und Wiesenökoton.

den Wiesen  in der Nähe des Waldes beobachtet wurde, am Laubwald in einer Entfernung von 100 m, und am Nadelwald in einer Entfernung von 50 m
vom Wald.

 Im Laubwaldökoton wurden 3740 Individuen gefangen, die 18 Familien
angehörten.  Die dominanten Familien waren (Abb. 2): *Scarabaeidae, Ela-*

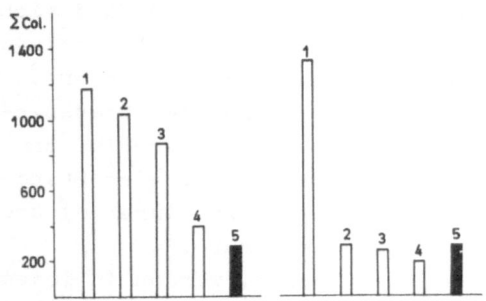

Abb.2. Dominante Familien  der  *Coleoptera*  (1969-1971):
I - Laubwald- und Wiesenökoton: 1.*Scarabaeidae*, 2. *Ela-*
*teridae*,  3. *Curcilionidae*, 4. *Carabidae*,  5. andere Familien; II - Nadelwald- und Wiesenökoton: $1_1$. *Elateridae*,
$2_1$. *Carabidae*, $3_1$. *Curcilionidae*, $4_1$. *Tenebrionidae*, $5_1$.
andere Familien.

*teridae, Curculionidae* und *Carabidae;* von anderen Familien waren vertreten: *Staphylinidae, Coccinellidae, Chrysomelidae, Silphidae, Alleculidae, Tenebrionidae, Dermestidae, Cantharidae, Lagridae, Anthicidae,
Byrrhidae, Dytiscidae, Hydrophilidae* und *Ipidae.*

Im Nadelwaldökoton wurden 2369 Individuen gefangen, die sich in 15
Familien gruppierten. Die dominanten Familien waren (Abb. 2): *Elateridae, Carabidae, Curculionidae* und *Tenebrionidae;* von anderen Familien
waren vertreten: *Staphylinidae, Silphidae, Chrysomelidae, Scarabaeidae,
Dryopidae, Cantathridae, Alleculidae, Coccinellidae, Cerambycidae, Dytiscidae* und *Byrrhidae.*

Das Zahlenverhältnis zwischen den dominanten und den übrigen Familien veranschaulicht die Abb. 2. Die prozentuelle Verteilung der dominanten Familien auf dem Boden des untersuchten Milieus illustrieren die
Abb. 3 und 4.

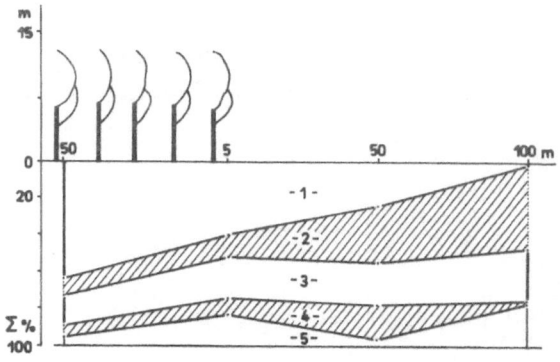

**Abb.3. Prozentuelles Vorkommen der dominanten Familien
der Coleoptera auf dem Boden des Laubwald- und Wiesen-
ökotons: 1. *Scarabaeidae,* 2. *Elateridae,* 3. *Curcilionidae,* 4. *Carabidae,* 5. andere Familien.**

Im Laubwaldökoton sinkt mit der Entfernung vom Wald die Dominanz
der *Scarabaeidae* und steigt jene der Elateridae. Der prozentuelle Anteil der *Curculionidae* an den einzelnen Standorten war gleichmässig;
die *Carabidae* lassen in einer Entfernung von 50 m vom Wald die Tendenz
zur Gruppierung auf der Wiese erkennen (Abb. 3).

Im Nadelwaldökoton dagegen behielten *Elateridae* auf allen Standorten die Hauptdominanz, besonders auf solchen, die 50 und 100 m vom
Wald entfernt waren. Der prozentuelle Anteil der *Carabidae* war auf den
einzelnen Standorten gleichmässig, die *Curculionidae* gruppierten sich
im Wald und die *Tenebrionidae* auf der Wiese in der Entfernung ab 5 m
vom Wald (Abb. 4).

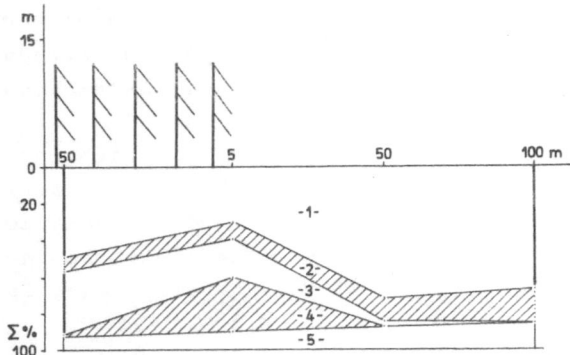

**Abb.4. Prozentuelles Vorkommen der dominanten Familien der *Coleoptera* auf dem Boden des Nadelwald- und Wiesen-Ökotons:** $1_1$. *Elateridae*, $2_1$. *Carabidae*, $3_1$. *Curcilionidae*, $4_1$. *Tenebrionidae*, $5_1$. andere Familien.

Der prozentuelle Anteil der auf den einzelnen Standorten der untersuchten Ökotone dominanten Arten ist auf den Abb. 5 und 6 dargestellt.

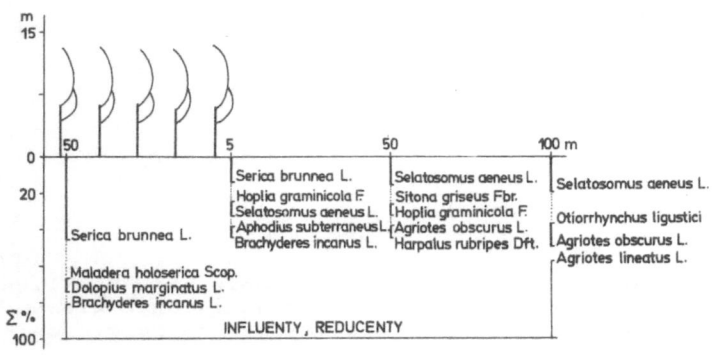

**Abb.5. Artendominanz der *Coleoptera* auf dem Boden des Laubwald- und Wiesenökotons.**

Im Laubwaldökoton war *Serica brunnea* L. die häufigste Art im Wald-inneren und in der Entfernung ab 5 m vom Waldrand, auf den übrigen Standorten der Wiese war es *Selatosomus aeneus* L. (Abb. 5 und 7). Mit der Entfernung vom Wald nahm die Anzahl von *Serica brunnea* L. ab, und die Anzahl von *Selatosomus aeneus* L. zu. Die häufigste Art der *Carabidae* war *Harpalus rubrifes* Dft. als dominante Art des 50 m vom Wald entfernten Standortes (Abb. 5).

Im Nadelwaldökoton war auf dem Standort im Walde *Dolopius marina-tus* L. die häufigste Art, dagegen in einer Entfernung von 5 m vom Wald

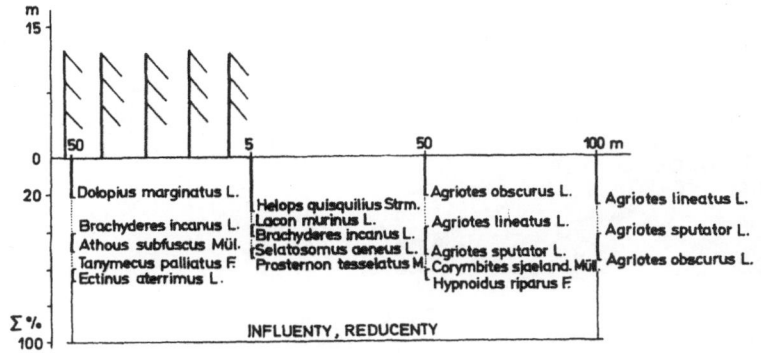

Abb.6. Artendominanz der *Coleoptera* auf dem Boden des Nadelwald- und Wiesenökotons.

die Art *Helops quisquilius* Strm. (einziger Vertreter von *Tenebrionidae)*; auf den 50 und 100 m vom Wald entfernten Standorten dominierten *Elateridae* mit der Gattung *Agriotes* Eschz.: *A. obscurus* L., *A. lineatus* L. und *A. sputator* L. (Abb. 6 und 8).


ZUSAMMENFASSUNG

*Scarabaeidae, Elateridae, Curculionidae* und *Carabidae* waren auf der Bodenoberfläche der Laubwald- und Wiesenökotone dominant. Mit der Entfernung vom Wald sank die Dominanz der *Scarabaeidae* und stieg jene der *Elateridae*. Die häufigste Art im Waldinneren und am Aussensaum des Waldes war *Serica brunnea* L. und auf der Wiese die Art *Selatosomus aeneus* L.

In Nadelwald- und Wiesenökotonen waren dagegen *Elateridae, Carabidae, Curculionidae* und *Tenebrionidae* dominant. Auf allen Standorten des Ökotons waren *Elateridae* die dominante Familie. Die häufigste Art auf dem Boden des Waldinneren war *Dolopius marginatus* L., 5 m vom Waldrand die Art *Helops quisquilius* Strm., und auf den übrigen Standorten der Wiese *Agriotes obscurus* L., *A. lineatus* L. und *A. sputator* L.

Die auf der Bodenoberfläche der untersuchten Wiesen- und Waldökotone vorkommenden dominanten Arten der *Coleoptera* zählen in ihrer Mehrzahl zu den ernsten Pflanzenschädlingen.

Abb.7. *Selatosomus aeneus* L.   die häufigste Art der Wiese am Laubwald.

Abb.8. *Agriotes lineatus* L.   die häufigste Art der Wiese am .Nadelwald.

Literatur

C y k o w s k i , R. K. (1973a): Der Einfluss des Laubwaldes *(Fageto-Quercetum)* auf *Coleoptera* im Wiesenboden. Ekologie Polska (im Druck, englisch).

- (1973b): Der Einfluss des Nadelwaldes *(Pineto-Vaccinietum)* auf *Coleoptera* im Wiesenboden. Ekologia Polska (im Druck, englisch).

G h i l a r o v , M. S. (1954): Art, Population und Biozönose. Zool. Z. 33, 4: 769 - 776 (russisch).

H o n c z a r e n k o , J. (1973): Der Einfluss des Waldes auf die Bodenentomofauna der angrenzenden Wiesengelände. Pol. Pis. Entomol. 43, 155 - 180 (polnisch).

Discussion

W. H ü t h e r : Haben Sie eine Hypothese, welche Faktoren die Unterschiede in der Käferfauna bedingen? Inwieweit spielen anthropogene Einflüsse eine Rolle?

R. K. C y k o w s k i : Die Untersuchungen wurden in naturnahen, durch den Menschen nicht weitgehender beeinflussten Pflanzengesellschaften durchgeführt.

# Rapport des populations de Carabidae et d'Elateridae dans les terres labourables de chernozem du nord-est de Yougoslavie

R. SEKULIĆ
Faculté d'Agriculture,
Novi Sad

C'est dans le but de prévoir l'apparition des insectes s'attaquant aux plantes cultivées qu'on fait en automne de prélèvements des échantillons de sol, en particulier dans les emblavures de blé et celles de betterave sucrière. Cela parce qu'en général le blé fait part de l'assolement avec des cultures sarclées étant sensibles aux attaques des fils de fer au début même de leur végétation, tandis que la betterave sucrière est en outre souvent menacée par de divers charançons et noctuides. Le matériel d'insectes recueilli de cette manière sert également de détermination de la composition quantitative et qualitative de la macroentomofaune dans des agrobiocoenoses différentes. Les superficies examinées s'élargissent d'année en année d'autant plus que s'augmente le nombre des organisations agricoles acceptant cette méthode de travail comme une mesure indispensable dans la protection des plantes contre les ravageurs.

Dans les terrains de nos recherches les populations d'Elateridae représentent un problème important et permanent en provoquant souvent le besoin des interventions chimiques. De l'autre côté, les populations de Carabidae sont les plus nombreuses parmi les insectes benéfiques y étant recueillis. Pour comparer leurs populations nous n'avons pris en considération que les données de deux localités portant sur les différences provoquées par de conditions spécifiques, étant donné qu'il s'agit de la même région géographique, du même type de sol, de la même époque de prélèvements et des mêmes agrobiocoenoses.

Ces deux localités (on en a enrégistrées comme A et B) se trouvent entre le Danube et la Tissa. Les données qu'on y traite portent sur une période de six années - de 1966 à 1971. Pendant ce temps on a examiné environ 4000 hectares de blé et 5500 hectares de betterave sucrière de localité A et 27000 hectares de blé et plus de 6000 hectares de betterave sucrière de localité B. Il faut dire tout d'abord qu'on y cultive du blé sur une superficie englobant de 40 à 50 % de terrains la-

bourables, tandis que la betterave sucrière seulement  sur 1/6 de localité A et 1/14 de localité B. Dans ces recherches on utilise la méthode
de prélèvement manuel, l'échantillon de sol mesurant 0,25 m$^2$  et  50 cm
de profondeur et leur nombre moyen étant un à deux hectares.

    Les résultats sont présentés  sur les trois graphiques.  C'est sur
le graph.1 qu'ils portent sur la densité moyenne des populations d'*Elateridae* et de celles de *Carabidae*  ou bien  sur  leur dynamique d'année
en année.  Comme on voit,  les populations  d'*Elateridae*  sont considérablement plus nombreuses  dans les cultures de blé  et celles de *Cara-*

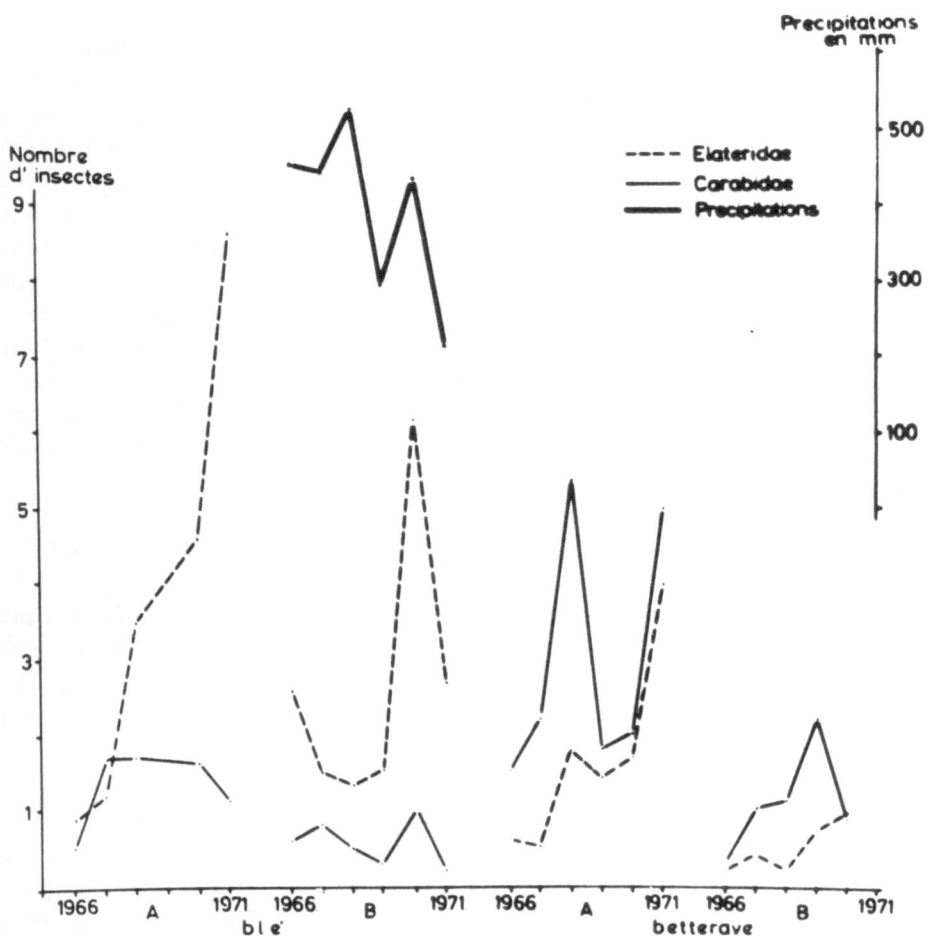

Graph 1. La densité moyenne des populations d'*Elateridae*
(----) et de celles de *Carabidae* (4.)

*bidae* dans les cultures de betterave sucrière. Il existe à cet égard aussi de différences nettes entre les deux localités. Tandis que les populations d'*Elateridae* s'augmentent d'une année à l'autre dans les emblavures de blé de localité A, on enrégistre une dépression de quelques années pour celles de localité B. C'est à peine en 1970 qu'on observe un acroissement de ces dernières-là atteignant la moyenne de 6,2 individus par m$^2$, leur nombre critique pour les cultures sarclées étant 3 à 5 sur les terrains examinés. L'augmentation des populations d'*Elateridae* enrégistré chaque année dans les cultures de blé de localité A est certainement due à l'humidité favorable grâce à l'irrigation des champs de cultures sarclées. Au fait, la végétation de blé à la suite de ces cultures est beaucoup plus forte en automne et plus résistente à la sécheresse, notamment bien accusée au printemps de 1971, en assurant par conséquant le microclimat plus favorable aux *Elateridae* dont le nombre moyen a atteint son maximum de 8,7 individus par m$^2$.

Pour expliquer la dépression dans les populations d'*Elateridae* de localité B où l'on n'applique pas d'irrigation, il faut parait-il en chercher les causes dans la période précédente. Cela d'autant plus qu'on prend en considération la quantité de précipitations - 400 à 500 mm dans le courant de la végétation de ces années-là, la moyenne de plusieurs années n'étant qu'environ 600 mm pour cette région. Dans la période précédente donc la superficie des cultures de blé était beaucoup plus restreinte presque jusqu'à 50 %. Cela veut dire qu'il a fallu s'écouler plusieurs années avant que cette culture revienne aux mêmes parcelles en suivant l'assolement. Or, l'élargissement considérable de la superficie des cultures de blé partie en monoculture ainsi que la quantité favorable de précipitations ont favorisé la ponte et l'existence des larves de stades initiales de développement en randant possible leur apparition en masse en 1970. Mais en 1971 déjà les précipitations pendant la végétation n'ont atteint que la moitié des années précédentes -220 mm en provoquant la sécheresse influençant mal les cultures de blé et les populations d'*Elateridae* aussi.

La densité moyenne des populations d'*Elateridae* dans les emblavures de betterave sucrière a été plusieurs fois plus faible que celle dans les emblavures de blé. Mais les populations de localité A ont quand même montré une faible tendence à augmenter, ce qui est certainement du à l'irrigation, en atteignant 4 individus par m$^2$, en 1971 malgré l'application des insecticides.

Quant aux populations de *Carabidae*, la situation a été tout différente. Leur densité moyenne dans les cultures de blé de deux localités a été tout faible pendant toute la période de six années. Par contre, leurs populations dans les cultures de betterave sucrière ont été plus

fortes avec les variations beaucoup plus accusées. Ainsi celles de localité A  ont eu tendance à augmenter  jusqu'en 1968  en atteignant 5,4
individus par m$^2$  et celles de localité B  jusqu'en 1969 avec 2,2 exem-
plaires par m$^2$ comme le maximum. (Cette différence  est la conséquence
possible de la plus ample application des insecticides sur les terrains
de localité B.)

En comparant les populations de deux familles dans les cultures de
betterave sucrière  on peut remarquer  une coïncidence - l'augmentation
ou la diminution des populations de Carabidae sont toujours suivies par
le même phénomène de celles d'*Elateridae*.  Ce rapport entre ces populations on peut considérer comme celui  de l'ennemi et de sa victime,mais
il est sans doute troublé  par  d'autres facteurs  du milieu extérieur,
anthropologiques notamment.  Parmi ceux-ci on pense en premier lieu aux
interventions chimiques.  Ainsi la diminution  des populations de *Carabidae*  dans les emblavures de betterave sucrière de localité A  en 1969
et 1970 a été provoqué par l'application des insecticides.  Dans ce cas
il s'agissait d'un traitement du sol par lindane contre des fils de fer
au début même de la végétation.  Si l'on prend en considération le taux
de lindane employé (100 kg avec 1,2 % de m.s. à l'hectare)  et sa haute
persistence, le résultat à l'égard de *Carabidae* est bien évident. L'application  des insecticides  contre  des fils  de fer étant souvent une
mesure  indispensable  sur les terrains examinés,  on doit tenir compte
de la méthode du travail au point de vue biologique  et économique à la
fois.  On l'a confirmé par les résultats de même localité  en 1971 déjà
où le traitement par thimet a été effectué en rangs de semis  de betterave sucrière.  De cette façon  on a enrégistré de nouveau,  malgré  le
traitement chimique, une progression des populations de *Carabidae*.

Quelle sera l'importance du traitement chimique  à l'égard des populations  de *Carabidae,*  effectué au début  de la végétation dans les
emblavures de betterave sucrière  contre de divers charançons par exemple,  dépend  de la superficie englobée  et de l'insecticide appliqué.
Ainsi en 1966  et  1967 toutes les emblavures  de betterave sucrière de
localité A  ont été traitées par l'huile de dieldrine.  On suppose que
c'était la cause principale y ayant diminué la densité  des populations
de *Carabidae*.  Dans d'autres cultures sarclées de cette localité on n'a
pas appliqué  du tout d'insecticides  de 1966  jusqu'en 1968.  C'était
aussi une des raisons possibles, avec l'irrigation, influençant l'abondance  des insectes  dans les cultures de blé, notamment celle des fils
de fer.

Quant aux agrobiocoenoses préférables aux *Elateridae,* il est connu
que ce sont les champs des céréales, selon de divers auteurs (Ghilarov,
1941;  d'Aguilar,  1962.;  Vukasović,  1964 etc.).  Mais, selon d'autres

(Scherney et Skuhravý cit. in Tischler, 1965), ces champs sont préférables aussi aux *Carabidae* par rapport aux cultures sarclées. Néanmoins, la situation à cet égard est tout différente dans nos localités de recherche. Elle peut, parait-il, être due aux migrations des adultes effectuées dans l'intervalle de deux mois s'écoulant entre la moisson des blés et notre prélèvement des échantillons de sol. Il sera intéressant de suivre dorénavant l'état des populations de *Carabidae* dans des cultures de blé après la moisson parce que l'on a commencé à brûler leurs restes ce qui va certainement diminuer davantage le nombre de ces insectes. Ce sont les données sur le pourcentage des larves et des adultes en populations de *Carabidae*, présentées sur le graph. 2, nous donnant une idée sur les migrations mentionnées. Comme on peut y bien voir c'est presque régulièrement dans les champs de betterave sucrière

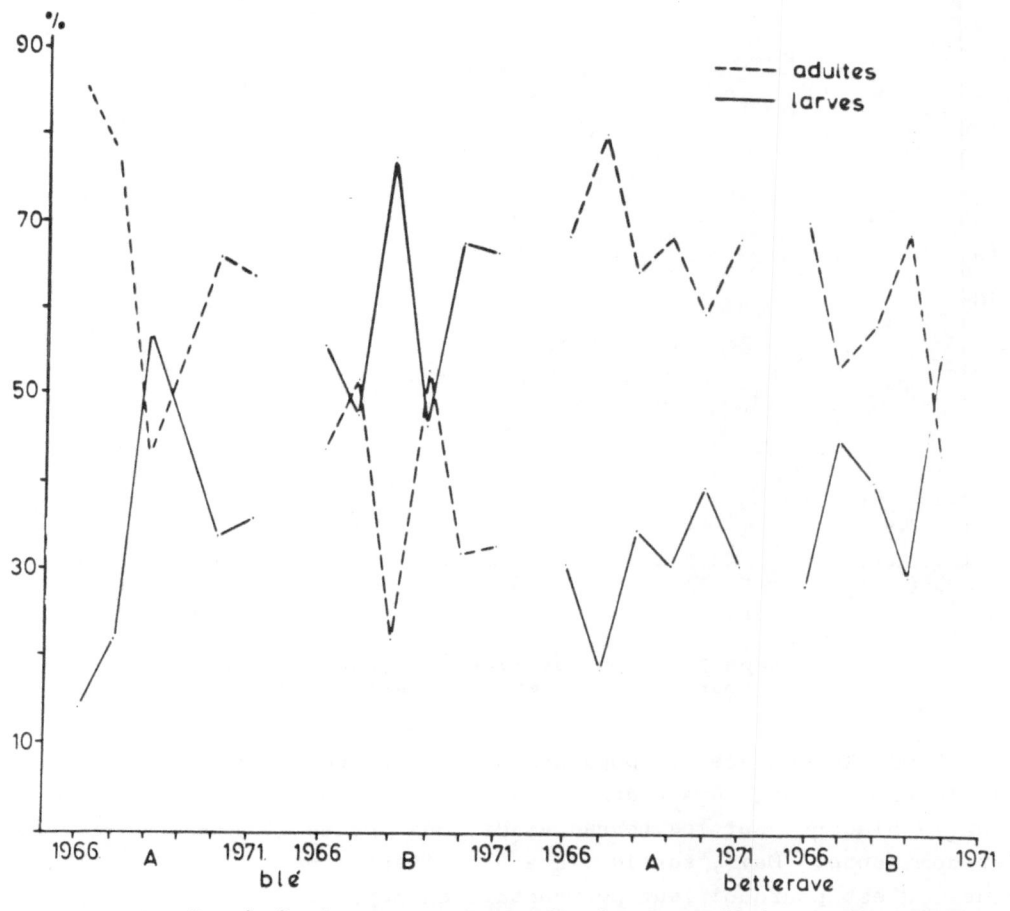

Graph 2. Le pourcentage des larves (4.) et des adultes
(----) en populations de *Carabidae*.

où dominent les adultes. Les larves sont habituellement dominantes dans les champs de blé, mais quelquefois les adultes y dominent aussi. Ce phénomène étant remarqué sur les terrains irrigués, on peut supposer qu'il s'agisse, grâce à l'humidité plus favorable, de leurs migrations moins intenses.

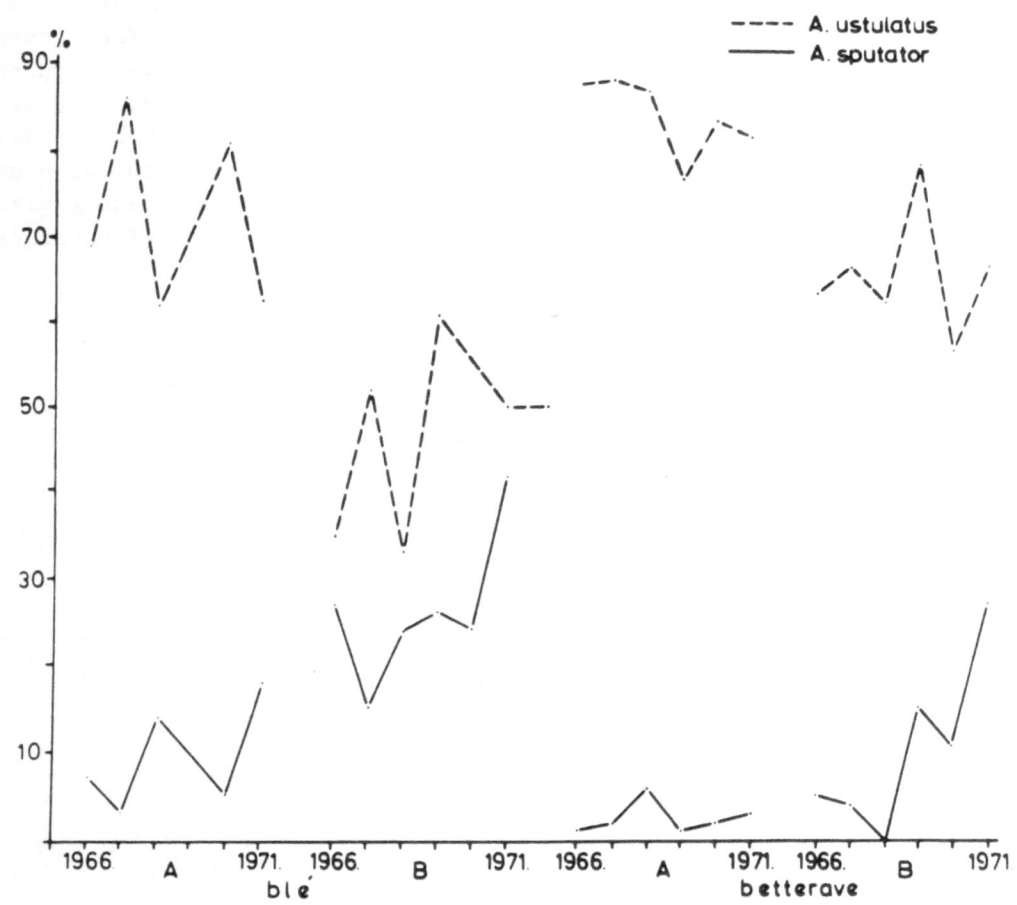

Graph 3. Le pourcentage d'*Agriotes ustulatus* (----) et *A. sputator* (4.) en populations d'*Elateridae*.

Parmi les espèces en populations d'*Elateridae* (graph. 3) *Agriotes ustulatus* domine, étant donné que les terrains examinés se trouvent pour la plupart sur les terrasses de loess où en général cette espèce est dominante. Mais, sur le plateau de loess c'est *A. sputator* qui domine. C'est pourquoi leur pourcentage en populations d'*Elateridae* dépend en premier lieu de la proximité de ce plateau (Djurkić, 1973). En outre, les différences existant entre les emblavures de blé et celles de betterave sucrière montrent que *A. sputator* parait plus sensible au changement d'agrobiocoenoses dans cette région. Aussi elle était toujours moins nombreuse dans les emblavures de betterave sucrière.

On n'a que commencé à déterminer les genres et les espèces de *Carabidae*. Selon les données à notre disposition jusqu'à présent, c'est le genre *Harpalus* qui domine sur les terrains de chernozem avec 68,8 %. Les espèces les plus abondantes de ce genre sont *H. calceatus, H. rufipes* et *H. distinguendus*. En second lieu est le genre *Pterostichus*, avec 15,9 % et avec l'espèce la plus nombreuse *P. vulgaris*. Parmi de plus importants on peut mentionner encore le genre *Amara*, avec l'espèce *A. eurinota*.

Bibliographie

Balachowsky, A. S., D'Aguilar, J. (1962): Entomologie appliquée à l'agriculture. Tome I, Coléoptères, Premier volume, Paris, 204 - 233.

Ghilarov, M. S. (1941): Über die Lokalisation der Schnellkäfer *A. lineatus* und *A. obscurus* während der Eiablageperiode auf den Fruchtfolgefeldern. C. R. Acad. Sci. U.R.S.S., 31, 725 - 728.

Djurkić, J. (1973): Observations sur les populations d'*Elateridae* dans le sol après la moisson des blés. Agriculture contemporaine 1, Novi Sad, 69 - 77.

Tischler, W. (1965): Agrarökologie, Jena, 156 - 167.

Vukasović, P. et al. (1964): Contribution à la connaissance de l'entomofaune coléoptère dans les terres labourables, luzernières, prairies et pâturages en Volvodina. Rev. Mat. Srpska Sci. Nat. 27, 84 - 100. Novi Sad.

Discussion

J. d'Aguilar : Y a-t-il une véritable relation entre les populations de *Carabidae* et les *Elateridae?* En effet les Harpalus, qui sont abondants, sont omnivores et pas seulement carnivores. D'autre part avez vous tenu compte des traitements herbicides?

R. Sekulić : Cette relation est à l'étude. Il n'a pas été tenu compte des traitements herbicides.

# Soil biological features
# of some alpine grasslands
# in Czechoslovakia

J. RUSEK, B. ÚHELOVÁ, J. UNAR

Entomological Institute of the Czechoslovak Academy of the Sciences,
Prague;

Botanical Institute of the Czechoslovak Academy of the Sciences,
Brno

The soil biological characteristics were studied of a number of plant
communities in the region of alpine meadows in the Tomanova dolina valley (West Tatra Mts.) Czechoslovakia, in 1969 - 1971. The main aim was
to study the relationships and correlations among the plant associations, soil types, the composition of soil arthropod fauna and the main
physiological groups of soil microflora. Also the dependence of biotic
components on different abiotic factors was considered. The dynamics
of soil arthropod fauna and soil microflora counts and activities were
investigated. This contribution, although containing only some of the
results, clearly demonstrates the qualitative as well as quantitative
diversity of the alpine biotopes.

THE CHARACTERISTICS OF THE AREA

The area under study is situated in the West Tatra Mts. forming a part
of the Tomanova dolina valley, the Tatra National Park. The Tomanova
dolina valley itself had been declared a strictly protected area. Its
length amounts to about 4.5 km and the total area to about 5 $km^2$. The
lowest point corresponds to 1225 m, the highest one (the summit of the
Mount Kresanica) to 2122 m above sea level. The valley is drained by the
Tomanový potok stream and by several small subterranean karst creeks.
A small glacial lake and some smaller bogs are situated at 1592 m ASL
in the upper part of the valley basin.

    The geology of the area is characterized by acidic rocks (granite,
pegmanites) south of the Tomanový potok stream and by limestone with
smaller outcrops of werfenic shales in the vicinity of the stream forming the northern part. The underlying rocks determine the soil types.
Ranker´s, brown soils, podzols and boggy soils are formed on the acidic
rocks and different types of mountainous rendzinas on the limestone.

The climate of the whole area  is of the cold alpine type  with an average annual temperature 0 $^{\circ}$C  and  average July temperature of 10 $^{\circ}$C and  with  yearly  precipitations  of 1750 mm  (average for 1951 - 1960; Fig. 1).  Frequent occurrence of strong winds blowing over the mountain ridge is typical.

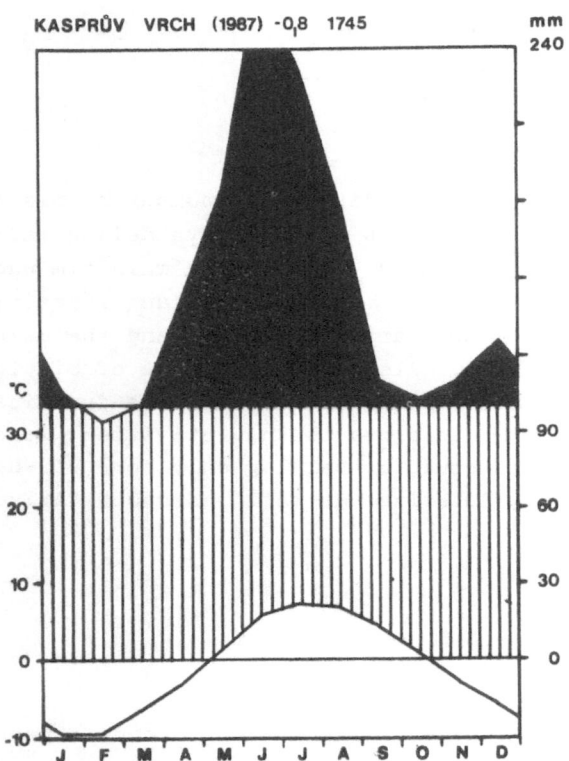

The flora of the Tomanova dolina valley is very rich and more than 500 species of vascular plants  have been found here. Phytosociological studies resulted in the identification and mapping of more than 50 plant communities. The upper boundary of the spruce forest is situated at the elevation of about 1400 m close to the mouth of the valley. The zone of dwarf pine  with the dominant  *Pinus mugo* Turra subsp. *pumilio* extends above the forest zone, forming a more or less compact  belt to the elevation of 1700 m.  Only islands  of dwarf pine  are present further up, the rest being covered by grassland biomass.

Very rich and strongly developed ecological differentiation of the Tomanova dolana valley makes this area highly preferential to the study of morphological and functional diversity of alpine biotopes.

PLANT COMMUNITIES

Typical examples  of selected localities  marked on two crossections of
the valley in Fig. 2 and 3  are described.  Five localities were chosen
in the area with underlying granite:

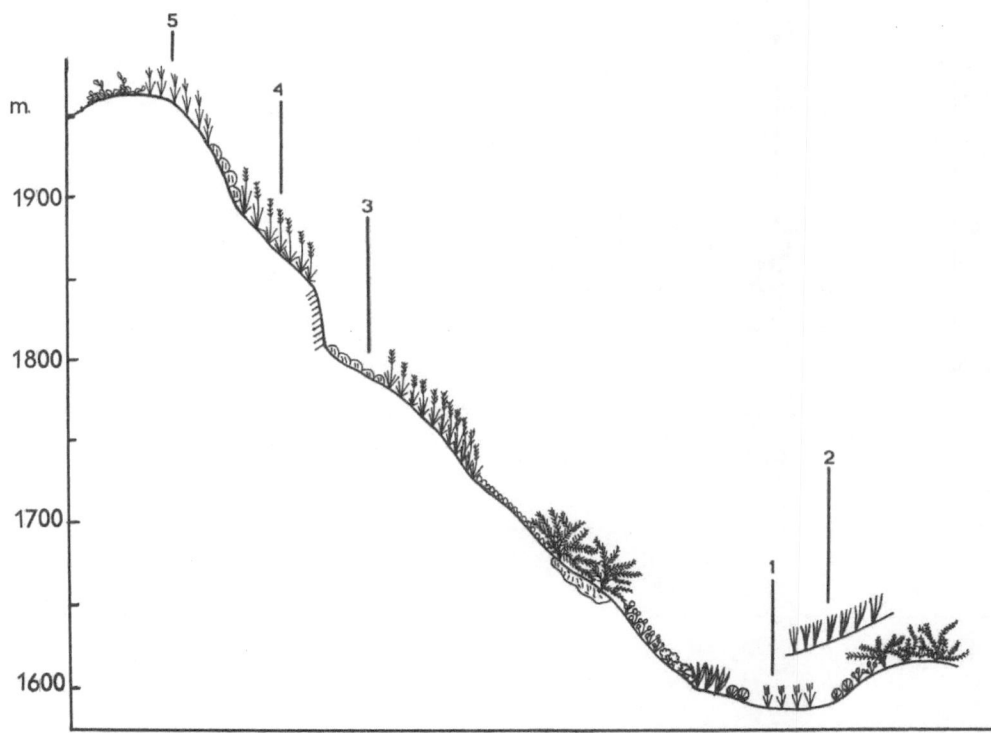

     1. The bottom of a former mountainous lake under the Tomanové sed-
lo pass.  The locality is 1592 m above sea level.  Its character is determined on the one side by the short time elapsed since its drying out
and, on the other side, by the thick layer of sediments, being now gradually eroded and transported away. High water content of the whole soil
profile presents another important ecological factor. The vegetation is
constituted  by isolated tufts  of *Eriophorum angustifolium*,  *Carex ca-
nescens*, and *Carex rostrata*, occasionally joined by *Deschampsia caespi-
tosa*. The soil type according to Kubiёna (1950)  is a  "Torfmoor". Some
properties of all soil types are given in Table 1.

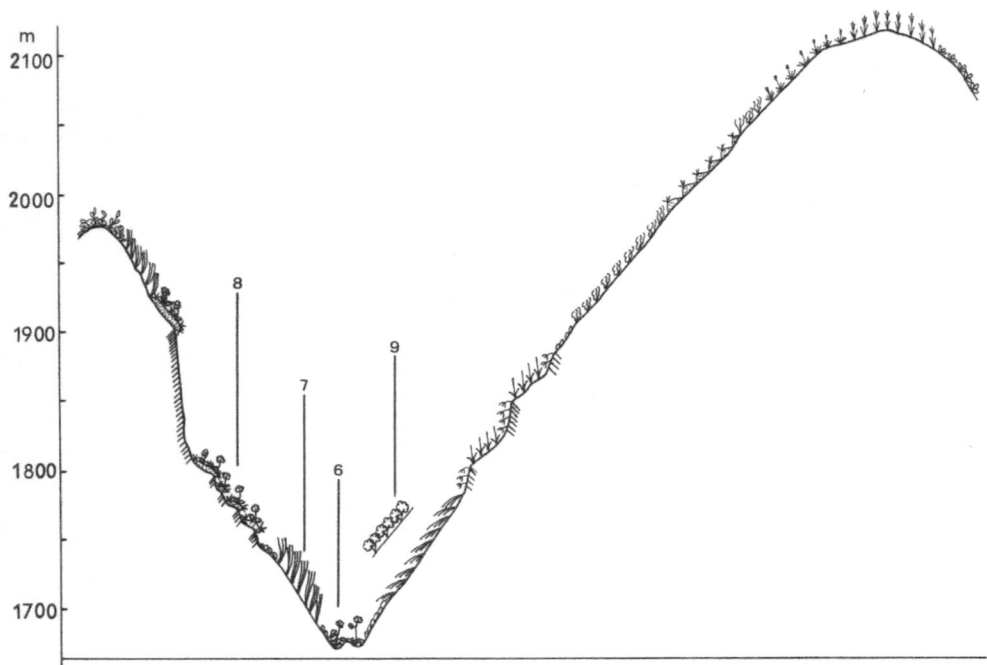

T a b l e   1.    Some soil properties  of  the  plant communities   under
study

| Plant community | Ca$^{++}$ mg.10$^{-2}$g | C % | N % | C : N | pH |
|---|---|---|---|---|---|
| 1 | 30 | - | 1.6 | - | 4.7 |
| 2 | 3 | 6.0 | 0.175 | 34 : 1 | 4.7 |
| 3 | - | - | - | - | 5.1 |
| 4 | - | - | - | - | 4.7 |
| 5 | 20 | 22.5 | 1.4 | 16 : 1 | 4.8 |
| 6 | 350 | 24.4 | 1.4 | 17 : 1 | 7.5 |
| 7 | 240 | 20.5 | 1.29 | 16 : 1 | 7.2 |
| 8 | 210 | - | 1.01 | - | 8.5 |
| 9 | 80 | 6.37 | 0.49 | 13 : 1 | 6.4 |

2. *Festucetum pietae* Krajina 1933

The occurrence of this plant community  is bound to shallow depressions
on the slopes, with a long lasting snow cover, and the resulting higher
soil moisture throughout the vegetative period. Typical examples are to

be found on the mild eastern slope  below the Tomanove sedlo pass (1590
to 1650 m ASL) on shale. The soil is thicker here than 0.5 m and it belongs to the "Alpine Rasenbraunerde".

3. *Sphagno - Vaccinietum myrtilli* Sillinger 1933

This conspicuous plant community occupies the slopes beneath the ridges
of the mountains  and  also on elevations of the northern slopes of the
Mount Polská Tomanová and  the Mount Liptovská Tomanová (1770 to 1950 m
ASL).   Sparse plants cover only about 50 per cent  of the soil surface,
the most important species being *Vaccinium myrtillus*  and *Vaccinium vi-
tis-idaea*. A greater part of their stems often protrudes through a dense,
up to 60 cm thick,  mat of *Sphagnum*. The existence  of this plant community  is conditioned mainly . by the northern orientation  and  by the
accompanying climatological and edaphic conditions, such as short duration of sunshine, long lasting snow cover, low rate  of litter decomposition, displacement of plant litter  to depressions  or lower situated
places. Depressed evaporation and the well-known ability of peat-mosses
to retain  water result  in the high moisture  content  throughout  the
whole profile during the year. The soil type is "Tundra Moos".

4. *Calamagrostidetum villosae tatricum* Pawlowski 1928

This plant community is typical of gullies and grooves on noncalcareous
substrates.  Ecological  factors  enabling it  to  establish  are: long
lasting snow cover,  adequate water supply  during the whole vegetative
period, and the accumulation  of small particles of parent rock  and of
not decomposed litter  in the densely closed cover.  In the area  under
study this plant community  is present on the slope of the Mount Polská
Tomanová between 1700 and 1900 m. The dominant species is *Calamagrostis
villosa*, providing  for about 70 per cent of soil cover.  The soil type
is "Alpiner mullastiger Ranker" with a high humus content.

5. *Juncetum trifidi*  (Szafer, Pawlowski et Kulczynski 1923) emend. Krajina 1933

This is a plant community of windward sides of the alpine ridges formed
by acidic rocks. Its occurrence is determined by the climate characterized by strong winds displacing  and drying out the soil, and often directly damaging the plants. Also in winter the snow cover is removed by
wind.  This results  in alternate thawing and freezing of the soil. The
soil type is "Eilagranker".

Common are species such  as *Juncus trifidus*  forming the main part
of the herbage stratum.  The most vigorous species  of the $E_0$ layer  is
*Cetraria islandica*.  The above-described plant community  can be found,
for instance, on the ridge between the Mount Liptovská Tomanová and the
Mount Polská Tomanová (1950 m. ASL).

In the area with underlying limestone the following plant associations have been chosen:

## 6. *Saxifragetum perdurantis* (Szafer et Kulczynski 1927)

This open plant community is developed in places of snow fields.  It is
noted for a short vegetative period. Even during the second half of July
the soil of this plant community is partly covered with snow and frozen.
The association is well developed on the bottom of deep karst sinkholes
and gullies in the small Hvíždalka valley (1650 m ASL). The soil belongs
to the group of snow-field soil types.

## 7. *Carduo glauci - Caricetum tatrorum*  Szafer et al. 1927

This plant community  comprises a high number of species.  It is formed
on sunny calcareous slopes,  covered  by a relatively shallow soil  interspersed by stones.  The soil type is greyish brown rendzina. The respective habitats  are often protected  from wind.  In winter  they are
covered by a deep snow layer, which, however, disappears rather early in
the spring.  Typical appearance provide the tussocks of *Sesleria tatrae*
and *Carex tatrorum*.  Large areas are covered by this plant community on
southern slopes of the Mount Opálená in the Hvíždalka valley.

## 8. *Caricetum firmae carpaticum* (Br.-Bl.) Pawlowski 1956

This is a typical calcicolous,  turf-forming plant community  formed on
sites strongly exposed to wind erosion.  Polygonal soils often occur on
such sites. The plants are adapted to the adverse conditions by reduced
growth.  Tuft and cushion forming species,  such as *Carex firma, Dryas
octopetala,* and *Silene acaulis* are typical. *Caricetum firmae carpaticum*
is a rather rare community  in the area  under  study.  It can be found
in the high situated Hvíždalka valley, between 1750 and 1925 m.

## 9. *Geranio - Alchemilletum crinitae*  Hadač 1969

This plant community, fairly rich in species, forms fully closed stands.
It occupies the gullies on calcareous rocks.  Important ecological fac-

tors are the deep snow cover  and an ample water supply provided by accumulated rainwater  bringing also dissolved nutrients.  The soil type
is  a turf rendzina.  *Geranio' - Alchemilletum crinitae*  is  limited  to
rather  smaller areas.  It is well developed  on eastern slopes  of the
Hvížďalka valley, between 1720 and 1770 m ASL.


SOIL ZOOLOGY


Soil *Arthropoda*  were sampled  by slow drying of soil samples according
to Tullgren.  On each site  10 soil samples  10 cm$^2$ times 5 cm  (depth)
were collected.  The complete experimental material has not been worked
up untill now and therefore the dynamics of soil arthropod fauna is not
dealt with in this contribution.  The classification  has been finished
for apterygots sampled during the first year of study, and here the data
from the July term will be discussed.  The coenological characteristics
is based only  on species  with an abundance  of more than 500 individuals . m$^{-2}$.


DIVERSITY OD *APTERYGOTA* IN THE PLANT COMMUNITIES UNDER STUDY


The total  of 49 forms  of *Apterygota*  have been found  in the selected
plant communities 1 to 9 (Table 2), the abundance of 25 species aurpas-
sing 500 individuals . m$^{-2}$.  When comparing the occurrence  of these 25
species  in plant communities 1 to 9  by means of Jaccard's quotient of
similarity (Table 3) it can be found that only *Festucetum pictae, Calamagrostidetum villosae tatricum*  and  *Juncetum trifidi* are similar (not
identical) so far the species composition of the *Apterygota* is concerned, nevertheless considerable differences exist in the dominance of individual species.  The *Apterygota* from  sites  on calcareous substrate
differ greatly, the corresponding Jaccard quotients remaining low.
     The following *Apterygota* species have been found in more than four
plant communities: *Folsomia quadrioculata*  (in 8 communities), *Onychiurus armatus* (6), *Isotomiella minor* (4) and *Friesea mirabilis* (4).  Ecological factors  do not appear to attain limiting levels for these four
species in the biotopes under study  and therefore they can be referred
to as ubiquites.
     Important species occurring in a single plant community  (and also
known for their ecological specialization  according  to the literature

T a b l e  2.  Abundance of Apterygota on 100 cm²

| Species        No. of loc. | 1 | 2 | 3 | 4 | 5 | 6 | 7 | 8 | 9 |
|---|---|---|---|---|---|---|---|---|---|
| Sminthurides aquaticus | 6 | - | - | - | - | - | - | - | - |
| Sphaeridia pumilis | 5 | - | - | - | - | - | 1 | - | - |
| Onychiurus armatus | 3 | 10 | - | 108 | 11 | - | 10 | 20 | 23 |
| Isotoma violacea | 2 | 32 | 1 | - | 37 | - | - | 54 | 42 |
| Tetrodontophora bielanensis | 1 | - | - | - | - | - | - | - | - |
| Folsomia quadrioculata | - | 165 | 6 | 391 | 27 | 24 | 48 | 54 | 42 |
| Tetracanthella arctica | - | 139 | 4 | 11 | 169 | - | - | - | - |
| Hypogastrura parva | - | 22 | - | - | - | - | 1 | - | - |
| Isotomiella minor | - | 10 | - | 8 | 16 | 5 | 1 | 2 | 2 |
| Isotoma notabilis | - | 8 | - | 17 | 3 | - | 2 | - | 2 |
| Lepidocyrtus lignorum | - | 4 | - | 15 | 32 | - | 3 | 2 | 3 |
| Friesea mirabilis | - | 2 | 16 | 2 | 13 | 3 | - | 5 | 5 |
| Tomocerus flavescens | - | 1 | - | - | - | - | - | - | - |
| Pseudanurophorus binoculatus | - | 1 | - | - | - | - | 2 | 14 | - |
| Proisitoma recta | - | - | 17 | - | - | - | - | - | - |
| Isotoma olivacea f. neglecta | - | - | 5 | - | - | - | - | - | - |
| Mesaphorura sensibilis | - | - | 1 | - | - | - | - | - | 5 |
| Ceratophysella sigillata | - | - | 1 | - | - | - | 1 | - | - |
| Tomocerus minor | - | - | 1 | - | 2 | - | 2 | 14 | - |
| Folsomia multiseta | - | - | - | 239 | - | - | - | - | - |
| Pseudisotoma monochaeta | - | - | - | 13 | - | - | - | - | - |
| Folsomia sensibilis | - | - | - | 8 | - | - | - | - | - |
| Ceratophysella granulata | - | - | - | 4 | - | - | - | - | - |
| Metaphorura bipartita | - | - | - | 1 | - | - | - | - | - |

Table 2 continued

| Species | No. of loc. 1 | 2 | 3 | 4 | 5 | 6 | 7 | 8 | 9 |
|---|---|---|---|---|---|---|---|---|---|
| Acerentomon sp. | – | – | – | – | – | – | – | – | 1 |
| Entomobrya juv. | – | – | – | – | 2 | – | – | – | 1 |
| Isotoma fennica | – | – | – | – | 21 | – | – | – | – |
| Ceratophysella succinea | – | – | – | – | 1 | 6 | – | – | – |
| Paratullbergia callipygos | – | – | – | – | – | 3 | – | – | – |
| Stenaphorura quadrispina | – | – | – | – | – | 1 | – | – | – |
| Anurida juv. | – | – | – | – | – | 1 | – | – | – |
| Tomocerus juv. | – | – | – | – | – | 1 | – | – | – |
| Pseudosinella zygophora | – | – | – | – | – | – | 9 | – | – |
| Mesaphorura sylvatica | – | – | – | – | – | – | 8 | 2 | – |
| Eosentomon bohemicum | – | – | – | – | – | – | 3 | – | – |
| Eosentomon larva I | – | 1 | – | – | – | – | – | – | – |
| Pseudachorutes subcrassus | – | – | – | – | – | – | 2 | – | – |
| Pseudosinella alba | – | – | – | – | – | – | 2 | – | – |
| Willemia intermedia | – | – | – | – | – | – | 2 | – | – |
| Megalothorax minimus | – | – | – | – | – | – | 1 | – | – |
| Isotoma sp. juv. | – | – | – | – | – | – | – | 8 | – |
| Folsomia alpina | – | – | – | – | – | – | – | 7 | – |
| Ceratophysella armata | – | – | – | – | – | – | – | 3 | – |
| Lepidocyrtus paradoxus | – | – | – | – | – | – | – | 1 | – |
| Micranurida forsslundi | – | – | – | – | – | – | – | 1 | – |
| Pseudachorutes asigillatus | – | – | – | – | – | – | – | 1 | – |
| Neanura sp. | – | 1 | 1 | – | – | – | – | – | – |
| Ceratophysella engadinensis | – | – | – | – | – | – | – | – | 32 |
| Willemia anophthalma | – | – | – | – | – | – | – | – | 4 |
| Suma | 17 | 396 | 53 | 817 | 334 | 44 | 98 | 128 | 131 |

ex.100cm$^{-2}$

T a b l e  3.    Index of Jaccard for species with abundance    500 indiv. on $m^2$

| 1 | 2 | 3 | 4 | 5 | 6 | 7 | 8 | 9 | |
|---|---|---|---|---|---|---|---|---|---|
| 100 | 0 | 0 | 0 | 0 | 0 | 0 | 0 | 0 | 1 |
|  | 100 | 12 | 83 | 100 | 33 | 28 | 20 | 37 | 2 |
|  |  | 100 | 9 | 25 | 20 | 16 | 28 | 28 | 3 |
|  |  |  | 100 | 71 | 25 | 22 | 17 | 17 | 4 |
|  |  |  |  | 100 | 28 | 25 | 33 | 57 | 5 |
|  |  |  |  |  | 100 | 20 | 12 | 12 | 6 |
|  |  |  |  |  |  | 100 | 28 | 28 | 7 |
|  |  |  |  |  |  |  | 100 | 37 | 8 |
|  |  |  |  |  |  |  |  | 100 | 9 |

as well as to our own previous records) are listed below: *Hypogastrura parva* occurring in *Festucetum pictae*, *Proisotoma recta* in *Sphagno-Vaccinietum myrtilli*, *Folsomia multiseta*, *F. sensibilis* and *Pseudisotoma monochaeta* in *Calamagrostidetum villosae tatricum*, *Ceratophysella succinea* in *Saxifragetum perdurantis*, *Pseudosinella zygophora* and *Mesaphorura sylvatica* in *Carduo glauci - Caricetum tatrorum*, *Pseudanurophorus binoculatus*, *Tomocerus minor* and *Folsomia alpina* in *Caricetum firmae carpaticum* and *Ceratophysella engadinensis* with *Eosentomon bohemicum* in *Geranio-Alchemilletum crinitae*. *Tetracanthella arctica* is typical of the mountain top  linked plant association *Juncetum trifidi* and for *Festucetum pictae*, reaching here the dominance of 50  and  35 per cent, respectively.  The number of *Apterygota* species occurring in individual plant communities is presented in Table 2. The greatest number of *Apterygota* species  with an abundance exceeding 500 individuals . $m^2$ are present  in  *Calamagrostidetum villosae*.  The joint occurrence  of species common to localities of lower altitudes  *(Folsomia multiseta)*  with the alpine type species  *(Folsomia sensibilis, Pseudisotoma monochaeta)*  is interesting.  This  corresponds  to the very rich floristic composition due to joint occurrence  of plant species  typical  of lower  altitudes with plants of alpine type (Jeník, 1961).

Plant communities  of  extreme type, i.e. *Sphagno-Vaccinietum myrtilli, Saxifragetum perdurantis, Carduo glauci - Caricetum tatrorum* and that with the dominant  *Eriophorum angustifolium*,  are the poorest with respect to the number of species.

The abundance of *Apterygota*
in plant communities under study

The abundance of *Apterygota* in sites 1 to 9 is given in Fig. 4. The highest abundance is seen to be connected with *Calamagrostidetum villosae tatricum* (81 700 individuals . m$^{-2}$). A relatively high abundance

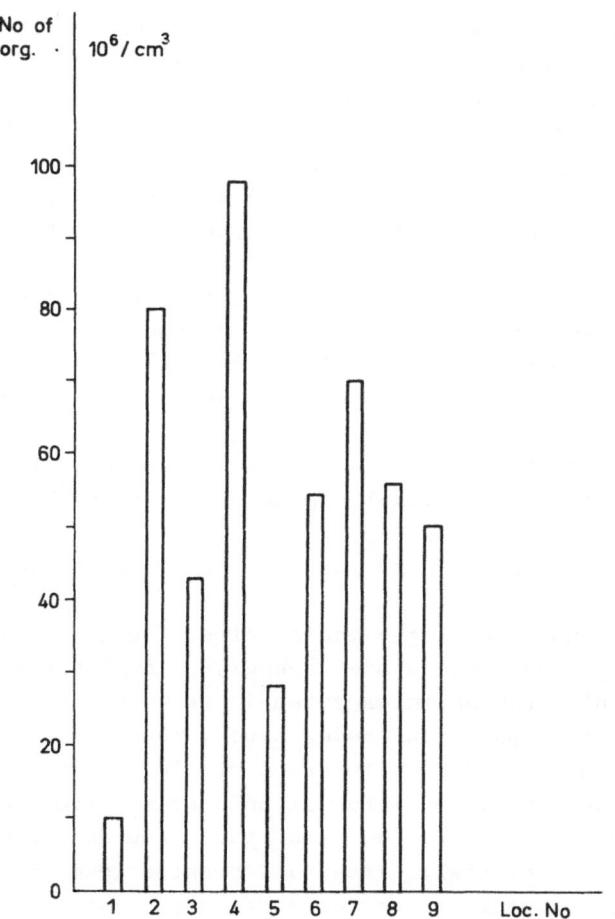

characterizes also the plant communities *Festucetum pictae* (39 600 individuals . m$^{-2}$) and *Juncetum trifidi* (33 400 individuals . m$^{-2}$). The lowest abundance has been found in the site with dominant *Eriophorum angustifolium* (1 700 individuals . m$^{-2}$). The abundance in the remaining five plant communities was also very low and exceeded 10 000 individuals . m$^{-2}$ in two cases only.

SOIL MICROBIOLOGY

Three methods were used for the study of soil microflora and its activities:

1. Soil samples were collected in monthly to six weeks intervals in selected biotopes during 1969 - 1971. Soil microbial counts were estimated using the plate count technique. Soil suspensions diluted to $10^{-6}$ were inoculated on Taylor (TA), Starch (SA), Meat-peptone (MPA) agar media, and diluted to $10^{-3}$ on Jensen (JA) and Ashby (AA) agar media.

2. The rate of cellulose decomposition was measured under natural conditions. Filter paper squares 1 $dm^2$ of size, placed in silon mesh bags, were exposed in five replicates on each habitat, 2 cm bellow the soil surface for about six months. Details of the method are given in Tesařová et Úlehlová (1968). This simple method makes it possible to compare the cellulolytic activities in different plant communities during the vegetative period.

3. The $CO_2$ production of soil samples from plant communities under study was measured under laboratory conditions using the method of Steubing (1965). The rates of potential microbial activities in soils from individual biotopes are compared by this method under optimal conditions.

Using these three methods we tried to obtain information about the composition and activities of soil microbial populations in the biotopes under study.

The preliminary results presented here pertain to the year 1969 only.

The counts on different nutrient agar media of soil microflora from the nine plant communities described above are given in Table 4.

Data contained in the table can be interpreted in terms of diversity and similarity of microbial populations under study. The microbial populations can be considered as homogeneous in cases of similar microbial counts on different nutrient media. Microbial population from the biotope with dominant *Eriophorum angustifolium* and that from *Juncetum trifidi* can serve as examples. Conversely, great differences in microbial counts on different nutrient media can be considered as indicative of high diversity of the respective microbial populations. The microbial populations from *Festucetum pictae* and *Calamagrostidetum villosae tatricum* can be quoted as examples. It can be also seen from the Table that soil microbial populations from the sites on calcareous substrate are more alike than those from the sites on acidic rocks.

Both the highly productive plant communities *Calamagrostidetum villosae tatricum* and *Carduo glauci - Caricetum tatrorum* posses the

T a b l e   4.    Microflora counts on nutrient agar media for soil samples from different plant communities

| Plant community | Nutrient agar medium | | | | |
|---|---|---|---|---|---|
| | MPA | SA | TA | JA | AA |
| | Number of org.$10^6$.cm$^{-3}$ | | | Number of org.$10^2$.cm$^{-3}$ | |
| 1 | 11 | 10 | 11 | 26 | 58 |
| 2 | 26 | 55 | 80 | 92 | 216 |
| 3 | 9 | 43 | 32 | 200 | 83 |
| 4 | 43 | 98 | 55 | 177 | 220 |
| 5 | 21 | 28 | 22 | 75 | 120 |
| 6 | 23 | 54 | 45 | 80 | 156 |
| 7 | 35 | 70 | 66 | 100 | 181 |
| 8 | 25 | 56 | 41 | 50 | 190 |
| 9 | 31 | 50 | 37 | 120 | 210 |

most diversified microbial population of the biotopes on the respective
geological substrates.   This points to a certain analogy  regarding the
structure, function and siting of the two plant communities.

The highest counts  of microorganisms  found  on all  the nutrient
media during 1969  are presented  as abundances  in Fig. 5.   The figure
shows  that  the microbial populations  of  *Calamagrostidetum  villosae
tatricum* and *Festucetum pictae*  (biot. 4 and 2)  were the most abundant
and that of the biotope No 1 with dominant *Eriophorum angustifolium* the
least  abundant.   A higher uniformity  of the sites  on calcareous substrate, regarding the abundance, is also shown in the Figure.

A comparison  of the data  on  abundance  of microbial populations
(Fig. 5)  with  those  on abundance  of *Apterygota*  (Fig. 4)  indicates
a rather close relationship between the two groups of organisms.

Table 5  presents  information  on the potentital microbial activities  of the soils  based  on data on $CO_2$ production  by  soil samples
under laboratory conditions,  as well  as information  on the cellulose
decomposition  under field conditions.   The data indicate that a correlation exists between the activities and microbial counts. The cellulose
decomposition  appears to be higher  on acidic substrates,  although it
can  be  occasionally  decreased  by other limiting factors, as e.g. in
*Juncetum trifidi*.   Somewhat higher respiration rate appears to be found
in samples from the sites on limestone.

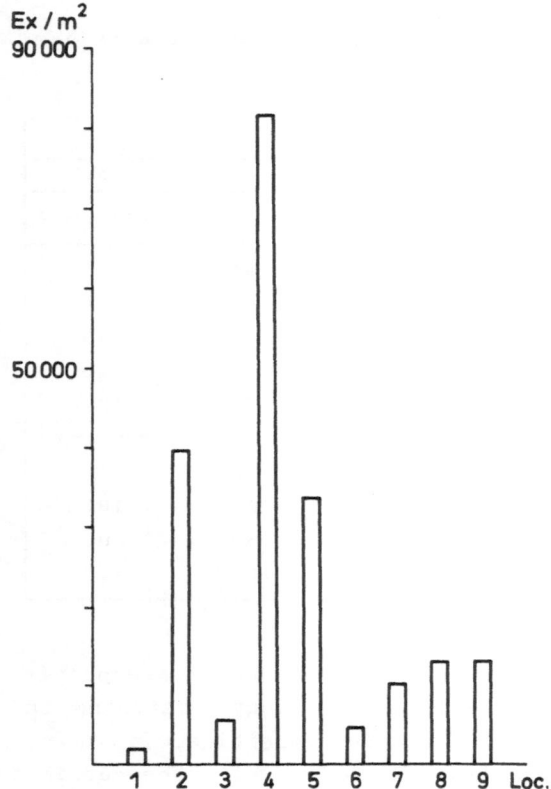

T a b l e   5.   Microbial activities in soils of plant communities under study

| Plant community | Soil moisture | $CO_2$ production $(mg.10^{-2}g.h^{-1})$ | Cellulose decomposition $(mg.g^{-1}.day^{-1})$ |
|---|---|---|---|
| 1. | 76.0 | 0.48 | 0.00 |
| 2. | 43.0 | 0.63 | 0.40 |
| 3. | 70.7 | 0.50 | 0.25 |
| 4. | 42.5 | 1.10 | 0.65 |
| 5. | 46.0 | 0.63 | 0.17 |
| 6. | 65.0 | – | 0.20 |
| 7. | 65.5 | 1.40 | 0.25 |
| 8. | 64.0 | 0.68 | 0.10 |
| 9. | 59.0 | 0.76 | – |

CONCLUSIONS

1. Nine selected alpine biotopes from the West Tatra Mts. are described as developed (a) on acidic substrate (the biotype with dominant *Eriophorum angustifolium, Festucetum pictae, Sphagno - Vaccinietum myrtilli, Calamagrostidetum villosae tatricum, Juncetum trifidi*) and (b) on calcareous substrate *(Saxifragetum perdurantis, Carduo glauci - Caricetum tatrorum, Caricetum firmae carpaticum, Geranio-Alchemilletum crinitae)*.

2. Preliminary results of the studies on *Apterygota* populations and on microbial populations of the respective biotypes are presented.

3. Three plant communities from acidic rocks, i.e., *Festucetum pictae, Calamagrostidetum villosae tatricum* and *Juncetum trifidi*, showed a certain similarity (not identity) as concerns the population of *Apterygota*. No such similarity could be found among the biotypes on limestone.

4. The highest abundance of *Apterygota* has been found in *Calamagrostidetum villosae tatricum* and in *Festucetum pictae*, the lowest one in the biotype with the dominant *Eriophorum angustifolium*.

5. The soil microbial populations of *Calamagrostidetum villosae tatricum* and *Festucetum pictae* are highly diversified, while those of the biotype with *Eriophorum angustifolium* and of *Juncetum trifidi* are rather similar.

6. The highest counts of soil microorganisms have been found in *Calamagrostidetum villosae tatricum* and *Festucetum pictae*, the lowest one in the biotype with *Eriophorum angustifolium*. This indicates that a relationship exists between the abundance of *Apterygota* and that of soil microorganisms.

7. The $CO_2$ production under laboratory conditions of soil samples from different biotypes appears to be related to the respective soil microbial counts. The cellulose decomposition under field conditions depends on the type of microbial population and on the limiting ecological factors.

R e f e r e n c e s

A a r o n s o n , S. (1970): Experimental Microbial Ecology. Academic
    Press, New York, 9, 236 pp.
F j o d o r o v , M. V. (1953): Příručka praktické mikrobiologie. SZN,
    Praha.

J e n í k , J. (1961): Alpinská vegetace Кrkonoš, Kralického Sněžníku a Hrubého Jeseníku. NČSAV, Praha, 409 pp.

K u b i e n a , W. L. (1950): Bestimmungsbuch und Systematik der Böden Europas. Enke-Verlag, Stuttgart, 392 pp.

K ň a z o v i c k y , L. (1970): Západné Tatry. VSAV, Bratislava, 211 pp.

N o s e k , J. (1969): The Investigation on the *Apterygotan* Fauna of the Low Tatras. Acta Univ. Carolinae, Biol., Praha, (5-6), 349 - 528.

S t e u b i n g , L o r e (1965): Pflanzenökologisches Praktikum. Methoden und Geräte zur Bestimmung wichtiger Standortsfaktoren. P. Parey, Berlin, 262 pp.

T e s a ř o v á , M., Ú l e h l o v á , B. (1968): Abbau der Zellulose unter einigen Wiesengesellschaften. Tagungsber., Nr.98, Mineralisation der Zellulose, DAL, Berlin, 377 - 388.

## D i s c u s s i o n

W. D u n g e r : Sie haben den Jaccard-Index zum Vergleich von Collembolen-Faunulae verwendet. Nach meiner Erfahrung ergeben viele gleichberechtigte Indizes sehr verschiedene Resultate. Welchen Wert messen Sie der Verwendung des Jaccard-Indexes bei?

J. R u s e k : Der Jaccard Index wird in dieser Arbeit nur für die Arten mit Abundanz $\geq$ berechnet. Man darf diesen Index nicht strikt verwenden, man muss mit einer Variabilität rechnen. Bei der Schlussauswertung des Gesamtmaterials werden auch andere Indices verwendet.

C. G r e g o i r e - W i b o : You notice a relationship between the abundance of *Collembola* and soil microorganisms. Is there also a relationship between the number of species of *Collembola* present and microbial activity?

B. Ú l e h l o v á : We did not look for it.

P. B e r t h e t : Les valeurs reprises dans vos tableaux correspondent-elles à des moyennes anuelles et avez-vous des informations concernant un parallélisme éventuel entre l'evolution saisonière de la faune et la production de $CO_2$ ou d'activité microbienne?

B. Ú l e h l o v á : The soil microflora counts are the average values for the vegetative period of 1969; the cellulose decomposition rates are average values for the whole year 1969 (winter and vegetative pe-

riod).  The whole material  has not yet been  fully evaluated;  also up
till now we have no idea about this parallelism.

G. P u g h :  Did your histograms  of microorganisms  refer to bacteria
only or to total microbial counts?  Could you tell us  something  about
your methods for estimating cellulose decomposition?

B. Ú l e h l o v á :  To bacteria only.  We did counts of bacteria, mi-
cromycetes and actinomycetes, but here are presented counts for bacteria
only.  The precise description of the method  is given in the paper: we
exposed 1 dm$^2$ of filter paper square  in silon mesh bog  in habitats in
five replications and measured the disappearence of cellulose.

S. K. S t e b a e v a :  In this report  there  are  many excellent interesting phenomena.  We are studying  the density of *Collembola* in the
high-altitude steppes  of Tuva, Sajan  and  Altai.  In those landscapes
there  are  similar pictures  of the *Collembola* distribution also,  for
example the low density of *Collembola* is registered in bags.  Other interesting phenomena - the low numbers of species  and  their high abundance in extreme conditions. It was a very interesting paper for me.

# Oribatoid mite complexes
# as the soil type bioindicator

D. A. KRIVOLUTSKY

Institute of Animal Evolutionary Morphology and Ecology,
Moscow

The study of the oribatoid mites ecology  and distribution in the soils
of the USSR  represents  a fragment  of pedozoological  investigations,
being carried out  under the guidance of M. S. Ghilarov.  The materials
were collected  in 1960-1972  from the tundra, taiga, mixed  and broad-
leaved forests, forest-steppe, steppe, desert  and  subtropical natural
zones. Samples were extracted in the modified Berlese´s funnels.  There
were some aspects of our investigations.

Ve r t i c a l   m i g r a t i o n s   of mites. There is close
correlation of the vertical distribution of oribatids, humus  and  root
systems in the soil profile (Fig.1). Mites can inhabit all parts of the
soil profile  to the depth 100-200 cm,  but their vertical distribution
varies widely at different seasons. For example, the vertical distribution of oribatids  in  the soddy-podzol, podzol  and  brown forest soil
types near Moscow differ especially in the winter,  as a result of different depth of the hard frozen soil horizon.  Oribatoid mites  inhabit
the forest litter  and all parts of the soil profile at a depth 115 cm,
in the soddy-podzol soil type of Piceetum-myrtillosum. There were daily
and seasonal migrations of the mites  in upper soil layer (0-10 cm)  at
all the seasons,  exept winter,  when  the soil  was frozen.  The mites
migrate in the inner parts of soil profile (10-115 cm)  at all the seasons  during  the periods  of long weather change.  They can react upon
a temperature gradient as low as 0.1 °C.

Some regularities  in the  z o n a l   d i s t r i b u t i o n  of
mites.  The oribatoid  mite population den. ity  is very low,  mites in-
habite only the litter  and upper soil part (0-10 cm).  In subarctic or
tundra zone  the population density  reaches 10000-20000 spec./m$^2$, they
penetrate in the soil to 5-30 cm depth. The population density in taiga,
where the podzol soils predominate, reaches 60000 -- 100000 spec./m$^2$. The
majority  of  the mites concentrate in the litter  and upper soil layer
(0 -- 10 cm), but many of the mites penetrate to 20-40 cm.  In chernozems

number of oribatids (in hundred of specimens per 1 dm²)

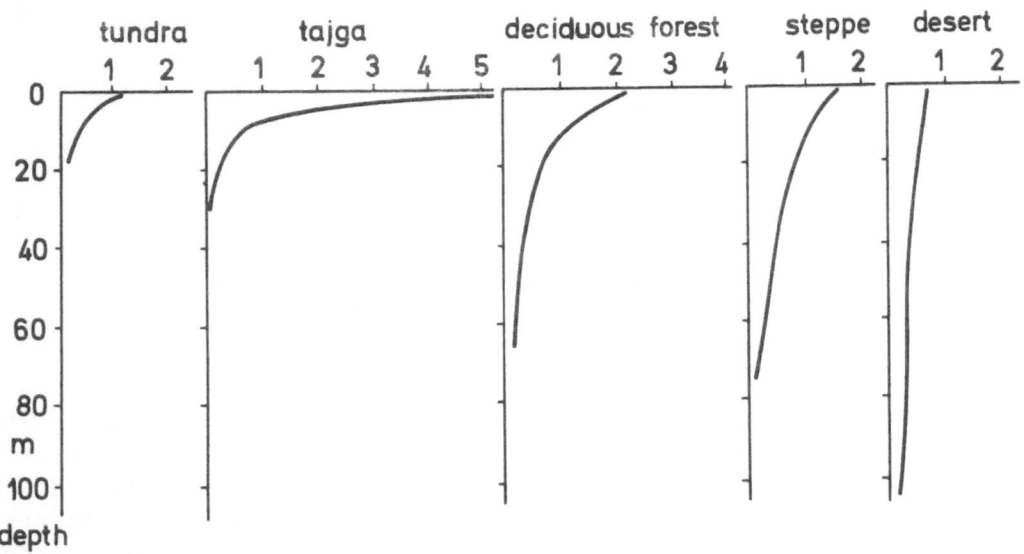

Fig.1. The oribatoid mite   vertical distribution   in the
zonal soil types of the USSR.

under  broad-leaved forests the oribatids penetrate to a depth of 1-2 m
and their population density is 40000 spec./m². The oribatid mite number in the arid territories  decreases gradually  in southern direction
(steppe,  semi-desert,  dry mediterranean regions,  desert), it reaches
20000 spec./m² in North steppe,  5000 in South steppe,  200 spec./m² in
deserts.  Oribatid mites occupy more than 1 m²  and  can migrate in the
inner soil layers in the summer period.  The population density  of the
oribatids in the natural habitats depends  on the soil moisture quality
of the fallen dead plant materials and intensivness of the decomposition
processes. Number of mites can be calculated by the coefficient of suitability (A)  by the formula $A = \frac{F+L}{F \cdot n} RK$. (R - index  of radiation balance,
F - annual amount of the fallen plant material,  K - coefficient of humidity,  L - weight of the litter,  n - number of the annual generation
of oribatids). The close correlation  between "A"  and real population

density of mites  in the natural zones  of the USSR  at the begining of
summer was registered.

Distribution in the  v e r t i c a l  z o n e s  in the mountains.
A study  of the oribatids quantitative distribution  in  different vertical zones  has been made in Terskey-Ala-Too ridge in Tian-Schan mountains in Central Asia. Materials were collected at 500-4000 m above sea
level in each vertical zone (desert, steppe, meadow, forest, subalpine·,
alpine, glacial).

The  oribatoid  mite  population  density  was the highest  in the
forest soils under *Picea schrenkiana*  and the least in the desert soils
(Fig. 2).  According to the present data it is possible to suggest that

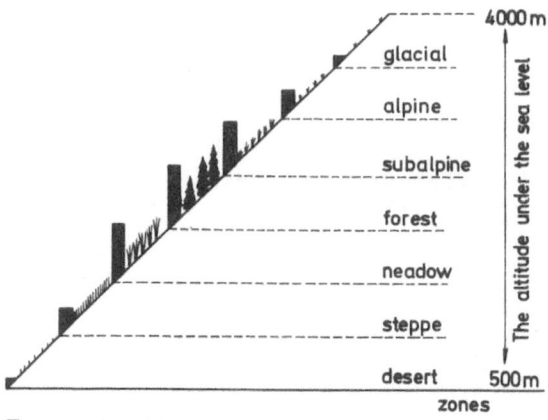

■  the oribatoid mite ponulation density

Fig.2. The oribatoid mite population density in the vertical zones of Tian-Schan.

the regularities of the quantitative distribution of *Oribatei* in vertical zones appears to be analogical to their distribution in the natural
zones of flat countries of Eurasia.

The number  of mites depends on the depth  of the hard frozen soil
horizon (in Eastern Europa),  or on the depth of the warm soil horizont
in the summer time (in northern Siberia)  (Fig. 3), where the other horizons are permanently frozen.

The oribatoid mites  are one  of the  most numerous soil dwellers,
the structure and dynamics of their communities' are considered in relation to soil types.  Phenology peculiarities, population dynamics, seasonal and daily vertical migrations of mites  have typical patterns for
any soil type. There are some faunistical complexes in the oribatid mite

fauna in every region, the ratio of these complexes varies widely and also is one of the indicators of soil properties (Fig. 4).

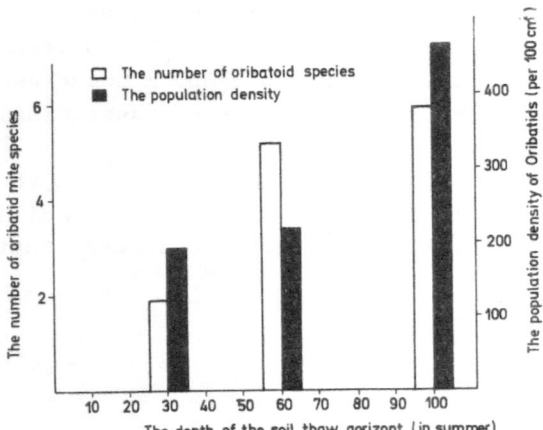

Fig.3. The ratio of the faunistic complexes of oribatoid mites in the forest soils near Moscow.

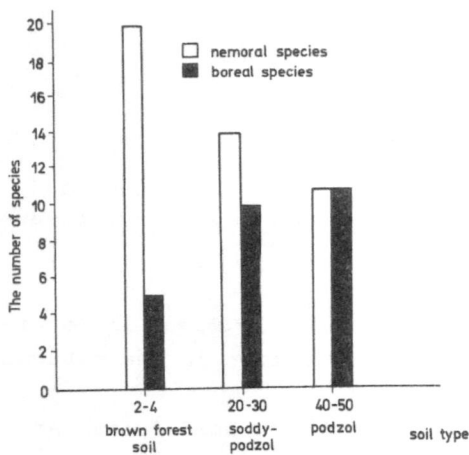

Fig.4. The species variability and number of oribatoid mites in the permanently frozen soils of Northern Siberia.

References

V t o r o v ,  P. P.,  K r i v o l u t s k y ,  D. A. (1968): Oribatoid
    mites of Western Kirgisia. Pedobiologia, Bd. 8, 123-133.
K r i v o l u t s k y ,  D. A. (1968):  Some regularities  in the zonal
    distribution of oribatid mites. Oikos, v. 19, 339-344.

# Seasonal variations of number and biomass of earthworms in grassland

S. TSURU
Fermentation Institute,
Chiba

There are various kinds of soil animals  in surface soils of grassland.
It was shown  that these soil animals  are related to the decomposition
process of cattle dung  and the turn-over of subsoils.  It is said that
the population density  and  the species composition affect  the decomposition rate and the humus formation in the soil. Howeverm it is still
important  to investigate such soil animal activities  establishing the
process of material cycle  in the grassland ecosystem.  There is little
known of such soil animals  in the grassland except for soil nematodes.
This paper  is dealing with the population density and the species composition, and furthermore their seasonal variations in 1966 - 1967.


STUDY AREA


The study area is situated in northern and eastern districts of Hokkaido. These have been developed into grazing land  and pasture during the
last decade  for  a large scale dairy project.  The area  includes many
types of soils, and has been subjected to various management conditions.
The soils  are generally divided  into bog soil,  alluvial soil,  heavy
clay soil and volcanic soil.  For a detailed description of these soils
see Sasaki (1960). Average rainfall is 940 mm, mean temperature 5.8 °C.
     The experimental plots  were chosen  to represent the various soil
types and different management conditions. A brief history is given for
the individual soil types:
     1. Bog soil identified  as  *Sphagnum*  peat,  altitude three meters
above sea level, not cultivated as yet.
     2. Alluvial soil  characterized  as  clay loam soil, altitude  8 m
above sea level.  In 1961, the plot  has been cleared  of Sasa bush and
red clover  and timothy grass  has been cultivated on it;  mineral fer-

T a b l e   1.    Chemical analysis of sampling area

| Surface soil | mois-ture /%/ | pH$_{H_2O}$ | Humus /%/ | C/%/ | N/%/ | C/N | CaO, mg/100g | MgO, mg/100g |
|---|---|---|---|---|---|---|---|---|
| Bog soil (Sphagnum peat) | 84.4 | 4.7 | 55.6 | 32.3 | 2.12 | 15.2 | 16.9 | 12.3 |
| Alluvial soil (clay loam) | 35.0 | 5.2 | 6.4 | 3.7 | 0.23 | 16.3 | 6.3 | 4.4 |
| Heavy clay soil (clay) | 18.4 | 4.9 | 4.8 | 2.8 | 0.16 | 17.6 | 3.4 | 1.9 |
| Volcanic soil (sandy loam) | 40.4 | 5.6 | 2.7 | 1.5 | 0.52 | 12.6 | 14.1 | 1.2 |
| Volcanic soil (sandy loam) | 34.4 | 6.3 | 7.7 | 4.5 | 0.38 | 11.8 | 11.7 | 2.0 |

tilizer (NPK = 30, 120, 40) and calcium carbonate (10 ton/ha) have been applied.

3. Heavy clay soil on a coastal hillside, 20 m above sea level. In 1957 cleared of Sasa bush and planted with red clover and timothy grass for grazing; soil fertilized with NPK = 6, 14, 6 and calcium carbonate (10 ton/ha).

4. Volcanic soil identified as sandy loam, at 20 m above sea level. In 1955, cleared of Japanese oak and Sasa bush and cultivated with oats and maize followed by orchard grass and red clover. Fertilized with composted manure (10 ton/ha) and neutralized with calcium carbonate (4.5 ton/ha).

5. Volcanic soil also identified as sandy loam, at 8 m above sea level. In 1955, cleared of Japanese oak and Sasa bush and cultivated with red clover and timothy at the time of sample collection. At other times, potatoes and oats were grown there. Application of mineral fertilizers, composted manure and calcium carbonate at the same rate as described under 4.

METHODS

For studies of earthworms populations, consisting of five doil repli-
cated samples of 0.25 $m^2$ each obtained from four well-mixed soil blocks
taken to a depth of 25 cm, were collected from eachplot. The pick up
methods were used in handling the annelids. The population density were

T a b l e   2.    Number and biomass in sampling area

|  | number /n/m$^2$/ | biomass /g/m$^2$/ |
|---|---|---|
| Bog soil | O | O |
| Alluvial soil | 25 | 1.75 |
| Heavy clay soil | 92 | 14.60 |
| Volcanic soil | 96 | 21.80 |
| Volcanic soil | 99 | 19.1o |

METHODS

For the study on earthworm populations, five soil samples (0.25 $m^2$ each)
obtained from four well-mixed soil blocks taken to a depth of 25 cm,
were collected from each blocks. The annelids were picked by hand.
Population density was calculated by the total number of annelids obtained from the well-mixed sampling soil blocks. The total annelid
catch was fixed with methanol and their biomass was weighed.

RESULTS AND DISCUSSION

The humus content of the sampling plots was below 10% except for the
bog soil (55%).
     The pH values ranged from 4.7 to 6.3, but did not influence the
population density of' the annelids. This was related to both humus
content and culture conditions.
     Calcium application had no effect on the distribution pattern of
the annelids. It was of interest that soil neutralization with calcium
had no effect on the population density. No distinct differences were
observed in population density in pasture- and arable soils. The spe-

T a b l e   3.    Distribution pattern of Annelids in Hokkaido

| | Bog soil | Alluvial soil | Heavy clay soil | Volcanic soil I | II |
|---|---|---|---|---|---|
| *Allolobophora japonica* | + | − | + | + | + |
| *A. caliginosa* | − | + | + | + | + |
| *Bimastus* sp. | + | − | + | + | + |
| *Dendrobaena* sp. | − | − | + | + | + |
| *Eisenia foetida* | + | + | + | + | + |
| *Pheretima agrestia* | − | − | − | + | + |
| *Ph. divergens* | − | + | + | + | + |
| *Ph. hilgendorfi* | − | + | + | + | + |
| *Ph. irregularis* | − | − | + | + | + |
| *Ph. phaselus* | − | − | + | + | + |
| *Ph. yunoshimensis* | − | − | + | + | + |

cies composition appeared to be rather simple in uncultivated soils
under Sasa vegetation as compared to that of the remaining soils. The
only species isolated from these uncultivated soils was *A. japonica*.
The species composition of grazing and pasture grass soils is characterized by *A. japonica* and *Bimastus* sp. (volcanic soils). Heavy clay
soils are characterized by *Dendrobaena* sp. and *A. caliginosa* isolated
from well-managed grassland soils. Macro-soil animals have been divided
into three groups, i.e. *Annelida, Mollusca, Insecta*. These were animals
with a body length of more than 2 mm.

Evidence has been obtained that population density is dependent
on the seasons. It increases gradually from the spring up to the autumn
when it attains its maximum and decreases again in the winter, being
lowest during the period of snow fall. A similar situation has been
observed for all three groups. Also changes in biomass were dependent
on seasonal disturbances. It is of interest that changes in annelid

T a b l e   4.    Population density $(n/m^2)$ and biomass $(g/m^2)$ of soil
fauna

| | number $(n/m^2)$ | % | biomass $(g/m^2)$ | % |
|---|---|---|---|---|
| *Annelida* | 1.924 | 74.1 | 596.81 | 97.40 |
| *Mollusca* | 20 | 0.4 | 2.10 | 0.03 |
| *Insecta* | 648 | 35.5 | 11.53 | 2.59 |

population density were identical to those observed with all other soil
animals.

The species composition  disclosed a marked dominance of *Allolobophora japonica* and *Pheretima yunoshimensis*

Grassland  soils  harbour various species  of  soil animals,  which
participate  in cycling processes of the material.  However, population
density of these soil animals  is low  in these soils  and  the species
composition  is simple in grassland ecosystems.  It is difficult to say
that the grassland fauna is poorer than the forest fauna.  According to
an annual observation by Nakamura (1968), population density and biomass
of annelids did increase.

Population density  is  influenced  by the seasons.  They increase
in number  from spring to autumn, and decrease again in the winter. The
same applies to the biomass.  The autumnal increase might be attributed
to an increase of hay. Management of grassland (grazing and pasture) is
one of the factors influencing seasonal changes in annelids.

SUMMARY

The present paper  deals with the seasonal variations of the population
density and biomass of annelids in grasslands of Hokkaido.  The population density and biomass  showed seasonal variations depending upon the
sampling periods in May and September of each year.

The main soil animals were divided into three categories, covering
*Annelida, Mollusca* and  *Insecta*.  Each of these groups  showed more or
less peculiar seasonal variations, in most cases the population density
was large in autumn and small in winter.

A c k n o w l e d g e m e n t:

The author is grateful to Dr.Y.Nakamura for his kind information on the
alluvial soil grassland in Hokkaido, and  to Dr.T.Matsuno  for his support in preparing our fields experiments.*


       * This study has been performed in the period of May and September
from 1961 to 1969.

R e f e r e n c e s

Agency of Development. Hokkaido (1960):   Report on grassland conserva-
     tion with soil animals. pp. 1-53.
N a k a m u r a , Y. (1968):  Studies  on  the ecology  of terrestrial
     *Oligochaeta*. appl. Entomol. Zool. 3, 89-95.
S a s a k i , S. (1960): Soil geography in Hokkaido. Japan, pp. 1-221.
T a m u r a , H.,  N a k a m u r a , Y.,  F u j i k a w a , T. and
     Y a m a u c h i , K. (1969):  An ecological survery of soil fauna
     in Hidaka Mombetsu, southern Hokkaido. J. Fac. Sci. Hokkaido Univ.,
     Ser. VI, Zool. 17, 17-57.
T s u r u , S. (1967):  On studies  of the  microbial decomposition of
     various litters and humus formation in volcanic soils. Ed. O. Graff
     and J. E. Satchell, Progress in Soil Biology,  Verlag Frieder Vie-
     weg & Sohn,  Braunschweig  and  North— Holland Publishing Company,
     Amsterdam, pp. 455-463.

# SECTION B

## INFLUENCE OF ABIOTIC AND BIOTIC FACTORS ON COMMUNITIES OF SOIL ORGANISMS

# Calorimetric studies on soil invertebrates and their ecological significance

J. A. WALLWORK
Department of Zoology, Westfield College, University of
London

Recent developments in the field of microbomb calorimetry have provided new approaches to the study of natural populations of soil animals. The use of this technique in the construction of energy budgets is well established. However, attention has also been drawn to the possibility that animals which have temporarily accumulated food reserves would be expected to show higher calorific values than animals which are moribund or undergoing food stress. Thus, calorific values can provide an index of the nutritional status of the population in the field.

The main objective of the present study, which was started in 1969 and is continuing, is to determine the extent to which calorific values vary from one animal population to another within a single community. Some of the questions I am attempting to answer in connection with this are:

1. Are there any broad differences in calorific values between trophic levels?

2. Are there seasonal variations in calorific values within the same species?

3. If seasonal variations exist, can these be related to the ecology of the population, notably its nutritional status?

The soil community chosen for this study is that of a beech forest floor in the Chiltern Hills of southern England. Thirty of the commonest and most conspicuous species of invertebrate have been investigated, 15 of which can be broadly classified as saprophages and 15 carnivores. Some of these populations were sampled regularly throughout the year, and data on seasonal variations in calorific values are available.

METHOD

The procedure for obtaining calorific estimates is straightforward, and
little time need be taken up by describing the method in detail. Briefly, the animals are collected by hand, and dried immediately for 3 days
in an oven at $60^{\circ}C$.  The dried material  is then ground  into a powder,
formed into a pellet, weighed, and burned in a microbomb calorimeter of
the Phillipson type The calorimeter is calibrated with standard benzoic
acid  and corrections are applied,  where appropriate,  for ash content
and high calcium carbonate content in the sample.


RESULTS

Average values of calorific equivalents were calculated for each of the
30 species studied,  and the frequency distribution  obtained  is shown
in Fig. 1.  Here, it may be noted that the majority of carnivores  have
calorific values in the range 5000-6000 cals/g, whereas the majority of
saprophages lie in the range 4000-5000 cals/g. The 15 species of carnivore  provided 66 samples  of material  for calorimetric determination,
and the average value  obtained  for all these samples  was:  $5219 \pm 81$
cals/g. The 15 species of saprophages provided 170 samples with an average value  of $4698 \pm 54$ calg/g.  These two estimates  are significantly
different at the 0.01 level of probability.
      The Figure  also  shows  that despite the fact  that the calorific
values  of saprophages  are significantly lower, on average, than those
of the carnivores,  there is  a broad overlap  between  the two trophic
groups, and the frequency distribution  of the saprophages is much less
strongly centred about the mean value than is the case  with the carnivores.
      The ecological significance  of these findings seems fairly clear.
In the first instance,  we must remember  that plant material  on which
saprophages feed  has a lower calorific value, in general, than  animal
tissue.  Conversion  from a low energy state  to a higher one occurs as
material passes along a food chain.  However,  the upgrading  of energy
requires work to be done and if the difference between the energy state
of the food consumed (i.e. its calorific value) and the energy state of
the animal tissue into which this food is converted  is too great, metabolic processes will not be able to achieve the conversion.  Thus, the
saprophage trophic level emerges  as an important intermediate stage in
the upgrading  of energy present, initially, in plant material, and fi-

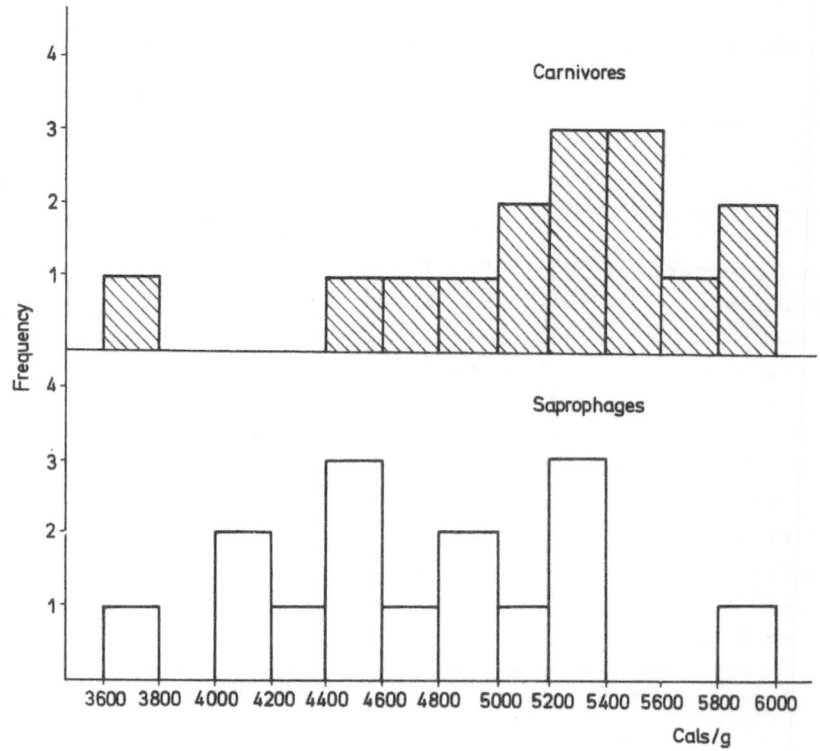

Fig.1.    Frequency distribution   of calorific equivalents
for   15 species   of saprophage   and   15 carnivores   from
a beech forest floor.

nally destined to find its way into the tissues of carnivores. Clearly,
calorific values   for saprophages   must be intermediate   between   those
for plant material   and those of top carnivores, and this   is obviously
borne out   by the results obtained.   Secondly,   the wide variability in
calorific value   shown   by the saprophages may be due to the fact that
their food sources are much more heterogeneous than those of carnivores.
We can now look at the individual species in more detail.

     Table 1   shows the results obtained for 5 species of millipede and
5 centipedes. Millipedes are saprophages and, as we might expect, their
calorific values   are lower than those of the predominantly carnivorous
centipedes.   Within the millipedes, as a group, the estimates   for different species   do not differ significantly   from each other, except in
one case   the value for *Cylindroiulus punctatus* is significantly higher
than that for   *Polydesmus angustus*.   This difference   may be related to
the fact   that the two species   have different ecologies.   *C. punctatus*
is an iuloid millipede capable of an active,   burrowing life,   while *P.*

T a b l e   1.    Calorific equivalents of myriapods   from a beech forest floor

| | Cals/g ($\bar{x} \pm SE$) | No. of samples (n) |
|---|---|---|
| *DIPLOPODA* | | |
| *Cylindroiulus punctatus* | 4563 ± 149 | 12 |
| *Ophyiulus pilosus* | 4516 ± 460 | 3 |
| *Blaniulus* sp. | 4147 ± 61 | 3 |
| *Polydesmus angustus* | 4109 ± 253 | 10 |
| *Glomeris marginata* | 4340 ± 149 | 4 |
| | | |
| *CHILOPODA* | | |
| *Lithobius curtipes* | 5377 ± 174 | 10 |
| *Lithobius variegatus* | 5167 ± 117 | 14 |
| *Lithobius calcaratus* | 5877 | 1 |
| *Brachygeophilus truncorum* | 5910 ± 240 | 5 |
| *Geophilus insculptus* | 5474 ± 278 | 4 |

*angustus*  is a "flat-back" millipede, having a sedentary, non-burrowing mode of life. As far as the various species of centipede are concerned, no significant differences  could be detected,  despite the rather wide range of values.  However, females of *Lithobius curtipes*  gave significantly higher values than males,  and this may serve to remind us  that calorific estimates for whole populations may be influenced  by the sex ratio.

Table 2 shows estimates  for annelids and molluscs. There is close agreement between the values for the two lumbricid  and one enchytraeid species.  The two slugs show values  which are comparable with those of the annelids, although  the estimate  for  *Arion hortensis*  is significantly lower than that for *Limax maximus*. Here again, this may indicate a relationship  between calorific content  and mobility, for *L. maximus* is a large grey slug which shows a high degree of mobility  in the laboratory.

Table 3  provides data for the isopod crustacean  *Trichoniscus pusillus* and various groups of insects. Although more samples are required to confirm many  of these estimates, they conform to what appears to be a general pattern emerging  from  this study.  Thus,  the highly mobile staphylinid beetles *Othius punctulatus*  and *Staphylinus* sp. show higher values than the less active saprophage *Trichoniscus pusillus*.

T a b l e   2.   Calorific   equivalents of annelids   and   molluscs   from a beech forest floor

| | Cals/g ($\bar{x} \pm SE$) | No. of samples (n) |
|---|---|---|
| *ANNELIDA* | | |
| *Dendrobaena sp.* | 5295 ± 68 | 17 |
| *Allolobophora chlorotica* | 5125 ± 57 | 2 |
| *Fridericia sp.* | 5207 ± 352 | 24 |
| | | |
| *MOLLUSCA* | | |
| *Arion hortensis* | 4828 ± 67 | 11 |
| *Limax maximus* | 5231 ± 136 | 8 |

T a b l e   3.   Calorific   equivalents   of   isopods   and   insects   from a beech forest floor

| | Cals/g ($\bar{x} \pm SE$) | No. of samples (n) |
|---|---|---|
| *ISOPODA* | | |
| *Trichoniscus pusillus* | 4488 ± 87 | 41 |
| | | |
| *INSECTA:PTERYGOTA* | | |
| *Othius punctulatus* | 5344 | 1 |
| *Staphylinus sp.* | 5594 ± 45 | 2 |
| *Feronia madida* | 4973 ± 45 | 3 |
| *Dolichopus sp. (larvae)* | 5658 ± 120 | 4 |
| | | |
| *INSECTA: APTERYGOTA* | | |
| *Entomobrya sp.* | 4993 ± 400 | 3 |
| *Tomocerus longicornis* | 5859 ± 16 | 3 |

To complete this general survey, we can examine estimates  for the various groups of Arachnida  (Table 4).  The 5 species  of mite (Acari) consists of 2 saprophages, *Damaeus onustus* and *Steganacarus magnus*, with relatively low calorific values,  and  3 predatory species,  *Pergamasus crassipes*, *Parasitus* sp. and  *Macrocheles submotus*.  It is worth noting that *M. submotus*, a predator  with relatively low mobility, has a lower

T a b l e   4.    Calorific equivalents of arachnids   from a beech forest
floor

|  | Cals/g ($\bar{x}$ ± SE) | No. of samples (n) |
|---|---|---|
| ACARI | | |
| Damaeus onustus | 4636 ± 130 | 15 |
| Steganacarus magnus | 3738 ± 161 | 14 |
| Pergamasus crassipes | 5241 ± 154 | 13 |
| Macrocheles sp. | 4541 ± 40 | 3 |
| Parasitus sp. | 5475 | 1 |
| | | |
| PSEUDOSCORPIONES | | |
| Chthonius ischnocheles | 4872 ± 485 | 3 |
| Neobisium muscorum | 5063 | 1 |
| | | |
| OPILIONES | | |
| Nemastoma lugubre | 3747 ± 230 | 4 |

calorific value   than   the very active P. crassipes   and   Parasitus sp.
The estimates  for the 2 pseudoscorpions - both very mobile predators -
are around 5000 cals/g, but more samples  are required to confirm this.
The value for the predaceous opilionid Nemastoma lugubre is surprisingly low, but there  may be  an explanation  for this  as we will see  in
a moment.

DISCUSSION

One idea  which  seems  to be repeated  throughout  this group-by-group
analysis is that the calorific value of a species appears to be directly related to its mobility.  Carnivores tend to be more mobile than the
saprophages  which form their prey, and we have already noted  that the
former have significantly higher calorific values, on average, than the
latter. However, even within the carnivore fauna, some species are more
mobile than others; similarly with the saprophages. In these cases, the
general pattern still applies.
    Taking this analysis  a stage further,  it may also be noted  that
highly mobile species  usually  have  high metabolic rates.  Also, many

carnivores show a seasonal periodicity in feeding activity, for example, various pseudoscorpions, opilionids, carabid beetles and centipedes. In contrast, many of the saprophages among the soil fauna, such as the oribatid mites, woodlice and collembolans, remain active and feeding throughout the year.

It may then be argued that species which feed throughout the year are receiving a constant supply of energy from their food, and are not under any selective pressure to build up food reserves. On the other hand, species which feed only during a limited period will tend to build up energy stores during their active period, and utilize these at other times of the year. From this it follows that "continuous" feeders will have relatively low calorific values, since they have no appreciable food reserves. Because of this, they are likely to be relatively sedentary species, with low metabolic rates. Periodic feeders, on the other hand, would be expected to build up food reserves, and their calorific values will be higher than those of the continuous feeders. The creation and mobilization of these reserves requires metabolic work, and this implies a higher metabolic rate. A relatively high metabolic rate also implies a relatively high degree of mobility.

This interpretation is supported by the results obtained in the present work. Continuous feeders, represented by the majority of the saprophages, have significantly lower calorific values, on average, than periodic feeders, represented by the majority of the predators. It has also been suggested that there is a direct relationship between calorific values and the degree of mobility of the species.

There are, however, certain species which do not appear to conform to this pattern. The opilionid *Nemastoma lugubre* is unusual in having a remarkable low calorofic value for a predator. However, this opilionid is one of the very few members of its group which is active throughout the year. Thus, the considerations which apply to continuous feeders may also apply in this case. Again, the collembolan *Tomocerus longicornis* has a remarkably high calorific value for a saprophage, and while this species may be a continuous feeder, it is highly mobile and may have a high metabolic rate.

A further line of enquiry has been suggested by the general findings outlined here. We can test the hypothesis that species which create and mobilize food reserves will show greater seasonal variation in calorific values than continuous feeders which do' not require such reserves, since they have a constant supply of energy. The seasonal data that have been obtained suggest as much, although the trends are not always clear-cut. Two extreme patterns are shown by *Arion hortensis*, a saprophage, and *Pergamasus crassipes*, a carnivore (Figs. 2, 3). Here,

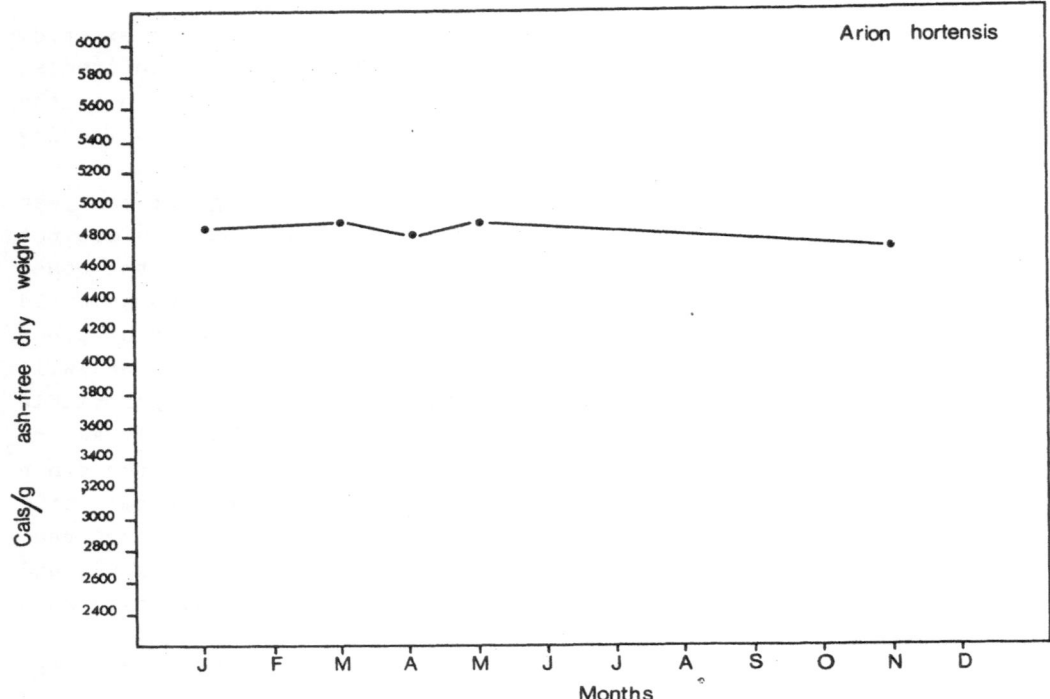

Fig.2.    Seasonal  variation   in   calorific  equivalents  of
Arion hortensis.

the lack of seasonal variation in the saprophage  (a continuous feeder)
contrasts sharply  with  the pattern shown  by the carnivore,  which is
probably a periodic feeder. However, we must be cautious about over-gen-
eralization here. Fig.4 shows the seasonal patterns for 2 lithobiomorph
centipedes, and they are obviously quite different. The marked seasonal
variation  in L. curtipes is lacking in L. variegatus.  This may be explained  by the fact  that  L. variegatus  tends  to feed  continuously
throughout the year, switching from a carnivorous diet  to a vegetarian
one in winter when prey is scarce. Clearly, a detailed knowledge of the
ecology of a species  is an important ingredient  in the interpretation
of calorific values.

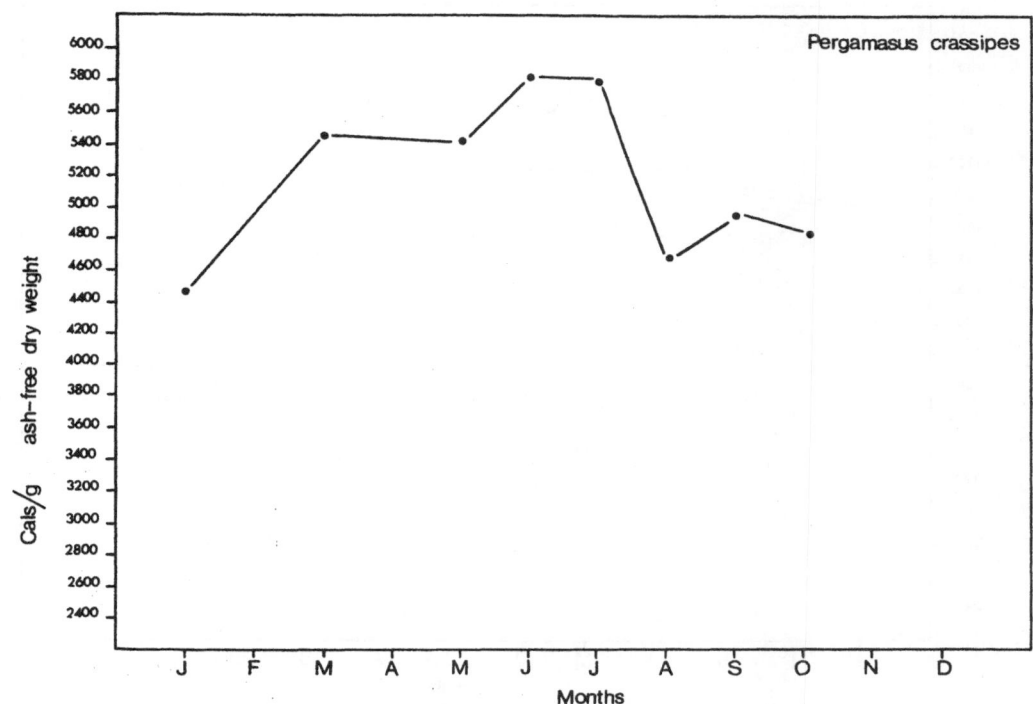

Fig.3.  Seasonal variation  in  calorific equivalents of *Pergamasus crassipes*.

Discussion

H. P e t e r s e n :  I wonder to which extent you have included microphytophages  in the saprophages.  You may except  that microphytophages
have higher calorific values than consumers  of dead plant material, as
I remember  Malcom Luxton  has shown  for oribatid mites. This may contribute to the explanation of the high calorific values for *Collembola*.
The high calorific values of *Collembola* is in agreement with results of
my own investigations.

J. A. W a l l w o r k :  I have used  the term "saprophage" here to include all feeders on plant material.But I agree with you that there are
differences in calorific values  within this group which may be related
to the type  of food material consumed.  This could explain  the rather
wide range of values obtained for the saprophages as a group.

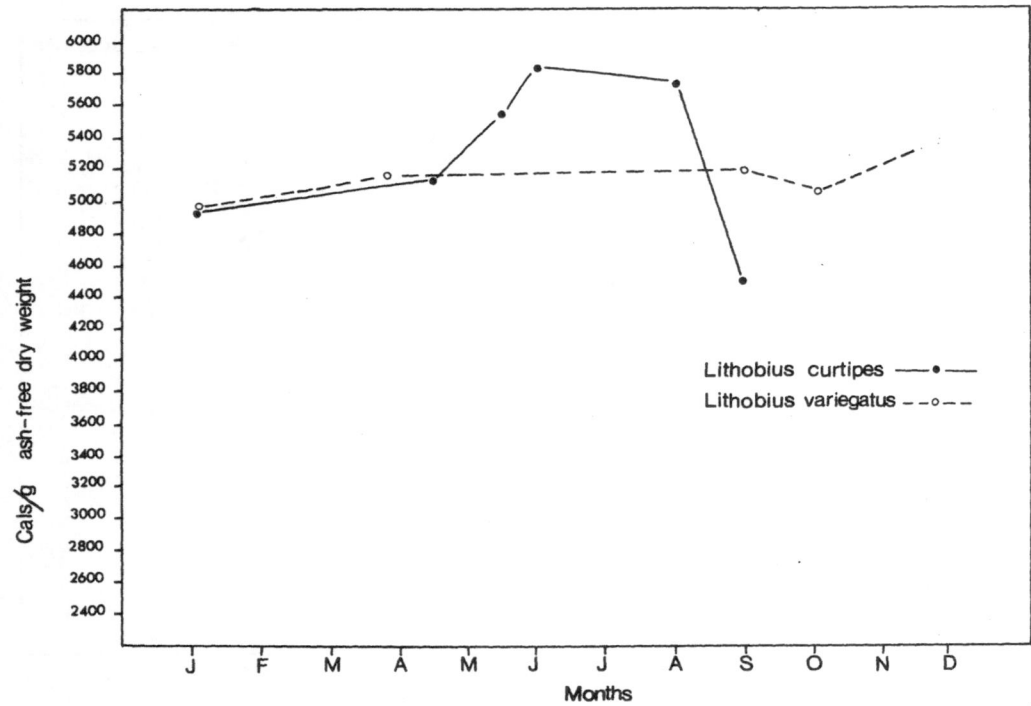

Fig.4.  Seasonal variation  in  calorific equivalents of
*Lithobius curtipes* and *Lithobius variegatus*.

# Influence d'une irradiation gamma chronique sur la microfauna d'un ecosystème méditerranéen à Cadarache

N. POINSOT *

Laboratoire d'ecologie, Faculté des Sciences
St. Jerome Traverse de la Barasse,
Marseille

Les résultats concernent l'étude de l'influence d'une source radio-active sur les communautés végétales et animales.

L'expérience est poursoivie au C.E.N. de Cadarache dans un vallon recouvert par une chenaie d'yeuses. La source est constituée par du [137] Ce de 1200 curies. Elle peut être comparée à celles de Woodwell et Sparrow qui, depuis 10 ans à Long Island étudient l'influence d'une irradiation gamma sur une forêt de chênes blancs et Pitchpins, les doses allant de plusieurs centaines de rads/jours à 1 rad/jour. Ils ont observé la destruction de la végétation avec une résistance particulière des lichens et des mousses et un changement dans la quantité d'éléments nutritifs dans le sol, cette diminution étant semblable à celle que l'on note après un incendie ou l'emploi d'herbicides.

L'autre expérience est celle de Whicker, qui depuis 1968, étudie l'influence d'une source radioactive de 8750 Curie (débit de dose de 650 r/h à $7,8 \cdot 10^{-3}$ r/h à 50 m de la source) sur une prairie du Colorado. Il est, là encore, difficile de définir le role exact des irradiations, car il y a d'abord destruction de la végétation. Le résultat le plus intéressant est l'explosion démographique des Pucerons.

A Cadarache, en ce qui concerne la végétation Fabries (1972) note une relation entre la répartition biogéographique des espèces et leur radiosensibilité, les ubiquistes étant les plus résistantes et les européennes les plus sensibles, Il observe aussi, une modification profonde des équilibres de la phytocénose: les ligneux disparaissent tandis que d'autres végétaux, comme les herbacés et les lichens, prolifèrent.

Pour la microflore, Tchernia (1971) constate après un and'irradiation qu'il y a une baisse de l'activité du sol, le nombre de germes de microflore totale diminue.

* Avec la collaboration de Grauby (A), Rougon (C et D).

Pour la microfaune et l'entomofaune une étude préliminaire nous avait fait choisir un deuxième vallon témoin parrallèle au vallon irradié. Dans chacun trois ou quatre niveaux ont été choisis. Le niveau inférieur, formé par une garrigue basse à Romarin, recoit un débit de dose de 14 r/h, le deuxième dans le Phillyrea 10,4 r/h, le premier niveaus du chêne vert 3,8 r/h et le dernier 1,4 r/h.

Parmi les réprésentants de l'entomofaune, notre attention a été attirée par le groupe des *Psocoptères*, dont nous avons dénombré 12 espèces. Après deux ans et demi d'irradiation, Bigot tire les conclusions suivantes:

- dans les deux niveaux supérieurs où les doses d'irradiation sont faibles (4,3 à 24,9 Kr) la croissance de la population est continue. En 1971 c'est la plus importante, l'effectif est passé de 7 individus en 1969 à 2 en 1970 et 247 en 1971.

- dans les autres niveaux, les valeurs les plus élevées des doses d'irradiation provoquent, dans un premier temps (1970), une diminution des effectifs.

- pour les doses moyennes (de 9,2 Kr à 53,5 Kr) les effectifs de Psoques diminuent moins de 1969 à 70 et atteignent des valeurs plus élevées en 1971 que pour les doses plus fortes (de 30 à 178,4 Kr).

Cette pullulation peut s'expliquer en partie par le régime alimentaire des Psoques dont certains se nourrissent d'algues vertes unicellulaires qui résistent très bien aux irradiations.

Chez les *Collemboles*, 56 espèces ont été inventoriées (Rougon C. et D., 1970).

A partir de Juillet 1969, nous avons comparé les fluctuations mensuelles des *Collemboles* dans le vallon témoin et dans le vallon irradié (fig. 1). Pour l'année 1969-1970, on constate qu'au mois d'août 1969, les *Collemboles* de la zone irradiée présentent une augmentation de l'effectif par rapport au vallon témoin. Cette augmentation est simplement due à une espèce d'*Isotomidae* qui pullule dans le vallon irradié, indépendamment de toute action radioactive.

De septembre à décembre, les fluctuations mensuelles sont sensiblement parallèles dans les deux vallons, mais l'effectif est légèrement inférieur dans le vallon irradié. Dans le vallon témoin, les *Collemboles* sont encore représentés en très grand nombre en décembre. Par contre, il semblerait que les radiations ionisantes commencent à agir dans l'aire d'irradiation. Dans ce vallon, ce n'est qu'en janvier que les *Collemboles* présentent une légère augmentation en nombre. A partir de ce moment-là, l'effectif des *Collemboles* du vallon irradié ne cesse de diminuer. Les radiations semblent même avoir supprimé le maximum d'abondance des *Collemboles*, qui a lieu en avril dans la vallon témoin.

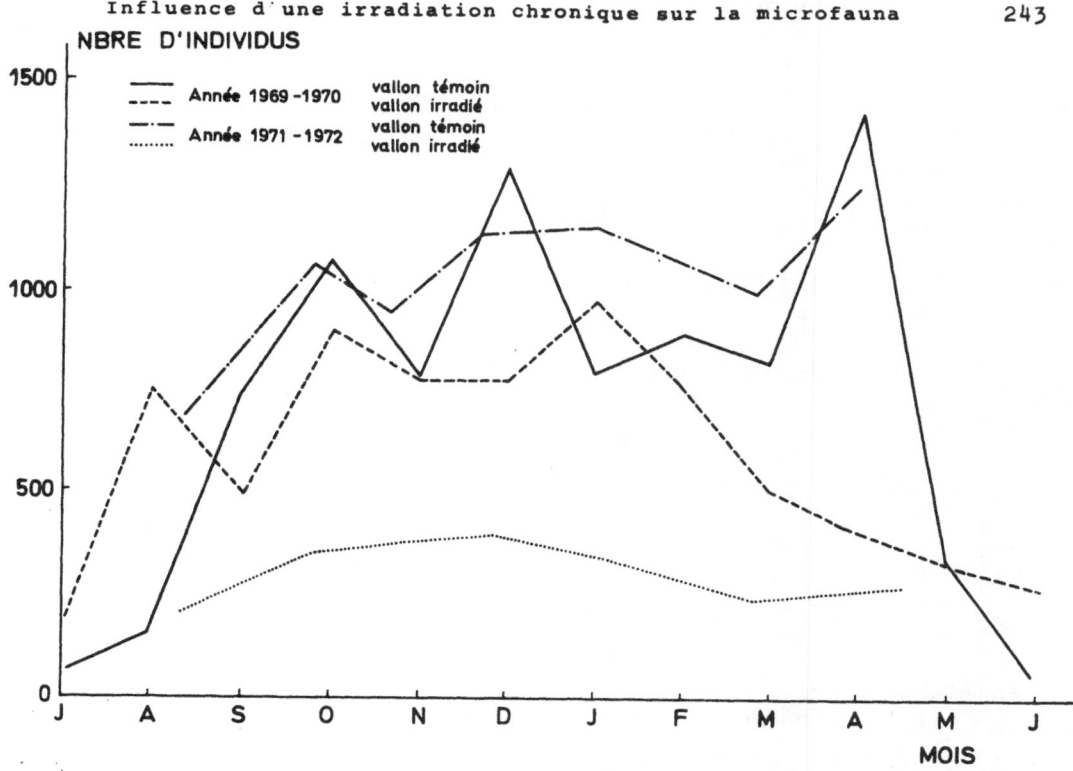

NBRE D'INDIVIDUS

Année 1969-1970   vallon témoin
                  vallon irradié
Année 1971-1972   vallon témoin
                  vallon irradié

MOIS

**Fig.1. Variations des populations de Collemboles dans le vallon temoin et le vallon irradié.**

Le même type d´étude comparative a été repris à partir de mai 1971 et les résultats sont reportés sur le même graphique. On constate une baisse très sensible des populations de *Collemboles* dans le vallon ir- radié. Cette courbe, bien qu´indicatrice, met cependant moins en évi- dence l´influence des rayons gamma que celles des variations des peu- plements dans chaque station.

En effet, dans le Romarin (niveau 1[+]), dans le vallon témoin, les *Collemboles* représentés en grand nombre accusent un maximum d´abondance en automne 1969 et 1971 (fig. 2A). Par contre, dans le vallon irradié, l´effectif des populations de *Collemboles* est nettement plus faible et le maximum d´abondance est reporté au moins de janvier 1970. En décembre 1969, la dose d´irradiation reçue par le Romarin est de 46,2 Krads. Sur les graphiques sont reportées les doses cumulatives d´irradiations tri- mestrielles. A partir du mois de février les peuplements de *Collemboles* diminuent progressivement dans l´aire irradiée alors que leur cycle se poursuit normalement dans le vallon témoin avec un maximum d´abondance en avril.

Cette diminution se retrouve à partir de mai 1971. On note en

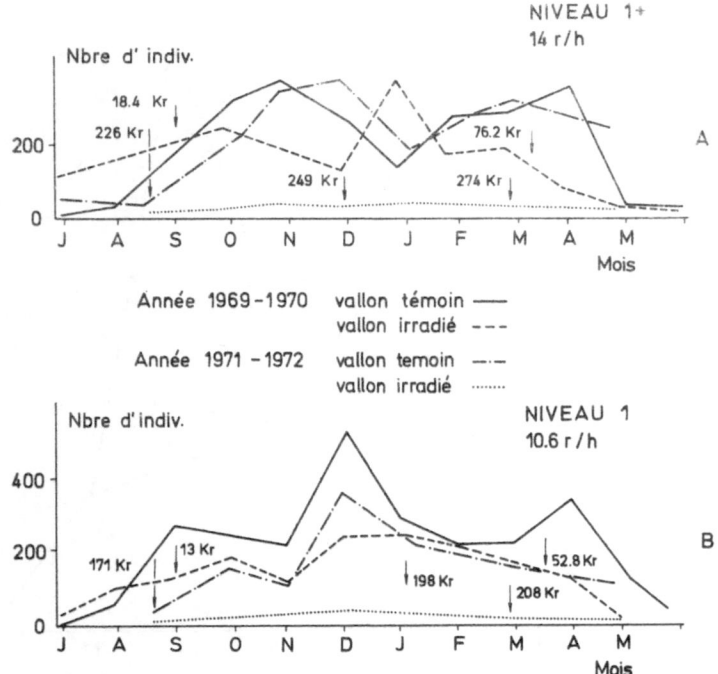

Fig.2. A: Variations des populations de Collemboles dans
le niveau 1⁺; B: Variations des populations de Collemboles dans le niveau 1.

octobre   seulement 4 *Collemboles*   au niveau 1⁺ et quelques Acariens; en
janvier 1972, malgré des conditions microclimatiques favorables,  on ne
récolte que 4 *Collemboles*, un oribate  et en avril une vingtaine appartenant tous à la même espèce.* Il n'y a plus de *Collemboles* arboricoles,
absence  qui peut s'expliquer, d'une part, par l'action directe des rayons et, d'autre part, par la disparition du feuillage complètement détruit par l'irradiation.

     Au niveau du *Phillyrea*, (niveau 1), jusqu'au mois de décembre 1969
les fluctuations  saisonnières  sont  sensiblement parallèles  dans les
deux vallons, mais l'effectif des populations de *Collemboles* est nettement inférieur  dans l'aire d'irradiation  (fig. 2B).  A cette date, le
*Phillyrea* a reçu  une dose de 36,6 Krads  et à partir  de ce moment-là,
soit 6 mois après le début de l'irradiation, le nombre  des *Collemboles*
ne cesse de diminuer.  En 1970-1972, leur nombre atteint  au maximum 10
dans les relevés. On notera dans ce niveau comme dans le précédent, une
raréfaction presque concomitante des Acariens.

     * Cette espèce, en élevage, mue et se reproduit normalement.

Dans le premier niveau du Chêne vert, par contre, les variations des populations de *Collemboles* sont presque parallèles dans les deux vallons. On note même, dans certains relevés, des peuplements plus importants dans la zone irradiée. Cette observation vient confirmer l'importance du débit de dose sur les invertébrés. En effet, en mai 1972, la dose reçue est de 80 Krads, dose qui avait provoqué dans les niveaux précédents la raréfaction des microarthropodes. De même, du fait de leur éloignement par rapport à la source, l'effet du rayonnement commence à peine à être sensible sur les feuillages qui jonchent le sol.

Enfin dans le troisième niveau, aucune influence des rayonnements gamma n'est actuellement détectable.

La diminution de l'effectif des *Collemboles* semble due en partie à l'effet direct des radiations ionisantes, le premier niveau étant le plus atteint. On peut y ajouter au moins deux effets indirects: augmentation du nombre des fourmis prédateurs de *Collemboles*, et particulièrement résistants aux radiations ionisantes (Whicker, 1970; Lemasne, 1972), défoliation complète dans le périmètre irradié de tous les arbustes; ce que l'on peut résumer sur la figure 3.

RÉSUMÉ

Les rayons gamma chroniques émis par un irradiateur disposé dans un écosystème forestier ont une action très marquée sur les organismes vivants que se soit l'entomofaune, la microfaune ou la microflore. On assiste en effet, au bout de trois ans, à une explosion démographique des Psoques, une disparition des *Collemboles* dans la zone la plus irradiée et à une diminution de l'activité du sol.

SUMMARY

Chronic gamma-irradiations of a forest ecosystem resulted in a significant action on living organisms, as far as entomofauna, microfauna or microflore are concerned. A demographic explosion of Psoque was observed three years later as well as a disappearance of *Collembola* in the nearest zone of irradiator and a reduction of the soil activity.

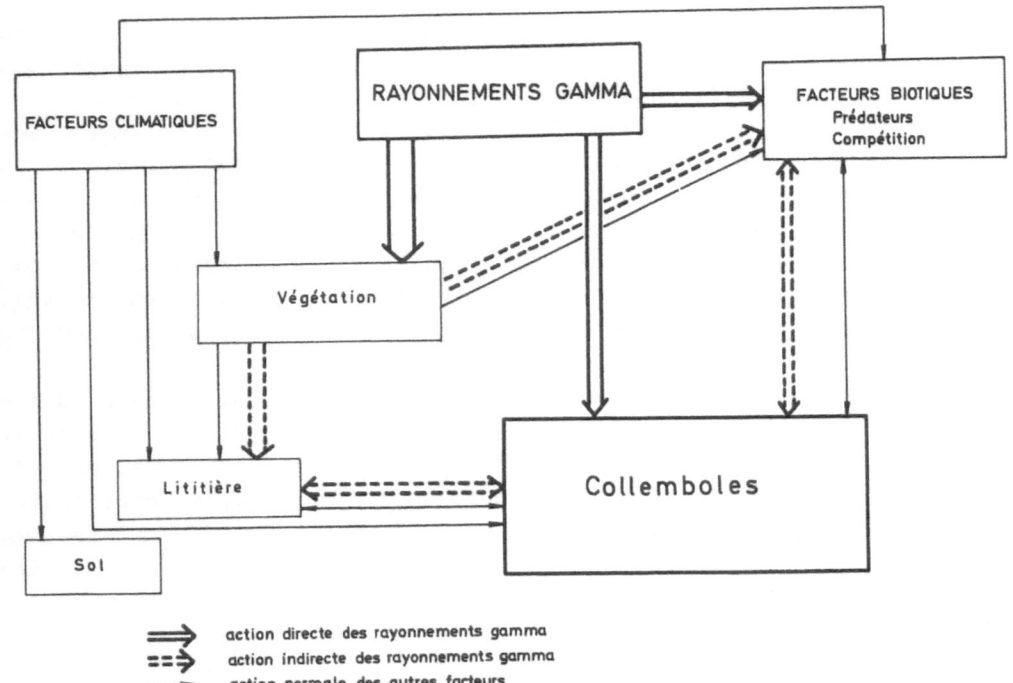

Fig.3. Interactions des facteurs  sur les populations de
Collemboles en milieu irradie (d'aprés C. Rougon).


B i b l i o g r a p h i e

F a b r i e s ,  M. (1972): Quelques aspects écologiques  et biogéogra-
    phiques de l'irradiation chronique  d'une phytocénose terrestre de
    type mediterranéen. Rapport C. E. A. R. 4 300 C. E. N. Saclay.

R o u g o n ,  C. et D. (1970): Premières données sur les insectes Col-
    lémboles et Coléoptères  de la forêt de Cadarache (B. d. R.). Ann.
    Fac. Sci., Marseille, 43, 27 - 34.

T c h e r n i a , F. (1971): Influence  d'une  irradiation  chronique
    gamma sur la microflore totale d'un sol méditerranéen. VIII. Symp.
    Int. Agronomica, Venezia, Maggio 1971.

W h i c k e r ,  F. W. (1970): Radioecology  of some  natural organisms
    and systems  in Colorado.  Eighth Technical Progress Report,  Dep.
    CFSTI, 40 p.

W o o d w e l l ,  G. M. (1962): Effects of ionizing radiation  on ter-
    restrial ecosystems. Science 138, 572-577.

W o o d w e l l ,  G. M.,  B r o w e r ,  J. H. (1967):  An aphid popu-

lation explosion induced by chronic gamma irradiation of a forest.
Ecology 48, 680-683.

W o o d w e l l ,  G. M.,  S p a r r o w ,  A. H. (1963): Predicted and
observed effects of chronic gamma radiation on a nearclimax forest
ecosystem. Rad. Bot. 5, 231-237.

D i s c u s s i o n

J. M. T h i b a u d :  Y-a-t'il des différences de résistance entre les
jeunes et les adultes? Que deviennent les oeufs?

N. P o i n s o t :  Lors  de cette expérience  je ne l'ai pas  observé,
mais d'après les observations  d'Edwards,  les jeunes  sont plus résistants.  Styron, lui, a montré que les oeufs  de *Sinella curviseta* sont
10 fois plus sensibles que les adultes aux rayonnements gamma.

P. L e b r u n :  La prolifération des psoques est-elle vraiment due au
développement des algues? Des recherches sur les microarthropodes corticoles  semblent  indiquer  que les animaux  deviennent  très abondant
lorsque leurs prédateurs,  parasites et compétiteurs ont disparus  dans
les zones fortement polluées au $SO_2$.

N. P o i n s o t :  Comme  pour les autres groupes  il est difficile de
doser l'influence exacte des facteurs biotiques  et abiotiques.  Il est
évident qu'en ce qui concerne les Psoques la disparition des prédateurs
ajoutée  à celle  de la prolifération  des algues  jouent  en faveur de
l' augmentation du peuplement.

J. d' A g u i l a r d :  Les Psoques  ont-ils  été  déterminés  jusqu'a
l'espèce? S'agit-il d'espéces algophages?

N. P o i n s o t :  Oui, les espèces  ont été déterminées  et parmi ces
espèces il y a à la fois des espèces saprophages et des algophages.

Z. M a s s o u d :  Est-ce que vous avez  un hypothèse  pour  expliquer
la résistance de deux espèces de Collemboles à la irradiation?

N. P o i n s o t :  Non.

# The role of pedobionts
# in biogeochemical cycles
# of calcium and strontium–90
# in the ecosystem

A. D. POKARZHEVSKY AND D. A. KRIVOLUTSKY

Institute of Animal Evolutionary Morphology and Ecology,
Moscow

Among pedobionts  many calcophil forms  are found,  nummerous groups of
active soil-formers  belonging to calcophils,  c.f. *Diplopoda*,  *Isopoda*
and  *Lumbricidae*.  However, quantitative studies on the contribution of
soil-dwellers  to the calcium turnover  have been  so far insufficient,
to say nothing of other alkaline-earth elements, i.e.barium, strontium,
radium.

Stable strontium  and  its radioactive isotopes are,  however, the
closest analogs of calcium, i.e. their behavior  and migration patterns
in living systems are similar to those of calcium.  Calcium is found to
affect the accumulation  of $^{90}$Sr in plants  and  vertebrates (Guliakin,
Judinceva, 1962;  Comar, Wasserman, 1964).  In this case quantative relationships are clear in general. The above problem for terrestrial and
soil invertebrates remains,  however,  open because  of the scarcity of
the available data. A quantative distribution of Ca and $^{90}$Sr has virtually not been investigated  with respect to the biomass of various animal groups in the ecosystem.

We have studied the $^{90}$Sr  and  Ca accumulation by pedobionts in an
$^{90}$Sr experimentally contaminated plot. The description of the plot under
investigation  and  some data on the $^{90}$Sr accumulation by animals  have
already been published  (Iljenko, 1970;  Ghilarov,  Krivolutsky,  1972;
Krivoluckij, Tichomirova, Turchaninova, 1972) and therefore are not presented here.

The content  of $^{90}$Sr  in relation  to Ca  in the  superficial soil
layer 5 cm thick  in the plot  under investigation amounted to $1.57 \cdot 10^7$
strontium units  (s.u.),  in the litter - $5.5 \cdot 10^7$ s.u.,  in leaffall of
birch *(Betula pendula)* - $2.75 \cdot 10^7$ s.u.. The $^{90}$Sr in relation to Ca content in pedobionts  and  their respective predators  was found to be as
follows:  earthworms  *(Allolobophora  calliginosa)* - $4.2 \cdot 10^3$ s.u., *Opi-
liones* - $3.7 \cdot 10^5$ s.u.,  spiders (Araneus) - $7.4 \cdot 10^5$ s.u.,  *Carabidae* -
$1.26 \cdot 10^7$ s.u.,  *Geotrupes stercorosus* - $3.7:10^6$ s.u.,  *Silpha  carinata*

*larvae* - $4 \cdot 10^6$ s.u., their imago - $3.1 \cdot 10^6$ s.u., small flying insects -
$1,03 \cdot 10^5$ s.u. (data on calcium were extrapolated after Clark, 1958,
Reichle et al. 1969), frogs *(Rana terrestris)* - $1.8 \cdot 10^6$ s.u., shrews:
*Sorex araneus* - $3.4 \cdot 10^5$ s.u., *S.caecutiens* - $4.97 \cdot 10^5$ s.u., *S.minutus* -
$3.57 \cdot 10^5$ s.u., *S.minutissimus* - $5.7 \cdot 10^5$ s.u. (frog gaster content -
$4.85 \cdot 10^6$ s.u., shrews one - $1.31 \cdot 10^6$ s.u.). Data on groups of animals
are presented in Table 1.

T a b l e  1.  Ca and $^{90}$Sr content in soil and suprasoil animals of the
ecosystem studied

| Group of animals | Ca ppm* | $^{90}$Sr Ci/g.$10^{-8}$* | s.u. $\cdot 10^5$ |
|---|---|---|---|
| *Lumbricidae* | 11800 | .05 | .042 |
| *Araneida* | 4000 | 2.22 | 5.6 |
| *Insecta* | 300 | 1.86 | 62. |
| *Vertebrata* | 35000 | 3.47 | 9.9 |

* - dry weight.

In the insects studied the values of $^{90}$SR in relation to Ca proved to
be about 10-fold higher than those in other species of soil fauna. Lower
means of $^{90}$Sr content in s.u. in small flying insects are accounted for
by a relatively small size of the contaminated plot and consequently by
large numbers of individuals outside of the "clean" areas. The low mean
of the radiostrontium content in s.u. in *Lumbricidae* appears to have
resulted from the self-purification of these animals in deeper "clean
layers" of the soil.

As is mentioned above the concentration of calcium affects the
accumulation of $^{90}$Sr in mammals. The quantitative characteristic of
this effect is an o b s e r v e d  r a t i o :

$$O.R. = \frac{^{90}Sr/Ca \text{ in animals}}{^{90}Sr/Ca \text{ in diet}} ,$$

e.g. in mammals the respective value is 0.25 (Comar, Wasserman, 1964).
The observed ratio can be estimated for the trophic links as follows:
(soil-layer) - *Lumbricidae* - $1.2 \cdot 10^{-4}$, *Lumbricidae* - *Carabidae* - =,000,
leaffall - *Cetonia aurata* - 0.16, animals - *Silpha carinata* lar. - 1.5,
animals - *S.carinata* im. - 1.2, small flying insects - *Araneus* - 7.2,
frog diets - frog - 0.37, shrew diets - shrew - 0.34.

Data on observed ratio for frogs and shrews are close to the published ones, which are mentioned above. There is scarce evidence con-

cerning the migration of the alkaline-earth elements through terrestrial invertebrates (Clark, 1958; Reichle et al, 1969; Van Hook et al,
1970) Therefore, it is premature to discuss the differences in observed
ratios in the food links of vertebrates and invertebrates. However,
there is ground to believe that in invertebrates, the observed ratio is
lower than 1 at the first trophic level, it being higher than 1 at the
second trophic level.

It is of interest to consider Ca and $^{90}$Sr distribution in ecosystem compartments and assess quantatively the role of different animal groups in the biogeochemical turnovers.

Let us firstly consider the data available presented in Table 2.

T a b l e  2.  The calcium and $^{90}$strontium distribution in the compartments of the ecosystem studied

| Compartment | Mass g/m$^2$ dry wt. | Calcium mg/m$^2$ | $^{90}$Sr  Ci/m$^2$ |
|---|---|---|---|
| Soil (layer 5 cm thick) | 25000 | 234000 | $3.20 \cdot 10^{-3}$ |
| Litter | 3200 | 5820 | $3.04 \cdot 10^{-4}$ |
| Leaffall | 720 | 4540 | $1.25 \cdot 10^{-4}$ |
| Carabidae im. | 1.16 | .322 | $4.06 \cdot 10^{-9}$ |
| Aranea | .10 | .412 | $2.29 \cdot 10^{-10}$ |
| Lumbricidae | .75 | 8.85 | $3.70 \cdot 10^{-11}$ |
| Frogs | .00145 | .0475 | $2.10 \cdot 10^{-11}$ |
| Shrews | .00085 | .0324 | $5.80 \cdot 10^{-11}$ |

The values for the mass of the litter and leaffall and calcium content
in them are quoted from the literature (Rodin, Bazilevich, 1965; Alek-
sakhin, Aleksakhina, 1971).

As is seen from Table 2, the role of animals in the calcium and
$^{90}$strontium accumulation in a terrestrial ecosystem is extremely lower
than that of soil and vegetation, the contribution of invertebrates
being much greater than that of vertebrates. The relative role of different groups of terrestrial animals is demonstrated in Table 3.

The biomass of animals per square unit is partly quoted from Krivolutsky, Shilova (1965). Although in that paper the correlation between
animal groups for a somewhat slightly different area was described,
however the order of values in correlation of groups is similar to that
at the plot studied. $^{90}$Sr content in vertebrate biomass was estimated
after Iljenko (1970), that for invertebrate biomass after Krivoluckij,

272 - 4383 - Sv.

Tichomirova, Turčaninova (1972). The amount of $^{90}$Sr accumulated in nematodes and protozoans is assumed to be equal to that in lumbricids, the above indicated groups being animals without endoskeleton dwelling in the same environment. It is evident that the invertebrates exceed vertebrates by an order not only in the biomass, but also in the calcium and radiostrontium content involved in this biomass.

However, the overall amount of calcium in animals constitutes only 1.13%, that of $^{90}$strontium is 5.6 $10^{-4}$% of their total content in the ecosystem. The facts that a number of animals reproduces a few times in the warm season of the year (protozoans as many as 10, collembolans nearly 5, rodents 2-4 and etc.) and the animal production may attain 110 to 160 kg/ha being taken into account, the animals are capable of consuming not more than 4 kg Ca and 45 $10^{-4}$ Ci $^{90}$Sr per ha annually.

The amount of $^{90}$Sr involved in biogenic migration by animals is 0.56 % of natural decay of $^{90}$Sr of the whole ecosystem. But considering the fact that $^{90}$Sr contaminated ecosystems are extensively incorporated by soils and involved in the biogeochemical turnover, the biogenic migration of this isotope can be assumed to be at least comparable to the values for the $^{90}$Sr migration leached off by the rainfall or dispersed by the wind from the ecosystem. As numerous animal groups are capable of migrating and covering large distances (especially insects, birds and ungulates) the role of animals in dispersion of radionuclides from ecosystem contaminated is not to be neglected (Tab. 3).

R e f e r e n c e s

A l e k s a k h i n , R. M. and A l e k s a k h i n a , M. M. (1971): Accumulation and distribution of Ca, Mg and stable Sr in the phyto-coenoses of hardwood and coniferous forests, in "Biological productivity and mineral cycling in the terrestrial plant communities", 232-235. Leningrad, Nauka (in Russian).

C l a r k , E. W. (1958): A review of literature on calcium and magnesium in insects. Ann. Entomol. Soc. Am., V. 51, 2, 142-154.

C o m a r , C. L. and W a s s e r m a n , R. H. (1964): Strontium. In: Mineral metabolism, 2A: 523-572, Acad. Press, N-Y, London.

G h i l a r o v , M. S. and K r i v o l u t s k y , D. A. (1972): Radioecological researches: in soil zoology. Colloquium Pedobiologiae, Dijon 1970, Paris 1971.

G u l i a k i n , I. V. and J u d i n c e v a , E. V. (1962): The radioactive decay products in soil and plants. Moscow. Gosatomizdat (in Russian).

T a b l e   3.     Ca and $^{90}$Sr   content in some compartments   of ecosystem

| Compartment | Mass kg/ha dry wt. | Calcium g/ha | Strontium Ci/ha |
|---|---|---|---|
| Soil (5 cm thick layer) | $2.5 \cdot 10^5$ | $25 \cdot 10^5$ | 32 |
| Vegetation | $10.0 \cdot 10^4$ | $5 \cdot 10^5$ | 5 |
| Litter & leaffall | $6.0 \cdot 10^3$ | $6 \cdot 10^4$ | 4.3 |
| Animals: | 79 | 470 | $23.5 \cdot 10^{-5}$ |
| *Protozoa* | 10 | 80 | $3.5 \cdot 10^{-7}$ |
| *Nematodes* | 5 | 40 | $2.0 \cdot 10^{-7}$ |
| *Enchytreids* | 5 | 40 | $2.0 \cdot 10^{-7}$ |
| *Lumbricids* | 7.5 | 88 | $3.7 \cdot 10^{-7}$ |
| *Microarthropods* | 15 | 50 | $6.0 \cdot 10^{-5}$ |
| *Insects* | 25 | 75 | $10\text{-}0 \cdot 10^{-5}$ |
| Other invertebrates (molluscs, spiders, etc.) | 10 | 50 | $4.0 \cdot 10^{-5}$ |
| All invertebrates | 77.5 | 420 | $20.0 : 10^{-5}$ |
| *Mammals:* | | | |
| *Rodents* | .85 | 30 | $30.0 \cdot 10^{-6}$ |
| *Ungulates* | .30 | 12 | $11.0 \cdot 10^{-6}$ |
| *Insectivores* | .056 | 2 | $.9 \cdot 10^{-6}$ |
| *Carnivores* | .028 | 1 | $.5 \cdot 10^{-7}$ |
| Other mammals | .056 | 2 | $.1 \cdot 10^{-6}$ |
| Birds | .056 | 2 | $.8 \cdot 10^{-6}$ |
| *Amphibians & Reptiles* | .056 | 2 | $1.7 \cdot 10^{-6}$ |
| All vertebrates | 1.4 | 50 | $3.5 \cdot 10^{-5}$ |

Annual $^{90}$Sr natural decay in the whole ecosystem.          .8

I l j e n k o ,  A. I. (1970):  The relationships of $^{90}$Sr and $^{137}$Cs migration in different links of food chains in zoocoenosis. J. Obshej Biologij (in Russian).

K r i v o l u t s k y ,  D. A.   and   S h i l o v a ,   S. A.   (1965): On the structure of biogeocoenosis of South Taiga.  Vestnik MGU, ser. V, geogr., 3, 72-75.

K r i v o l u c k i j ,  D. A.,   T i c h o m i r o v a ,   A.  L.   and T u r c h a n i n o v a ,  V. A.   (1972):  Strukturänderungen des Tierbestands  (Land- und Bodenwirbellose)   unter   dem Einfluss der Kontamination des Bodens mit Sr$^{90}$. Pedobiologia, Bd. 12, 374-380.

R e i c h l e ,   D. E.,   S h a n k s ,   M. N.   and   C r o s s l e y , D.
    A., jr. (1969): Calcium, potassium and  sodium content of forest
    floor arthropods. Ann. Entomol. Soc. Am., V. 62, 1, 57-62.
R o d i n ,   L. E.   and   B a z i l e v i c h ,   N. J. (1965): ˙Dynamics
    of the organic matter  and the biological turnover of ash elements
    and nitrogen in the main types of the world vegetation. Moscow-Le-
    ningrad, Nauka.
V a n   H o o o k ,   R. I., jr.,   R e i c h l e ,   D. E. and   A u e r -
    b a c h ,   S. J. (1970): Energy and Nutrient Dynamics  of Predator
    and Prey Arthropod Populations in a grassland Ecosystem. ORNL-4509,
    UC-48 - Biology and Medecine. Oak Ridge Nat. Lab.

## D i s c u s s i o n

D. E. R e i c h l e : Yours is a commendable effort in summarizing what
little  is known  about the incorporation of calcium  in various components of the ecosystem — including  the decomposer invertebrates.  For
comparison to the extensive data  in your Table,  I would like to offer
the following values  for  our deciduous forest IBP site  in the United
States. Calcium mass $(mg/m^2)$ in ecosystem components are: soil to 60 cm
$3.9 \times 10^5$ mc $Ca/m^2$ above and belowground vegetation $2.7 \times 10^3$ mg $Ca/m^2$,
litter 48 mg $Ca/m^2$;  all decomposer invertebrates 8.6 mg $Ca/m^2$;  decomposer microflora 0.5 mg $Ca/m^2$. No reply is necessary.

# Decomposition processes of swampy plants in peat soils

L. S. KOZLOVSKAJA
Carelian Branch, Academy of Sciences USSR,
Petrozavodsk

Decomposition of swampy plants and the role of soil invertebrates and microorganisms in the formation of different kinds of peat are under our investigation.

A litter-fall of herbs (mixture of *Spirea,* reedgrass and fern), birch, bogbean, sedge, cedar, *Sphagnum* and other plants placed in nylon medium, were placed in peat soil of a flat bog. The succession of soil microorganisms and invertebrates was traced in the process of mineralization of litter-fall and the loss of litter-fall weight recorded. Earthworms were placed into sacks with litter-fall (6 specimens of *Eisenia ukrainae* Malev. and about 20 *Enchytraeidae*) in order to disclose the participation of invertebrates in decomposition processes.

The experiments enabled a classification of plants into hardly and easily decomposed forms. Hardly decomposed forms were blocked from attacking organisms by phytoncides and began to disintegrate quickly after their destruction only (Kozlovskaja, Dick, Melentjeva 1969). *Sphagnum* mosses are reckoned among the hardly decomposed forms. According to data by V.A. Baturo and V.E. Rakovsky (1957) they contain many phenols. A great deal of green mosses and cedar needles, the latter with a large quantity of resins, also belong to the hardly decomposed forms. Such plant species as bogbean, totentilla, avens, fern etc. are decomposed best of all.

Sedges occupy an intermediate position among them. Referring to a comparison of the analyses, easily decomposed species contain a lot of proteins, readily-hydrolysed carbohydrates and Ca what makes them attractive for invertebrates and microorganisms in the absence of resistant phytoncides (Kozlovskaja, Dick, Melentjeva 1969).

The dependence of decomposition trend upon the contents of different organic matters of such types as tannins, phenols, terpens, organic acids was evidenced by Minderman and Daniels 1966, Edwards, Satchell and Low 1967, Chernova and Karpachevski 1969 et al.

The investigation  of decomposition patterns  of swampy plants allows to elucidate three types of cycles as regards their mineralization
and humification.  Plants poorly blocked by organic compounds  from the
attack  of invertebrates  and  microorganisms fall into the first type,
called by us the "overground-litter cycle of mineralization". According
to their chemical composition investigated  by E.I. Skobeleva  and S.N.
Tyuremnov (1966)  they are rich in proteins, Ca readily hydrolysed carbohydrates.  Such plants  as  bogbean, potentilla, calls, avens, ferns,
spirea etc. belong to this type.

Second type: Destruction of plants begins in autumn from the stage
of surface decomposition.  Rust fungi, molluscs and insect larvae dwelling on the soil surface, take the active part in these processes, which
result  in desiccation  of the plants.  Epiphytic  and saprophytic soil
microorganisms  seem to be especially significant  in  the second stage
of a litter decomposition.  The plant fall  is decomposed  by larvae of
*Diptera (Bibionidae, Sciaridae)* (Table 1).  Fluorescent  and  sporous
bacteria predominate among saprophytic microorganisms.

This is followed  by the soil stage,  during  which  soil forms of
*Collembola*  and mites, dipterous larvae *(Sciaridae, Chironomidae)*, lumbricids and enchytraeids  are especially active.  Ammonificators, fluorescent bacteria, actynomycetes, sporous and cellulose-destroying microorganisms  play the important role.  Miroshnichenko et al. (1972)  gave
a similar scheme for herb decomposition. Since they supposed that plant
destruction  was completed  at the litter stage,  they did not describe
the participation of animals in this process.

Roots of high plants participate  in the soil stage  of decomposition.  We distinguished a third type of decomposition  called the "litter-soil cycle of humification". It appears to have the least number of
stages and the most lasting.  Mosses blocked  by phenolic and other organic compounds are generally decomposed in this way. The disappearance
of the lower parts burried in the soil, and the growth of new shoots is
characteristic of mosses.

The stage of moss decomposition caused by the epiphytic microflora
in the litter layer  is short and proceeds jointly with the soil stage.

The purely soil stage of moss decomposition  seems to be the main
one.  Fluorescent bacteria and other ammonificators  play a significant
role at the beginning  of this stage.  Oribatid mites participate first
of all in the decomposition of oligotrophic *Sphagnum* mosses, oligochaets
in that of mesotrophic and eutrophic *Sphagnum* mosses.  Green mosses are
decomposed by larvae of Diptera, mainly of Tipulidae.

The additional inoculation of animals resulted in the more complete
transformation of organic matter  (Table 2).  In the presence of earth-

T a b l e   1.    The abundance of soil animals in the samples of the decomposed fall-litter placed in the soil (per 100 g of absolut dry weight)

| Animal groups | Herbs | | | Birch leaves | | | Carex lasiocarpa | | | Cedar needles | | | Sphagnum | | | Bogbean | | |
|---|---|---|---|---|---|---|---|---|---|---|---|---|---|---|---|---|---|---|
| The period of exposition in days | 40 | 65 | 360 | 40 | 65 | 360 | 40 | 65 | 360 | 40 | 65 | 360 | 40 | 65 | 360 | 30 | 95 | 36 |
| Enchytraeidae | 5.0 | 20.8 | 10.2 | 125 | 5.0 | - | 2.4 | 12.2 | 14.5 | - | - | 16.8 | - | - | 30 | - | - | - |
| Lumbricidae | - | - | 2 | - | - | 1 | - | - | - | - | - | - | - | - | - | - | - | - |
| Nematoda | - | 2.6 | 4 | - | - | - | - | - | - | - | - | - | - | - | - | - | - | - |
| Acari | 48 | 416 | 6.8 | 2 | 520 | 60 | 15.4 | 168 | 25 | 36 | 316.8 | 5.6 | 20 | - | - | 500 | 200 | - |
| Collembola | 70.6 | 1204 | 262 | 135 | 230 | 360 | 46 | 254 | 260 | 36 | 86.4 | 246.4 | 26 | - | 150 | - | - | - |
| Coleoptera larvae | 2 | - | - | 8 | - | - | - | 4.2 | - | - | - | - | - | - | 10 | - | - | - |
| Diptera larvae | 2 | 6.6 | 3.4 | 5 | 25 | 48 | - | 25 | 10 | - | 19.2 | 11.2 | - | - | 10 | 570 | 300 | - |

T a b l e  2.  Microflora variations in fall-litter deposited in soil (mlm/l g of absolute dry matter)

| Investigated object | Nutrient agar, Meat-pepton agar | Starch-ammonia agar | Ashby-agar | Meat-pepton + mash | Environment of Hetchinson | Mash-agar |
|---|---|---|---|---|---|---|
| *Herbs (without earthworms)* | | | | | | |
| 40 days | 46.95 | 8.16 | – | 0.008 | 1.51 | – |
| 65 days | 67.52 | 105.92 | 257.42 | – | 1.90 | – |
| 365 days | 197.28 | 34.11 | 42.33 | 0.12 | 0.6120 | 0.032 |
| *Herbs (without earthworms)* | | | | | | |
| 40 days | 204.8 | 8.67 | – | 1.05 | 2.66 | – |
| 65 days | 50.40 | 115.2 | 537.6 | 0.048 | 1.688 | – |
| 365 days | 117.3 | 53.04 | 58.65 | 0.46 | 0.039 | 0.056 |
| *Sphagnum (without earth-worms)* | | | | | | |
| 40 days | 107.87 | 7.97 | – | 0.48 | 6.48 | – |
| 65 days | 103.78 | 122.30 | 85.47 | 1.83 | 1.221 | – |
| 365 days | 159.28 | 29.27 | 45.63 | 0.60 | 0.047 | 30.13 |
| *Sphagnum (with earth-worms)* | | | | | | |
| 40 days | 957.32 | 18.93 | – | 0.96 | 4.21 | – |
| 65 days | 103.95 | 288.15 | 232.37 | 2.81 | 1.83 | – |
| 365 days | 197.52 | 58.43 | 79.01 | 1.11 | 0.061 | 41.15 |

worms and enchytraeids, the large numbers of ammonificating microorganisms including sporous bacteria develop in the plant remains.  The dominance of *Bac. megaterium* and *Bac.virgulus* indicate intensive mineralization processes.  Parallel  with the carbohydrates  and  proteins increases the rate of cellulose decomposition. The role of microorganisms fixing atmospheric nitrogen  increased significantly.  A further intensification  of microbiological processes  was observed two months after animal inoculation;  it was demonstrating by an increase of microorganisms decomposing nitrogen mineral compounds  and among them - *Actinomycetae.*

Schemes of cycles in humification  and  mineralization of swampy plants

1. Ground-level cycle  of mineralization of herbaceous plants.
   Above-ground  stage.  Parasitic fungi and surface invertebrates take part.
   The stage of litter  or surface soil decomposition. Epiphyllous and  saprophytic microorganisms and  litter invertebrates  take part (complete mineralization).
2. The cycle  of  litter-soil moss humification.
   Litter  stage.  Saprophytic microorganisms  and invertebrates participate.  The participation of parasitic microorganisms  is unimportant.
   Soil stage. Soil microorganisms and invertebrates take part.

3. Ground-level-soil cycle  of humification of herbaceous plants and litter-fall of tree species. Ground — level stage.  Parasitic fungi and surface invertebrates take part.
   The stage of rags or of surface litter decomposition.
   Epiphyllous, parasitic  and saprophytic  microorganisms  and top — litter  invertebrates  are taking part.
   The stage  of litter decomposition. Epiphyllous and saprophytic microorganisms  and  litter invertebrates  are taking part.
   The soil stage. Soil saprophytic microorganisms,  invertebrates and  partly  roots  of  higher plants take part.

Discussion

E.  v o n  T ö r n e :  Gibt es Ihrer Meinung  nach funktionelle Zusammenhänge zwischen den von Ihnen beobachten mikrobiellen  und zootischen Phaenomenen?

L. S. K o z l o v s k a j a : Die Tätigkeit der Mikroorganismen und Wirbellosen ist eng verbunden. Die Anwesenheit von Lumbriciden und Enchytraeiden steigert die Aktivität von Mikroorganismen und fördert ihre Vermehrung. Als Folge geht die Zersetzung der Pflanzenreste intensiver und in einer günstigeren Richtung für die Bodenfruchtbarkeit vor.

E. v o n T ö r n e : Welchen Einfluss hat Ihrer Meinung nach der Wechsel des hygrothermischen Regimes auf den raumzeitlichen Ablauf der metabolischen Prozesse; ist mit zeitweiligen vertikalen Verschiebungen der Aktivitäts-Strata zu rechnen oder ist anzunehmen, dass Schwankungen des hygrothermischen Regimes nur modifizierend wirken?

L. S. K o z l o v s k a j a : Die höheren Temperaturen steigern die Lebensintensität von Organismen. Die Moorböden sind durch breitere Temperaturamplituden der oberen Bodenschichten auch während des einzigen Tages charakteristisch. Auch die Verminderung der Feuchtigkeit wirkt positiv auf die Prozesse der Zersetzung der Pflanzenreste durch Bodenorganismen. In entwässerten Moorböden verläuft die Destruktion der organischen Substanz viel schneller und vollkommener weil sich die Zahl der Organismen, deren Artenzahl und Individienzahl, erhöht.

R. C o v a r r u b i a s : Est-ce que vous avez obtenu des correlations precises qui demontrent des lieus entre la mezofaune et les microorganismes?

L. S. K o z l o v s k a j a : Ja, wir haben solche Korrelationen studiert. Unter dem Einfluss von Lumbriciden und Enchytraeiden steigern sich im Boden die Prozesse der Ammonifikation, der aeroben Fixation des Stickstoffes und der Zersetzung der Zellulose. Die saprophagen Dipterenlarven beeinflussen die Ammonifizierung. Die Anwesenheit von Diplopoden steigert die Lebensaktivität der pektin- und zellulosezersetzenden Mikroorganismen.

# Colonisation par les Microarthropodes du sol de cinq types de litière en décomposition

**R. MIGNOLET¹ ET Ph. LEBRUN**

**Laboratoire d'Ecologie générale,
Louvain-la-Neuve**

Le présent travail s'inscrit dans le cadre d'une large étude de la décomposition des litières en relation avec leur colonisation par la Microfaune et la Microflore. Le point de vue développé ici sera celui des réactions de la faune vis-à-vis des caractéristiques des six substrats.

## DONNÉES PHYSICO-CHIMIQUES

### Matériel et méthodes

Au début du mois de novembre 1972, cinq x 50 sacs (tulle de nylon, mailles 3 $mm^2$) contenant respectivement 10 g de feuilles de chêne *(Quercus robur)*, 10 g de noisetier *(Corylus avellana)*, 10 g de charme *(Carpinus betulus)*, 10 g d'érable *(Acer pseudoplatanus)* et 5 g de papier Kleenex, ont été placés sous la litière fraîche d'une chênaie à charmes de la forêt de Meerdael (Moyenne Belgique).

Tous les mois, à dater du 7 décembre 1972, cinq sacs de chaque type ont été repris, coupés en deux et une moitié a été placée à l'extraction au Berlèse; par la même occasion, 5 quadrats de litière naturelle (témoin) étaient prélevés et semblablement traités. La seconde moitié a servi aux recensements des microorganismes. Pour chaque "litière" les analyses suivantes ont été réalisées: teneur en eau, perte de poids sec, teneur en cellulose et en élements minéraux (sodium-potassium-magnesium-calcium). Ces dernières ont été mesurées au spectrophotomètre tandis que la teneur en phosphore a été estimée à l'aide d'un colorimètre. Les valeurs du pH (eau) ont été déterminées à l'aide d'un pHmètre électronique.

Au cours de l'exposé des résultats, nous ferons parfois appel à des données recueillies au cours de l'année 71-72 dans une expérience

¹ Boursier de l'Institut pour l'Encouragement de la Recherche Scientifique dans l'Industrie et l'Agriculture (IRSIA).

analogue réalisée sur trois litières   (coudrier, chêne et litière naturelle: Mignolet, Priemen et Gregoire-Wibo, en prép.; Mignolet, en prép.).


## Caractérisation des litières
*a) L'acidité*

L'estimation des pH  nous permet  de classer  les litières monospecifiques dans l'ordre d'acidité croissante:  noisetier (5.68 $\pm$ .03), érable
(5.02 $\pm$ .06),  chêne  (4.7 $\pm$ .13),  charme  (4.43 $\pm$ .04).  Pour l'année
1971-1972, nous avions obtenu 5.65 $\pm$ .08 pour le noisetier et 4.47 $\mp$ .11
pour le chêne.


*b) Teneurs en éléments minéraux*

Les figures 1 (a,b,c)  illustrent l'évolution des teneurs en Magnesium,
Calcium et cendres des substrats étudiés.  On constate  que l'érable et
le noisetier  sont  les plus riches  en ces éléments.  Par contre,  les
teneurs  en Phosphore  sont  nettement plus faibles  dans  les feuilles
d'érable que dans les autres substrats (fig. 1d).  A partir  de concentrations en Potassium très différentes,  on arrive après 10 mois  à une
homogénéité de teneurs qui l'exclut comme élément distinctif (fig. 1e).
Les concentrations en Sodium  étaient très faibles (5-10  ppm) et à peu
près équivalentes  dans  les quatre substrats;  aucune évolution claire
n'a pu être mise en évidence pour cet élément.
    Signalons qu'au cours de l'étude de l'année précédente, nous avions
observé  le même facies d'évolution  des teneurs  en minéraux  du chêne
et du noisetier.


*c) Teneurs en cellulose*

Nous avons constaté  que les teneurs  en cellulose du charme, du chêne
et de l'érable  différaient relativement peu.  Par contre, le noisetier
est nettement moins riche  en cet élément (fig. 2),  comme  nous avions
déjà pu  le constater précédemment (Mignolet, en prép.).

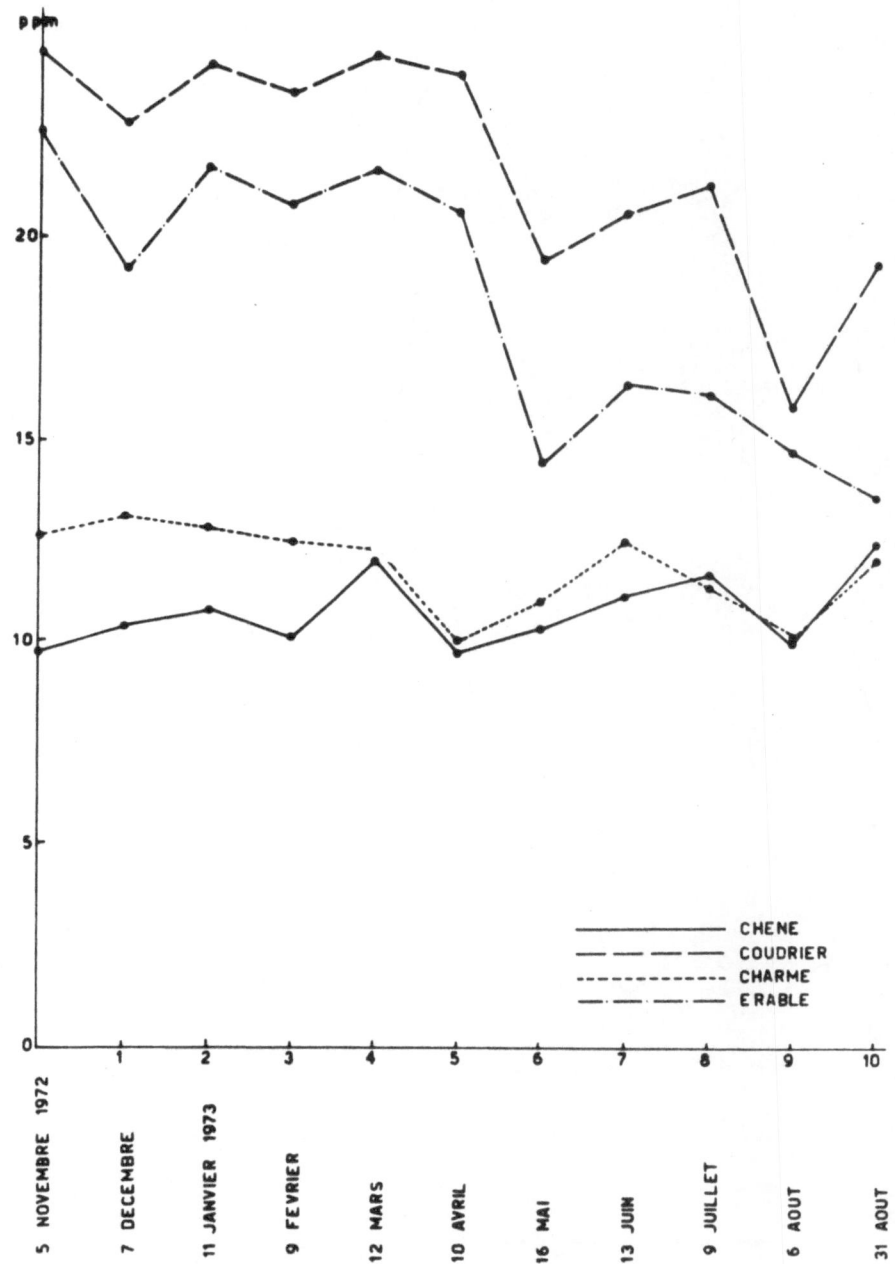

p pm

20

15

5

10

CHENE
COUDRIER
CHARME
ERABLE

0

1  2  3  4  5  6  7  8  9  10

5 NOVEMBRE 1972

7 DECEMBRE

11 JANVIER 1973

9 FEVRIER

12 MARS

10 AVRIL

16 MAI

13 JUIN

9 JUILLET

6 AOUT

31 AOUT

Fig.1a) Evolution des teneurs en Magnesium.

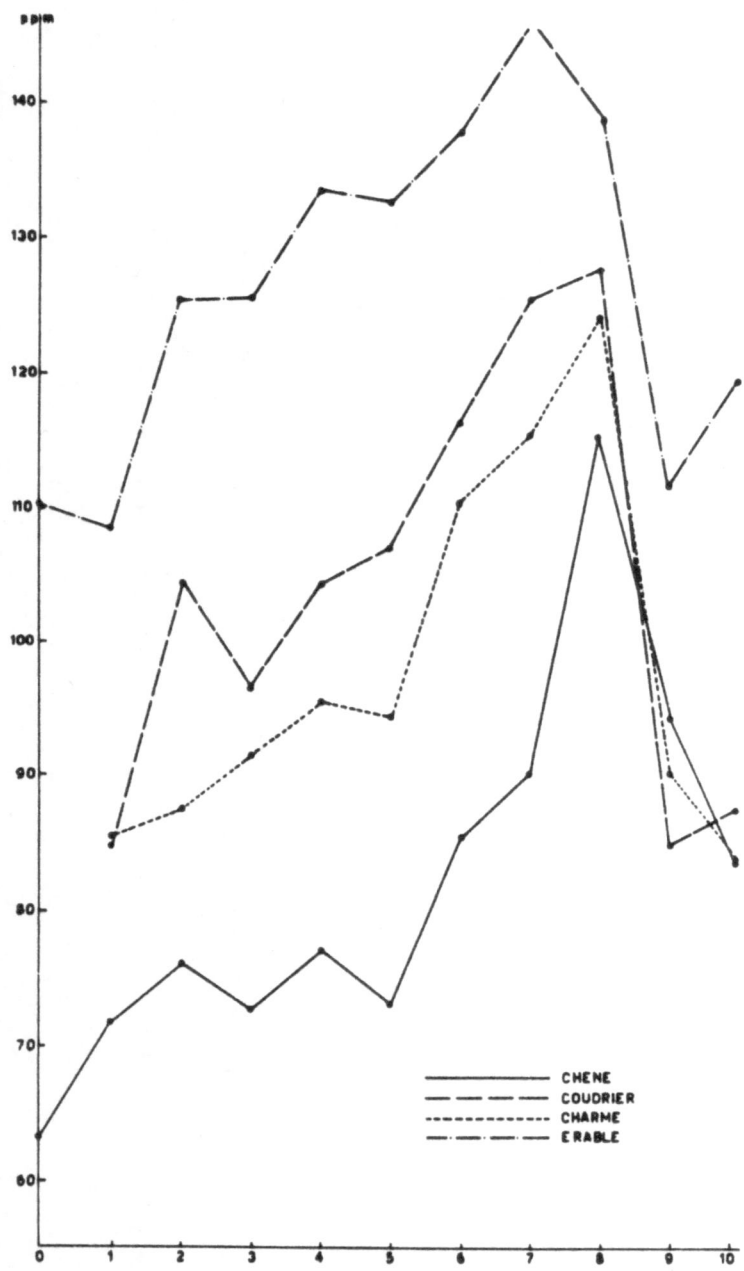

Fig.1b) Evolution des teneurs en Calcium.

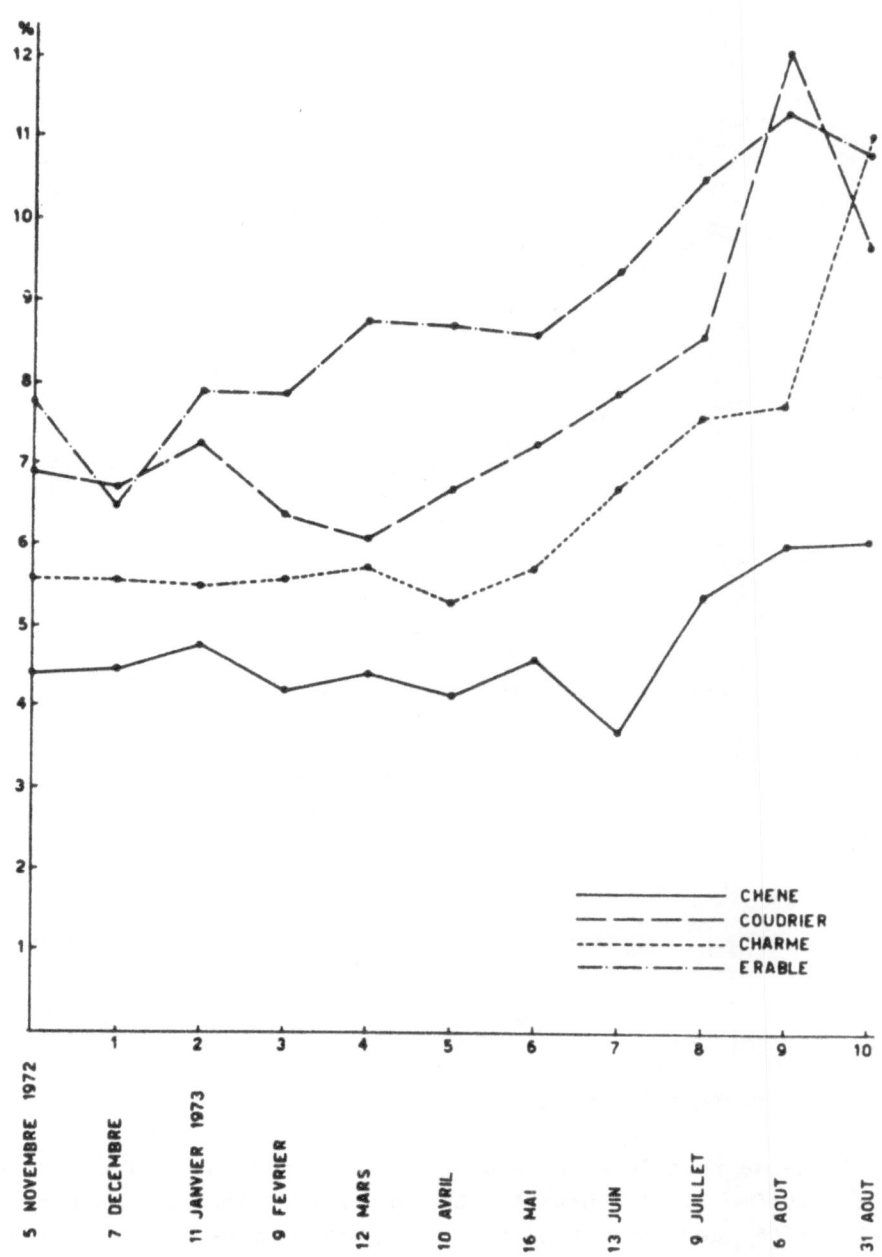

Fig.1c) Evolution des cendres totales.

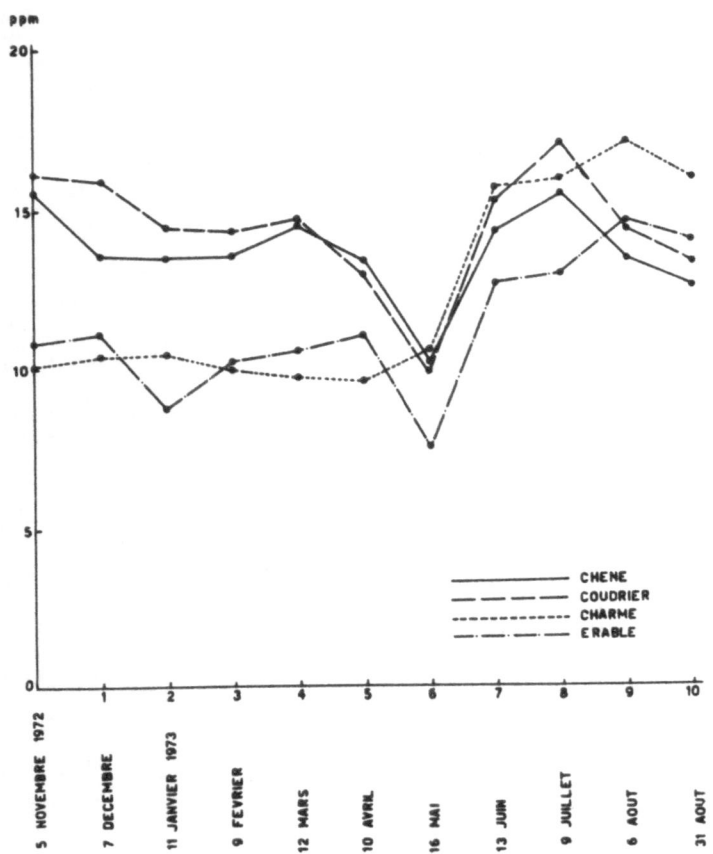

Fig.1d) Evolution des teneurs en Phosphore.

## d) Réactivité au facteur hydrique

Au cours d´étude de relations entre densités fauniques et taux d´hydra-
tation de la litière, on a constaté  (Lebrun, 1971; Mignolet, en prép.)
que la variabilité  des teneurs en eau  est généralement plus élevés en
période d´assèchement. Une façon de vérifier ce phénomène est de mettre
en relation la moyenne  et  l´écart type estimés  aux dates de prélèvements. L´examen de la fig. 3 (a,b) nous permet de constater une augmentation effective  de l´écart-type  avec la baisse  de teneur en eau. On
remarquera la similitude  des pentes des droites des litières  étudiées

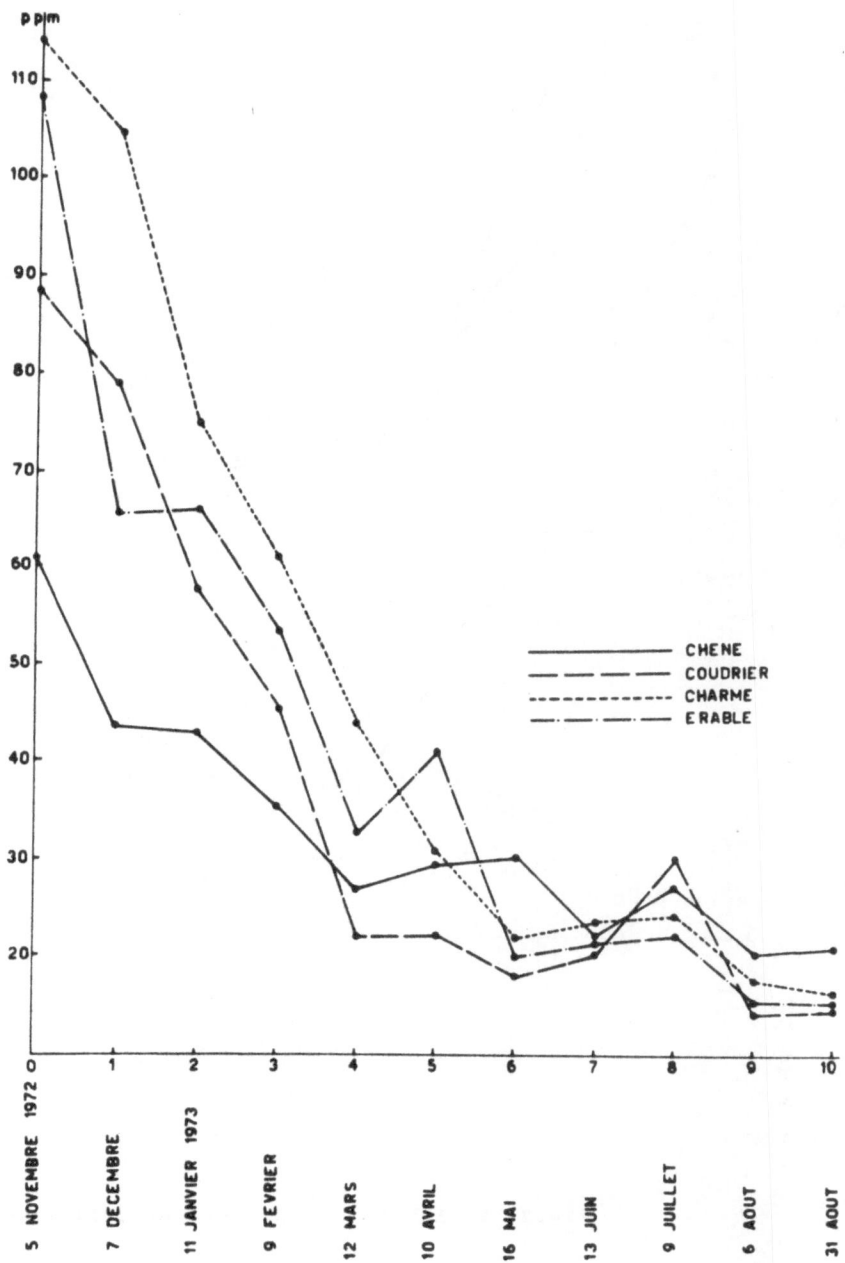

Fig.1e) Evolution des teneurs en Potassium.

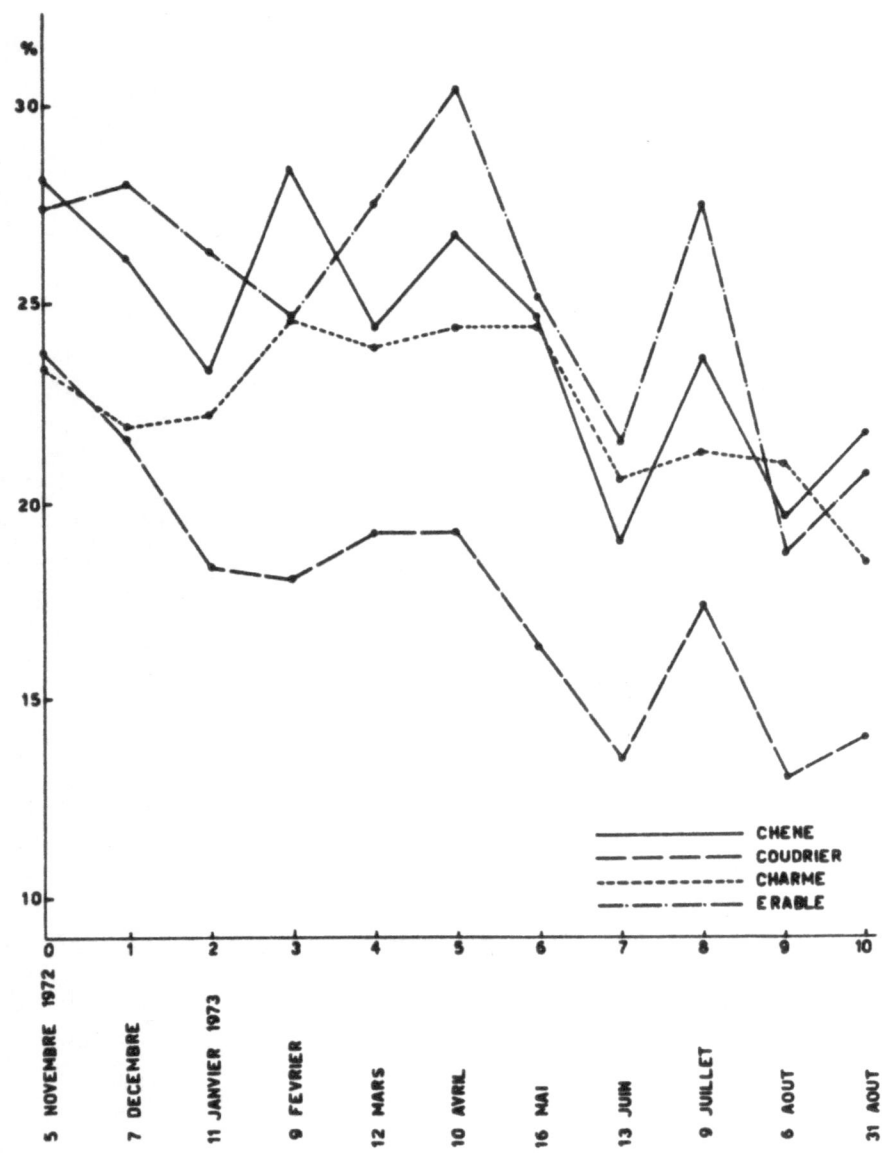

Fig.2. Evolution des teneurs en cellulose.

à la fois en 1971-1972 et 1972-1973. Dans le cadre du présent travail, cette méthode nous permet de grouper les litières en fonction de leur réactivité au facteur hydrique: le mode d'hydratation de l'érable et du noisetier est distinct de celui du charme et du chêne (fig. 3b).

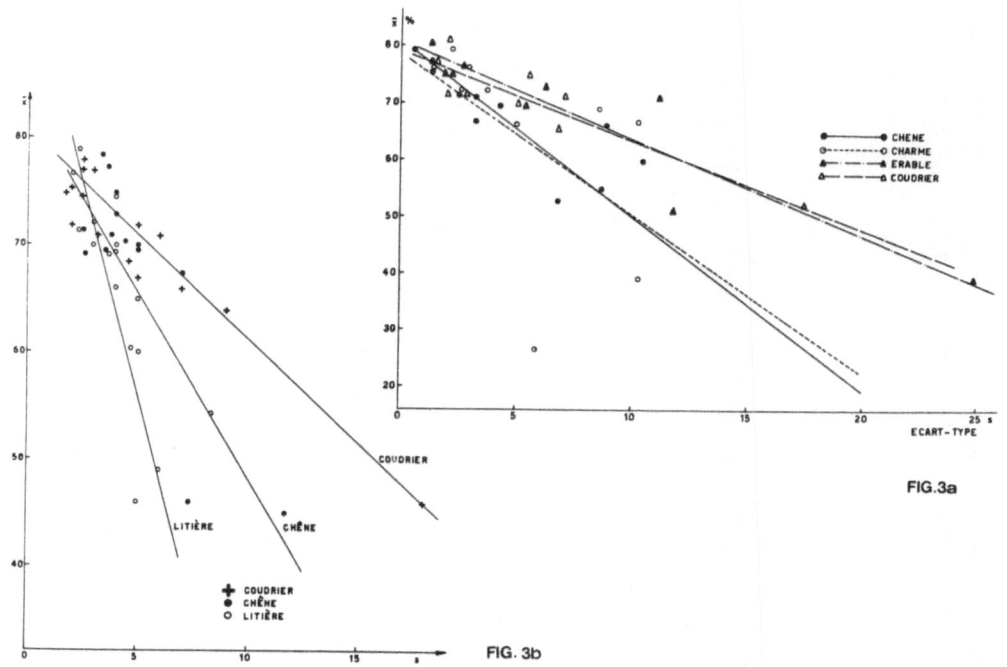

FIG.3a

FIG. 3b

Fig.3a) Relations entre moyennes  et écart-types des teneurs en eau dans les litières monospecifiques (3a 1972-
1973· 3b 1971-1972).

Par ailleurs, une analyse  de la variance  a montré que les degrés
d´hydratation des 4 litières monospécifiques n´étaient  que légèrement
différents;  l´effet temps  était nettement plus important  que l´effet
litière.


*e) Vitesses de décomposition*

La figure 4 (a,b)  décrit  l´évolution  des poids secs  de l´érable, du
chêne  et du  papier Kleenex (a), du charme  et  du noisetier (b), pour
l´année 72-73 ainsi que du chêne et  du noisetier  au cours  de l´année
71-73 (intervalle de confiance: 0,95) Une analyse de la variance a montré un effet litière et un effet temps très significatifs, un effet interaction  légèrement significatif.  On constate  que c´est  la litière
d´érable  qui perd  son poids  le plus rapidement;  le noisetier  et le
charme  évoluent  a peu près  à la même vitesse,  le chêne  se décompose
plus lentement.

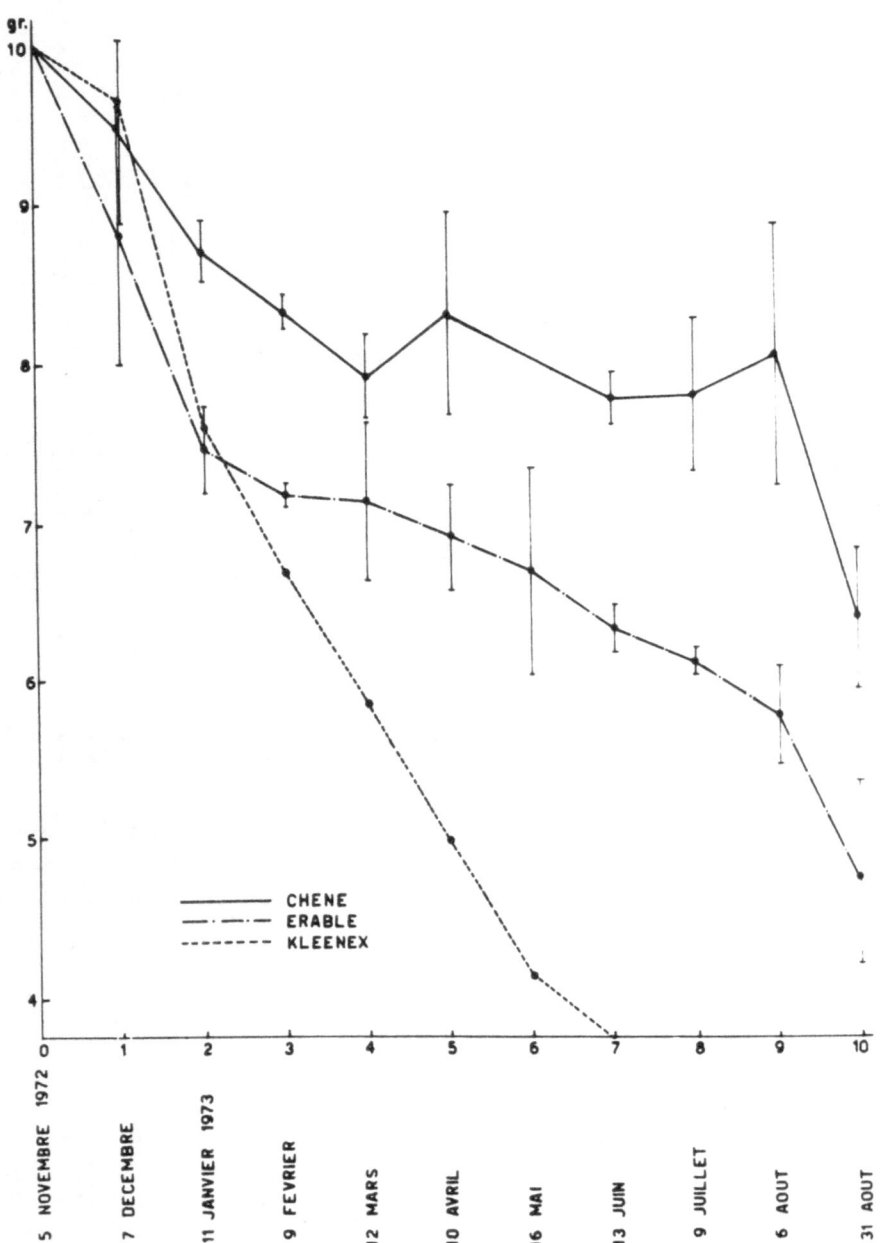

Fig.4a) Evolution des poids secs (chêne, érable, Kleenex, 1972-1973).

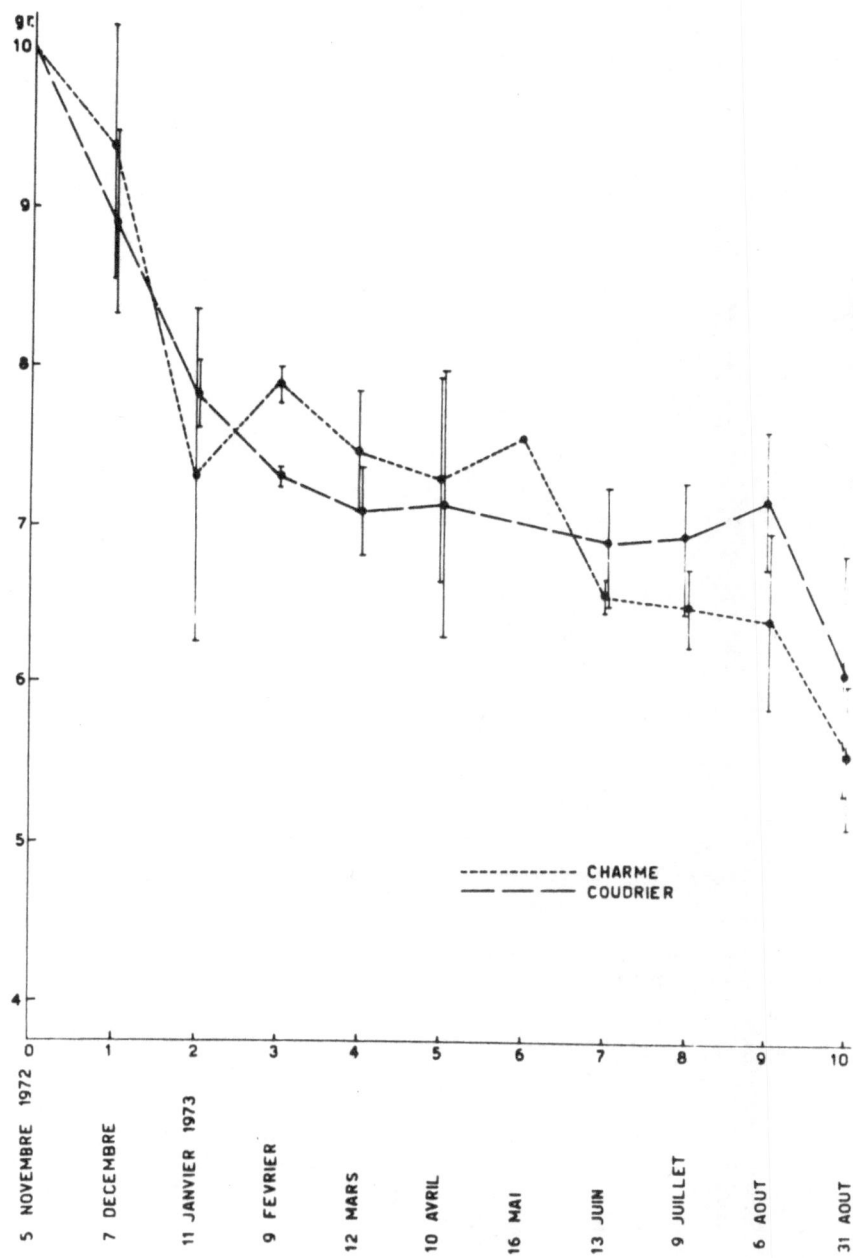

gr.

CHARME

COUDRIER

5 NOVEMBRE 1972

7 DECEMBRE

11 JANVIER 1973

9 FEVRIER

12 MARS

10 AVRIL

16 MAI

13 JUIN

9 JUILLET

6 AOUT

31 AOUT

Fig.4b) Evolution des poids secs (charme, noisetier, 1972-1973).

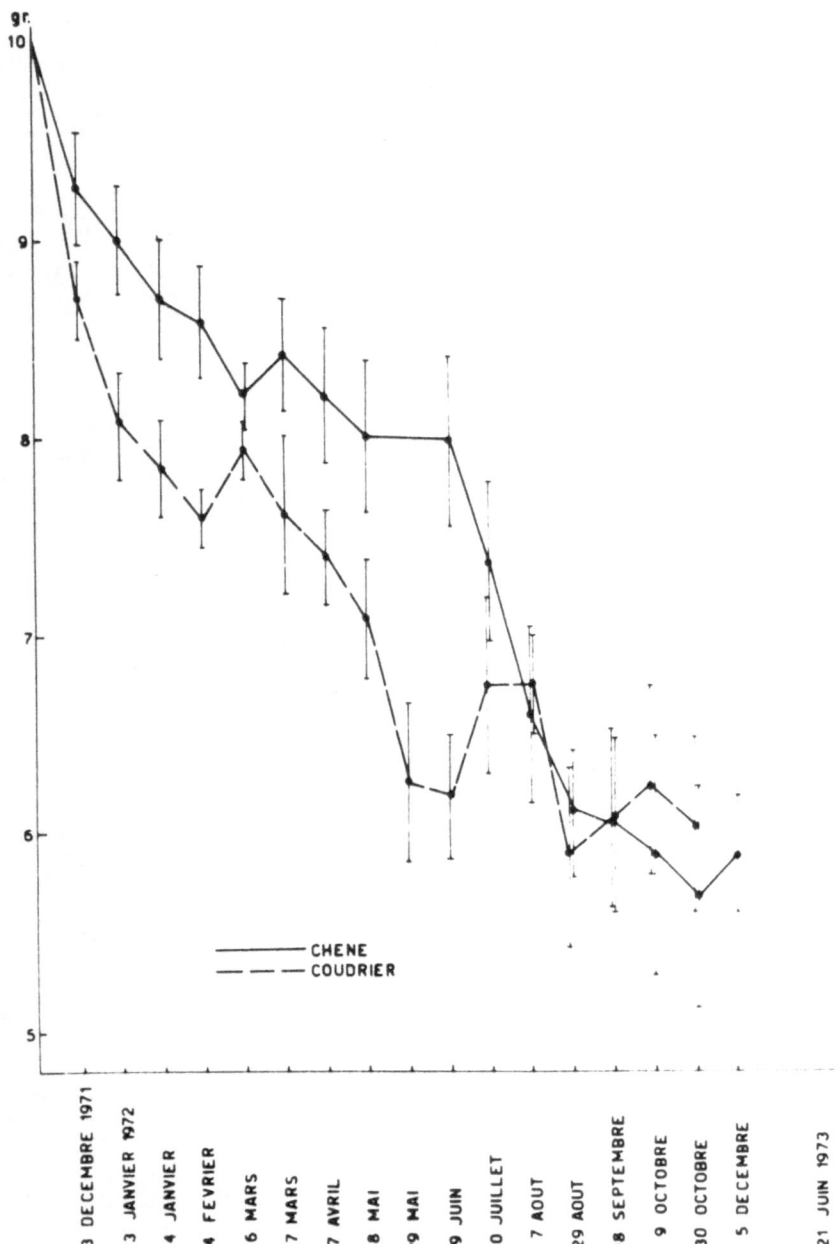

Fig.4c) Evolution des poids secs (chêne, noisetier, 1971-1972). Interval de confiance: 95.

Nous avons attribué le décalage dans les vitesses de décomposition du chêne et du noisetier entre les années 71-72 et 72-73 à une pluviosité moindre dans la seconde année (Mignolet, en prép.).


D i s c u s s i o n

L'examen de ces données nous permet de construire le tableau 1: en donnant à chaque litière un rang déterminé par la taille des onze caractéristiques étudiées, nous pouvons faire un regroupement des substrats

T a b l e a u  1.    Caractérisation des litières monospécifiques

| Espèces<br><br>Caractéristiques | Acer pseudo-<br>platanus | Coryllus<br>avellana | Carpinus<br>betulus | Quercus<br>robur |
|---|---|---|---|---|
| Vitesse de<br>decomposition | I | (2) | (3) | 4 |
| Intensite de<br>la rel.X/S des<br>teneurs en eau | I | 2 | 4 | 3 |
| pH | 2 | I | 4 | 3 |
| Cellulose | I | 4 | 3 | 2 |
| Magnesium | 2 | I | 3 | 4 |
| Calcium | I | 2 | 3 | 4 |
| Cendres totales | I | 2 | 3 | 4 |
| Cendres insolubles | I | 2 | 3 | 4 |
| Potassium initial | 2 | 3 | I | 4 |
| Sodium | pas de différence | | | |
| Phosphore | (2) | (I) | 3 | 4 |

(les chiffres entre parenthèses indiquent l'absence d'écarts significatifs entre les moyennes observées). A l'analyse de ce tableau on constate une majorité de "1" dans le colonne de l'érable, une prédominance de "2", "3" et "4" respectivement dans celles du noisetier, du charme et du chêne. Ce mode de caractérisation purement utilitaire ne préjuge bien sûr en rien de l'importance relative des éléments étudiés.

LA FAUNE DES ORIBATES

Les questions qui nous ont préoccupé  étaient de savoir dans quelle mesure  les caractères physico-chimiques  des litières artificielles pouvaient influencer la faune des Oribates.

Nous n´évoquerons ici  que des considérations basées  sur le total
des Oribates, l´analyse spécifique  devant être abordée ultérieurement.


## Vitesse de colonisation

Nous avons constaté  que sur l´ensemble de la période d´observation (de
décembre à août)  le nombre d´Oribates  colonisant les litières artificielles est très inégal. A titre exemplatif, la figure 5 montre les valeurs obtenues pour la litière de noisetier et celle d´érable.  Afin de
pouvoir comparer les vitesses de colonisation et comme une simple relation linéaire semble suffisante pour décrire cette évolution temporelle,
on a comparé les pentes des droites des moindres carrés. Celles-ci don-
nent donc une idée de la vitesse de colonisation des différents milieux.
Comme il apparaît au tableau 2,  le noisetier est l´essence dont la vitesse de colonisation  est la plus rapide.  Nous obtenons ensuite  dans
l´ordre:  le charme, l´érable et le chêne, les vitesses de colonisation
dans ces deux dernières litières  devant être considérées  comme  à peu
près identiques.

En complément  de la vitesse de colonisation, on pourrait ne s´in-
téresser qu´au bilan faunique, c´est-à-dire le nombre moyen d´individus
occupant un substrat déterminé.  C´est ce qui est présenté au tableau 2
également.  Il apparait  que si les milieux érable  et chêne  sont sensiblement  de même capacité  en faune d´Oribates,  celui  de charme et
surtout  celui de noisetier  sont plus riches  parmi les essences pures
tandis que  la litière reste le milieu  le plus peuplé et le kleenex le
moins.

De plus, cet ordre de capacité faunique se répète assez bien à toutes les époques de prélèvement.  Le tableau suivant montre en effet une
bonne  répétition  de l´odre  litière-noisetier-charme-chêne-érable  et
kleenex (voir tableau 3).

Un test de concordance de Kendall  (voir Siegel 1956) effectué sur
les rangs montre d´ailleurs que cet accord est statistiquement significatif au seuil 1 % ($W$ = 0,88). Notons que nous n´avons effectué ce test
que sur les cinq prélèvements  pour lesquels tous  les 6 substrats  ont
été prélevés et la faune dénombrée.

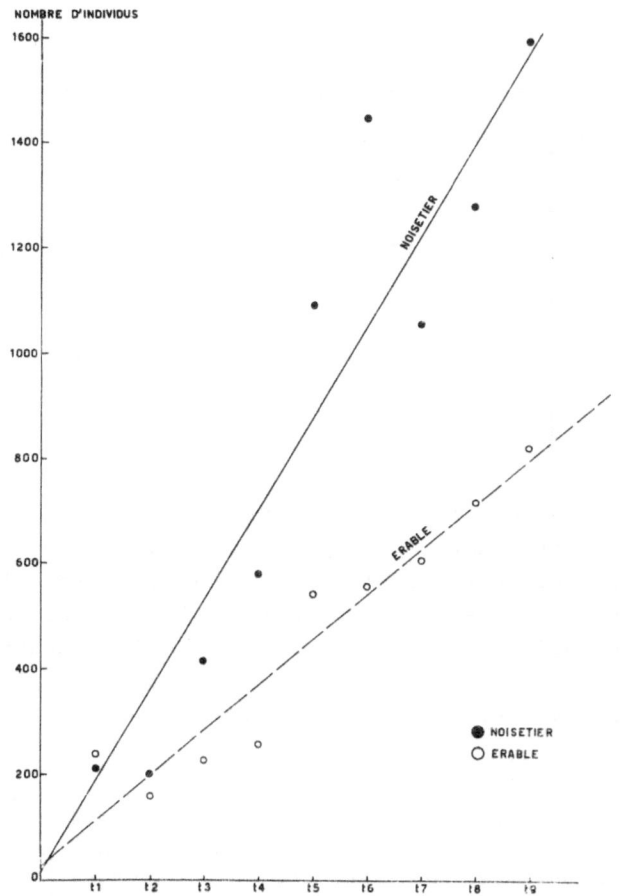

Fig.5. Vitesse de colonisation  par les Oribates des litières pures d'érable et de noisetier.


## La diversité spécifique

La "diversité spécifique" des litières  a été étudiée par la fonction $H$
de Shannon (Shannon   et   Weaver 1949).   Cette fonction donne une mesure
globale,  très synthétique,  de la structure numérique des communautés,
c'est-à-dire de la répartition  de l'abondance relative des différentes
espèces. Le tableau 2 monter que c'est dans la litière de noisetier que
la répartition des individus entre les espèces est la moins hétérogène.
Nous avons ensuite  la communauté des feuilles  de charme  et nous con-
sidérons, enfin, que les autres communautés  diffèrent peu.  Cependant,

T a b l e a u  2.    Vitesse de colonisation bilan faunique et diversité spécifique

|           | Vitesse (pente de la droite) | Bilan (Nombre moyen d'Oribates par sac) | Diversité (H. Shannon) |
|-----------|------------------------------|-----------------------------------------|------------------------|
| Noisetier | 1,80                         | 173                                     | 4,44                   |
| Charme    | 1.28                         | 123                                     | 4,27                   |
| Érable    | O,87                         | 88                                      | 4,05                   |
| Chêne     | O,81                         | 96                                      | 3,87                   |
| Litière   | -                            | 214                                     | 3,82                   |
| Kleenex   | -                            | 40                                      | 3,95                   |

on est surpris de constater que c'est la litière naturelle qui présente la valeur la plus faible. D'autre part, l'analyse de l'ordre de diversité établi à 5 temps de prélèvements confirme l'ordre précédent, la communauté de la litière naturelle étant la plus faible. Ici encore, un test de concordance a établi que cet ordre n'est pas dû au hasard ($W = 0,6$, valeur significative au seuil 5 %). Il en est de même, d'ailleurs, pour le degré de parité spécifique ("evenness" *sensu* Pielou, 1966) ($W = 0,7$).

T a b l e a u  3.    Test de concordance nombre total d'oribates

|        | Litière | Chêne | Charme | Noisetier | Érable | Kleenex |
|--------|---------|-------|--------|-----------|--------|---------|
| $t_3$  | 1       | 3     | 4      | 2         | 5      | 6       |
| $t_4$  | 1       | 5     | 3      | 2         | 4      | 6       |
| $t_5$  | 2       | 4     | 3      | 1         | 5      | 6       |
| $t_7$  | 1       | 4     | 3      | 2         | 5      | 6       |
| $t_8$  | 3       | 5     | 2      | 1         | 4      | 6       |

## Affinités entre les litières

L'indice d'affinité utilisé pour établir les regroupements entre les différentes litières et celui dérivé de la fonction $T$ de Kullback (1959) utilisé notamment par Wauthy, Lebrun, Mercier et Sougnez (1973) Si l'on

considère deux biocénoses, trois fonctions d'indétermination peuvent
être définies:

$H_{(x)}$,     soit l'indétermination qu'un individu appartienne à une
espèce donnée;

$H_{(y)}$,     soit l'indétermination qu'un individu appartienne à l'une
ou l'autre des biocénoses;

$H_{(x,y)}$,    soit l'indétermination qu'un individu appartienne à l'une
ou l'autre des biocénoses sachant qu'il appartient à une
espèce donnée.

De là, on définit $T_{(xy)} = H_{(x)} + H_{(y)} - H_{(xy)}$ et l'indice de divergence $D = \dfrac{T_{(xy)}}{H_{(y)}}$ pouvant varier de 0 à 1.

Pour la facilité, on utilisera une fonction décroissante de $D$ ($A =
= 1 - D$). Cet indice compare globalement le degré de concordance entre
le spectre d'abondance relative des espèces de deux communautés.

Le tableau 4 montre la matrice de coeficients d'affinité établis
par couples de milieux (total des prélèvements). D'emblée, on constate
que toutes les valeurs sont très élevées et très proches puisque comprises entre 0,83 et 0,98. Ceci est normal car les sacs de litière ar-
tificielle ne sont pas des espèces étrangères et sont enfouis dans la
litière naturelle. C'est à partir de celle-ci que la colonisation survient. Il serait donc surprenant que les effectifs spécifiques soient
très tranchés comme il serait simpliste de supposer qu'une espèce très
abondante dans la litière naturelle soit très peu représentée dans les
autres substrats.

D'autre part, on remarquera que la communauté du noisetier s'in-
dividualise le mieux puisque toutes les affinités qu'elle manifeste
sont inférieures à 0,9.

C'est à partir de la matrice du tableau 4 que l'on opère une série
de regroupements. On réunit tout d'abord le chêne et l'érable et à ce
premier groupe viennent s'adjoindre successivement le charme, la litière, le kleenex et, en dernier lieu, le noisetier. Ce réseau d'affinité met l'accent sur une structure en continuum et sur la forte pa-
renté de la litière naturelle avec le charme, l'érable et le chêne.
La position du papier kleenex est moins marginale que prévu et on pourrait s'interroger sur la valeur de ce substrat comme milieu de référence hébergeant des espèces balladeuses arrivant au hasard.

Quant à la communité du noisetier, elle est encore ici très à part
des autres.

T a b l e a u   4.   Diagramme en treillis des valeurs d'affinités

|  | Chêne (oak) | Érable (mapple) | Charme (hornbeam) | Litière (litter) | Kleenex (paper) |
|---|---|---|---|---|---|
| Chêne (oak) | - |  |  |  |  |
| Érable (mapple) | 0,98 | - |  |  |  |
| Charme (hornbeam) | 0,93 | 0,97 | - |  |  |
| Litière (litter) | 0,92 | 0,92 | 0,93 | - |  |
| Kleenex (paper) | 0,88 | 0,90 | 0,90 | 0,90 | - |
| Noisetier (hazel-tree) | 0,83 | 0,88 | 0,89 | 0,84 | 0,84 |

## Discussion et conclusion

Ces observations nous incitent à rechercher dans les caractéristiques physico-chimiques des litières les critères qui justifieraient ce classement. Deux caractéristiques apparaissent: il s'agit d'une part de la teneur en cellulose et d'autre part du *pH* qui individualisent le substrat noisetier et réunissent le chêne et l'érable comme c'est le cas pour les critères faunistiques densité, vitesse de colonisation, diversité et affinités avec les autres communautés. Ceci pourra se confirmer par l'analyse des corrélations que nous aborderons ultérieurement.

En ce qui concerne la densité des Oribates on soulignera qu'elle n'est nullement en relation avec les pertes de poids. Par là on veut dire, si on se rappelle que la perte de poids va de la plus rapide à la moins rapide dans le sens érable-charme-noisetier et chêne que la vitesse de colonisation ne suit nullement un ordre inverse. La quantité pondérale de matière organique n'apparaît donc pas comme un élément prépondérant sur le nombre d'Oribates colonisateurs.

D'autre part, la composition pondérale des différentes essences de la litière naturelle va dans le sens chêne, érable et charme ce qui est loin de correspondre à l'ordre d'affinité que présente la communauté d'Oribates de la litière vis-à-vis de ces essences.

Il est bien évident qu'une litière naturelle représente bien plus qu'un simple mélange de diverses feuilles bien que cela n'ait jamais été démontré. En tout cas, nos résultats le confirment pour la faune des Oribates.

# Remerciements

Cette recherche a été partiellement subsidiée par l'Institut pour l'En-couragement à la recherche scientifique  dans l'Industrie et l'Agricul-ture  (I.R.S.I.A.).  Les analyses  ont été réalisées  au laboratoire du Prof. J. Lambert;  nous remercions Mr. Denudt  pour ses conseils et en-couragements.

# Bibliographie

K u l l b a c k , S. (1959):  Information  theory  and  statistics. J. Wiley, New York, 395 pp.

L e b r u n , Ph. (1971): Ecologie et biocénotique de quelques peuple-ments d'Arthropodes édaphiques. Mém. Inst. Roy. Sci. Nat. Belg., 165, 203 pp.

S h a n n o n , C. E. et W e a v e r , W. (1949): The mathematical theory of communication. The University of Illinois press, Urbane, 117 pp.

S i e g e l , S. (1956): Nonparametric statistics  for the behavioral sciences. MacGraw-Hill, London, 312 pp.

W a u t h y , G., L e b r u n , Ph., M e r c i e r , N. et S o u g n e z , N. (1973): Comparaison  de communautés d'Oribates de litière de chênaies. C.R. Vième Coll. Zool. Sol., Praha, Sept. 73.

# Discussion

C. G r e g o i r e - W i b o : Vous attendiez-vous à obtenir des va-leurs d'affinité aussi elevées et aussi semblables étant donné les dif-ferences relativement importantes mises en evidence pour  les litières etudiées?

P. L e b r u n : Non à priori- cependant, l'étude des affinités a cha-que période de prélèvement pourra nous donner (par une éventuelle répé-tition de l'ordre d'affinité) une idée de la valeur de notre classement. D'autre part,  l'analyse spécifique  donnera  des précisions concernant les différences observées.

E. v. T ö r n e : Könnte Ihrer Meinung nach der Index der Diversität als ein orientierendes Mass zur Einschätzung der Optimalität des biogenen Stoff- und Energie-Kreislaufes dienen?

P. L e b r u n : Il est prémature de vouloir assigner à l'indice de Shannon une signification (ou une corrèlation avec un quelconque facteur) aussi précise- dans l'état actuel cet indice mesure la structure numérique de la communauté et donc le degré d'hétérogénéité de répartition des individus entre les espèces.

E. v. T ö r n e : Ist Ihrer Meinung nach prinzipiell zu erwarten, dass Informationen über den Verlauf der Freizetzung von mineralischen Substanzen aus der verrottenden Streu Hinweise für eine optimierte Forstdüngung erbringen könnten?

R. M i g n o l e t : Peut être à longue échéance. Ce type d'étude nous semble de toute façon pouvoir déboucher sur une meilleure caractérisation d'essences végétales dites "améliorantes".

Z. S. K o z l o v s k a j a : Quelle forme de phosphore était etudiée: totale ou libre?

R. M i g n o l e t : Nous analysions le phosphore total.

H. E i j s a c k e r s : What kind of preparation (drying out) was used for the natural litter. Have you got any idea of the effect of the bag itself on the colonisation by acarids, for instance by comparing "free" litter and litter in "bags"?

P. L e b r u n : The leaves were air-dried during three weeks. It seems that there is no inhibiting effect of bags on *Acarids* and *Collembola*.

H. E i j s a c k e r s : Why was cellulose chosen as a parameter, instead of studying more simple sugars?

R. M i g n o l e t : L'année précédente nous avions effectué des analyses de sucres réducteurs. Il nous est apparu que les teneurs variaient trop rapidement, peut-être d'une heure à l'autre, en fonction des facteurs climatiques: une forte pluie pourrait lassiver ces sucres. Nous avons préféré étudier un élement plus stable.

M. B. B o u c h é : Il est intéressant de constater que le coefficient *H* de Shannon est constamment faible pour la litière naturelle. Vous semble-t-il lié au fait que l'on est en présence d'une faune en équilibre avec le substrat offert par opposition à celle des litières monospécifiques.

Ph. L e b r u n : Je pense que c'est effectivement le cas: la diversité spécifique et la parite élevées dans les litieres monospecifiques seraient l'indice d'une "désorganisation". Celle-çi se marque par une

répartition moins hétérogène  de l'abondance  entre les différentes espèces.

Nous avons constaté  le même fait  pour des communautés  de Microarthropodes corticoles vivant  dans diverses conditions de salubrité de
l'air: les communautés les plus soumises à la pollution par le $SO_2$ sont
les plus diversifiées, les communautés de référence  sont les moins diversifiées.

# Ecological energetics of decomposer invertebrates in a deciduous forest[1] [2] and total respiration budget

E. REICHLE, J. F. McBRAYER[3], AND S. AUSMUS

Environmental Sciences Division, Oak Ridge
National Laboratory,
Tennessee, USA[4]

INTRODUCTION

The analysis of ecosystems has recently received the increasing attention of investigators as they attempt to characterize the functional
dynamics of natural systems by quantifying their component processes
and the interconnections of these processes. The ecosystem concept
implies that an understanding of the processes and their interactions
can provide the ability to predict changes in ecosystem behavior resulting from changes imposed on those processes. While many attributes
of ecosystems can be characterized, all are inextricably related to the
carbon and mineral element metabolism by organisms composing that system.

    The decomposer community of the deciduous forest is responsible
for one of the major ecosystem processes, decomposition and elemental
cycling, and contributes the major heterotrophic respiration $CO_2$ flux
from the forest. It has been estimated that greater than 36 % of res-

[1] Research supported by the Eastern Deciduous Forest Biome US-IBP,
funded by the National Science Foundation under Interagency Agreement
AG-199, 40-193-69 with the Atomic Energy Commission, Oak Ridge National
Laboratory.

[2] Contribution No. 111 from the Eastern Deciduous Forest Biome,
US-IBP.

[3] Present address: Laboratory of Nuclear Medicine and Radiation
Biology, University of California at Los Angeles, Los Angeles, California, USA.

[4] Operated by the Union Carbide Corporation for the U.S. Atomic
Energy Commission.

piration losses  and greater than 95 % of the heterotrophic respiration losses of forest ecosystems result from metabolic activities of the decomposer biota. (Reichle et al, 1973).  Nevertheless, the relative contributions of the various decomposer groups to the total carbon dioxide efflux from the forest floor have not been quantified for any ecosystem.

In a mesic deciduous forest  in  Eastern Tennessee,  USA,  we have quantified the biomass  of the major decomposer flora  and faunal populations.  Annual respiratory  losses  from each group  were established using measures of mean individual weight,  field estimates  of standing crops, laboratory determined  respiration rates  (and  respiratory quotients), and regressions of respiration rates  on body size (Reichle, 1971).  Respiration  rates  of individual populations  were  integrated throughout the year using a mathematical function  of daily temperature and a $Q_{10}$ = 2. Annual summations of the decomposer populations respiration rates  were compared to in situ $CO_2$ efflux measurements  using infrared gas analysis (Edwards and Sollins, 1973).

SITE DESCRIPTION

The forest ecosystem studied  is a second-growth,  mesophytic deciduous stand established on deep alluvial silt loam soil underlaid by dolomitic limestone.  *Liriodendron tulipifera*  is  the dominant canopy species in the stand.  Other canopy species include: *Quercus velutina, Q. alba, Q. coccinea, Q. rubra, Q. prunus, Pinus echinata,* and  *Carya tomentosa. Cercis canadensis* and *Cornus florida* are the primary understory species. Over 90 % of the field layer  of the stand  is found  in three species: *Parthenocissus  quinquefolia, Hydrangea arborescens,*  and  *Polystichum acrostichoides.*  Mean age of the canopy species is 48 years; basal area is approximately 22 $m^2$ ha $^{-1}$.

The litter  is  a typic mull,  with a mean annual standing crop of 560 g dry weight $m^{-2}$.  The A horizon of the soil is well developed; exchangeable potassium averages 0.31 meq/100 g  and  calcium averages 8.4 meq/100 g.  The organic matter  of the  soil averages 4.5 %  in the top 35 cm.  Mean annual soil temperature is 13.3 C and mean annual precipitation is 126.5 cm. Precipitation is rather evenly distributed throughout the year and soil water tension rarely exceeds 3 bars.  Temperature is the primary abiotic factor influencing the rate  of decomposition in litter and soil strata.

METHODS

Biomass estimates   of litter macroarthropods   were determined   from bi-
weekly 0.1 m$^{-2}$   ring samples   of the   forest floor O1   and   O2 horizons
(Moulder and Reichle, 1972).   Annual biomass of litter   and   soil mesofauna   were calculated   from monthly means of soil cores   to a depth of
15 cm (McBrayer, 1973).   All litter and soil substrates   were extracted
on a monthly basis (Reichle et al. 1973).   Nematode standing crops were
estimated monthly using Baerman funnel extraction   and   separated   into
trophic groups by direct microscopic observation (Ferris, 1972;   Ausmus
et al. 1973).

Microfloral   populations   were estimated using   the ATP assay and,
chloraphyll A extraction. The ATP values are apportioned into bacterial
and fungal (including actinomycetes)   active biomass using respirometry
with   and   without antibiotics (Ausmus, 1973).   Standing crops of roots
< 0.5 cm diam. are after Harris et. al (1973)   who separated roots from
soil by washing and sifting with a 2-mm nesh screen.

Empirical   determinations   of nematode   and   microbial respiration
were performed   on a Gilson differential respirometer,   and oxygen consumption values   were converted to $CO_2$ respired using an R.Q.   of 0.80
(Ausmus, 1973;   Ferris, 1972).   Arthropod respiration rates   were   calculated from body size relationships (Reichle, 1971), using the regression of log oxygen consumption ($Y = \mu l\ O_2\ hr^{-1}$) on log live body weight
($X$ = mg): of the form:   $Y = 0,339\ X^{0.808}$.   Modifications   of parameters
in this equation   were used to estimate respiration rates for soil Acarina and Collembola (McBrayer, 1973). Respiratory quotients of 0.82 for
arthropods and nematodes, and 0.85 for earthworms   were used to convert
$O_2$ uptake values to $CO_2$ evolved.

Metabolic   and population statistics   for the decomposer community
are presented   in   Table 1.   Standardized laboratory   respiration rates
were extrapolated   to the forest environment   using a $Q_{10}$ = 2 relationship   and   a computer-generated   sine wave   annual temperature function
(mean annual temperature = 13.3 $^{\circ}$C).   In situ measures of $CO_2$ evolution
from the forest floor have been determined by Edwards and Sollins (1973)
using an infrared gas analysis technique.

RESULTS AND DISCUSSION
Biomass of. the Decomposer Biota

The majority of litter invertebrates (77 % by number and 88 % by weight)
belong to ten taxonomic groups: *Araneàe, Chilopoda, Coleoptera, Collem-*

T a b l e  1.  Calculation of annual $CO_2$ respiration by forest soil and litter decomposers as calculated from body size, biomass and oxygen respiration rate data. Column five gives temperature of respirometry data (13.3° C annual temparature mean was used to convert laboratory data to field values). R.Q. values of 0.80 (microflora and *Nematoda*), 0.85 (*Enchytraeidae*) and 0.85 (all other taxa) were used to convert $O_2$ consumption to $CO_2$ efflux

| Decomposer Taxon | Mean Individual Body Weight (Indiv.$^{-1}$) | Mean Annual Biomass ($m^{-2}$) | $O_2$ Uptake Rate (mg $O_2$ $g^{-1}$ dry wt $day^{-1}$) | Temperature Of $O_2$ Determinations (°C) | Annual $CO_2$ Efflux Population (g $CO_2$ $m^{-2}$ $yr^{-1}$) (Totals) |
|---|---|---|---|---|---|
| *Microflora* | | | | | |
| Algae | 100 ng | 0.31 g | 612.6 | 10 | 72.1 |
| Bacteria | 1 | 5.40 | 751.6 | 10 | 1541.1 |
| Fungi | 100 | 5.93 | 651.3 | 10 | 1466.5 |
| | | | | | 3079.7 |
| *Nematoda* | 0.3 g | 0.89 g | 9.912 | 10 | 3.35 |
| | | | | | 3.35 |
| *Pulmonata* | 37.08 mg | 222.46 mg | 2.400 | 13 | 0.161 |
| | | | | | 0.161 |
| *Arthropoda* | | | | | |
| *Phalangida* | 0.30 mg | 5.79 mg | 5.760 | 13 | 0.009 |
| *Pseudoscorpionida* | 0.415 | 9.3 | 2.136 | 15 | 0.005 |
| *Chilopoda* | 0.69 | 32.32 | 22.120 | 13 | 0.214 |
| *Diplopoda* | 1.08 | 29.2 | 71.712 | 15 | 0.543 |
| *Araneae* | 0.856 | 50.9 | 37.020 | 10 | 0.733 |
| *Acarina* | | | | | |
| *Gamasina* | 0.280 | 2836.3 | 2.856 | 10 | 0.131 |
| *Uropodina* | 0.182 | 237.7 | 0.120 | 10 | 0.011 |
| *Orbatei* | 0.092 | 3678.2 | 0.045 | 10 | 0.067 |
| *Prostigmata* | 0.004 | 59.3 | 0.002 | 10 | 0.00003 |

Table 1 continued

| Decomposer Taxon | Mean Individual Body Weight (Indiv.$^{-1}$) | Mean Annual Biomass (m$^{-2}$) | $O_2$ Uptake Rate (mg $O_2$ g$^{-1}$ dry wt day$^{-1}$) | Temperature Of $O_2$ Determinations (°C) | Annual $CO_2$ Efflux Population (g $CO_2$ m$^{-2}$ yr$^{-1}$) | (Totals) |
|---|---|---|---|---|---|---|
| Pauropoda | 0.006 | 10.5 | 2.160 | 18 | 0.005 | |
| Symphyla | 0.084 | 104.1 | 10.464 | 15 | 0.282 | |
| Protura | 0.003 | 8.3 | 1.272 | 18 | 0.003 | |
| Diplura | 0.081 | 58.3 | 17.640 | 15 | 0.267 | |
| Insecta | | | | | | |
| Collembola | | | | | | |
| Onychiuridae | 0.005 | 37.6 | 1.272 | 18 | 0.010 | |
| Poduridae | 0.008 | 12.4 | 1.872 | 18 | 0.005 | |
| Isotomidae | 0.017 | 150.4 | 2.688 | 18 | 0.087 | |
| Entomobryidae | 0.045 | 295.6 | 4.752 | 18 | 0.303 | |
| Sminthuridae | 0.008 | 6.3 | 0.984 | 18 | 0.002 | |
| Orthoptera | 6.73 | 26.91 | 3.360 | 13 | 0.027 | |
| Psocoptera | 0.080 | 1.9 | 2.688 | 18 | 0.0009 | |
| Coleoptera | 0.075 | 134.4 | 16.896 | 15 | 0.589 | |
| Hymenoptera | 0.110 | 16.92 | 4.800 | 13 | 0.243 | |
| Lepidoptera (larvae) | 0.100 | 5.73 | 4.800 | 13 | 0.008 | |
| Diptera (larvae) | 1.800 | 849.6 | 1.687 | 13 | 0.460 | 4.005 |
| Annelida | | | | | | |
| Enchytraeidae | 0.080 mg | 0.5 g | 3.769 | 20 | 0.368 | |
| Lumbricidae | 0.167 g | 10.5 g | 0.011 | 20 | 0.225 | 0.593 |

*bola, Diplopoda, Diptera, Hymenoptera, Lepidoptera* (larvae) *Orthoptera,*
and *Pulmonata* (Moulder and Reichle, 1973). Mean annual biomass is 842 mg
dry wt m$^{-2}$. Microarthropods in litter average 5.9 x 10$^4$ indiv. m$^{-2}$ with
a biomass of  342 mg m$^{-2}$  (McBrayer and Reichle, 1971).  Annelids, pri-
marily *Octolasium* sp.  earthworms (10.5 g m$^{-2}$)  constitute the bulk of
the soil invertebrate biomass. *Oribatei* (3.7 g m$^{-2}$) and *Gamasina* (2.8 g
m$^{-2}$) mites  and  larval *Diptera* (0.8 g m$^{-2}$)  are  the dominant soil-inhabiting microarthropods (McBrayer, 1973). Nematode population averaged
annually 1.2 x 10$^6$ indiv. m$^{-2}$ (0.89 g dry wt m$^{-2}$). Of this total almost
50 % were phytophagous,  46 % predators,  and  the remaining fungivores
and bacteriovores (Ferris, 1972).

   Microbial populations  varied throughout the year  with changes in
temperature  and moisture, with moisture limiting at temperatures above
20°C.  Densities per gram  of substrate for bacteria  were 2.3 x 10$^8$ in
litter, and 5.6 x 10$^6$ in A1 soil to 5 cm.  Fungal densities  were 2.7 x
x 10$^6$ for O1 litter,  1.24 x 10$^8$  for O2 litter, and  0.3 x 10$^6$  for A1
soil to 5 cm  (Edwards,  1972).  Metabolically active microbial biomass
(that part  of the ATP pool  which  mathematically  corresponds  to the
microbial populations which are respiring  and  catabolizing substrates
at maximum rates) for the entire O1, O2  and  A1 profile averaged: bacteria 5.4, fungi 5.9  and algae 0.3 g dry wt m$^{-2}$ (Ausmus, 1973).  Total
microbial standing crop averaged 10 times these values,  and were  composed of inactive tissues, e.g., rhizomorphs, spores and fruiting bodies
(Ausmus, 1973).

## Decomposer respiration

Annual  contributions  of decomposer taxa  to the carbon dioxide efflux
from the forest floor  are  summarized in Table 1.  Column 1 enumerates
the decomposer taxa examined;  both mean individual body sizes and mean
annual biomass are given in following columns. Biomass was approportion-
ed as 26.7 % *Arthropoda,* 33.9 % *Annelida,* 2.7 % *Nematoda* and 0.7 % *Pulmonata* (Fig. 1).  Collectively,  invertebrate biomass amounted to 20.8 g
dry wt. m$^{-2}$,  whereas microbes  (bacteria, fungi, algae)  accounted for
11.6 g dry wt m$^{-2}$. Temperatures are given for the respective laboratory
oxygen consumption rates, which are converted to mean annual forest $CO_2$
efflux (g $CO_2$ m$^{-2}$ yr$^{-1}$) according to procedures outlined in the methods
section. Of the total of 3088 g $CO_2$ m$^{-2}$ yr$^{-1}$ respired by the decomposer
community, invertebrates  were  responsible  for approximately 0.3 % --
*Arthropoda* 0.1 %, *Annelida* 0.02 %, *Nematoda* 0.1 % and *Pulmonata* 0.005 %.

Microflora contributed toward 99.7 % of the decomposer respiration, although they consituted only 36 % of the standing crop (Fig. 1).

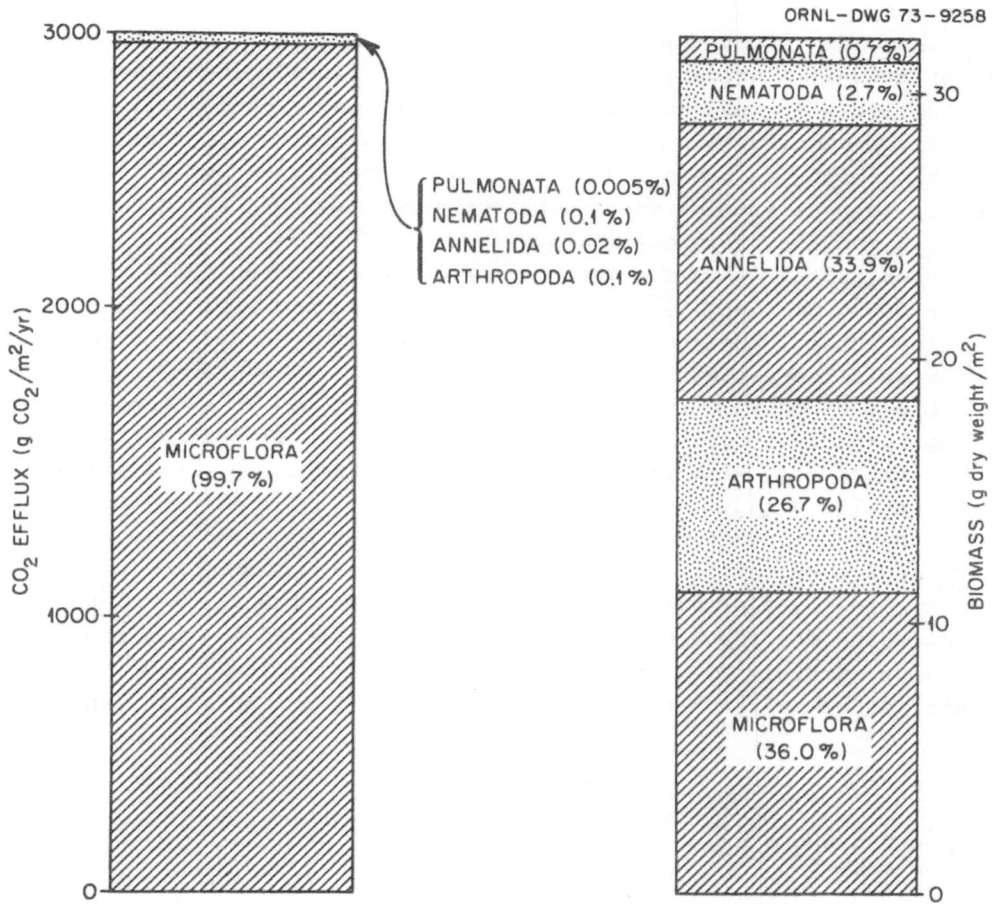

Fig.1. Comparative biomass and respiratory $CO_2$ fluxes for the major decomposer biota of the soil and litter strata in a mesic, deciduous forest.

## Forest floor respiration

Total biomass and respiration of the invertebrate decomposer fauna are consistent with similar estimates previously developed for forest soils with mull organic layers (Macfadyen, 1963; Kitazawa, 1967; Edwards, Reichle and Crossley, 1970). Data collected by European investigators (reviewed by Macfadyen, 1963) have demonstrated that, while the various faunal components of the decomposer community may vary considerably in biomass and respiratory contributions, finite substrate resources dictate a remarkably stable total metabolic activity. If this hypothesis can be substantiated, then our ability to predict total metabolic activity of natural substartes from faunal and floral composition should eventually allow us to understand (1) the basic ecological processes underlying substitution of similar functional roles (niches) by different biotic components and (2) the consequences of alteration of species composition of the decomposer community on heterotrophic respiration and the decomposition process.

Continuous, 24-hr measurements of $CO_2$ efflux from the forest floor (Edwards and Sollins, 1973) yielded data on diurnal variation, seasonal baseline rates, and the limiting effects of temperature and moisture variables. The estimate of total annual $CO_2$ evolved from the floor of this mixed mesophytic forest, based on integration of seasonal values, was approximately 3.8 kg $CO_2$ $m^{-2}$ $yr^{-1}$. Edwards and Sollins (1973) have shown that roots, with their associated rhizoflora in the top 15 cm of soil annually accounted for between 22 and 36 % of the total $CO_2$ evolved from the forest floor. Ausmus (1973) has independently estimated that root respiration contributes ~ 21 % and rhizosphere microflora approximately 9 % of the annual $CO_2$ efflux from the floor of the same forest. Data for the same forest ecosystem presented in this paper show that approximately 71 % of the efflux of $CO_2$ from the forest floor can be attributed to the heterotrophic decomposer biota of litter and soil.

R e f e r e n c e s

A u s m u s , B. S. (1973): The use of the ATP assay in terrestrial decomposition studies. In: Modern Methods in the Study of Microbial Ecology (T. Rosswall, Ed.). Bull. 17 Ecological Res. Com. Swedish Nat. Sci. Res. Council, 511 pp.

A u s m u s , B. S. (1973): Litter and Soil Microbial Dynamics in a Deciduous Forest Stand. PhD. Thesis, University of Tennessee, Fall 1973.

A u s m u s , B. S., F e r r i s , J. M., R e i c h l e , D. E., and W i l l i a m s , E. C. (1973): The role of primary consumers in forest root process. The Belowground Ecosystem: A Synthesis of Plant Associated Processes. Ft. Collins, Colorada, September 5-7, 1973.

Edwards , N. T. (1972): Decomposition Processes, pp. 26-37. In: Oak Ridge IBP Research Site Annual Progress Report. EDFB Memo Rep. 71-92. Oak Ridge National Laboratory.

Edwards , C. A., Reichle , D. E., and Crossley , D. A., Jr. (1970): The role of soil invertebrates in turnover of organic matter and nutrients, pp. 147-172. In: Analysis of Temperate Forest Ecosystems (D. E. Reichle, Ed.). Springer-Verlag, Berlin-Heidelberg-New York, 304 pp.

Edwards , N. T. and Sollins , P. (1973): Continuous measurement of carbon dioxide evolution from partioned forest floor components. Ecol. 54, 406-412.

Ferris , J. M. (1972): Terrestrial decomposition processes: Population and energy dynamics of selected invertebrates. EDFB Memo Rep. No. 72-166. Oak Ridge National Laboratory.

Harris , W. F., Kinerson , R. S., and Edwards , N. N. (1973): Comparison of belowground biomass of natural deciduous forests and loblolly pine plantations. In: The Belowground Ecosystem, A Synthesis of Plant Associated Processes, Ft. Collins, Colorado, September 5-7. 1973.

Kitazawa , Y. (1967): Community metabolism of soil invertebrates in forest ecosystems of Japan. In: Secondary Productivity of Terrestrial Ecosystems (K. Petrusewicz, Ed.), Panstwowe Wydawnictwo Naukowe, Warsaw, 879 pp.

Macfadyen , A. (1963): The contribution of the microfauna to total soil metabolism. In: Soil Organisms (J. Doeksen and J. van der Drift, Eds.), North-Holland Pub. CO., Amsterdam, 453 pp.

McBrayer , J. F. and Reichle , D. E. (1971): Trophic structure and feeding rates of forest soil invertebrate populations. Oikos 22, 381-388.

McBrayer , J. F. (1973): Energy Flow and Nutrient Cycling in a Crytozoan food web. PhD. Thesis, University of Tennessee, Fall 1973.

Moulder , B. C. and Reichle , D. E. (1972): Significance of spider predation in the energy dynamics of forest floor arthropod communities. Ecol. Monog. 42, 473-498.

Reichle , D. E. (1971): Energy and nutrient metabolism of soil and litter invertebrates. In: Productivity of Forest Ecosystems (P. Duvidmeaud, Ed.), UNESCO, 707 pp.

Reichle , D. E., Crossley , D. A., Edwards , C. A., McBrayer , J. F. and Sollins P. (1973): Organic matter and 137Cs turnover in forest soil by earthworm populations: application of bioenergetic models to radionuclide transport. In:

Radionuclides in Ecosystems (D. J. Nelson, Ed.), US AEC CONF-710501,
U.S. Document Printing Office, Springfield, Virginia.

R e i c h l e , D. E.,  D i n g e r ,  B. E.,  E d w a r d s ,  N. T.,
H a r r i s ,  W. F.  and  S o l l i n s ,  P. (1973): Carbon Flow
and Storage in a Forest Ecosystem. In: Carbon  and  the Biosphere
(G. M. Woodwell, Ed.), US AEC CONF-720510, U. S. Document Printing
Office, Springfield, Virginia.

# D i s c u s s i o n

A. P a l i s s a :  Die Lichtstärke ist ein wichtiger Umweltfaktor, der
die Aktivität beeinflusst. Beleuchtungstärke und $O_2$ Verbrauch sind kor-
reliert, Abweichungen von cca 100 % sind möglich.  Unter welchen Licht-
verhältnissen  wurden  die Zahlen gewonnen? Ergeben sich  Hinweise  auf
rhythmische Prozesse?

D. R e i c h l e :  To avoid bias  in our respiration estimates  and to
obtain realistic values  for  extrapolation  to natural environments of
the forest  we have employed hourly means  of continuous 24-hr respirometry under total darkness  for soil-inhabiting microarthropods  and 12
hr light at < 50 footcandles for litter-inhabiting species.

E. v o n  T ö r n e :  Do you feel  that  comparative  respirometry  is
a realistic measure  of the importance  of organisms  in the ecosystem?

D. R e i c h l e :   Carbon dioxide evolution,  or any other measure of
respiratory energy flux,  while a useful measure  of the energetic cost
of metabolism  do not indicate  the functional roles performed by these
organisms as they metabolize.  Respiration is a measure  of the "costs"
of doing business, but not the object of this energy exploitation.

J. M. A n d e r s o n :  Have you made independent estimates  of microbial biomass for comparison with ATP estimates?  Were the $Q_{10}$ relationships of individual soil animal species  and  microorganisms integrated
with temperature measurements made on the site?

D. R e i c h l e :  Yes, we have independent measures of microbial numbers  and mean isolate weights  from  serial dilution plate counts-this
of course is not a reliable measure of "active" biomass.  Metabolically
active  microbial  biomass  was estimated  by standardizing ATP  assays
against $CO_2$ respiration rates  for which empirical $CO_2$ efflux/mass correlations had been established for active cells in the culture. Yes, we
used a computer-generated continuous  temperature  function  which  was
a sine wave with an annual mean of 13.3 $^{\circ}C$ and amplitudes corresponding
to seasonal temperature maxima and minima.

# The significance of the millipede Glomeris marginata (Villers) for oak-litter decomposition and an approach of its part in energy flow

VAN DER DRIFT J.

Institute for Nature Management (R. I. N.)
Arnhem

*Glomeris marginata* (Villers) is a widespread inhabitant of litter layers in temperate deciduous forests with both mor and mull conditions. In oak forests in The Netherlands it occurs frequently, at densities of 5-10 mature individuals per square meter. *G. marginata* feeds mainly on moderately decomposed litter. Under laboratory conditions, about 90 % of the ingested leaf litter is returned to the soil as faecal pellets measuring approximately 1.5 - 2.0 x 1.0 - 1.5 mm and consisting mainly of leaf fragments with a surface area of up to 1 mm$^2$. The decomposition of pellets of *G. marginata* on a diet of hazel leaves was studied by Nicholson et al. (1966).

The aims of the present study were 1. to compare the faeces production of adult *Glomeris* under constant laboratory conditions with defecation in the field; 2. to obtain quantitative data on the faeces production and weight increase of mature *Glomeris* in the field over a three-year period; and 3. to evaluate the role of *Glomeris* in litter breakdown and its share in energy dissipation in an oakwood.

## Defecation under constant laboratory conditions and in the field

Adults varying in weight between 60 and 250 mg, were kept individually in cylindric perspex cells (diameter and height 5 cm) provided at both ends with lids (mesh size 1 mm) and filled with unselected litter material collected in the field. In the laboratory the cells were kept in the dark in a container with a relative humidity of 95 % and at constant temperature of 15°C, and in the field they were dug into the forest litter layer. The pellets were picked out and weighed (oven-dry) weekly. About once a month the litter in the cells was refreshed. Fig.1 shows two representative graphs. The upper graph (natural conditions)

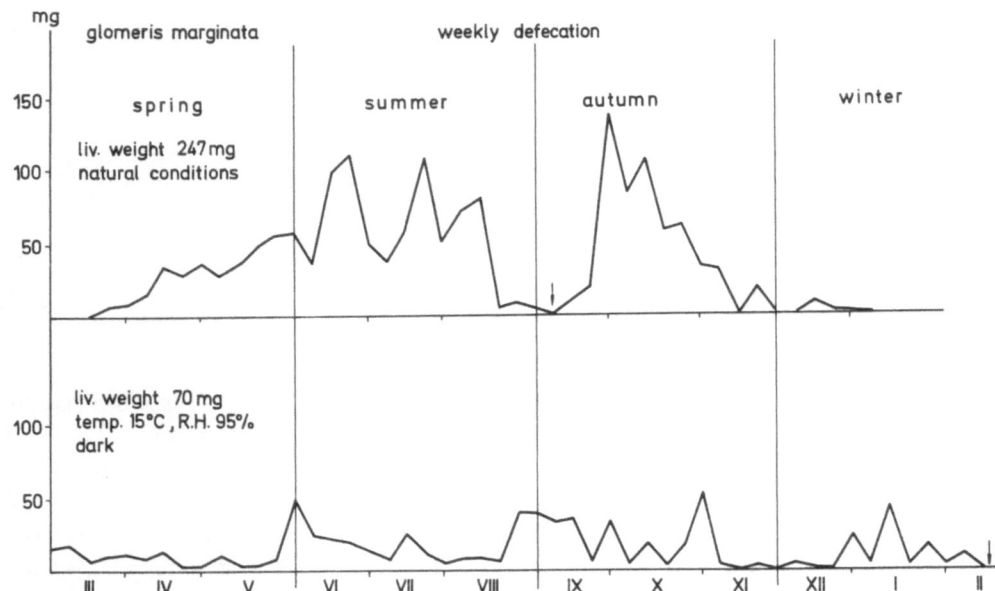

Fig.1.    Weekly productions of faecal pellets    (mg. oven-
dry)    by a mature specimen of *Glomeris marginata*.    Upper
graph: Natural conditions.    Lower graph: laboratory conditions.

shows, not withstanding the wide fluctuations, a clear pattern: defecation begins in March, increases regularly  in April and May, fluctuates
on a rather high level up to mid-August,  is then completely  to almost
completely  suppressed  until mid-September,  increases sharply  to top
values in October, and decreases gradually until it ceases in December.
        At the zero point  of defecation  the animal moults  (indicated by
an arrow), which is evidently related  to the suppression of defecation
and the subsequent sharp increase.   In some cases moulting occurred one
or two weeks earlier or later, and with the same increase in defecation.
        The lower graph (laboratory conditions)  shows a rather capricious
sequence of higher and lower defecation values  and no sign of the pattern found  under natural circumstances.   In this case moulting did not
occur until February, but in other cases it occurred at other times and
once even with an interval of a few months. It seems evident that laboratory conditions disturbed the organism's internal regulation.
        The data collected in this way clearly cannot serve as a basis for
estimation of the performance in the field.

**Feaces production and weight increase of *Glomeris* adult  in the field**

In ten cells of the same type as described above and also provided with unselected litter material  and dug into the litter layer, *Glomeris* was kept individually  over  a three-year period.  The faecal pellets  were collected, dried, and weighed, the litter  was refreshed  once a month, and the live weight of the specimen was recorded.  Fig. 2 shows the re-

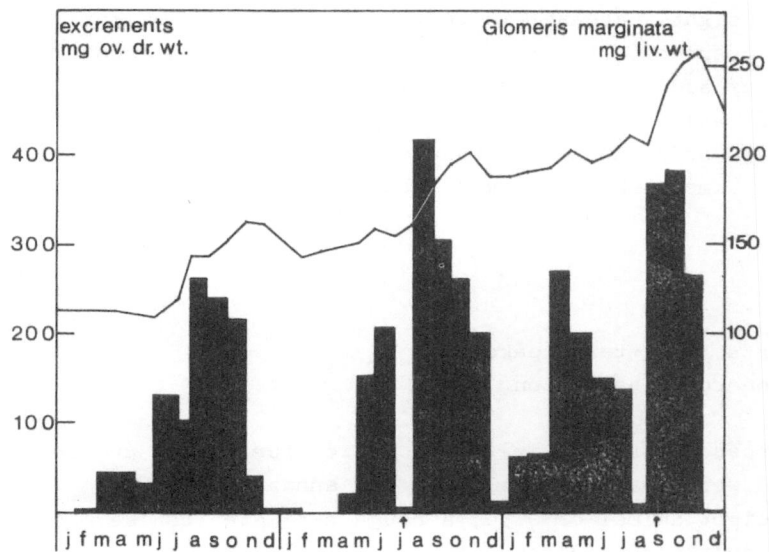

Fig.2.  Monthly productions of faecal pellets by one mature *Glomeris marginata* during 1969, 1970 and 1971 (diagram).  Upper line indicates the live weight of the millipede. Arrow: moulting observed.

sults of a representative case.  The diagram gives a less detailed picture than Fig. 1,  being based on monthly observation,  but the general
pattern is the same.  Moulting (indicated by an arrow)  was accompanied
by a decrease in defecation,  although there is some overlapping due to
the long intervals.

     In the mild winter of 1970/1971, with litter temperatures of up to
$7^{\circ}C$, pellets were produced as early as January and February.  It is not
certain whether this was also the case for free-living animals.  Bocock
and Heath (1967) found in winter most of the *Glomeris* population in the
soil at depths below 5 cm.  The annual defecation was calculated by addition of the monthly pellet production.

     It is clear that the weight increase occurs shortly after moulting;
at that time the integument  is soft and food intake,  like  defecation,

is at the highest level.  The annual weight increase  can be determined
by comparison of the weight levels  before  moulting  in the successive
years.  If the weight of the skin cast at moulting is added, individual
production can be estimated.

For seven animals that survived the three-year period of the field
investigation and whose initial weight varied from 60-150 mg, the oven-
dry weight of the total faecal production during the three years amounted 10.8 10.5 and 9.3 times the live weight of the animals before moulting.  The average weight increase  in percentage  of these weights  was
31, 26, and 27 %, respectively.  The average production  in each of the
years - obtained  by adding the average weights  of the cast skins - is
2 % higher: 33, 28, and 29 % of the live weight.

On average, *Glomeris*  annually produces an amount of faeces (oven-
dry) equalling ten times its own weight  and  increases its body weight
by 30 %.


**The rôle of *Glomeris* in litter breakdown
and its share in energy dissipation**

On the basis  of these averages,  giving a ratio  for faeces production
(oven-dry) to live weight animal of 10 and for annual production  (live
weight) to live weight animal of 0.30, a rough estimate  can be made of
the share of the adult *Glomeris* population  in litter breakdown  and in
the energy dissipation  by the soil òrganisms involved in decomposition
in the oakwood where the culture experiments were done.

This 150-years-old oakwood with a 10-15 cm thick litter  and humus
layer, was taken as object in an I.B.P. study, the results of which will
be published elsewhere; a summary can be found in the Netherlands I.B.P.
Report 1973. For our purpose it is sufficient to mention that the average annual decomposition in the soil  was estimated at 3300 Kcal/$m^2$, of
which only  about 150 Kcal  was due to the activity  of the soil fauna.

Field sampling and quantitative surveys indicated that the average
density of adult  *Glomeris*  (mean live wt. 100 mg)  may be put  at five
per $m^2$. Average annual leaf-litter production over the 1967-1970 period
was 367 g/$m^2$. In laboratory feeding experiments with oak-litter, defecation came to 91 % of consumption.

The annual defecation of *Glomeris* calculated from the population's
live weight per $m^2$ (5 x 100 mg)  and the ratio  of 10 is 5000 mg/$m^2$. To
derive  the corresponding consumption  from this value  by applying the
Defecation:  Consumption ratio of 0.91,  the defecation value  must be

corrected  by the subtraction of 5 %,  which is  the difference between
the ash content of the faeces in the culture cells  and the ash content
of the pellets  in the oak-litter  feeding  experiments,  in which sand
grains were removed from the litter. Comparison of the resulting annual
consumption value  of 5.22 g/m$^2$ with the leaf-litter production  of 367
g/m$^2$ shows  that 1.4 % of the average annual leaf-litter production is
consumed by Glomeris.

The difference between consumption and defecation gives the amount
of assimilated food, in this case 470 mg/m$^2$.  Part  of this  amount  is
stored  in production, i.e., according to the above calculations,  30 %
of the live weight. Since the average dry weight of Glomeris is 25 % of
the live weight,  dry production  is  0.30 x 500 x 0.25 = 40 mg/m$^2$. The
remaining part is respired: 470-40 = 430 mg/m$^2$. If the caloric value of
the respired material  is put at 4.8 Kcal per gram, about 2 Kcal per m$^2$
is dissipated per year.  Thus, on the basis  of the total annual decomposition of 3300 Kcal/m$^2$  the share of Glomeris  in the energy flow  is
only 0.06 %, or about 1.3 % of the total respiratory activity  of soil
fauna, which was put at 150 Kcal per m$^2$ for this wood plot.

It must be admitted that both of these estimates  of the share  of
the mature Glomeris population  in litter breakdown  and  in the energy
flow through soil are rough. However, they indicate the order of magnitude of the part  taken  by Glomeris in these processes  in the wood in
which the study was performed.

It may be remarked that the population density referred to is low.
Local  concentrations  in other wood plots  showed densities  upto four
times larger.  If the indicated ratio  is also valid under these conditions and the litter production is of the same order, Glomeris may consume about 5 % of annual leaf fall,  an estimate which is in agreement
with that of Thiele (1968).

SUMMARY

From field-and laboratory observations  on the feeding pattern  of Glomeris marginate  it appears that the characteristic pattern observed in
the field  is completely lacking  at constant laboratory conditions. On
base of litter consumption and weight increase in the field the role of
an adult Glomeris population in breakdown of oak litter  and  its share
in the energy flow through soil is estimated.

R e f e r e n c e s

B o c o c k ,   K. L. and   H e a t h ,   Y. (1967):   Feeding   activity of
      the millipede Glomeris marginata (Villers) in relation to its ver-
      tical distribution in the soil. In: Progress in Soil Biology, 223-
      240 pp.,   ed. O. Graff   and   J. E. Satchell.   Vieweg & Sohn Braun-
      schweig.
F i n a l   R e p o r t   I B P ,   The Netherlands (1973): To be published
      by North Holland Publishing Company, Amsterdam-London.
N i c h o l s o n ,   P. B.,   B o c o c k ,   K. L. and   H e a l ,   O. W.
      (1966):   Studies   on the decomposition   of the faecal pellets of a
      millipede   Glomeris marginata (Villers).   J.Ecol. 54, 755-766.
T h i e l e ,   H. U. (1968): Die Diplopoden   des Rheinlandes Decheniana
      120, 343-366.

Discussion

J. A. W a l l w o r k :   There is some evidence that populations of *Glo
meris* in grassland   can consume   much higher proportions of litter than
you have suggested here.   Do you think that there is a significant difference   between   the role   of *Glomeris*   in grassland and woodland ecosystems?

J. V a n   d e r   D r i f t :   I have   no experience   with   *Glomeris* in
grassland   but it is known   that *Glomeris'* consumption and defecation is
highly dependant   on type of food.   So hazellitter   is consumed in much
greater amounts than oaklitter and perhaps this is the case with grass-
landlitter too.

# Consommation annuelle d'une population naturelle de vers de terre (Millsonia anomala Omodeo, Acanthodrilidae: Oligochetes) dans la savane de Lamto (Côte d'Ivoire)

P. LAVELLE

Laboratoire de Zoologie de l'E. N. S.,
Paris

*Millsonia anomala* est l'espèce la plus importante du peuplement en Vers de terre des savanes de Lamto. Constituant de 40 à 80 % de la biomasse, elle est représentée, dans un faciès herbeux à *Loudetia simplex*, par 215.000 individus pesant en poids frais (tube digestif plein) 250 kg et produisant annuellement plus de 400 kg de matière vivante par hectare (Lavelle, 1972).

C'est une espèce géophage qui se nourrit dans les premiers centimètres du sol. On a cherché à mesurer la consommation de l'animal de façon à établir le bilan d'énergie d'un individu puis d'une population naturelle.

METHODE D'ETUDE

La technique employée est fondée sur la modification des propriétés physiques de la terre excrétée par le Ver, et tout particulièrement de sa structure dont on sait qu'elle est rendue plus stable par la digestion. Afin d'accentuer ce phénomène, on a utilisé comme milieu d'élevage un sol artificiel à structure particulaire: on prend de la terre humide que l'on fait passer au travers d'un tamis à mailles de Imm; les déjections du Ver placé dans ce milieu se présentent sous forme d'unités compactes -cylindres allongés et petits conglomérats- que l'on sépare aisément du reste par un nouveau tamisage.

On a étudié trois variables déterminantes de la quantité de terre ingérée par l'animal: l'humidité du sol, sa température et le poids du Ver. Par la suite, la consommation d'une population naturelle a été calculée, connaissant chaque mois la température du sol et sa teneur en eau moyennes, ainsi que la structure pondérale de la population.

VARIATIONS DE LA CONSOMMATION EN FONCTION DE L'HUMIDITE DU SOL

L'humidité du sol  est le principal régulateur de l'activité des Vers de
terre  dans la savane de Lamto; son influence  sur la consommation de M.
anomala est très nette (fig.1a). Aux valeurs du pF supérieures à 4,2 les

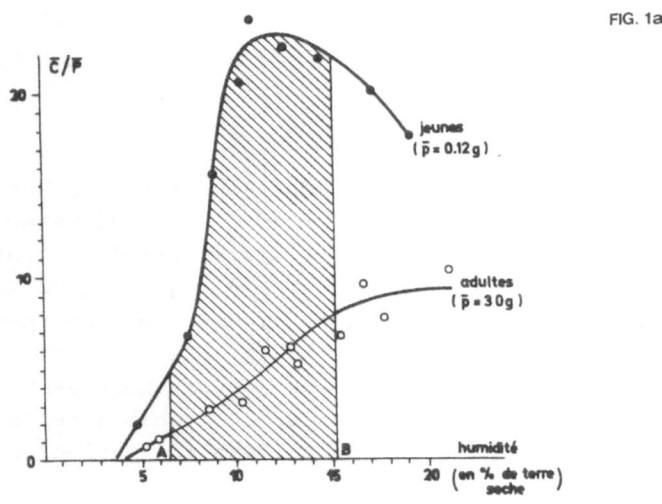

Fig.1.  Variations de la consommation de terre par Millsonia anomala  en fonction de l'humidité du sol (1a), de
la température moyenna (1b)  et du poids du ver (1c). C:
consommation  moyenne  journalière  en g  de terre sèche;
P: poids frais moyen du ver.

Vers sont quiescents et n'ont donc aucune activité. Lorsque la teneur en
eau augmente, la consommation de terre s'accroît.  Les valeurs optimales
de l'humidité -à 25,6° C- sont de 11 % pour les jeunes (≈pF 2,5) qui ingèrent alors 24 fois leur propre poids de terre  chaque jour  et de 20 %
pour les adultes  (pF < 2),  dont la consommation journalière  est alors
égale à 9 fois leur propre poids.
     Pour des valeurs  plus élevées  de l'humidité du sol, les Vers con-
somment moins.

VARIATIONS DE LA CONSOMMATION DE TERRE EN FONCTION
DE LA TEMPERATURE DU SOL

On a élevé des M. anomala,  jeunes et adultes,  dans une terre contenant
12 % d'eau, à quatre températures moyennes -16,9° C, 24,8° C, 29,8° C  et

33,3° C- recouvrant l'ensemble des valeurs observées sur le terrain entre 0 et - 10 cm.

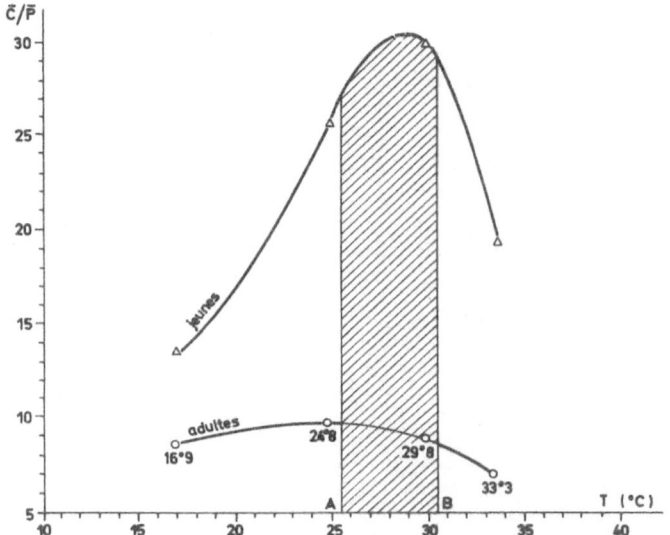

La consommation des jeunes Vers est minimale à 16°9 alors qu'ils
ingèrent en moyenne 13,9 fois leur poids de terre journellement. Elle
augmente jusqu'à 29°8 où elle est maximale (30,5 poids/jour) puis décroît à 33°3 (19,5 poids/jour).

A 16°9 les adultes ont une bonne activité; ils consomment déjà 7,5
fois leur propre poids chaque jour. C'est à 24°8 qu'ils se nourrissent
le plus (9,4 poids/jour). Au delà de cette valeur leur consommation
décroît lentement (8,0 poids/jour à 29°8) puis plus brutalement (3,8
poids/jour à 33°3).

Il apparaît ainsi que les jeunes et les adultes ont des consomma-
tions très différentes et des réactions dissemblables aux facteurs du
milieu: les valeurs optimales pour les jeunes sont plus sèches et plus
chaudes et les adultes consomment dans tous les cas relativement moins
de terre.


VARIATIONS DE LA CONSOMMATION AU COURS DU DEVELOPPEMENT

On a mesuré la consommation de l'espèce tout au long du développement,
à une humidité moyenne de 12 % et rapporté les résultats à la température de 25°6 (fig. 1c).

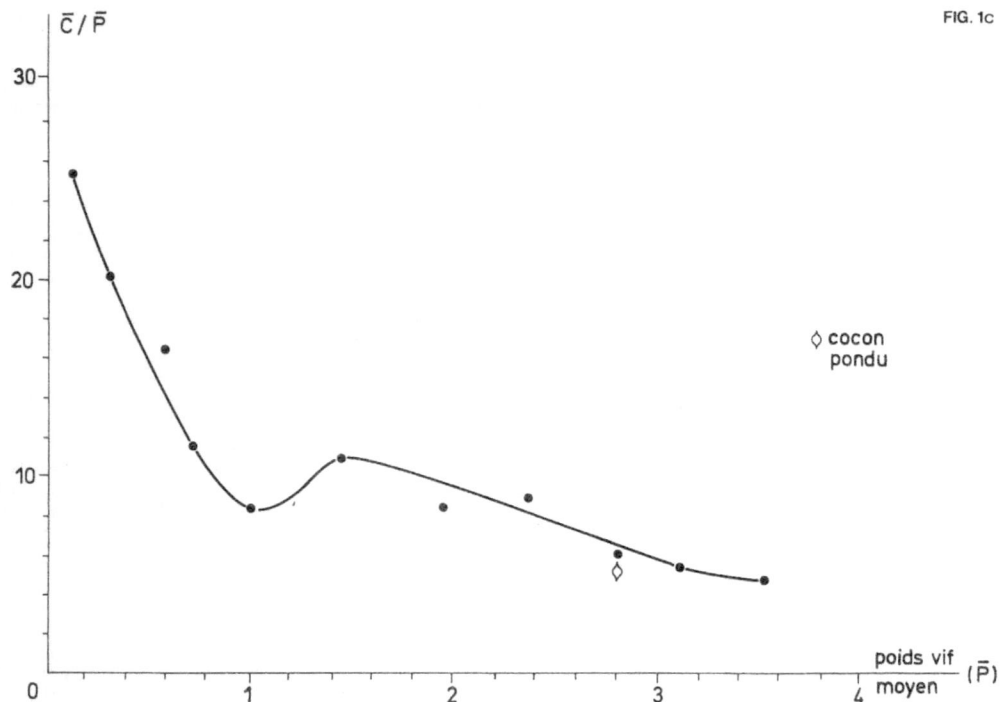

FIG. 1c

Le stade adulte a été atteint au bout de 240 jours et la consommation a été de 2 kg de terre sèche pendant ce temps (moyenne sur 4 individus). Maximale à la naissance (24,5 poids/jour), elle décroît jusqu'à
la puberté, remonte légèrement au cours de celle-ci (10,9 poids/jour)
puis diminue régulièrement tout au long de la vie adulte jusqu'à des
valeurs inférieures à 4 fois le poids de l'animal chaque jour.


CONSOMMATION ANNUELLE D'UNE POPULATION NATURELLE

En utilisant les relations précédentes, on peut calculer la consommation
de terre d'une population pendant un temps donné, connaissant la température moyenne et la teneur en eau du sol ainsi que la structure pondérale de cette population. Ce calcul a été réalisé pour une population
de savane herbeuse à *Loudetia simplex* pendant une année, de août 1971
à juillet 1972. Le tableau 1 donne pour chacun des mois la température
et la teneur en eau moyennes du sol entre 0 et -10 cm ainsi que la structure de la population.

La combinaison des courbes des figures la et lc donne le système de courbes de la figure 2. Ainsi, connaissant le poids d'un individu ou d'un groupe d'individus (l'ensemble des Vers d'une classe de poids par exemple), et par ailleurs l'humidité moyenne du sol, on peut calculer le taux de consommation théorique à une température de 25°6. Les courbes de la figure 3, tirées de la figure lb, permettront d'effectuer une cor-

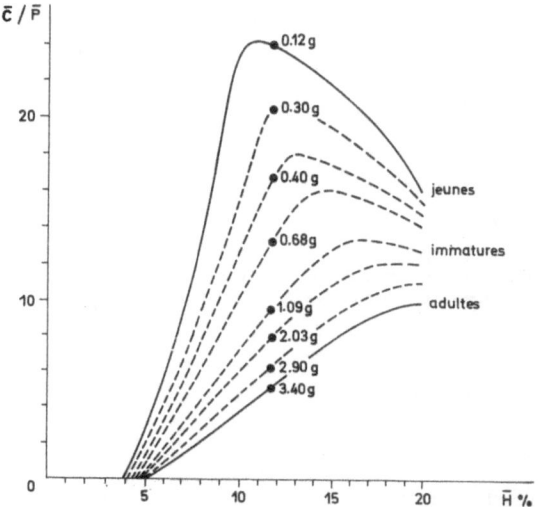

**Fig.2. Variations de la consommation** de terre en fonction de son humidité par les individus de différents poids chez *Millsonia anomala*.
c: consommation moyenne journalière en poids de terre sèche;
p: poids frais moyen du ver;
H %: teneur en eau de la terre.

rection tenant compte de la température réellement enregistrée sur le terrain. De proche en proche, classe après classe, on calcule ainsi la consommation totale de terre par la population considérée.

La valeur de consommation obtenue est de 507 tonnes de terre sèche par hectare dans l'année étudiée par la population définie (215.000 Vers/ha pesant en poids frais 250 kg). Un tel résultat s'explique certainement par la très faible teneur en matière organique des sols de Lamto (de 1 à 2 % en moyenne) et la valeur également faible du taux d'assimilation de cet aliment par l'animal.

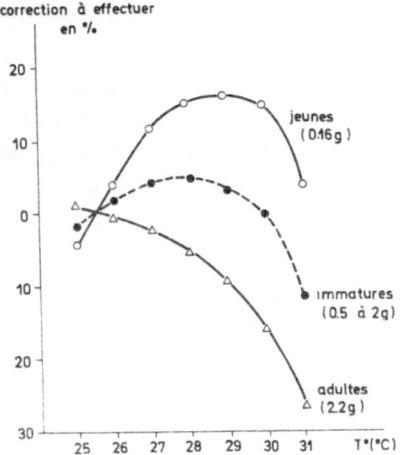

Fig.3. Système d'abaques permettant d'effectuer des corrections en fonction des températures observées.

SUMMARY

Consumption  of the earth-feeding worm  *Millsonia anomala*  (Oligochaeta-
*Acanthodrilidae*)  has been measured  for individuals  of different sizes
bred in different temperature and soil moisture conditions.

The total consumption of a field population  can then be calculated
if one knows,  on a given period of time,  the weight structure  and the
mean values of temperature and soil moisture.

In one year,  a field population  of  215.000 worms  (250 kg  fresh
weight) ingests 507 t/ha oven dry earth.

B i b l i o g r a p h i e

L a v e l l e ,  P. (1972):  Peuplement et production des Vers de terre
de la savane de Lamto (Côte d'Ivoire). Ann. Univ. Abidjan, sér. E
(Ecologie), V, 3, 37-51.

- (1973):  Consommation annuelle de terre par une population naturelle de Vers de terre: *Millsonia anomala* Omodeo(Oligochètes-*Acanthodrilidae*) dans la savane de Lamto (Côte d'Ivoire). I. Coll. Soc.
Ecol. (sous presse).

# Contribution à la définition d'une zone prairiale ecotone du domaine du Pin-au-Haras (Normandie) par la gamasofaune édaphique (Arachnides, Parasitiformes)

C. ATHIAS-HENRIOT
Laboratoire de Faune du Sol, I. N. R. A.,
Dijon, France

## INTRODUCTION*

Situé aux confins du Perche (61-F), dans le Domaine du Pin-au-Haras, le pâturage permanent *P.EX* a servi de station expérimentale pour les travaux de la section PT (prairie permanente) de la participation française au P.B.I. Un échantillon de microarthropodes édaphiques y a été prélevé durant 2 ans (1969-1971).

Au nord, la prairie *PP* est limitée par le bord de chemin *L* (fig.1). La bande prairiale la plus septentrionale *Poo* forme une zone écotone entre *P.EX* et *L*. Cette zone *Poo* est encore appelée à servir de station d'études écologiques; par conséquent, les prélèvements de microarthropodes édaphiques effectués sous quelques types tranchés de couverture peuvent être exploités pour contribuer à définir cette zone *Poo* par rapport aux autres sites du Domaine, principalement ceux qui la flanquent, *P.EX* et *L*.

## MATÉRIEL ET TECHNIQUES

Le Domaine du Pin-au-Haras est situé en climat tempéré océanique; les sols y sont de texture argileuse; le paysage actuel (distribution des parcelles forestières et prairiales; tracé des chemins) a des origines séculaires (Ricou, 1972).

Chaque prélevat (Bouché, en prépar.) utilisé pour l'extraction des microarthropodes édaphiques est formé de litière et de terre superficielle. Ces prélevats ont été soumis à l'extraction éthologique (variantes de l'appareil de Tullgren).

* Les symboles et sigles sont définis en annexe.

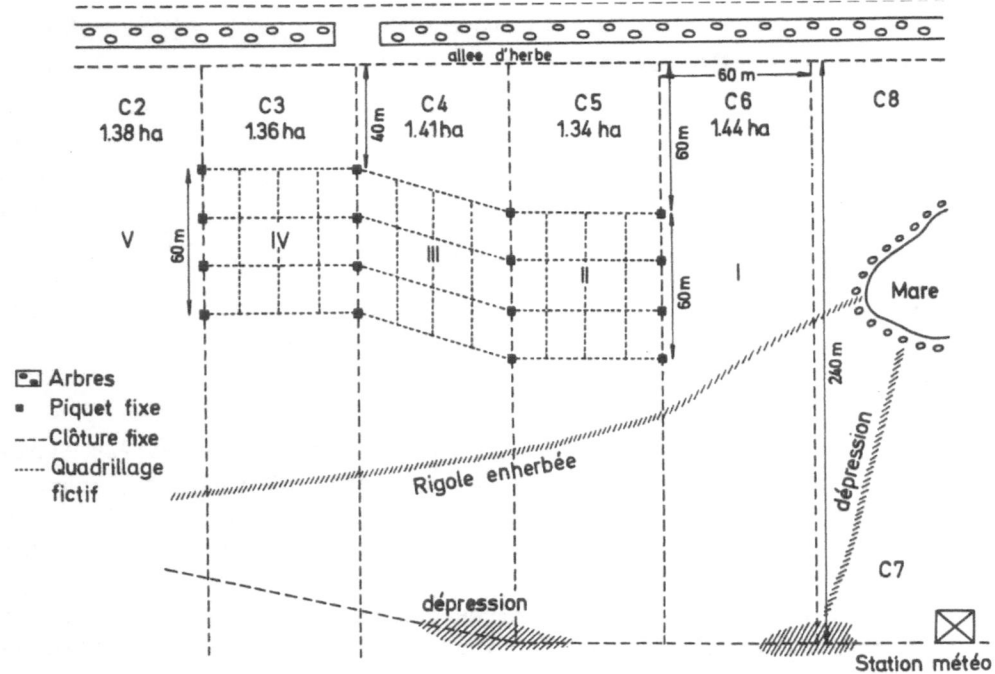

Fig.1. Domaine du Pin-au-Haras, lieu-dit Borculo (d'apres Ricou, 1972); III - *P.EX*; $C_4$ = *Poo*; allée d'herbe et arbres = *L* (voir Annexe).

Le groupe zoologique choisi pour l'analyse est celui des gamasides (à l'exclusion des uropodes), en raison des possibilités de reconnaissance spécifique.

Les sites de récolte sont *F. L. Poo* et *P.EX*. Comme on n'a pas procédé à des relevés quantitatifs en *L* et en *F*, la comparaison des compositions faunistiques entre les 4 sites se base sur les données de présence/ absence; l'indice calculé est *r*. L'utilisation d'un coefficient de similitude faunistique se justifie par le fait que, dans les conditions où ont eu lieu les relevés, la variabilité biogéographique n'intervient pas et que les écarts à l'hypothèse d'indépendance peuvent entièrement être rapportés aux caractéristiques propres à chacun des sites de récolte. L'indice *r* a été retenu parce qu'il est lié au $x^2$ (Daget, 1971).

Dans le calcul des *r*, la valeur de *p* a été fixée de la manière suivante:

1. Pour la comparaison de la macrodistribution d'un site donné, *p* est le nombre total d'espèces de gamasides récoltées dans ce site.

2. Pour la comparaison des compositions faunistiques des sites entre eux, $p$ = 86, nombre d'espèces de gamasides récoltées dans le Domaine.

A titre comparatif, les chiffres obtenus pour la hêtraie $F$ sont également donnés ci-après.

Il est implicite, vu l'emploi du coefficient $r$, que le statut spécifique des animaux utilisés soit établi avec rigueur (même dans le cas où un binôme linnéen ne peut encore être attribué à chacune des espèces). En outre, le terme espèce est ici utilisé sans considération de la nature du taxon: espèce biologique (= sexuée) ou "espèce" uniparentale (De Bach, 1969).


RÉSULTATS

Ceux-ci comprennent le dépouillement faunistique des gamasides non uropodes, le calcul d'indices et leur comparaison.


## Compositions faunistiques

Les individus récoltés en $F$, $L$ et $PP$, entre janvier 1969 et septembre 1971, appartiennent à 86 espèces de gamasides; plus de la moitié d'entre elles a déjà reçu un nom d'espèce et le dépouillement faunistique a donné lieu jusqu'à présent à 3 notes (Athias-Henriot, 1972, 1973 et 1973a).

Le tab. 1 présente les données des 4 sites, leur "originalité" ($E$ %) et leur "richesse" ($K$). La parcelle prairiale.$P.EX$ est la plus "originale", bien que faiblement appauvrie; la prairie écotone $Poo$ se distingue de $P.EX$ et de $L$, qui la flanquent, par une "originalité" et une "richesse" réduites.


* La détermination des espèces de gamasides nécessite fréquemment un long travail de relevés morphologiques, de comparaison et de révision; pour le matériel zoologique recueilli dans le Domaine du Pin-au-Haras, cette opération n'est pas encore terminée.

**Comparaison des macrodistributions dans chacun des 4 sites** (tab. 2)

Les coefficients $r$ ont été calculés entre prélevats de chacun des 4 sites successivement; leur distribution est légèrement asymétrique (fig. 2)* D'après le tab. 2, on voit que le pâturage $P.EX$ présente, par rapport à la forêt, une nette homogénéisation ($\bar{r}$ élevé et $\frac{100\ s}{\bar{r}}$ faible). La macrodistribution du bord de chemin $L$ est un peu plus disparate que celle de la forêt. La bande prairiale écotone $Poo$ est intermédiaire entre $L$ et $P.EX$ et les chiffres correspondants sont voisins de ceux de la forêt.

T a b l e a u  1.  Composition  spécifique  des gamasides édaphiques de 4 sites  dans le Domaine du Pin-au-Haras
(pour la signification des lettres, consulter l'annexe)

| Site | F | L | Poo | P.EX | Total |
|------|------|------|------|------|-------|
| $p$ | 32 | 53 | 27 | 48 | 86 |
| $p'$ | 6.66 | | 17,96 | | |
| $c$ | 8 | 38 | 5 | 28 | 79 |
| $E(\%)$ | 18,75 | 22,64 | 7,41 | 25,00 | |
| $K$ | 28,06 | 29,36 | 11,62 | 20,75 | |

T a b l e a u  2.  Tableau comparatif des similitudes faunistiques  entre prélevats de 4 sites du Domaine pour les espèces de gamasides édaphiques
(pour  la signification des lettres, consulter l'Annexe)

| Biotopes | F | L | Poo | P.EX |
|----------|---------|---------|---------|---------|
| $\bar{r}$ | + 0,150 | + 0,063 | + 0,140 | + 0,534 |
| $v$ | 0,031 | 0,036 | 0,024 | 0,018 |
| $s$ | 0,177 | 0,190 | 0,155 | 0,134 |
| $v/\bar{r}$ | 0,206 | 0,571 | 0,171 | 0,033 |
| $100s/\bar{r}$ | 118 % | 302 % | 111 % | 25 % |
| Mode | + 0,10 | - 0,10 | + 0,15 | + 0,50 |

    * Ces distribution  n'ont pas été analysées du point de vue mathematique.

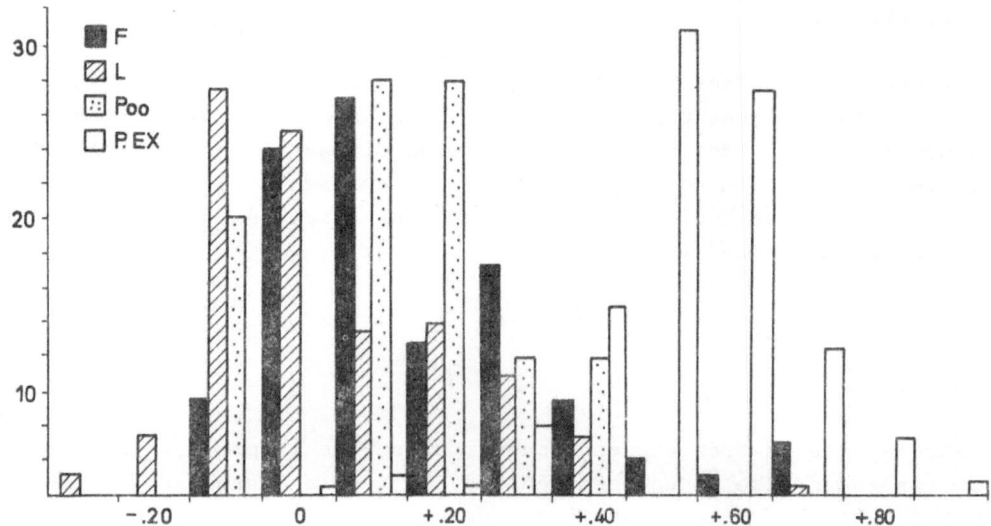

Fig.2. Frequences des indices de similitude r, calcules
entre prelevats d'un meme site (F, L, Poo, P.EX et: voir
Annexe).

Comparaison des compositions entre sites (tab. 3)

Les coefficients r ont été calculés d´après les nombres totaux d´espè-
ces par site, en faisant p = 86.

Seul L conserve un caractère forestier. L´affinité prairiale de
Poo est mise en évidence; cependant cette zone écotone ne partage pas
la dissemblace de P.Ex vis-à-vis de L.

T a b l e a u   3.   Diagramme des similitudes
faunistiques entre les espèces de gamasides édaphiques de 4 sites du Domaine du Pin-au-Haras
(pour la signification des lettres, voir annexé)

| r | P.EX | Poo | L | F |
|---|---|---|---|---|
| P.EX | - | + 0,50 | - 0,32 | - 0,28 |
| Poo | + 0,50 | - | + 0,02´ | - 0,21 |
| L | - 0,32 | + 0,02 | - | + 0,31 |
| F | - 0,28 | - 0,21 | + 0,31 | - |
| Total | - 0,10 | + 0,31 | + 0,01 | - 0,18 |

RECAPITULATION

Par rapport  aux sites $L$ et $P.EX$  qui la flanquent, la zone écotone $Poo$
est,  faunistiquement parlant,  moins "originale"  et moins "riche"; la
macrodistribution  des espèces y est  plus disparate  qu'en $P.EX$,  mais
moindre qu'en $L$.  Du point de vue faunistique, son affinité est principalement prairiale,  bien qu'elle ne présente pas·,  à l'instar de $P.EX$,
de dissemblance vis-à-vis de $L$.


OBSERVATIONS

Il semble  que la très ancienne substitution  de la prairie  permanente
à la forêt conduise  à une homogénéisation  de la macrodistribution des
gamasides édaphiques et à un léger appauvrissement du spectre faunistique, lequel, cependant,acquiert une "originalité" propre. Le seul effet
notable observé en $L$  est un léger accroissement  de l'hétérogénéité de
la macrodistribution.

  L'écotone $Poo$  est une prairie exploitée;  de par son contact avec
$L*$ et $P.EX$, elle présente - pour la faune de gamasides  et sa macrodistribution - des caractères  intermédiaires  entre ceux  de ces 2 sites;
cependant,  il n'y a pas,  dans  cette bande écotone,  d'enrichissement
faunistique; le contraire semble même vraisemblable.

  Les résultats obtenus  pour les gamasides édaphiques de la hêtraie
$F$ ne sont pas représentatifs  d'un état  de climax;  celui-ci n'existe
plus  au Pin-au-Haras;  ces données  ont  cependant semblé correspondre
à l'état de référence  le moins mauvais, pour le Domaine, des points de
vue biogéographique et biocénotique.  Son remplacement  par une couverture laissée à elle-même (site $L$) ne s'accompagne pas d'appauvrissement
faunistique ni d'homogénéisation  de la macrodistribution,  au moins en
ce qui concerne les gamasides édaphiques.


A n n e x e

$a$   = nombre d'espèces communes à deux unités de référence (prélevats
        ou sites, ...).


  * Ce contact va bien au delà de la continuité edaphique; par exemple, $Poo$ reçoit une partie de la nécromasse foliaire produite  par  les
arbres ou arbustes de $L$.

$b$   =   nombre d'espèces présentes  dans la première des deux unités de référence (prélevats ou sites) comparées (voir $c$).

$c$   =   nombre d'espèces présentes  dans la seconde  des deux unités de référence (prélevats ou sites) comparées (voir $b$).

$C$   =   nombre de prélevats.

$d$   =   $p - (a + b + c)$.

$E$   =   proportion d'espèces exclusives.

$F$   =   Hêtraie mixte à sous-bois de charmes *(Fagus - Carpinus)* plus ou moins dense située  dans la Domaine à quelques centaines de mêtres de L et PP.  Cette forêt est soumise à l'exploitation traditionnelle sans importations artificielles  (au moins jusqu'en 1971).

$K$   =   $\dfrac{P - p'}{\log C}$.

$L$   =   bord sud du chemin  de terre longeant au nord la prairie $PP$. Ce bord  de chemin  a plusieurs mètres de large; il est fortement déclive et ne subit ni exportations ni importations volontaires. Il porte une épaisse couverture herbacée, principalement graminéenne, ainsi que des arbres et buissons épars.

$P$   =   nombre  d'espèces  trouvées  dans un espace  et un temps  donne dans un prévelat, dans un site ...).

$p'$  =   nombre moyen d'espèces par prélevat.

$P.EX$ =  parcelle expérimentale  de prairie permanente pâturée,  implantation d'un programme de recherche dans le cadre de la participation française  au  Programme Biologique International  (parcelle CIII in Ricou, 1972).  La prairie reçoit de temps à autre des scories.  La terre nue y est estimée occuper 8 % de la surface (Hedin et Lecomte, comm. pers.).

$Poo$ =   bande nord de $PP$, située entre $P.EX$ et $L$.

$PP$  =   pâturage permanent composé de Poo et $P.EX$.

$r$   =   $\dfrac{ad - bc}{(a+b)\ (a+c)\ (b+d)\ (b+c)}$.

$s$   =   $\sqrt{v}$.

$v$   =   variance.


R é f é r e n c e s

A t h i a s - H e n r i o t ,  C. (1972): Sur  la systématique  et la géographie de la lignée *Truncus* du sous-genre *Anidogamasus* Athias, 1971 (Genre *Paragamasus* Hull, 1918: Arachnides,  Gamasides tocospermiques, Parasitina). Bull. Mus. Natl. Hist. Nat.,  3ème série, n° 33, Zool. 27, 317-325.

- (1973): Observations sur les genres *Neojordensia* Evans et *Ortha-denella* n.g. en Europe occidentale (Gar..sides, Dermanyssina, Asci-dae). Acarologia (sous presse).

- (1973a): *Arctopsis inexpectatus* n.g., n.sp., gamaside litiéri-cole nouveau de Normandie *(Arachnides, Dermanyssina, Ascidae)*. Bull. Soc. Entomol. France (sous presse).

B o u c h é , M. B. (en préparation): Discussion d'écologie. I. Intro-duction. II. Quelques modes d'études en rapport avec l'uniformisa-tion spatiale.

D a g e t , J. (1971): Les modèles mathématiques en écologie (Cours professé à la faculté des Sciences de Paris). Laboratoire de Zoo-logie de l'Ecole Normale Supérieure, 105 pp. ronéotées.

D e  B a c h , P. (1969): Uniparental, sibling and semi-species in re-lation to taxonomy and biological control. Israel J. Entomol. 4, 11-28.

R i c o u , G. (1972): Prairie permanente pâturée, Pin-au-Haras (Fran-ce). Rapport présenté à la Réunion P.B.I. "Prairie-Toundra", Fort-Collins, 14-26 août 1972, 57 pp. ronéotées.

## D i s c u s s i o n

H. F r a n z : La prairie expérimentale est-elle pâturée? Nous ⟨avons fait l'expérience en Styrie, que les pâturages qui reçoivent une fumure ou même seulement les excréments des vaches, ont une faune caractéris-tique "nitrophage" qui est très répandue dans la zone tempérée. Cela peut être la cause du fait que la prairie expérimentale a une faune très caractéristique.

C. A t h i a s : La pâture expérimentale ne reçoit aucun soin agrono-mique, si ce n'est un apport de scories tous les 3 ou 4 ans. Un des facteurs de l'originalité faunistique (Tab. 1) pourrait être le rejet des bouses de vaches.

F. A t h i a s : Est-ce que l'échantillonnage est comparable dans tou-tes les parcelles? Ce serait important dans le cas de la zone écotone dont l'apparante pauvreté est assez curieuse.

C. A t h i a s : Le plus grave défaut des prélèvements est d'ordre chronologique. Cependant, l'unité chronologique de l'echantillon tient à ce que tous les prélèvements ont eu lieu au cours des deux phases de l'éxpérience. Le rythme des prélèvements qualitatifs (en *F,L* et *Poo*) a seulement été moins régulier et moins dense que celui de l'échantil-lon quantitatif effectué en *P.EX*.

M. S. G h i l a r o v : Les Gamasides de vos prélèvements à quels groupes écologiques appartiennent-ils?  Sont-ils libres, nidicoles ou parasites?  La répartition de ces groupes écologiques doit être différente.

C. A t h i a s :  Les prélèvements  en prairie *P.EX*  ont été  conformes
à un protocole casualisé rectifié préétabli. L'opérateur sur le terrain
ignorait s'il allait rencontrer une habitation d'animal. Ce fut rarement
le cas.  Par contre, les dépouillements des Gamasides ont montré que la
densité des nids était plus élevée dans la couverture *L*;  cette circon-
stance est certainement - et pour une part - à l'origine de l'hétérogénéité faunistique élevée de la gamasofaune observée en *L*.

# Feeding activity and availability of food in Collembola

ELS N. G. JOOSSE

Department of Zoology, Free University,
Amsterdam, The Netherlands

INTRODUCTION

Although collembolan populations often consist of many thousands of individuals per $m^2$, and despite the fact that they occur in nearly all
soils and probably play a major role in the formation and maintenance
of the soil, only little is known of their feeding biology and especially about the variation in their food supply, about the question if
food shortage ever occurs. It is suggested by Anderson & Healey (1972)
that decomposers live in an excess of food and that this may be the
cause of the low food specialization found among these groups of species.

From laboratory investigations performed by Usher et al. (1971) it
appeared that the availability of food is very important both in regu-
lating the rate of population growth and in determining the maximum
population density. Only one study was performed on the effects of the
availability of food in field conditions namely by Van der Kraan &
Vreugdenhil (1973). They demonstrated that *Hypogastrura viatica*, which
lives in an enormous excess of food (it feeds rather exclusively on
unicellular algae), is at the same time limited in its food supply. It
cannot burrow in the soil and as a result it food is limited to the
algae growing at the surface of the soil. This is sometimes dried out
or inundated, causing periods of starvation, which sometimes lead to
a migration of masses of animals searching for food (Van der Kraan, unpublished). This similtaneous feeding activity gives rise to a synchronization of the moulting rhythm. In the laboratory it could be observed
that when animals are reared with insufficient food, so when a starvation period is given, a synchronization of moulting and reproduction
can be induced after subsequent food supply (Joosse & Veltkamp, 1970:
Joosse, 1971).

The purpose  of this study  was to investigate  if a food shortage
occurs in collembolan species living in the soil.  Therefore the number
of individuals of *Orchesella cincta* found with and without gut contents
in the field was analyzed.

Several authors  (Christiansen, 1964;  De With & Joosse, 1971; Anderson & Healey, 1972;  Bödvarsson, 1970, 1973)  have seen  that  among
animals  taken  from the field  a high percentage  of empty guts  occur.
And although Christiansen  suggested  that individuals  with empty guts
might have been feeding on bacteria,  which would not be visible in gut
content preparations, it seemed probable to us that empty guts indicate
lack of feeding.  De With demonstrated that this lack of feeding is associated with moulting.

Concerning the relation  between feeding activity and moulting, we
analyzed  in the laboratory  the number  of empty guts  which occur  at
different temperatures  in cultures  with an excess  of food.  The food
consisted of *Pleurococcus* (green algae),  taken from the bark of trees,
mixed in water  and offered in drops.  One could expect that empty guts
occurring in those cultures,  were caused by activities needed  for the
preparation  of a new cuticle and gut wall  and not by a food shortage.

Knowing  these data  one can go  to the field  and collect animals
to investigate the percentage of gut contents with the idea that a comparison  with the laboratory data should demonstrate  an eventually periodical food shortage, which should be expressed  in the occurrence of
significantly more empty guts in field animals in comparison to laboratory animals.


LABORATORY OBSERVATIONS

The laboratory observations on the feeding activity  of *Orchesella cincta* were made in two ways:
1. by observations on 10 separately reared individuals at three temperatures 10, 15 and 20 °C.
2. to have data  on more individuals  under more or less natural conditions where the animals live  in interaction  with each other, observations  were  made on mass cultures of 100 animals.  These observations
were performed  on three populations  collected at different periods of
the year, to get an impression about a possible varying feeding activity
during the year. The first population was collected in April. when most
animals were not yet full-grown and very active in reproduction (Joosse,
1969), the second in June, the animals being  at the end of their life;

the third in August, a population born in spring, active in reproduction and not yet full-grown.

The results of the individual observations are presented in Fig.1.

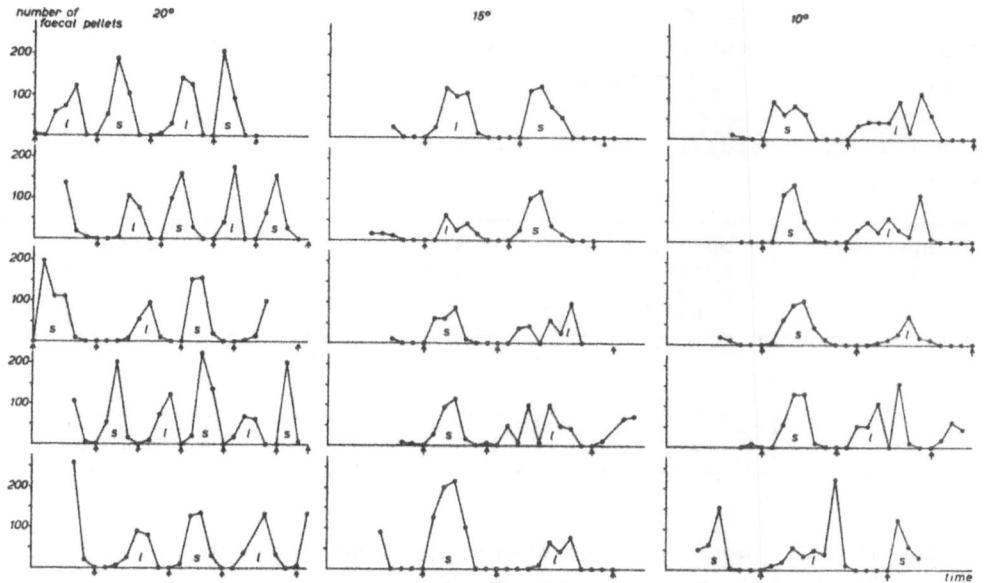

Fig.1. Feeding activity in relation to moulting and temperature in *Orchesella cincta*.

The production of the number of faecal pellets is shown for 5 individuals at each temperature. The first thing we noticed was a definite pattern in the feeding activity indicating that some days around the moult no faeces is produced, that means that empty guts occur. Secondly, alternating long and short instars can be distinguished. This was known before (Joosse & Veltkamp, 1970). The long instars are associated with reproduction, one can see that in the first days of those instars no faeces is produced. Solitary females which were offered spermatophores at different moments of their reproductive instar, appeared to be receptive for reproductive activities from 4 till 33 hours after the moult, just the period in which they do not feed and in which they walk around searching for spermatophores or, when not solitary reared, for places to lay eggs.

From these observations we have calculated how many animals were active in feeding and how many were engaged with moulting preparation, namely by counting - at each observation - the number of animals that

did not produce faecal pellets. The resulting data are presented in
Table 1, where the mean duration of the moulting preparation at the
different temperatures is given in days and as a percentage of the
instar duration. One can see that the relative time taken for the preparation of a new cuticle is independent of the temperature and so is
the percentage of empty guts. It appears that about 40-50 % of guts are
empty, 50-60 % have gut contents.

T a b l e  1.   Feeding activity of *Orchesella cincta* in relation to
temperature

| Temperature (°C) | Duration moulting preparation (days) | As % of instar time | Mean % empty guts per observation |
|---|---|---|---|
| 10 | 4.2 ± 0.8 | 43.5 ± 10.1 | 45.7 ± 13.5 |
| 15 | 3.8 ± 1.3 | 44.8 ± 9.1 | 49.5 ± 16.5 |
| 20 | 2.3 ± 1.0 | 42.8 ± 14.2 | 42.4 ± 17.3 |

The mass cultures in the laboratory showed somewhat different
results. First the observation method was different. The feeding activity could not be established by counting the faecal pellets in this
situation. We were helped by the remarkable and very practical phenomenon that feeding animals were nearly always present on the food,
contrary to non-feeding animals which sat never on the food. We counted
during three weeks 3 times per day the number of animals present on the
food. At the end of the three weeks the animals were fixed in an alcohol
mixture according to Gisin (1960) and cleared in 10% KOH to determine
the gut contents. A small difference appeared between the percentage
feeding activity and the percentage of gut contents (0.8), at higher
temperature somewhat more (1.3) than the feeding activity indicated.
The results are given in Table 2.

First, it has to be noticed that the animals of experiment II
(June) are comparable to the animals from the preceding individual
observations, being taken from the same population. A difference in
feeding activity becomes apparent: the animals in the mass cultures
feed less continuously than the solitary animals, the percentage of gut
contents being much lower. This may be caused by disturbance effects
of the animals among each other, but especially by disturbance effects
after the observations for which the lids of the glass boxes had to be
taken off, which was not necessary in the individual observations.

T a b l e   2.   Feeding activity  in experimental populations of *Orche-sella cincta*  of a different age structure  in relation  to temperature

| Temperature (°C) | f.a. | I (April) % gut contents | f.a. | II (June) % gut contents | f.a. | III (August) % gut contents |
|---|---|---|---|---|---|---|
| 2 | 18.7 | 44.6 | -- | -- | -- | -- |
| 5 | 30.8 | 40.3 | 5.3 | 7.2 | 3.7 | 5.3 |
| 1o | 39.7 | 39.7 | 18.8 | 22.2 | 18.5 | 18.5 |
| 15 | 42.2 | 39.7 | 21.0 | 27.4 | 21.5 | 29.8 |
| 20 | 40.3 | 33.7 | 27.1 | 27.1 | 23.9 | 22.3 |

A comparison  between  the various mass cultures  shows large dif-ferences.  Population I,  taken  in April,  had  a significantly higher feeding activity,  than populations II and III.  This is an interesting phenomenon  and  we suppose  that it has something to do  with the fact that this population  has hibernated.  It is adapted to low temperature conditions and is stimulated intensively by a temperature rise. In field populations  it was established (Joosse & Veltkamp, 1970)  that spring populations show  a very rapid growth  and  a simultaneous reproduction activity  after  a temperature rise.  These  favourable conditions were artificially introduced  in the laboratory:  higher temperatures and an excess of food.  They induced a high metabolic activity. The summer po-pulations were less  stimulated  especially at low temperatures. Mortal-ity was even so high at $2^{o}$, that these data were unreliable.

Observing these results one can expect  that the percentage of gut contents in field populations may vary depending on the food supply but also  on a varying feeding activity  of the population  as  a result of preceding conditions.

FIELD OBSERVATIONS

During the period February  till  September 1973,  100 animals were col-lected each month  at moments  of particular climatic conditions. These particular conditions  were  dry circumstances,  when  the animals con-centrate  in a few places  where survival  is  possible.  Food competi-tion  may  occur  in these limited spaces  and  one would expect a food shortage.  In Fig. 2 the percentage of gut contents in the various sam-

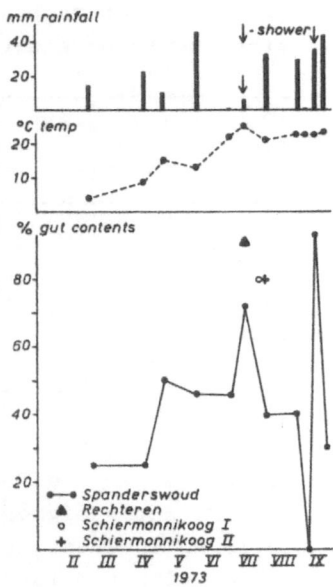

Fig.2. Percentage of gut contents in mass cultures of *Orchesella cincta*. Data about the mean maximum temperature of 7 preceding days and total rainfall of 14 preceding days are obtained from the nearest weather stations.

ples is shown. To give an impression of climatic conditions, the course of the maximum temperature of the 7 preceding days and the total rainfall of the 14 preceding days are given. An arrow indicates that rainfall occurred in one shower about 24 hours before the sample was taken.

Many points have a value which does not indicate a food shortage and which can be interpreted by moulting phenomena. In spring the values were somewhat lower, as the temperature was below $10^\circ$. This is in agreement with the laboratory observations. Surprisingly high values in the gut contents became apparent at the end of June and at the end of August. These points are clearly caused by the particular climatic conditions. The phenomenon appeared at several sampling sites at about the same time as is indicated in Fig.2. These samples were taken one or two days after heavy rainfall after a long period of drought. In August the phenomenon is most clear: after a period of drought just before rainfall the concentration of animals in places convenient for survival was so high that they could hardly be found. The guts were empty for a 100 %. This indicates a food shortage. Two days afterwards, after heavy rainfall, the same population had gut contents for 93 %. The same is true in June, only the sample was taken a longer time before the rainfall.   We suppose the rainfall   caused a very rapid growth   of soil algae and fungal spores.   This occurs within some hours   as becomes apparent   from soil respiration studies   (Tamm & Krzysch, 1963),   showing a very high $CO_2$ production   after rainfall as a result of an increasing metabolic activity of soil microorganisms.   The gut contents of the August animals appeared to be filled with yeasts.

This periodical food shortage is, in fact, a similar phenomenon as described   by   Van der Kraan & Vreugdenhil (1973) for *Hypogastrura viatica* as was shown above, and by Birch (1971) for the rabbit in semi-arid areas of Australia.   The simultaneous feeding activity   after those unfavourable periods which also occurred in *Hypogastrura viatica*   is considered to be of important value   for the population.   These conditions lead to a synchronized reproduction,   because the number   of encounters between many animals   being concentrated   and in the same active reproduction phase results   in an explosion of juveniles.   This happens also in spring when after severe conditions (drought, cold) many local populations become extinct, except for the few that happen to be in favourable places   (Joosse,   1969).   When   favourable weather returns,   these places act   as foci   from which   the region is recolonized.   And these founder populations clearly have a high reproductive value.   A last important point   seems to be the fact   that   the very frequent occurrence of empty guts   is the result of moulting   and of a varying food supply, and   that   locomotory activity   is   variably   associated   with moulting (Joosse, 1971);   this must be regarded   in estimations of the soil biological importance of these animals.

SUMMARY

From a study   of the percentages of gut contents   in populations of the collembolan species   *Orchesella cincta* (Linné),   it appeared that empty guts   are present frequently and varyingly during the year.   Laboratory experiments indicated   the relation   of the percentages of gut contents to temperature and moulting. With this standard measure it was tried to gain insight into the feeding conditions of *O. cincta*   in their natural habitats.   Food shortage   could be demonstrated,   after long periods of drought. Subsequent rainfall induced a high simultaneous feeding activity.   It was stated that this synchronization effect, leading to simultaneous reproduction, has an important value for the restoration of the population density after unfavourable periods.

References

A n d e r s o n ,   J. M. and  H e a l e y ,   I. N. (1972): Seasonal and
    inter-specific variation   in major components   of the gut contents
    of some woodland *Collembola*. J. Anim. Ecol. 41, 359-368.
B i r c h ,   L. C. (1971): The role  of environmental heterogeneity and
    genetical heterogeneity in determining distribution and abundance.
    Proc. Adv. Study Inst. Dynamics Numbers Popul. (Oosterbeek, 1970),
    109-128.
B ö d v a r s s o n ,   H. (1970):  Alimentary   studies   on seven common
    soil-inhabiting *Collembola*  of Southern Sweden.   Eı tomol. Scand.1,
    74-80.
    - (1973): Contributions to the knowledge of Swedish forest *Collem-
    bola*. Inst. For Skogs Zool. Res. Note, nr. 13, 1-43.
C h r i s t i a n s e n ,   K. (1964): Bionomics of *Collembola*. Ann. Rev.
    Entomol. 9, 147-178.
G i s i n ,   H. (1960): Collembolenfauna Europas. Genève, 312 pp.
J o o s s e ,   E. N. G. (1969):  Population structure   in  some surface
    dwelling  *Collembola*  in a coniferous forest soil.   Neth. J. Zool.
    19, 621-634.
    - (1971):  Ecological aspects  of aggregation in *Collembola*.   Rev.
    Ecol. Biol. Sol 8, 91-97.
    - and  V e l t k a m p ,  E. (1970): Some aspects of growth, moult-
    ing and reproduction  in five species  of surface dwelling *Collem-
    bola*. Neth. J. Zool. 20, 315-328.
K r a a n ,   C. van der, and  V r e u g d e n h i l ,   A. P.  (1973):
    Presence and accesibility of food  for *Hypogastrura viatica* Tullb.
    1872 *(Collembola-Hypogastruridae)*. Neth. J. Zool. 23, 125-129.
T a m m ,   E. and  K r z y s c h ,  G. (1963):  Der Einfluss  der Boden-
    temperatur und der Bodenfeuchte auf die $CO_2$- Produktion eines leh-
    migen Sandbodens. Z. Acker-Pflanzenbau 117, 359-378.
U s h e r ,   M. B.,    L o n g s t a f f ,   B. C.   and   S o u t h a l l,
    D. R. (1971): Studies on populations of *Folsomia candida (Insecta:
    Collembola)*.  The productivity of populations  in relation to food
    and exploitation. Oecologia 7, 68-79.
W i t h ,   N. D. de and  J o o s s e ,   E. N. G. (1971): The ecological
    effects of moulting in *Collembola*. Rev. Ecol. Biol. Sol 8, 111-117.

Discussion

P. L a v e l l e :  En faisant des expériences comparables sur les vers de terre, j'ai observé que la remplissage du tube digestif variait au cours de la journée, le maximum se produisant la nuit. Avez-vous remarqué un tel rythme chez les collemboles?

N. G. J o o s s e :  The occurrence of food in the guts of *Collembola* seems to be related to the moulting stage only. Observation on individual specimens make clear that the animals feed during the day.

G. M a r c u z z i : Asks whether illumination can influence the rhythm of feeding activity of *Collembola* (epiedaphic species).

N. G. J o o s s e :  Light conditions were constant in the experiment namely darkness from 18 till 6 o'clock. No experiments were performed with respect to the influence of light on feeding activity.

H. P e t e r s e n :  I would like to mention that the proportion of specimens with empty guts varies considerably from one collembolan species to another. The smaller forms living deeper in the soil seem to use a much shorter proportion of their life in these nonfeeding moulting phases. Have you calculated the transport of food through the gut for the species you have studied?

J. S a t c h e l l :  Dr Petersen told us that *Collembola* from deeper soil layers generally have their guts full while empty guts are often found in surface-living species. There is a parallel between surface feeding and deep burrowing earthworms. Is it possible that in *Collembola* the difference is related to the nutritional quality of the food available in the two horizons?

H. P e t e r s e n :  I think that there is a parallel to what we find in the earthworms. Some surface-living *Collembola*, which have a high proportion of empty guts. most often have their guts filled with selected food components: pollen, spores or hyphae, while the deeper living forms have gut contents which are mixtures of mineral particles, organic particles and microfloral compounds.

N. G. J o o s s e :  It would be very interesting to analyse the nutritional value of the food eaten by these different species. We have no information yet either about differences in the quality of food taken by *O. cincta*. We only know that *Pleurococcus* (green algae) are a very good food substance and that it is present in the guts during the whole year.

A. S z e p t y c k i :   How long   is the life cycle of the studied species?   Did the author make experiments   wit   very young instars   of the
species under study?

N. G. J o o s s e :   The life cycle of *O. cincta* depends on the temperature. In field conditions we established that the time from egg to egg
lasts   about   3 months.   The differences in gut contents   between young
and adult individuals have not yet been studied.

C. G r e g o i r e - W i b o :   Qu'entendez-vous   par "reproduction activity"?   Observez-vous   les exuvies   sur le terrain?   Observez-vous la
croissance de la population après la pluie?

N. G. J o o s s e :   Reproduction act   is found in alternating moulting
instars in which eggs and spermatophores are produced. This occurs only
several hours   after   the moult.   Exuviae cannot be found again   in the
field   although the laboratory   results demonstrate that   this species
does not eat its exuviae.   We only assume that the population increases
rapidly for a rainfall   is often followed by the occurrence of many juveniles.

# Influence of brushwood and undergrowth upon distribution of litter beetles in poor pine forests

A. SZUJECKI

(in co-operation with J. Szyszko and S. Perlinski)

Forest Institute,
Warszawa

The influence of environmental differentiation upon the occurrence of animals permanently attracts notice of biocenology, mainly in view of the economically important knowledge of this phenomenon. This is why the ameliorative role of deciduous undergrowth (mainly oak and beech), introduced artifically to poor pine forests, should be studied comprehensively. Considering the role of undergrowth, we cannot overlook the influence of brushwood on the soil fauna, which can be restricted by the undergrowth.

1. Studies on the distribution of adult *Staphylinidae* and their ecological groups with regard to the presence of oak undergrowth in poor pine forest and to the form of their appearence (in clusters or singly).

2. Studies on the distribution of litter beetles (mainly *Staphylinidae* and *Elateridae*), with regard to the distribution of brushwood patches in the richer and poorer parts of a habitat of the fresh pine forest type.

3. Studies on the influence of oak and beech undergrowth upon the distribution and activity of *Carabidae* under various geographical conditions of Poland.

The first and the second tasks were realized in a period from 1967 to 1972 in the vicinity of Warsaw. Experimental plots were set up in pine forests of mean age classes (32 - 67 years) in the association *Peucedano-Pinetum* Mat. 1962. The forests are situated on podsol soils - loose sands of fluvioglacial origin, characterized by a considerable thickness. Due to inappropriate forest manegement the oligotrophic habitats underwent degradation with the result that the brushwood showing a predominance of *Endoton schreberi*, *Vaccinium myrtillus*, *Vaccinium vitis idaea*, *Calluna vulgaris* and *Festuca ovina* belongs to the moss facies of degenerative young stand phases. Under these circumstances the phyto-ameliorative role of the 20 - 40 years' oak undergrowth is to be expected.

The third task  was executed in 1970-1971  under similar, but geographically different, habitat conditions: the experimental plots  were placed  in forests  of Gościeradów  in South Poland,  and in forests of Smolniki in North Poland.


RESEARCH METHODS

Task 1. We selected nine experimental plots  and  marked them  with the symbols I, II and III.

Symbol I - Forests with a dense undergrowth of *Quercus robur* forming clusters from 6 - 10 to 17 x 22 m  in diameter, height up to 3.5 m, covering approximately 33% of the area.

Symbol II - an occasional oak tree - approximately 10% of the area.

Symbol III - No undergrowth.

We collected 513 samples  in order  to obtain  a picture on conditions below the canopy of the undergrowth (A), at its edge (B) and outside the canopy (C).  Insect samples  were collected from the litter of an area of 0.25 m$^2$ with a metal frame  and  sieve of the Reitter type. Sample collection  proceeded  from  May till October;  the samples were taken to the laboratory for classification.

Task 2. We selected two experimental plots displaying considerable differences in their brushwood components. These were:

Plot "84": *Vaccinium myrtillus*-1 %, *Entodon schreberi*  and *Dicranum undulatum*-89 %, oak litter-10 %.

Plot "161": *Vaccinium myrtillus*-88 %,  *Entodon schreberi*  and *Dicranum undulatum*-11 %, oak litter about 1 %.


A total of 300 samples  (40 for  every  vegetation patch  in each plot) was taken from May till October.

Task 3. We selected 13 plots  in the fresh-forest habitat and collected 286 litter samples. In addition, we obtained another 117 samples from ethylene-glycol traps  (bottling glass jars)  and 228 samples from trapping furrows measuring 1.0 x 0.3 x 0.3 m. The representative sample were *Carabidae* caught  in  the traps  within 14 days  or in the furrows within 7 days.

RESEARCH RESULTS

Task 1. We collected 3,123 individuals belonging to 128 species. Basing
on results of a previous investigation (Szujecki 1966), 8 of these species appeared to prefer the habitat of fresh pine forest (class $F_3$),
30 to be accompanying species (Class $F_2$), which are less abundant in the
pine forest than in the other forest biotopes or which do not prefer
any type of forest habitat. Species of class $F_1$ were those found in
various forest- and non-forest habitats; species of class $F_0$ were those
not present in the fresh-forest habitat, but typical of other types of
forests or of non-forest habitats. Species of class $F_1$ and $F_0$ will be
referred to as non-forest species.

The principal representatives of our material were *Othius myrme-*
*cophilus* (25.4%) and *Sipalia circellaria* (12.5%). The dominant species
of the individual fidelity classes were these: $F_3$ - *Sipalia circellaris*
(68.4%), $F_2$ - *Othius myrmecophilus* (57.4%), $F_1$- *Philonthus fuscipennis*
and *Ph. varius* (16.1 % each), *Xantholinus linearis* (14%), *Tachyporus*
*hypnorum* (9.8%), $F_0$ - *Gabrius pennatus* (23.8%).

There is no evidence that the undergrowth affects the density of
the individuals of all *Staphylinidae*. In plots I and II the density
amounted to 6/0.25 $m^2$, in plots III - 6.2 on the average. The density
of individuals of the most important species was analogous: *Othius myr-*
*mecophilus*. 1.5 - 1.7 and *Sipalia circellaris* 0.7 - 0.8. The same concerns the individual ecological groups ($F_3$- $F_2$). Apart from the general
lack of the wanted dependence within plots I and II (with oak undergrowth) it was observed that the distribution of representatives of the
individual species and ecological groups of *Staphylinidae* was irregular.
Within plots I the density of the forest species decreased form zone A
towards zone C from 4.7 - 3.6 - 2.6, that of non-forest species increased respectively from 1.1 - 1.5 . 4.2 indiv. 0.25/$m^2$. In the forest
obejcts examined this picture was proportional for all groups of plots
I - III, though it did not concern the individual species. Analogous
were also the proportions of distribution of forest species versus
non-forest ones, i.e., in the successive zones A - C it was 0.73 - 0.74 -
0.52.

In plots II the unit admixture of oak affects only slightly microclimatic conditions determining the distribution of the forest and non-
forest *Staphylinidae*. On the average, under the oak crown canopy the
density amounted to 5.8 at the edge to 6.8, outside the canopy to 5.3
individuals/0.25 $m^2$. However, variation of analogous values in the individual forest objects did not render itself to generalisation, although the density of representatives of forest species in zone C was

lower than that under the oak canopy and, at the edge, differences were observed  in the density  of representatives of the non-forest species.

The relation between forest  and non-forest species in zones A - C was 1.0 - 0.85 - 0.53.

In plots  without  undergrowth  this dependence  was equal to 0.71 thus proving  that non-forest species grow in forests  with undergrowth with the same intensity as in those without undergrowth;  in sun-heated places, however, their density increases,  whereas  in forests  without undergrowth their distribution is more regular.

Task 2. We collected a total of 2,035 specimens  of *Staphylinidae*. We observed no direct dependence  between vegetation patch and density. In each plot, density was highest in patches with a dominant vegetation. In plot no.84, in each vegetation patch, the dominant species was *Sipalia circellaris*, the subdominant species *Othius myrmecophilus*. The situation was reversed in plot no.161, where the dominant species was *Othius myrmecophilus*.  It was a regular feature  in both plots  that the least advantegeous relation in the number of forest-  and  non-forest species occurred in the moss patches:  0.50 in plot no.161, 0.23 in plot no.84. This may be related  to microclimatic conditions,  because moss patches are, generally, less shaded than the remaining vegetation patches.

The distribution and density of *Staphylinidae* larvae  were similar to those of the adults. On the other hand, *Elateridae* larvae, with *Athous subfuscus*  and  *Dolopius marginatus*  as the dominants  among the 3, 787 species, preferred oak litter on plot no.84 (oak litter - 15.3, huckleberry - 10.8, mosses - 9.6),  and  huckleberry on plot no.161  (huckleberry - 23.1, mosses - 10.0, litter - 9.1).

It is of interest  that the distribution  of the beetles  examined was more regular  in the poorer plot no.84  than in the richer plot no. 161, in which they preferred the dominant huckleberry patches.

Task 3.  We collected a total of 2,802 specimens of *Carabidae*  belonging to 38 species.  In the fresh-forest habitat  (without an undergrowth)  the average coefficient  of individuals in the traps  was 6.8, in forests  with  an oak undergrowth 7.8;  the density  was 0.2 and 0.2 respectively on 0.25 m$^2$.  In the forest of Smolniki,  analogous  values were these: trapping jars - 5.4 and 11.6 respectively; litter samples - 0.3 and 0.4 respectively.  This indicates that the density of *Carabidae* was not affected  by the undergrowth.  The higher  catching coefficient in traps in plots with an undergrowth should be ascribed to an increased activity of the insects  in places where the occurrence of brushwood is limited by a more intensive shading of the ground.  The species composition  of *Carabidae* communities  depends  on geographical conditions and  local edaphic possibilities.  Artificially  introduced undergrowth

may modify only the structure of existing associations.   At Góścieradów
the dominance of *Pterostichus aethiops* attaining 88.2 and 89.2% respectively in forests with and without undergrowth was typical of all plots.
At Smolniki, *Pterotichus oblongopunctatus* dominated   in plots without
undergrowth (34.5%);   *Pterostichus niger*   (27.7 %)   dominated   in plots
with an undergrowthm   while   *Pt. oblongopunctatus*   was   represented   by
18.2% only.  This species, together  with the remaining species  of the
ecological group  belonging to the spring type  of development,  escape
from shaded sites.   In plots without undergrowth, species of the spring
type   were represented by 79,9%,   those of the autumn type by 21,1%.  In
forests  with an undergrowth covering 70% of the area   under consideration, their representation was 40.4%   and   59.6% respectively.  The non-
existence of this dependence in plots at Góścieradów may be ascribed to
the low percentage of undergrowth (20 - 40%) and to the high percentage
of one dominant species only.


DISCUSSION


The results obtained demonstrate that oak (or beech) undergrowth, artificially introduced to pine forests in poor habitats, does not increase
the density of litter beetles.  On the other hand,  the undergrowth affects the distribution of both the species  and the individuals  in the
habitat, and their activity.  Besides the markedly defined influence of
the undergrowth on litter beetles due to changes in lighting conditions,
the other possibilities  of the undergrowth vary according to differentiated environmental  and  geographical conditions even within the same
habitat type of the forest.  Here, the influence  of the undergrowth is
reflected  in  the litter properties,  its composition  and  thickness,
humidity, acidity, contents of organic carbon  and nitrogen, as well as
in their proportions indicating the quality  of the course of humification processes.
      Generally, the most advantageous (lowest) ratio  of C : N occurred
below the oak canopy;  it was slightly higher at the edge  and  highest
outside the canopy. Having regard to these values, an inversely proportional correlation  in the density of individual forest species of *Staphylinidae* could readily be observed  in each plot.  However, this correlation cannot be presented  in a general mathematical formula in view
of the fact  that C : N values change within the corresponding zones of
the various plots  and  within various zones of the same plot.  It may,
however, be possible  to elaborate  a rectilinear relationship  between

the contents of organic carbon in soil compounds(calculated by Tiurin's
method) and the density of *Staphylinidae* by the equation:

$$y = 16.9 - 0.3 x,$$

where $y$     - density (on 0.25 $m^2$),

$x$     - contents of $C$ in layer $A_{oL}$.

In this case, the correlation coefficient (0.68) is high.

The verifiableness of this regression equation for other materials
from similar habitats is as yet doubtful. It evidently cannot be used
for materials taken from richer soils. The principle of inverse propor-
tionality in the density of *Staphylinidae*, and that of the percentage
of carbon in organic soil·compounds has been observed by Szujecki (1966,
1972) under different environmental conditions. Although this correlation indicates that such dependence does, in fact, exist in nature,
I have as yet been unable to find a satisfactory explanation of this
phenomenon. Apparently, a lower carbon percentage indicates a higher
decomposition rate of the litter (not necessarily correlating to its
thickness). In the fresh pine forest, the biological activity of the
soil in the saprophage horizon is based on the mesofauna the members
of which constitute the principal source of food for the most abundant
*Staphylinidae* species, i.e., *Sipalia circellaris* and *Othius myrmecophilus*. As the undergrowth changes, an important role is played also by
other factors influencing the life of soil beetles, e.g., humidity,
acidity, and modifying the influence of this relationship.

In this way. the undergrowth differentiates considerably the homogeneous poor pine forest which, consequently, affects the fauna. There
is no evidence as yet that this may be responsible for an increase in
the number of animals examined, by contrast, it may reduce forms which
develop under conditions of stronger light. Considering the role of the
undergrowth in the development of *Carabidae* under various forest- and
soil conditions, J. Szyszko (in press) inferred that if the undergrowth
is a natural component of a forest on richer soils, it affects posi-
tively the frequency of these insects. On the other hand, the density
of *Carabidae* is reduced in a fresh forest by the elimination of species
of the spring developmental type. Since the percentile participation of
these species varies in the different geographical zones, ultimate results from an introduction of undergrowth can be obtained only on the
basis of an earlier faunistic analysis of forests assigned fro phyto-
amelioration, and of an analysis of both soil and phytosociological
conditions.

R e f e r e n c e s

S z u j e c k i ,  A. (1965):  Preferendum wilgotnościowo-glebowe niek-
    tórych gatunków kusakowatych  (Coleoptera, Staphylinidae) żyjących
    w sciołce leśnej i na obrzeżach jezior. Zesz. nauk. SGGW, Leśnictwo,
    Warszawa.
S z u j e c k i ,  A. (1966):  Zależność między wilgotnością wierchniej
    warstwy  gleb leśnych  a  rozmiesneniem kusakowatych  (Coleoptera,
    Staphylinidae)  na przykładzie  nadleśnictwa Szeroki Bór w Puszczy
    Piskiej. Folia forest. Pol., ser. A, Warszawa.
S z u j e c k i ,  A. (1972):  Staphylinidae (Col.)  kak pokazateli ne-
    kotorych svojstv počvy  i  razvitija sosnovych drevostojev.   XIII.
    Int. Entomol. Congr. Proc. III, Moscow.
S z y s z k o ,  J.  (in press):  The dependence  of the occurrence  of
    Carabidae  upon  some soil properties  and  species composition of
    forest stand. Ekologia Pol., Warszawa.

# Influence of humidity and specific interactions on collembolan populations in a pine forest

M. KACZMAREK
Ecological Institute,
Warsaw

Both, continous ability to reproduce and the activity period extended throughout the year, account for the potentially rapid development of the collembolan population almost at any time of the year. This is supported by the results of many studies conducted in various environments.

Among the environmental factors influencing population numbers the most important are soil moisture, air humidity and temperature, food resources and possible competitive interactions (Weis-Fogh 1948, Sheals 1957, Poole 1961, Kaczmarek 1963 and others).

The study was carried out in a large complex of the lowland pine forests on slightly podsolized soils with eluvial level at the maximum depth of 25 - 30 cm. Five plots have been selected at a distance of 2 - 3 km from each other, and marked with numbers from I to V according to the increase in mean soil moisture (I - 6 %, II - 7.7 %, III - 9.4 %, V - 13.4 %). From each plot 50 randomly distributed samples of a size of 10 $cm^2$, 10 cm deep, were taken generally once a month from April to December. Each sample was divided into a surface layer - the litter, and a deeper layer - the soil. Fauna was extracted with a simplified Tullgren apparatus for 3 days.

To characterize temperature, relative air humidity and the amount of precipitation, a period of 10 days prior to sampling was considered.

The object of this report is to compare collembolan population dynamics under different conditions of soil moisture in a pine forest. In order to determine precisely the effect of habitat moisture, the changes in the number of insects were analysed with regard to two aspects - annual cycle and habitat gradient. These autecological data provide a basis for the interpretation of interspecific relationships. To estimate the possibility of competitive interactions, the annual changes in both numbers and mutual distribution of particular species were compared with the expected ones on the basis of habitat preference of these species.

A detailed analysis  of  7 most abundant species  is  presented in
this report.  These are  *Isotomiella minor, Isotoma notabilis, Folsomia*
*quadrioculata, Anurophorus laricis,  Lepidocyrtus lanuginosus, Willemia*
*anophthalma,  Tullbergia krausbaueri,  Anurida pygmaea,  Orchesella bi-*
*fasciata.* They constitute about 90 % of the total number of individuals,
and all of them were found in the first 100 samples for each plot. This
indicated that the same collembolan community occurred everywhere.

The highest mean annual collembolan numbers occurred in the middle
of ·the moisture gradient for all the plots.

During successive periods the peak of numbers moves  from one plot
to another  in  a very characteristic way  (Fig. 1).  At the lowest air

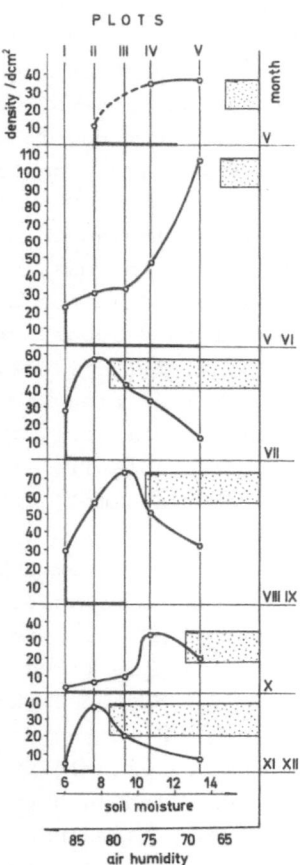

Fig.1. Comparison of the density  of *Collembola*  in five
plots: during successive periods.

humidity recorded, the peaks occurred in the plots with the higher
moisture content, and at higher air humidity in the drier plots respectively. In each case, a regular decline in numbers was observed
from the peak toward extreme points of the gradient. The coincidence
of the peak of numbers for successive plots with relative air humidity
indicates that humidity was the main factor determining habitat optimum
for the whole community. The plot with the lowest moisture content was
beyond optimum humidity throughout the study period. There was no peak
of total numbers there and average numbers were generally the lowest.
A quite exceptional situation was observed only during a period of
heavy rains and high relative humidity, when the lowest numbers were
recorded in plot V, in which the moisture content was highest.

No marked relation has been recorded, however, of seasonal variations in average numbers to air temperature or humidity, except for the
fact that the total number for the growing season was distinctly higher
than that for the period from late autumn to early spring. A somewhat
different presentation of seasonal changes in the distribution of the
peaks shows characteristic differences in collembolan response to variations in humidity conditions between these two periods (Fig. 2). During
the warm season of the year the upper limit of numbers went up dis-

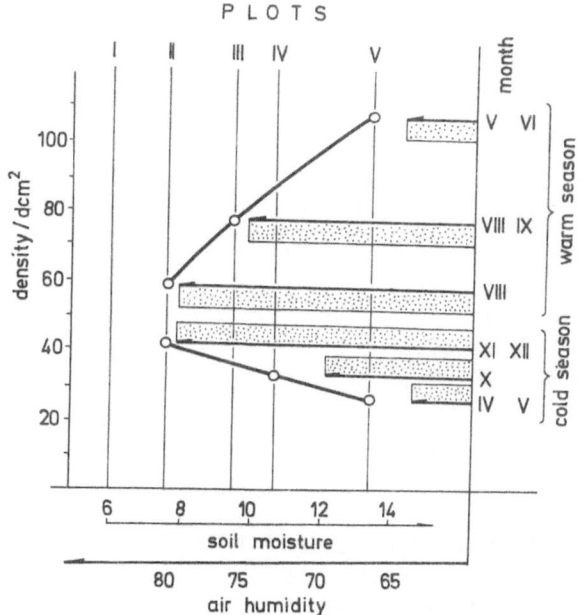

**Fig.2. Density peaks of** *Collembola* **under different conditions.**

tinctly with increasing soil moisture. During the cold period it declined slightly. In other words, soil moisture at a similar air humidity has even opposite effects on maximum numbers during these two periods.

Density of a particular species with different habitat preferences made it possible to assess their habitat specialization under study conditions (Fig. 3). The curves of density distribution of particular

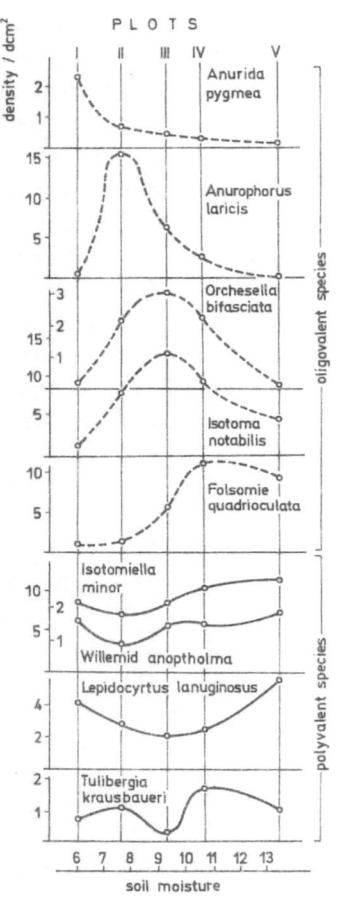

Fig.3. Density of particular species with different habitat preferences.

species along the gradient of soil moisture generally follow two patterns - 1) the peak falls into one of the plots, these are the species with high habitat specialization, which we call oligovalent species, and 2) the peaks of numbers fall into the plots with extremely different conditions, these are the species with low habitat specializa-

tion,which we call polyvalent species. A distinct symmetry of the curve
for oligovalent species  in relation  to the peak  of numbers  suggests
that this peak  indicates the point of optimal moisture in the gradient
of habitats.  The extreme soil moisture  requirements recorded here for
A. laricis and F. quadrioculata  are in accordance with the results obtained by Agrell under laboratory conditions (Agrell 1941). The highest
numbers of polyvalent species at the extreme plots,  which are situated
outside  the optimal moisture conditions  for  the total community, indicate  that their tolerance  of moisture conditions  is high, but does
not indicate the actual position of optimal moisture conditions. It has
been found in studies under laboratory conditions that this optimum for
I. minor falls  into  high moisture values, and that  L. lanuginosus is
a species comparatively resistant to drying of the substrate.

An analysis of seasonal variations  in numbers  along the gradient
generally indicates that there  is a regular shift in the peaks of numbers toward drier plots  during the periods of higher precipitation and
air humidity,  and  toward  the plots  with higher moisture  during the
periods of low precipitation  and  air humidity.  At the same time, the
peak for A. laricis oscillated between plots II and III, for F. quadrioculata between plots III and V,  for I. notabilis  between plots II and
V, for I. minor  between plots I and V (Fig. 4).  This is in accordance
with the assessed ranges  of habitat requirements, since  A. laricis is
defined as a xerophilous species and F. quadrioculata as a hygrophilous
species; the highest range of moisture tolerance is observed for I. minor, an intermediate one for I. notabilis.  These regularities indicate
that the observed seasonal variations in numbers of each of the studied
species  are related  to the moisture of the habitat.  The character of
the density distribution,  however,  indicates that there are different
degrees of this relation.  A. laricis has the most regular density distribution.  A steady decrease in numbers  was observed for this species
from the most densely populated plot  toward both ends  of the gradient
during all periods.This indicates that the observed changes were mainly
dependent on habitat conditions.  The density distribution for I. notabilis and F. quadrioculata was slightly less regular, which may indicate
that  also other habitat factors  than soil moisture  and  air humidity
contribute to the differences in population dynamics.  Finally, regular
density distribution  for the polyvalent species I. minor  was observed
only in spring. But the general trend remains – maximum number for plots
with a higher moisture  content  appear  during  dry periods, for plots
with a lower moisture content during humid periods. In sum, the changes
in I. minor numbers, though to some extent dependent  on soil moisture,
were not conditioned by its variability and  must have been a result of
the effect of other ecological factors.

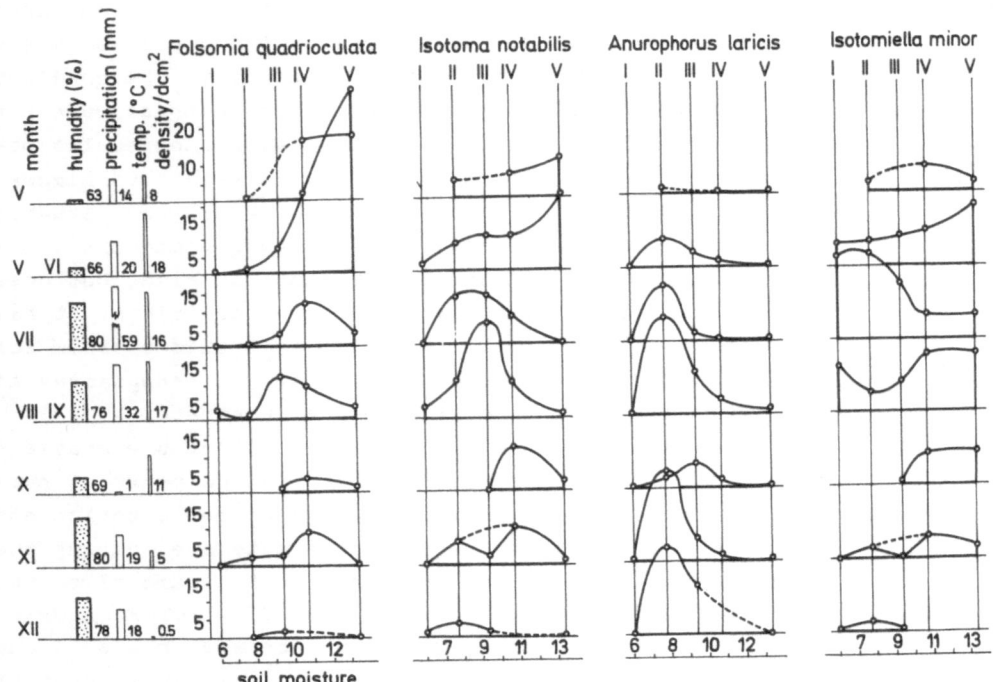

Fig.4.   Density   of   particular   species   under   different
conditions during successive periods.

When mean annual density distribution of the species are compared,
it is possible   to conclude that the differences   in population numbers
which cannot be explained by differences in humidity conditions between
the plots,   may be due   to interspecific relationships   (Fig. 5).   This

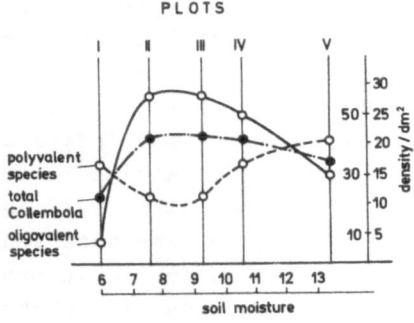

Fig.5. Comparison of density distribution   of oligo- and
polyvalent species.

366a-    4383 - Sv.

comparison indicates that the distribution of oligo- and polyvalent species is characterized by compensatory interactions. These inter- actions result in a considerable levelling of their joint numbers under optimal conditions of soil moisture and air humidity, that is to say in plots II - V.

The situation indicates that the reduction in the number of poly- valent species in the intermediate section of the habitat gradient is probably related to a great density of oligovalent species, which have their ecological optimum in these habitats. Accordingly, large numbers of polyvalent species in the plots with extreme habitat conditions, outside the optimum for oligovalent species, could be related to a lower impact of the latter. Therefore, we probably have here a limiting ef- fect of specialized species on the numbers of unspecialized species.

R e f e r e n c e s

A g r e l l , J. (1941): Zur Ökologie der Collembolen. Untersuchungen
    im schwedischen Lapland. Opuscula Entomol. Suppl. III, 236 pp.
K a c z m a r e k , M. (1963): Jahreszeitliche Quantitätsschwankungen
    der Collembolen verschiedener Waldbiotope der Puszcza Kampinoska.
    Ekologia Polska, A, 11, 127-139.
P o o l e , T. B. (1961): An ecological study of the *Collembola* in
    coniferous forest soil. Pedobiologia 1, 113-137.
S h e a l s , J. G. (1957): The *Collembola* and *Acarina* of uncultivated
    soil. J. Anim. Ecol. 26, 125-134.
W e i s - F o g h , T. (1948): Ecological investigations on mites and
    *Collembola* in the soil. Nat. Jutland. 1, 137-270.

D i s c u s s i o n

H. F r a n z : Diese ausgezeichnete Untersuchung zeigt deutlich, dass man eindeutige Schlussfolgerungen nur auf Grund von Untersuchungen zie- hen kann, die das Verhalten der einzelnen Arten genau erfassen. Auf Grund solcher Untersuchungen können auch menschliche Eingriffe in das Ökosystem in ihren Auswirkungen genau erfasst und beurteilt werden. Auf Grund von derartigen Untersuchungen kann man auch die Praxis von der Bedeutung bodenbiologischer Untersuchungen überzeugen.

F. A t h i a s :  In this study,  you consider  two factors,  i.e. soil
humidity and air humidity.  My question is: are the types of soils similar in all the five plots?  It is important  for the evaluation of the
gradient of soil moisture on different plots.

M. K a c z m a r e k :  In all plots there was slightly podsolized soil.
The gradient of soil moisture, which we had in plots I - V was connected
as we suppose  with the difference  in water run off and soil porosity.

G. M a r c u z z i :  The author  has spoken  on interspecific competition - does she think  that may be there  is not sufficient food supply
for all species and for all specimens?

M. K a c z m a r e k :  In  field conditions  we can observe  only  the
result  of competitive relationships  among  collembolan species,  when
they change their density.  We can suppose  that the basis for such relationships may be the environmental physical factor in connection with
food supply.

W. D u n g e r :  Ist es denkbar,  dass  in Bezug  auf die Feuchtigkeit
oligovalente Arten z.B. hinsichtlich der Ernährung (relativ) polyvalent
sind  und umgekehrt, so dass  Abb. 5  teilweise, auf der Basis  anderer
Faktoren,  auch reziprok zu denken wäre.

M. K a c z m a r e k :  Es gibt nur sehr kleine Wahrscheinlichkeit, dass
die untersuchten  polyvalenten Arten  eine  enge Nahrungsspezialisation
haben, weil ihr Abundanzmaximum  in extrem verschiedenen Feuchtigkeitsbedingungen lag.  Die Feuchtigkeitsbedingungen, besonders für saprofage
Arten  beeinflussen  auch die Nahrungsbedingungen  und die Nahrungsqua-
lität.

C. G r e g o i r e — W i b o :  Are your species  present  together in
the same plots?

M. K a c z m a r e k :  Yes, they are.

# The influence of acclimation and soil moisture on the temperature preference of Eisenia rosea (Lumbricidae)

A. J. REINECKE

Department of Zoology, Potchefstroom University,
South Africa

INTRODUCTION

Earthworms are capable of vertical migration in the soil  which is frequently  the result  of temperature difference  in the soil  (Satchell,
1967).  Most soil-dwelling  invertebrates exhibit a preference temperature (Wallwork, 1960;  Madge, 1969;  Edwards, 1961) but live and reproduce  at other temperatures,  depending  upon the availability  of soil
moisture.

It can be accepted that, due to a natural interaction between factors  and the resulting complexity of the factors  operating in the natural environment,  an animal  will not always select the optimal grade
of a certain factor. Animals are governed in nature by the quantity and
changeability of factors for which they have a minimum requirement. The
limits set by one operating factor  will therefore influence the selection of the optimal condition for another factor.

Knowledge  of the temperature preferences  has been gained through
the work of various authors. El Duweini & Ghabbour (1965) found that *A.
caliginosa*  occurred in a temperature gradient  between  2 $^\circ$C and 37 $^\circ$C
with high frequencies at 8 $^\circ$C,  19 $^\circ$C and 28 $^\circ$C.  *Pheretima californica*
occurred  between 26 $^\circ$C and 35 $^\circ$C  in a temperature gradient while *Alma*
sp. showed  a  definite preference  for  temperatures of $\pm$ 25 $^\circ$C. These
authors investigated the temperature preferences by using a gradient of
0 - 40 $^\circ$C.  They did not investigate the possible influence of acclimation on their results.

Grant (1955)  found that most of the *Pheretima hupeiensis* individuals preferred a temperature  between 15 $^\circ$C and 23 $^\circ$C.  *Eisenia foetida*
occurred mostly between 15.7 $^\circ$C and 23.2 $^\circ$C  while *A. caliginosa* exhibited a wider range between 10.0 $^\circ$C and 23.2 $^\circ$C. According to this author
all three species preferred temperatures above 23 $^\circ$C  and  below 10 $^\circ$C.
All specimens were acclimated at a temperature of 22 $^\circ$C.

Wolf (1938) studied the directing influence of temperature upon *Eisenia foetida* and concluded that a temperature of 27.5 $^\circ$C resulted in a "thermotropic" reaction with the earthworms moving to cooler areas.

The present investigation was aimed at determining the temperature preference of *Eisenia rosea* and investigating the influence of two different acclimation categories as well as soil moisture on the preference temperature.

The author is indebted to the C.S.I.R. for financial support.

## APPARATUS, MATERIAL AND METHODS

A temperature gradient trough (Fig. 1) was constructed consisting of two metal water baths A and B (20 cm x 20 cm) and a copper trough con-

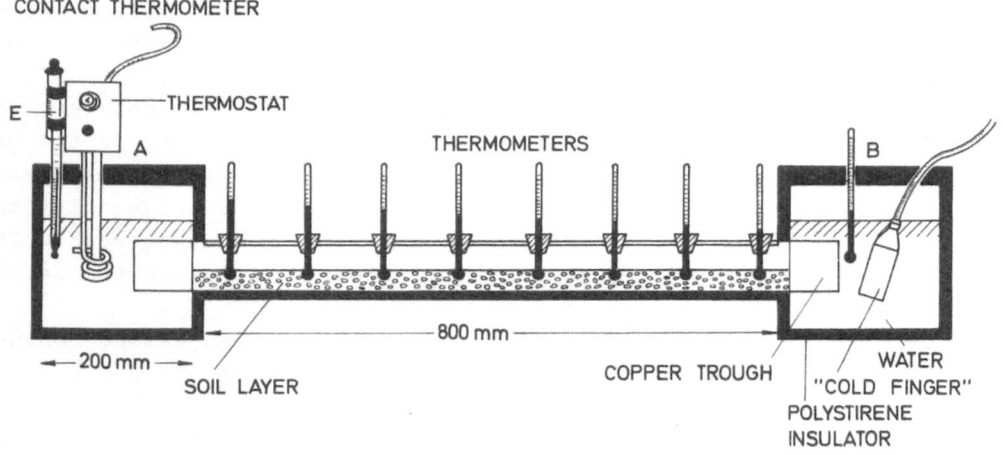

Fig.1. Temperature gradient was constructed of two metal baths and a copper trough.

necting the two baths. One end of the trough was permanently fastened to bath A while the other end protruded into bath B. This construction made it possible to vary the distance between the two baths. The trough was provided with a perspex lid with openings 10 cm apart through which rubber stoppers with thermometers were placed. The baths and trough were isolated by polystirene foam.

Testing of the apparatus over periods of 24 hours showed that a stabilised gradient could be maintained in spite of diurnal changes in air temperature. Small fluctuations of 0.3 $^\circ$C to 0.6 $^\circ$C between the day and night average were recorded by a thermograph.

A soil layer  5 - 6 cm deep  was  placed  in the copper trough and wetted thoroughly. Condensation of water occurred on the perspex lid at the warm end  of the trough  which posed the problem of a moisture gradient  being established across the length of the trough.  Soil samples were taken  from different zones  in the trough after each test period. Soil moisture  determinations  with  an Ultramat moisture-meter  showed a slight  difference  between  the two ends  of the trough  although no gradual gradient was established.

To prevent evaporation and subsequent condensation of water on the lid in the warm end  of the trough, cotton wool  was placed on the soil surface. Moisture determinations were carried out for all the different temperature zones after each experiment.

Fifteen worms  were placed evenly distributed  in the trough after wetting the soil.  The lid was closed and sealed and the experiment remained  for 30 to 60 minutes  at room temperature.  This period  proved long enough for the worms to dig into the soil.

The temperature gradient  was established by gradually raising the temperature in bath A and cooling the water in bath B.  With the use of the thermostat the water could be heated very slowly until the required maximum  was reached.  This procedure  was of great importance  because a gradient with a steep slope was required. Cooling was also done slowly. The required temperature gradient  was reached after 24 hours which gave the worms time  to move away  from extreme temperature conditions. The possibility of lethality due to a thermic shock was thus prevented.

To prevent the influence of different structural and textural conditions in different parts  of the trough, all soil  was homogenised by thorough mixing before being used in the experiment.

Grant (1955)  has shown that acclimating of worms  at certain temparatures  higher than normal  leads  to increased  lethal tamperatures which poses the possibility that the same phenomenon might occur in the case of preference temperatures. All worms used in the present investigation  were sampled at soil temperatures ranging from 19 $^{\circ}$C to 23.8 $^{\circ}$C and kept in stock  in a temperature  regulated breeding chamber.  Worms used in the experiments  were kept at a temperature of 20.0 $^{\circ}$C  for one week prior to the experiments.

Fresh specimens  were  used  in each experiment.  Where worms died during an experiment all soil was recovered to prevent the possible influence of toxic substances.

Worms were sampled  and counted  in each zone at the end of an experiment. Due to the migration of earthworms when the soil is disturbed, it was deemed necessary  to use  metal pieces to partition off the different zones before analysing the soil.

RESULTS

## The temperature preferences of *E. rosea*

The results of the investigation are given in table 1.  From table 1 it
can be seen  that the worms acclimated at 20.0 °C, as shown by the histogram  in Fig. 2,  tend to aggregate  between  25.0 and 26.9 °C  which
includes the mode and median of the frequency distribution.

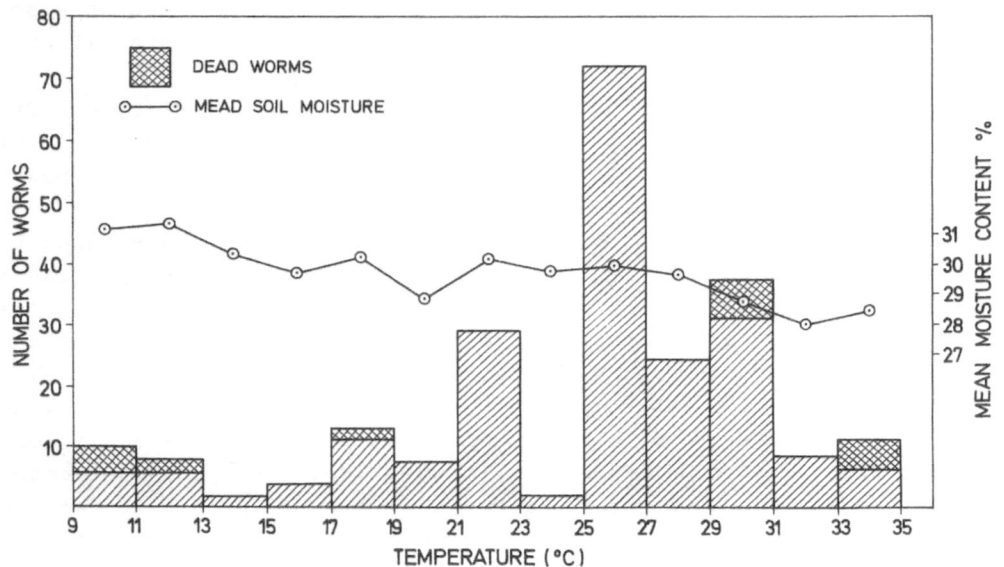

Fig.2.  Histogram illustrating  the overall distribution
of *Eisenia rosea*  in a temperature gradient.  Worms were
previously acclimated  at  20,0 ± 0,5 °C.  Soil moisture
content: 28 - 31 %.

For a 95 % confidence limit the "preference temperature" lies between 24.1 and 25.6 °C.  A total of 227 worms were used of which 19 died
during the experiments. Death occurred mainly at the extremities of the
temperature gradient.

A heterogeneity occurred  in the frequency distribution  which  is
illustrated  by the coefficient  of variance  of 21.3 %  which was calculated.

The "preference temperature"  as used in this paper indicates that
temperature at which most of the worms were found and which can be considered as a centre around which the worms  are most likely to be found
when acclimated at the same temperature.

T a b l e   1.    The compiled results  of different series of experiments
performed to determine the preference temperature of *Eisenia rosea*

| Temperature °C | Mean soil moisture % | | | Number of worms | | |
|---|---|---|---|---|---|---|
| | Series 1 | Series 2 | Series 3 | Series 1 | Series 2 | Series 3 |
| 9,0 - 10,9 | 31,2 | 31,9 | 17,3 | 6+(4) | 4+(3) | O |
| 11,0 - 12,9 | 31,4 | 31,7 | 17,0 | 6+(2) | 3+(1) | 16 |
| 13,0 - 14,9 | 30,3 | 30,7 | 16,6 | 2 | 1 | 4+(1) |
| 15,0 - 16,9 | 29,7 | 30,0 | - | 4 | 3 | - |
| 17,0 - 18,9 | 30,3 | 31,2 | 15,9 | 11+(2) | 8 | 28ᵣ |
| 19,0 - 20,9 | 28,9 | 29,4 | 15,7 | 7 | 12 | 16 |
| 21,0 - 22,9 | 30,2 | 28,9 | 14,8 | 29 | 22 | 42 |
| 23,0 - 24,9 | 29,7 | 29,1 | 14,7 | 2 | 11 | 19 |
| 25,0 - 26,9 | 29,9 | 29,3 | 14,8 | 72 | 66+(1) | 2+(1) |
| 27,0 - 28,9 | 29,8 | 28,9 | 13,0 | 24 | 34 | 1+(1) |
| 29,0 - 30,9 | 28,7 | 29,5 | 13,4 | 31+(6) | 35+(2) | O |
| 31,0 - 32,9 | 28,0 | 29,7 | - | 8 | 9 | - |
| 33,0 - 34,9 | 28,4 | 28,3 | - | 6+(5) | 6+(6) | - |

Series 1: Relatively high moisture content; acclimation 20°C.
Series 2: Relatively high moisture content; acclimation 25°C.
Series 3: Relatively low moisture content; acclimation 20°C.
(Dead worms are given in brackets.)

     The preference temperature as determined here cannot be considered
without  mentioning the moisture conditions  due to the close relationship in the action of these two factors  in a soil environment.  A mean
soil moisture content varying between 28.0 and 31.4 % for the different
zones, was determined.  The lack of a definite gradient  made the consideration of the moisture factor in the calculations unnecessary.
     The results of the experiments  to determine the temperature pref-
erences of *E. rosea*  after  acclimating the worms at 25.0 ± 0.5 °C  are
given  in table 1.  A prominent peak occurred  between 25.0 and 26.9 °C
(Fig. 3) which includes both the mode  and median of the frequency distribution.  Thirteen worms  died during the course of these experiments
mainly at the high and low temperatures.
     For a 95 % confidence limit  the preference temperature  lies bet-
ween 24.8 and 26.0 °C.  A mean moisture content of 28.3 and 31.9 % prevailed in the trough. A coefficient of variance of 18.5 % was calculated.

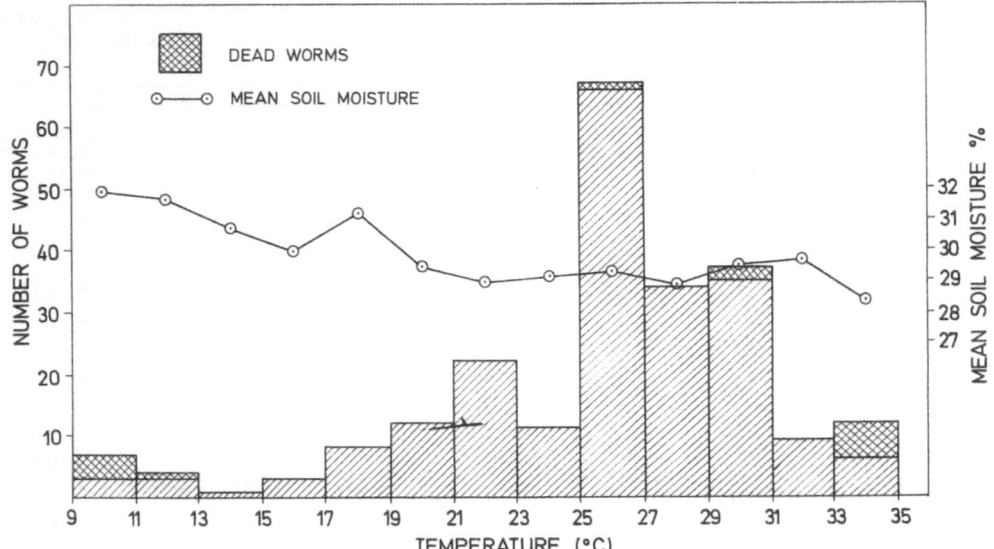

Fig.3.  Histogram  illustrating  the overall distribution
of *Eisenia rosea*  in a temperature gradient.  Worms were
previously  acclimated  at  25,0 ± 0,5 °C.  Soil moisture
content: 28,3 - 31,9 %.

## The influence of acclimation upon the preference temperature

If the results obtained for the two acclimation groups are compared, it
is clearly seen that the preference temperature of the higher acclimated group  is of a higher order  than  that  of the group  acclimated at
20.0 °C.  The difference  in the $\bar{x}$ - values  was 25.39 - 24.86 = 0.53 °C.
On an average the group acclimated at 25 °C  revealed a preference temperature 0.53 ° higher than that of the 20 °C group.

Testing the two variance estimates

$$28.24 - 22.27 = 5.97$$

and     $F = \dfrac{28.24}{22.27} = 1.2$

(with $DF^1 = 207$ and $DF^2 = 214$).

For p = 0.05 the critical values of the variance ratio F show that values of F  equal to or bigger than 1.00  will occur less frequently than
p = 0.05.  Or, any ratio smaller than 1.00  is compatible with the nul-
hypothesis of equal variances.

When the $S.$ $S.$ of both groups are pooled:

$$s^2 \text{ (pooled)} = \frac{SS + SS^1}{(N-1) + (N^1-1)} = 25.2.$$

Then since each sample has a different size, we compute an estimate of the variance of the difference between means by

$$s^2(\bar{x} - \bar{x}^1) = s^2(\frac{1}{N} + \frac{1}{N^1})$$

$$= 25.2 \ (\frac{1}{215} + \frac{1}{208})$$

$$= 0.23$$

$$s(\bar{x} - \bar{x}^1) \quad = 0.23 = 0.45$$

$$t = \frac{(\bar{x} - \bar{x}^1)}{s(\bar{x} - \bar{x}^1)} = \frac{1.2}{0.45} = 2.6$$

For $DF$ = 200 at $p$ = 0.05 $t$ must exceed 1.97 to give the required significance. Where $t$ is determined as 2.6 we can therefore conclude that the difference in the preference temperatures of the two differently acclimated groups were significant.

An increase of 5 °C in the acclimating temperature has raised the preference temperature of E. rosea by 0.5 °C. This finding accentuates the necessity for the previous history of the factor to be known before temperature preferences can be determined correctly. Due to this fact a direct comparison of the results of the various researchers who have done experiments in this field without considering the role of acclimation is not possible.


TEMPERATURE PREFERENCE AT LOW MOISTURE CONDITIONS

The results of the investigation at low moisture conditions of 13 – 17 % soil moisture are given in table 1 and illustrated in Fig. 4. A preference temperature between 17.6 °C and 21.8 °C ($p$ = 0.05) was calculated. If this result is compared with that obtained for the 20 °C-acclimated group at 28 - 31 % soil moisture it is clearly seen that the preference temperature difference under the different moisture conditions. The difference in the $\bar{x}$-values of the two groups was 24.8 – 19.7 = = 5.1 °C. Testing the two variance estimates, pooling and computing an estimate of the variance of the differences between means as was done

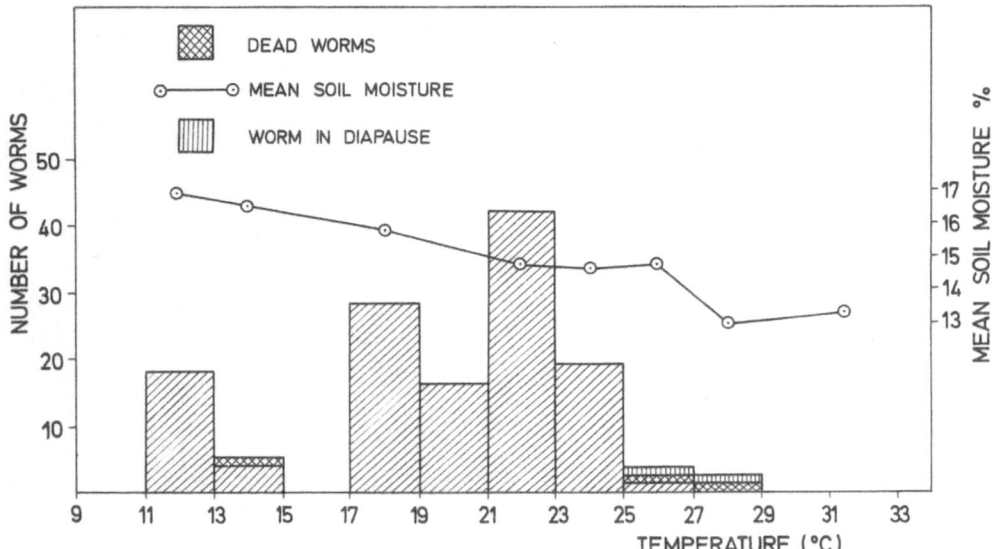

Fig.4. Histogram illustrating the overall distribution of *Eisenia rosea* in a temperature gradient with a relatively low moisture content: 13-17 %, worms were previously acclimated at 20,0 ± 0,5 °C.

in the previous section $t = 3.9$. We can therefore conclude that the difference in the preference temperatures at the two moisture categories were significant ($p = 0.05$).

It then follows that under the wetter conditions *E. rosea* preferred a temperature which was 5 °C higher than the preference temperature under the drier conditions. The difference in the soil moisture content was approximately 14 %.


DISCUSSION

The results of this investigation clearly underlined the close relationship between soil moisture and temperature in a soil environment and the influence of its interaction on the migration of the earthworms in a temperature gradient.

The physiologically non-static character of the preference zone is subsequently illustrated.

SUMMARY

*Eisenia rosea (Lumbricidae)*,  tested in a temperature gradient,  showed a preference temperature of 25.4 $^\circ$C (acclimation at 25 $^\circ$C)  and 24.9 $^\circ$C (acclimation at 20 $^\circ$C). At comparable moisture conditions a significant difference (*p* = 0.05) in preference temperatures of 0.5 $^\circ$C was obtained. Worms  tested at ± 30 % soil moisture  showed  a preference temperature 5 $^\circ$C higher than those tested at ± 16 % soil moisture after acclimation at 20 $^\circ$C.

R e f e r e n c e s

Edwards , C. A. (1961): The ecology of *Symphyla:* Pt III. Factors controlling soil distributions.  Ent. Exptl. and Appl. 4, 239-256.
El Duweini , A. K. & Ghabbour , S. I. (1965): Temperature relations  of three Egyptian oligochaete species.  Oikos 16, 9-15.
Grant , W. C. (1955): Temperature relationships in the meguscolecid earthworm *Pheretima hupeiensis*. Ecol. 36, 412-417.
Madge , D.S. (1969): Field and laboratory studies on the activities of two species of tropical earthworms. Pedobiol. 9, 188-214.
Satchell , J.E. (1967): *Lumbricidae*. In: Soil Biology (Ed. A.F. Raw), Academic Press, New York.
Wallwork , J. A. (1960):  Observations on the behaviour of some oribatid mites in experimentally controlled temperature gradients. Proc. Zool. Soc. London, 135(4), 619-629.
Wolf , A. V. (1938): Notes on the effects of heat in *Lumbricus terrestris* L.  Ecol. 19, 346-348.

D i s c u s s i o n

J. S a t c h e l l :  Did you find any problem  in the batches of worms you introduced  in each experiment  aggregating  together  from tactile stimulation?

A. R e i n e c k e :   I used  only ,15 worms in each run  and  did not notice it in the different runs but it could possibly be an explanation for  the relatively high heterogeneity  in  the frequency distribution.

# SECTION C

## INFLUENCE OF HUMAN ACTIVITIES ON COMMUNITIES OF SOIL ORGANISMS

# Influence of soil conditions
# on the distribution of diplopods
# in Southern Polesje (Byelorussia, USSR)

E. I. CHOTKO AND B. R. STRIGANOVA

Department of Zoology and Parasitology
Acad. Sci. Byelorussian SSR,

Institute of Evolutionary Morphology
and Ecology Acad. Sci. USSR,
Moscow

Diplopods as well as other members of the mesofauna can be most useful when solving different problems of soil diagnostics. These myriapods are important components of soil communities in mixed and broad-leaved forests. They also can inhabit arable soils, to which they migrate from forests and littoral habitats (Tischler 1965). The humidity of the soil and its calcium contents seem to be the main factors determining the distribution and population density of diplopods (Ghilarov 1957). This report presents data on diplopod communities in the Polesje lowland (Southern Byelorussia).

Polesje represents a vast swampy woodland including the whole basin of the river Pripjat'. Soddy podzolic soils are the zonal soil types there. They develop on loamy sands or loams with a comparatively high soil acidity. A great part of the Polesje lowland is occupied by swamps and peatbogs representing the entire scale of bog soils. The regular long-term drainage of this area caused essential changes of the soil regime. The soil fauna of the drained sites of Polesje was found to differ considerably from that of bog soils. Diplopods were found to be numerous in bog-reclaimed soils. Their population density reached 300 spec./m$^2$ in some regions of Polesje (Kipenvarlitz 1961). Peculiarities of soil animal communities in improved bog soils provided information about the change in soil formation processes from azonal bog patterns to zonal soddy podzolic ones which are characteristic of Polesje (Ghilarov 1965).

Quantitative investigations of diplopods were carried out in 8 sites with a different soil- and vegetation cover. These sites are the following: 1. *Pinetum caricoso-sphagnosum* in high bog with stagnant water. 2. *Pinetum myrtillosum* in a depression on the soddy podzolic sandy moist soil. 3. *Quercetum myrtillosum* on soddy podzolic loamy soil. 4. *Glutinoso-alnetum urticosum* on the mud-humus soil. 5. *Glutinoso-alnetum filicosum* on peat-humus soil. 6.-8. Crops of timothy (I-V years), oats and barley on the bog-improved soil.

20 soil samples, 0.25 m$^2$ in area, were taken from each forest habitat.  Soil cores in field habitats  were extracted  by means of a soil auger  from an area of 0.02 m$^2$ (100 cores from each field). The list of diplopod species and their population density are presented in Table 1.

Diplopods are the most numerous in alder (166-50 spec./m$^2$). *M. laeticollis* appeared to be the dominant species.  *Chr. sjaelandicus* and *P. complanatus*  were recorded in all samples.  *P. germanicum* and *Chr. projectus* found in the sedge alder  were not distributed in other investigated habitats.

Diplopods in pine forests were scarce.  Their density in the moist *Sphagnum* forest reached 7.4 spec./m$^2$.  Two species - *P. complanatus* and *P. fuscus* - were recorded  in both pine sites, the latter predominating. It was not found in other habitats.

The density of diplopods  in the oak forest  was as low as that in the pine forest.  Two species were recorded there.  Diplopods were comparatively numerous in arable soils of Polesje under perennial crops of timothy, where *Chr. sjaelandicus* predominated.  They were absent in the fields in the first 3 years, appeared after the fourth year and reached their maximum density in the fifth year in these fields. Diplopods concentrated  in arable habitats on boards of drainage trenches - up to 17 spec./m$^2$. *P. complanatus*, *J. scanicus*, *M. laeticollis* and *Chr. sjaelandicus* were found there.

The diplopod fauna  of  Bieloviezhskaja Pushcha  being  continuous with the Polesje region seems to be richer and more varied. Records are available on 15 species  of diplopods, with a population density  of up to 70 spec./m$^2$ (Jawlowski 1949,  Lokshina 1964,  Ghilarov et al. 1971). *Glomeris connexa*, widely distributed in Bieloviezhskaja Pushcha averaged in this area 50-85 % of the whole number  of diplopods  in forest habitats.  Glomeridae are known to be associated with forest brown soils in Europe also characteristic of the Bieloviezhskaja Pushcha. These myriapods  are absent  in regions  where  soddy podzolic processes  prevailed in the soil formation. They were not recorded in our samples on podzolic and bog soils.  On the territory of Polesje they were found to be scarce in an ash-oak forest on humic gley soil only  (Vorontzov and Zinovjeva 1951). Diplopod communities in Southern Polesje are represented by species widely distributed in different forest types  or in association with alders.  Rich humic soils of alders with a comparatively high Ca content (2-4 %)  and low actual pH (4.7)  appear to be most suitable for diplopods.  The Ca  content increases  after  cultivation  of these soils This creates prerequisites for migration of diplopods from alders to open sites  and, therefore, diplopods  are  numerous  in alders  and arable soils.  Low densities of diplopods  in forest podzolic soils re-

T a b l e   1.   Distribution of Diplopods in Byelorussian Polesje

| Diplopoda | Pinetum | | Quercetum myrtillosum | Glutinoso-alnetum | | Crops | | |
|---|---|---|---|---|---|---|---|---|
| | caricoso-sphagnosum | myrtillosum | | urticosum | filicosum | Timothy | oats | barley |
| | 1 | 2 | 3 | 4 | 5 | 6 | 7 | 8 |
| Mean density | 7.4 | 1.6 | 0.5 | 16.6 | 50.0 | 22.7 | 2.6 | 1.6 |
| Polyzonium germanicum | | | | + | | | | |
| Polydesmus complanatus | + | + | + | | + | | | |
| Proteroiulus fuscus | + | + | | | | | | |
| Microiulus laeticollis | | | + | + | + | + | | |
| Leptoiulus proximus | | | | | | | + | |
| Chromatoiulus projectus | | | | + | | | | |
| Chromatoiulus sjaelandicus | | | | | + | + | + | + |
| Julus scanicus | | | | | | + | | |
| Sarmatiulus vilhense | | | | | + | + | | |

corded in Polesje are characteristics of other regions of the forest zone. Both the Ca content and pH value appear to be the main factors determining the distribution of diplopods in the Byelorussian Polesje.

R e f e r e n c e s

G h i l a r o v , M. S. (1957): Black wireworms (Juloidea) and their role in the soil formation. Pochvovedenie 6, 74-80 (in Russian).
G h i l a r o v , M. S. (1965): Zoological method of soil diagnostics. Moscow, 1-276 (in Russian).
G h i l a r o v , M. S., P e r e l , T. S. and U t e n k o v a , A. P. (1971): Use of invertebrates for the characteristic of soils in Bieloviezhskaja Pushcha. In: Bieloviezhskaja Pushcha, Minsk, 4, 193-212 (in Russian).
J a w l o w s k i , H. (1949): Myriapods of the National Park Bialowiecza. Ann. Univ. M.C. Sklodowska, Lublin, Sec. 15, 309-323.
K i p e n v a r l i t z , A. F. (1961): Changes of the soil fauna of marsh-bogs under the influence of land-reclaiming and cultivation. Minsk, 1-197 (in Russian).
L o k s h i n a , I. E. (1964): Die Diplopoden in den Waldböden der Belovežkaja Pušča. Pedobiologia 4, 4, 299-309 (in Russian).
T i s c h l e r , W. (1965): Agrarökologie. Jena, 1-300.
V o r o n t z o v , A. I. and Z i n o v j e v a , L. A. (1951): To the characteristic of the invertebrate fauna in forest soils of Polesje woodland. In: About the forests of Polesje, Minsk, 125-140 (in Russian).

# Studies on the influence of industrial pollution on soil animals in pine stands, aims and methods of the soil-block model experiment

M. GÓRNY
**Forest Institute**
**Warszawa**

## BACKGROUND

During previous studies (Górny, 1971, 1972) it was inferred that industrial air pollution exerts a marked influence on soil animal populations in pine stands. It was also stated that an initially slight positive influence of pollution marked by a greater reproduction of some species of soil animals, changed into a distinctly negative influence and might have lead to the death of some populations. The intensity of the negative influence was being reduced with the distance from the source of pollution, in this case nitrate industry. This fact enabled to distinguish zones of different soil animals communities, similar to those of plant communities.

This is but a situation in which polluting gases and smoke are distributed with more or less constant intensity from one source for a prolonged period, resulting in both acute and slowly progressing changes, the first due mainly to gases, the second to a cumulation of solid particles in the soil. The results of our studies show the influence of pollution on forest soil animals, and indicate the animal species to be used as ecological indicators of changes in the environment.

For purposes of the practice, evidence should be available on the boundary of negative influences from which negative changes are still reversible. This knowledge may be important in suggesting the recultivation programme in areas affected by industrial pollution.

## IDEAS

Transformation of matter and energy in the soil are important for the entire ecosystem. In the areas under consideration transformation processes have been changed radically by industrial pollution.

    The object of this study  was to obtain   an answer   to these ques-
tions: 1 - what is the influence of pollution on the disorganisation or
destruction  of populations  of soil animals?  2 - How can pollution be
restricted  and what is the period   required   by the communities to re-
generate?  We have tried to estimate the time   and degree of changes in
soil animal communities  due to intensive or limited air pollution, and
to compare the reaction of soil animals to pollution  with that of animals living on tress and other forest plants.

    Under conditions  of acute industrial pollution,  community stress
may be encounter  as a result of a weakened  or destroyed capability of
selfregulation  in the community.  Its causes  may not be identical  to
those responsible  for individual stress  (according to Selye 1956) and
therefore, they should be studied at the community level.

    Having regard to all the aspects  to be studied  we decided to use
models, i.e., soil blocks in undisturbed condition. Of the soil animals,
we selected three groups of small soil invertebrates: enchytraeid worms,
oribatid mites and springtails (Collembola).  We assumed that the route
from the source of pollution to the soil  examined could either be prolonged or shortened by an interchange of soil blocks,  and that the influence  of pollution  could either be weakened  or intensified  in the
same way. Thereby, we were aware of the fact that the testing of a model
could not be compared  with  the testing  of a hypothesis  and  that it
could never seriously be considered  to be a true representative  of an
ecosystem, because no quantitative model is perfect. While constructing
the model we tried to collect a representative set of data for testing.
In order to compare conditions  of the model soil blocks  with those of
natural soil, we set up a provisional experiment on April 1969. In this
we used 10 soil blocks  (25 x 25 x 30 cm)  which we dug up, placed into
big, bottomless pots and returned to the soil. From each block, we took
twice two samples (100 cm$^3$), i.e., one month  and  5 months a:ter digging,  and extracted from them mites and springtails. In both instances,
differences in populations of mites and springtails were similar in the
model and natural soil and, hence, statistically insignificant. On these
grounds we hoped to obtained  from the soil-block model experiment performed for the purpose  of studying the influence  of industrial pollution on various groups  of soil animals,  information on the components
of litter  and  soil fauna, and on a set of processes such as reproduction or death, motility, decomposition activity  (feeding) etc.  In addition, the tests indicated that the species composition  and number of
specimens  were similar in both the blocks and natural soil in spite of
the fact that this similarity  was more marked  at the first inspection
(one month after the digging of the blocks)  than at the second inspec-

tion (5 months later). Our results  suggested  that soil blocks of the
described volume may be useful for experimental purposes and for extrapolation and also, that the experiment should be carried out in a limited period not exceeding one year.


METHODS

Our provisional test  confirmed  that soil blocks  are suitable experimental models.  They were used also by Stebaeva (1967)  for testing the
reaction of soil invertebrates under different geographical conditions.
        Our experiment  was performed in three local zones selected on the
basis of the risk-scale for forest trees and other vegetation.  We distinguished a control zone  with good living conditions  for the plants,
a zone  showing evidence of pollution  on the vegetation,  and a forest
death zone. Blocks of undisturbed soil should have been taken from each
zone  and placed in big, bottomless pots.  For transport purposes, however, these blocks  would have to be  of a limited size  but big enough
to represent the soil animal community from which a representative number of samples  could be taken  for a statistical evaluation of the material.
        Taking  these difficulties  into  consideration  we decided to use
soil blocks  with a litter cover,  measuring 25 x 25 cm  on the surface
and  approximately  30 cm in depth.  The volume of the soil sample from
which  any  of the three groups  of animals  were to be extracted,  was
100 $cm^3$, its diameter  was 5.1 cm  (surface area  approximately 20 $cm^2$,
depth 5 cm).  It was accepted that the total of surfaces of the samples
used in the experiment should not exceed 25% of the soil block surface,
in order to forego any larger environmental changes  in the soil blocks
and, consequently, in the soil animal community.
        This indicated that:
        1. no more than 2 samples  should be taken from each soil block at
any sampling occasion;
        2. for the duration of the experiment, the total number of samples
taken from one soil block should not exceed 6 - 8;
        3. consequently, the experiment conducted for a period of  3 years,
had to be divided into two parts which provided an additional advantage,
i.e., the comparison  of two sets  of soil blocks  taken  at  different
times.
        Considering the fact that springtails and mites might be extracted
simultaneously, it was possible to take each time two samples from each

block, i.e., one for the estimation of the two groups, the other for enchytraeid worms.

The next question to be solved was the number of samples to be taken each time and, subsequently, the number of soil blocks to be dug up from each zone. The number of samples had to be determined by the smallest area still capable of characterising the whole community and the pattern of distribution of the different groups or species of animals. All animals of the three groups, particularly enchytraeids and springtails, are distributed in patches. In view of this fact attention had to be given to the number of samples enabling the determination of the number of these irregularly distributed species.

Although we have not yet concluded our study on quantitive methods enabling an evaluation of enchytraeid populations in forest soils, our present results indicate that the smallest number of samples from which sparsely distributed populations of enchytraeids can be obtained in these soils and plant patches, is 5-10; for the entire area with different plant patches, the number of samples should be 20-30. Our data have been obtained from calculations of the average number of specimens per sample and the quantity of standard deviation. The distributions of quantity in the different samples close to average density indicates the actual distribution of worms in the soil.

According to these calculations, 5 samples are. adequate for the extraction of enchytraeids, and another 5 samples for the extraction of both mites and springtails. This, in turn, influences the number of blocks used and their distribution (Fig. 1).

For every zone it was decided to dig up 5 blocks each from the two other zones, and two blocks of soil from the same zone. This makes a total of 12 soil blocks for each zone.

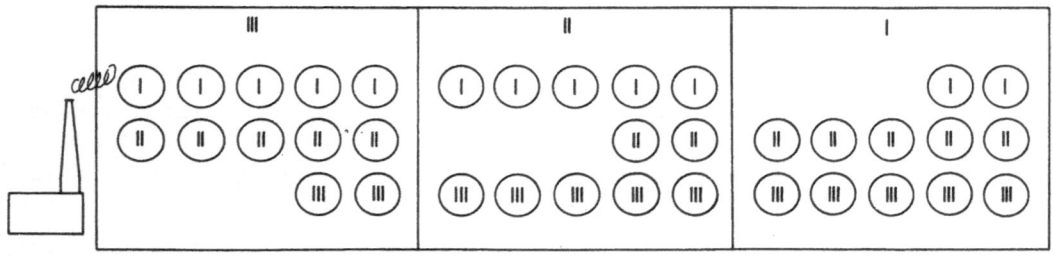

I     control zone
II    zone of visible influences of pollution on forest vegetation
III   woods death zone
O     soil-block

Fig.1.  Scheme  of the soil  block  distribution  in different zones of pine forest.

At the time  of sampling  it was agreed  to take  two samples from
each block and three additional samples from natural soil to supplement
the two samples taken from the soil blocks of the same zone.

When samples  are taken from blocks or natural soil in undisturbed
condition, they should be divided in segments in accord with their natural layers  (litter, humus, mineral soil, etc.)  and  then transported
separately  to the laboratory  for extraction.  Mites  and  springtails
should be extracted within 3 days,  enchytraeid worm within five hours.


CONCLUSION


It is rather difficult  to form ecological models.  Normally it is only
possible to maintain their identity  within a limited range of environmental data and a limited range of organisms.

Using models  for testing a hypothesis  it should be necessary not
to consider  the statistical significance.  Some phenomena  may be predicted by ecologists  without mathematics, which in turn  does not deny
the value  of mathematical models  in ecology.  The statistical significance helps in testing and evaluating conclusions, but does not replace
the experience  of ecologists.  The last  may be of still greater value
in view  of radical changes  in  our environment.  An important part of
testing models is also the confidence in extrapolation.

The last is,  among others,  the case  of our experiment  in which
soil blocks were to be used as models.  In this experiment it, was proposed to take  into account all the possible precautions to afford extrapolation.  If done cautiously the zoocenological observation of soil
blocks may  predict other situations  which are likely to occur in nature.  The model experiment, although not without weak points, may offer
just now some future predictions dealing with the consequences of these
environmental negative changes.


R e f e r e n c e s

G ó r n y , M. (1972):  Badania  zoocenologiczne gleb  borów sosnowych
     w sąsiedztwie  Zakładów  Azotowych w Puławach.  XIX.  Ogólnopolski
     Zjazd Naukowy  PTG - Ochrona  Środowiska  Glebowego, Puławy,  PTG,
     216-218.

G ó r n y , M. (in print): Badania kierunków i tempa zmian w zooceno-
    zach gleb leśnych  pod wpływem  narastania lub ograniczania wpływu
    emisji przemysłowych. Materiały I. Ogólnopolskiego Sympozjum – Eko-
    logiczne Aspekty Chemizacji, Łódź, PWN.

S e l y e , H. (1956): The stress of life. New York: McGraw Hill.

S t e b a e v a , S. K. (1967): Pedobiologische Experimente  mit aus-
    getauschten Bodenblöcken  im  südöstlichen Altai-Gebirge  und der
    Severnaja Baraba. Pedobiol. 7, 2/3, 172-191.

# D i s c u s s i o n

G. J o s e n s :  You have given a list of pollutants; is it known what
happens with some of these substances which are strongly reactive after
some days, weeks or months?

M. G ó r n y : No, it is not known.

M. S. G h i l a r o v :  The method of soil block transplantation (containing animals  in their natural milieu) was already used by S.K. Stebaeva when studying the *Collembola* dependence on soil  and ecosystem as
a whole (published in "Pedobiologia" some years ago) This method proved
to be adequate.  The method is promising in soil pollution studies too,
as Dr Górny has shown.

K. H. D o m s c h :  If the aim of the studies was to use soil invertebrates as indicators of soil pollution  by industrial emissions, a comparison of a biological method  with a direct chemical analysis  of the
pollutants should be considered.

M. G ó r n y : Of course it would be. But discussing the soil-block-ex-
periment  I don't mention all the environmental studies which should be
performed.

H. F r a n z :  Wenn man Bodenblöcke  in Töpfe gibt, verändert man  die
natürliche Wasserbewegung im Boden  und damit auch den Transport chemi-
scher Substanzen im Boden.  Es wäre besser  die Bodenblöcke  ohne Töpfe
in die Versuchsfläche zu verpflanzen, aber selbst dann würde der natürliche Wasserhaushalt verändert.

M. G ó r n y :  Natürlich kann kein Model den natürlichen Umgebungsver-
hältnisse entsprechen.  Wir können und müssen aber die Modelexperimente
benützen, wenn wir schon keine andere Möglichkeiten haben.  Dann müssen
wir alles machen, um die Modelverhältnisse  am meisten  den natürlichen
anzunähern.

# Response of ants to environment pollution*

J. PETAL, H. JAKUBCZYK, K. CHMIELEWSKI, A. TATUR
Institute of Ecology P. A. S.,
Dziekanow near Warszawa

In this report  we present preliminary results on adaptation of ant populations to an environment heavily polluted by a nitrogen plant. These adaptations consist of changes in the number and development of ant populations and modification of the microhabitat within ant nests.

Attention is given  to modifications  in the content of one of the main pollutant, i.e., mineral nitrogen compounds and microfloral groups involved in the transformation of these compounds.

Ants modify their microhabitat  by changing the structure of their nests, which is followed  by the changes in aeration,  humidity pH conditions.  It is accompanied by modifications  in the development of the soil microflora  and  in  chemical processes  which  occur  in the soil (Czerwiński et al. 1971, 1972; Jakubczyk et al. 1973).  Under unfavourable conditions this behaviour is to the advantage of the ants.
of the ants.

STUDY AREA

The study areas  are  situated  in the vicinity  of a nitrogen-plant in Puławy, Lublin district.

The soil  in this area  developed  from  dune sand  and is covered with a dead pine forest.  In 1971 the maximum emission of dust  was 271 $t/km^2/year$, the pH ranged  from 3.6-7.2 (Tab. 1).  Ammonium and nitrate nitrogen participated with about 25 per cent in the dust.  Maximum pollution  was recorded in summer. The effect of pollution is  especially harmful to plants of sandy little buffered soil. Efforts are being made now to restore  the soil.  To neutralize  the acidifying effect of pol-

* Project "Man and Environment" supported  by the Committee of the Polish Academy of Sciences.

lutants the soil  is fertilized with calcium, phosphorus and potassium.
Also experiments with plant cultivation are conducted.

Studies on the ants were carried out in the most polluted zone:a)in
totally degraded areas sparsely covered with clusters of *Calamagrostis,*
*Agrostis* and *Senecio silvaticus,* and b) in restored plots.


METHODS


The following problems  were studied:  a) population of ants, including
species composition,  density of ant nests,  size of communities within
ant nests and production;  b) soil respiration in ant nests (N)  and in
the same soil layers uninhabited by ants (C). The same samples were used
to determine: c) the number of some microfloral groups, and d) the content of mineral nitrogen.

a) The size of ant population  was estimated by counting the nests
in series  of  transects  the width  of 0.5 m.  Several nests  were excavated in order to estimate the number of individuals  and  to distinguish old individuals  from those produced  in the current year (Pętal,
1972).

b) Respiration rate as an index of total soil activity was measured with a Wartburg respirometer at 30 $^\circ$C,  both in the presence of glucose and without glucose.

c) The number of microflora was estimated for the following groups
by  the plate  dilution  method:  ammonyfying bacteria,  microorganisms
utilizing mineral nitrogen, fungi, and *Actinomycetes*. The nitrification
rate was measured in vitro.

d) Mineral nitrogen  was measured by the colorimetric method after
extraction from soil with 5% solution of $H_2SO_4$.


RESULTS

C h a n g e s   i n   a n t   p o p u l a t i o n s

In heavily polluted areas a significant simplification  was observed in
ant species composition  and  decreases occurred  in the density of ant
nests, in standing crop of biomass and in production (Tab. 2).

These areas  are  mainly  inhabited by *L. niger, L.* accompanied by
*Myrmica ruginodis* Nul.  and  *M. schencki* Em. in a less devastated zone.
The density of nests was significantly lower than in cultivated meadows

T a b l e  1.    Total  amount  of sediments  and
principal pollutants in 1971

| Total amount of sediments | Nitrogen plant |
|---|---|
| ton/km$^2$/year | 271.00 |
| N-NH$_4$ ton/km$^2$/year | 37.20 |
| N-NO$_3$ ton/km$^2$/year. | 31.70 |
| S-SO$_4$ ton/km$^2$/year | 6.30 |
| S-SO$_2$ adsorbtion in mg SO$_2$/ 100 cm$^2$ PbO$_2$/day | 0.38 |
| pH of sediments | 3.6-7.2 |

T a b l e  2.   Changes in ant population under influence
of nitrogen plant pollution

| Population parameters | Devastated area | Restored area |
|---|---|---|
| Number of species | 2 | 3 |
| Density of nests/m$^2$ | 0.008 | 0.034 |
| Myrmica ssp. | 0.001 | 0.020 |
| Lasius niger | 0.007 | 0.007 |
| Area occupied by ant nests cm$^2$/m$^2$ | 0.006 | 0.027 |
| Standing crop mg dw/m$^2$ | 2.05 | 9.67 |
| Myrmica ssp. | 0.46 | 9.42 |
| Lasius niger | 0.50 | 0.25 |
| Production mg dw/m$^2$ | 0.94 | 2.29 |
| Myrmica ssp. | 0.08 | 2.18 |
| Lasius niger | 0.86 | 0.11 |
| P/$\overline{B}$ | | |
| Myrmica ssp. | 0.17 | 0.23 |
| Lasius niger | 0.54 | 0.44 |
| By/Ny | | |
| Myrmica ssp. | 0.62 | 0.50 |
| Lasius niger | 0.80 | 0.63 |

(Jakubczyk and al., 1973). The difference between the minimum values is about 1000 times  and that between the maximum values  is 2500 times as compared with meadows.  This is probably due to the sandy soil with the minimum content of humus,  which is unable to buffer harmful effects of nitrogen compound accumulation.  When calcium, potassium and phosphorus are introduced to the soil in restored areas, the population of *Myrmica* increases considerably:  the density  of ant nests  increases 20 times, standing crop of biomass 19 times and production about 27 times. *Lasius niger* population, however, are reduced, although  the number  of nests does not change.  The young generation  growth  rate ($B_y/N_y$ ratio)  for both the species  is  higher  within the devastated zone, which  may be a result of biological adaptation, since larval development  is limited to the period of maximum plant  and  animal food supply  (Pętal, 1967). In  the devastated zone  it is finished  by the end of June, in the restored zone it lasts slightly longer.

Of the two ants populations, *L. niger* is the more resistant one; its number in the most polluted areas is higher than the number of *Myrmica* ssp.

Higher adaptability of *L. niger* is also shown in the modifications of microfloral activity and pollutant contents.


A d a p t a t i o n   o f   s o i l   e n v i r o n m e n t

a) **Mineral nitrogen compounds in soil** (Tab. 3).
The compounds  of mineral nitrogen  accumulated  in the soil cause a decrease in pH, which approximates 4.  The pH increases in ant nests. It is higher as its value in the environment decreases.

Mineral nitrogen  occurs mainly in nitrate form.  Its  content  is lower in *L. niger* nests, and seems to remain at the some level independently of the gradient of pollution. In *Myrmica* ssp. nests it is higher than in the environment but decreases along this gradient.  The content of nitrite form  of nitrogen ($N/NO_2$), however, is higher  in ant nests.

Ants seem to reduce  in their nests the content  of pollutants accumulated in large quantities in soil (Fig. 1).

b) **Soil microflora**
Each of the two species  of ants influence microfloral activity in different ways. In the nests of *Myrmica* the number of all microorganisms decreases.  In the nests of *L. niger*  there is a remarkable increase in the number of organisms using mineral nitrogen and ammonifying bacteria.

(in ml $O_2$/g d.w./hour)

| Area | Species | | control soil | | | ant nests | | |
|---|---|---|---|---|---|---|---|---|
| | | | without glucose | with glucose | % of increase | without glucose | with glucose | % of increase |
| Nitrogen plant | Myrmica | V | 25,0 | 52,3 | 210 | 23,2 | 112,3 | 480 |
| Devastated area | Lasius niger | V | 18,8 | 46,3 | 250 | 17,0 | 76,6 | 450 |
| Restored area | Myrmica | V | 33,0 | 75,8 | 230 | 32,0 | 95,9 | 300 |
| | | VI | 56,6 | 140,0 | 247 | 17,6 | 87,6 | 500 |

T a b l e  4.   PH and mineral nitrogen in control soil and in ant nests

| Area | Species | Deep of control sample cm/ml in cm | pH in KCL | | N mineral in p.p.m. | | | | | |
|---|---|---|---|---|---|---|---|---|---|---|
| | | | | | $N/NH_3$ | | $N/NO_3$ | | $N/NO_2$ | |
| | | | C | N | C | N | C | N | C | N |
| Nitrogen plant Devastated area | Myrmica | 0-10 | 3.8 | 5.2 | 6 | 5 | 15 | 180 | 0 | 0.4 |
| | Lasius | 0-5 | 4.2 | | 7 | | 500 | | 1 | |
| | niger | 5-10 | 3.9 | 4.3 | 6 | 6 | 250 | 320 | 0.5 | 0.45 |
| Restored area | Myrmica | 0-10 | 4.3 | 5.0 | 32 | 34 | 360 | 300 | 3.3 | 0.60 |
| | Lasius | 0-5 | 4.4 | | 4 | | 500 | | 0.1 | |
| | niger | 5-10 | 3.2 | 4.5 | 50 | 6 | 500 | 285 | 0.1 | 0.27 |

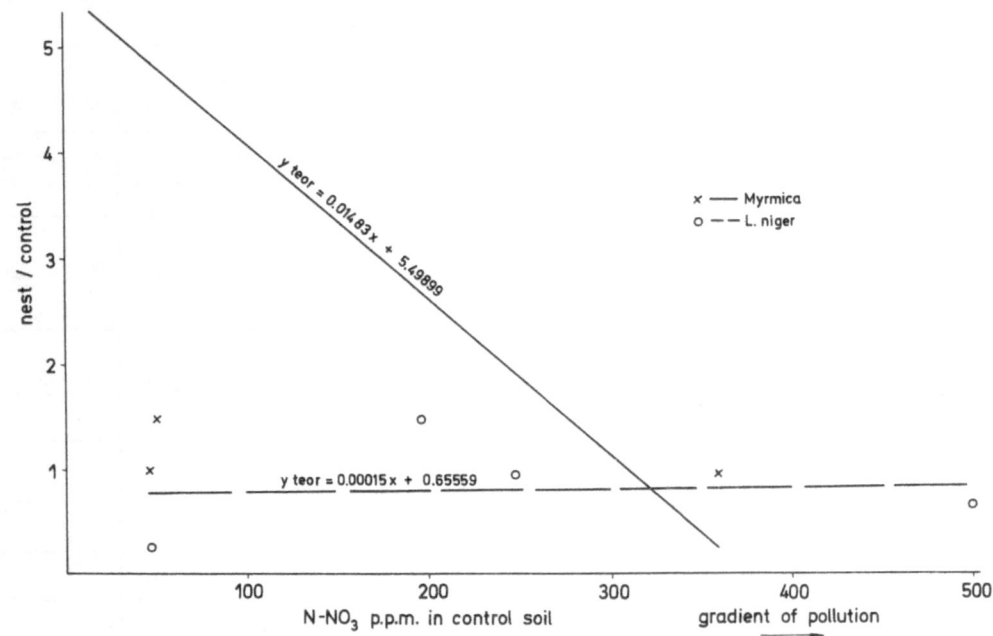

Fig.1. Relationship between the content of nitrate nitrogen in environment and in ant nests.

Also the percentile participation of resting spores of ammonifying bacteria decreases.

The number of microorganisms utilizing mineral nitrogen, ammonifying bacteria, is inversely correlated with habitat contamination.

There is a difference in the distribution of the nests of both species along this gradient (Fig. 2 and 3). *L. niger* occurs where the number of these microfloral groups is the lowest and, in an inversed proportion, changes their activity in the nest. *Myrmica* ssp. rather constructs nests where the number of ammonifying bacteria and organisms utilizing mineral nitrogen is higher, and causes the decrease in their activity in the nests along the environment gradient. These processes are stronger in *L. niger* nests than in *Myrmica* ssp.

Nitrification is in inversed proportion in ant nests and in the environment; it is highest in *Myrmica* ssp. nest and lowest in *L. niger* (Fig. 4).

There is compensation of all these microbiological processes in the environment and in ant nests; the higher it is in the environment the lower it is in ant nests and vice versa.

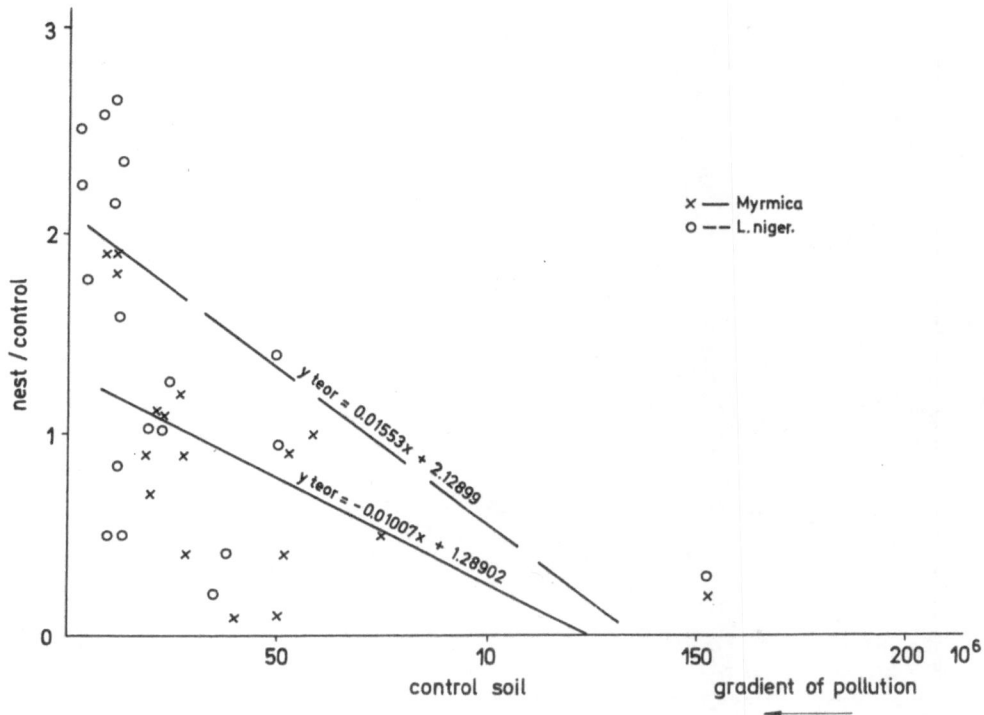

Fig.2. Relationship between the number of microorganisms utilising mineral nitrogen in the environment and in the ant nests.

c) **Soil respiration** (Tab. 4).

The respiratory rate usually decreases in the nests of both species. The addition of glucose results in an increase in the respiratory rate which is manyfolds higher in the ant nests than in the control soil. This indicates that the inhibition of microbiological activity is higher in ant nests.

CONCLUSIONS

The population size of ants is adapted according to the capacity of the habitat; the higher the contamination, the lower the number of species and population size. An improvement of habitat conditions in restored sites is followed by a rise in the population due to an increase in

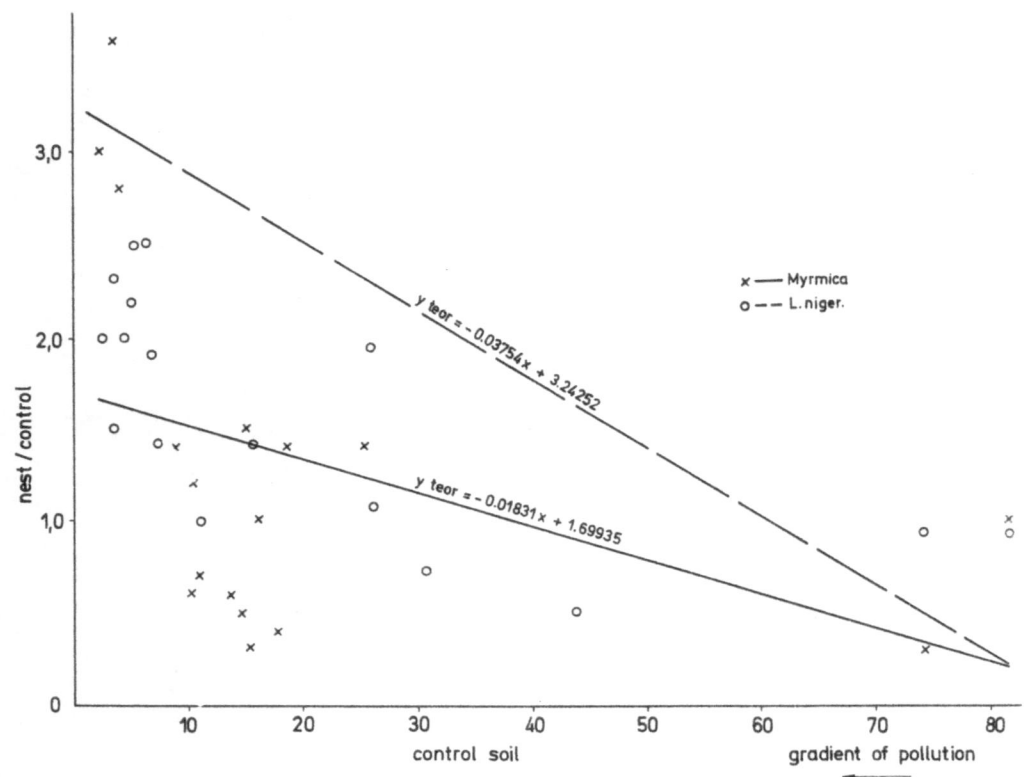

Fig.3. Relationship between the number of ammonifying bacteria in the environment and in ant nests.

production, which is subjected to the most significant changes. Then changes in the standing crop and in the number of nests are observed.

Mortality of adult individuals of *Myrmica* ssp. seems to be lower in the most heavily contaminated habitat. This is indicated by the bio-mass turnover, which is 2 times lower than that in the meadows.

Modifications in population parameters are followed by changes in the soil habitat in the ant nests. The content of pollutants, i.e, nitrate nitrogen is reduced in the nests. It seems that microbiological processes may be responsible for this.

The increase in the activity of microorganisms utilizing mineral nitrogen and ammonifying bacteria could result in a decrease in the nitrate nitrogen content.

Total soil respiration is reduced in ant nests but the great increase of respiratory rate after glucose addition indicates that inhibition of the microflora is higher in ant nests than in the environment.

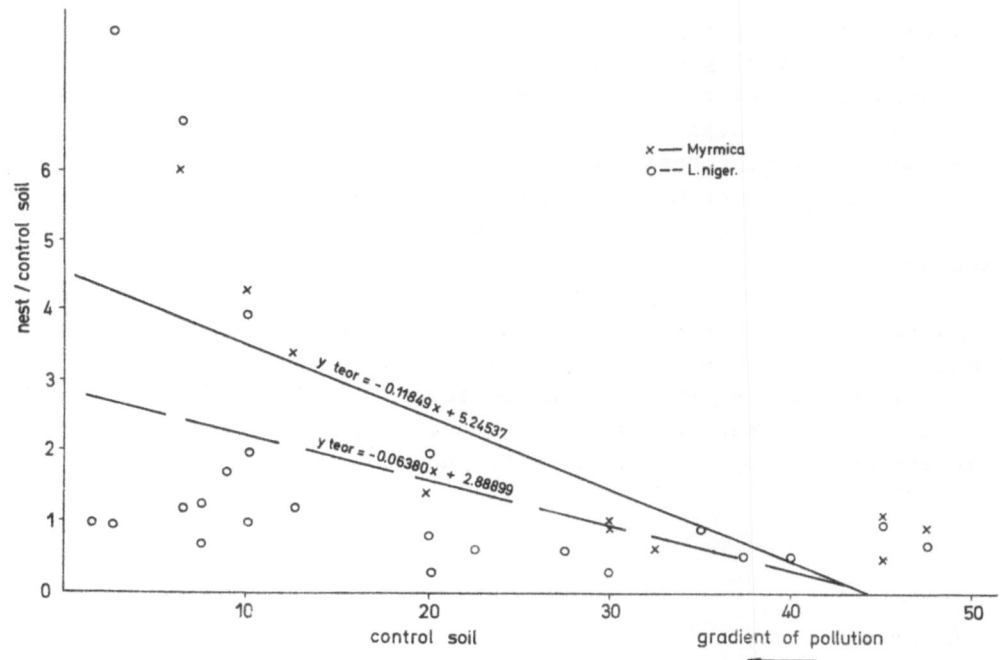

Fig.4. Relationship between the nitrification in the environment and in ant nests.

It is possible that a significant inhibition of the microflora
living in ant nests results in a slower biological decomposition of organic matter there. Higher pH within ant nests also retards chemical
decomposition of organic matter. Consequently, harmful pollutants may
be more effectively buffered in ant nests than in the surrounding environment.

SUMMARY

Response of ants to habitat pollution by dust from nitrogen-plant has
been presented. The heavier the habitat pollution, the lower the number
of species and the size of ant populations. The first phase of regulation of the population size consists of the changes in production, followed by changes in standing crop and in the number of nests. In the
most heavily contaminated habitats the mortality of *Myrmica* ssp. adults
is lower, which indicates a very low turnover of biomass (2 times lower
than in cultivated meadows). The decreases in the population size re-

lated to heavy habitat contamination were the lowest for *L. niger*. This is a better adapted species than *Myrmica*.

There are changes in the content of mineral nitrogen in ant nests; the content of pollutants is reduced. This may be due to a high increase in the number of microorganisms utilizing mineral nitrogen.

It seems that the microflora exhibits changes in activity to suppress the ris in the content of nitrate nitrogen which occurs in large quantities in the environment.

Respiration of the soil in ant nests is reduced but it increases considerably after glucose addition. This indicates that the inhibition of microflora in ant nests is very high. Due to this the decomposition of organic matter in ant nests may be lower. Higher pH in ant nests also reduces chemical decomposition. Consequently, the buffering of harmful pollution may be more effective in the nest of ants.

R e f e r e n c e s

C z e r w i ń s k i , Z., J a k u b c z y k , H., P ę t a l , J. (1971): Influence of a.._ hills on the meadow soils. Pedobiologia 11, 277-285.

J a k u b c z y k , H., C z e r w i ń s k i , Z., P ę t a l , J. (1972): Ants as agents of the soil habitat changes. Ekol. Pol. 20, 153-161.

J a k u b c z y k , H., P ę t a l , J., C z e r w i ń s k i , Z., C h m i e l e w s k i , K. (1973): Influence of ants on changes of the microbial activity of the soil in environment polluted by mineral nitrogen compounds. Int. Conf.: Indicatores Deterioratiae Regions, Czechoslovakia, Most, July 1973.

P ę t a l , J. (1967): Productivity and the consumption of food in the *Myrmica laevinodes* Nyl. population. In: K. Petrusewicz (ed.), Secondary Productivity of Terrestrial Ecosystems, Warszawa-Krakow, p. 841-857.

P ę t a l , J. (1972): Methods of investigating the productivity of ants. Ekol. Pol. 20, 9-22.

Zanieczyszczenia atmosfery oraz rekultywacja gruntów zdewastowanych w rejonie Puław. Instytut Uprawy, Nawożenia i Gleboznawstwa, Puławy, 51, 1-221, 1973.

Discussion

M. S. G h i l a r o v :   Probably the smaller changes  in bacterial activity in ant nests  to that  in surrounding soil  of polluted areas is
connected  with the fact that in ant nests  (with living ants) contamination of matter with poisonous substances was  lower (even occasional).

J. P e t a l :   The ant nests  were found  in completely degraded areas
and the control  samples of the soil  were taken in the proximity of the
nests. It seems  that ants exhibit changes in activity of certain groups
of microflora.   It can suppress the rise  in the content of nitrate nitrogen which occurs in large quantities in the soil  and  in industrial
dusts.

G. J o s e n s :   Vous avez parlé de respiration  des nids  de fourmis;
comment avez-vous mesuré cette respiration?

J. P e t a l :   Au laboratoire, à l'aide d'un appareil de Wartburg.

# Traits généraux du peuplement animal endogé de la savane de Lamto (Côte d'Ivoire)*

F. ATHIAS, G. JOSENS, P. LAVELLE

Laboratoire de la Faune du Sol,
Dijon;

Chargé de Recherches du F. N. R. S.,
Université de Bruxelles

Laboratoire de Zoologie, Ecole Normale Supérieure,
Paris

ETUDE DU MILIEU

## Localisation

La station d'Ecologie Tropicale de Lamto (5°02'W, 6°13'N) est située entre les savanes soudaniennes du Nord et la forêt ombrophile du Sud de la Côte d'Ivoire (15,16).

## Climat

Le climat est de type tropical humide à deux saisons: saison sèche de novembre à mars et saison humide d'avril à octobre, parfois interrompue par une petite saison sèche en août.

La pluviosité moyenne annuelle (sur 9 ans) est de 1290 mm, mais varie beaucoup d'une année à l'autre.

Les températures mensuelles moyennes sont stables: elles ` varient de 25,1 °C (août) à 28,3 °C (mars). Les amplitudes thermiques quotidiennes sont comprises entre 9 °C (août) et 13 °C (mars). L'humidité relative moyenne de l'air est toujours relativement élevée; elle varie de 72 % (saison sèche) à 85 % (saison humide).

## Biotopes

La savane de Lamto est située dans une mosaique de galeries forestières

* La présente étude entre dans le cadre des recherches poursuivies a la station d'Ecologie Tropicale de Lamto (Côte d'Ivoire), installée avec l'aide du Centre National de la Recherche Scientifique (R.C.P. n°60) et de l'Université d'Abidjan, dans le but d'analyser la structure et la vie d'une biocénose terrestre (P.B.I.).

et de savanes à Palmiers Roniers (*Borassus aethiopum*) avec une strate
arbustive de densité très variable. Les sav nes sont percourues chaque
année par des feux de brousse (27).

T a b l e a u   1.    Caractéristiques physico-chimiques d'un sol ferrugineux tropical de plateau en savane arbustive ouverte, entre 0 et 10 cm
de profondeur

| Granulomètrie (% de Terre seche) | | | |
|---|---|---|---|
| Argiles | (  2 | )  : | 6,9-7,1 |
| Limons fins | (  2 -   20 | )  : | 8,0-8,8 |
| Limons grossiers | ( 20 -   50 | )  : | 10,5-12,0 |
| Sables fins | ( 50 -  200 | )  : | 20,0-35,5 |
| Sables grossiers | (200 - 2000 | )  : | 35,0-43,0 |
| **Caractères chimiques** | | | |
| Carbone | 0,6  -  1,2 % | C/N = 13-17 | |
| Azote | 0,04 -  0,07 % | | |
| Matière organique | 1-2 % | | |
| pH | 5,8-6,7 | | |
| **Potentiel capillaire** | | Teneur en eau | |
| pF 5 | | 1,6 % | |
| pF 4,7 (point d'hygroscopie maximale | | 2,3 % | |
| pF 4,2 (point de flétrissement) | | 4,0 % | |
| pF 3 | | 8,2 % | |
| pF 2,5 (capacité au champ) | | 11,8 % | |
| pF 2 | | 15,6 % | |

## Sols (cf. tableau 1)

Les sols sont de type ferrugineux tropical à texture sableuse; ils comportent fréquemment un lit  de graviers.  Pauvres en matière organique,
ils sont plus ou moins lessivés  et hydromorphes suivant leur situation
topographique. Leur structure dispersée les rend très sensible à l'érosion (8).

Les températures mensuelles moyennes  dans les sols sont comprises entre  26 et 33 °C.  Les amplitudes thermiques quotidiennes  sont inférieures à 1 °C à 40 cm  de profondeur,  mais  elles dépassent 25 °C  en surface après le passage du feu (2,3,27).

L'humidité du sol atteint ou dépasse la capacité au champ (pF 2,5) pendant 12 à 23 semaines par an;  en saison sèche, le point de flétrissement permanent (pF 4,2)  est atteint pendant 6 à 12 semaines (23,18).

La microflore  est caractérisée  par l'abondance  relative des Actinomycètes  et par la faible densité des Bactéries, particulièment des bactéries minéralisatrices de l'azote (25,26).


STRUCTURE DU PEUPLEMENT

Le tableau 2 regroupe les principaux résultats déjà obtenue sur le peuplement animal du sol.  Ce tableau  appelle  quelques remarques  et restrictions:

a) les données mentionnées  sont celles qui ont été réellement observées,  elles  sont entachées,  dans chaque groupe animal,  d'une erreur qui dépend d'une part  de la technique d'extraction  utilisée (cf, colonne 3).  et d'autre part de la taille des échantillons (cf. colonne 4 et 5);

b) plusieurs facteurs  concourent en outre à accentuer l'hétér généité des résultats:

c) l'hétérogénéité du milieu;  les données  du tableau  ne concernent que la savane arbustive ouverte, le faciès le plus répandu à Lamto;

d) l'hétérogénéité du climat; la pluviosité notamment  est susceptible de varier du simple au double d'une année à l'autre;

e) la distribution fortement agrégative  de  certains organismes (Termites, Fourmis, Microarthropodes, ...).

Cette  hétérogénéité  explique  pourquoi  les données  moyennes du tableau 2 ne comportent que deux chiffres significatifs  et sont accompagnées de valeurs limites parfois très écartées de la moyenne.

Comparées aux peuplements  des milieux herbacés tempérés, la mésofaune  des savanes  de Lamto est pauvre, particulièrement en  Enchytréides, en Collemboles et en Acariens.

En revanche, les *Symphyles*, les *Protoures*, les *Diploures* et les Cochenilles  sont relativement abondants.

La macrofaune  est dominée par la biomasse relativement importante des Vers  de terre. Les Termites  caractérisent  les milieux tropicaux à Lamto,  ils sont relativement moins abondants  que dans d'autres éco-

T a b l e a u  2.  Composition et caractéristiques écologiques de la faune endogée dans une savane arbustive ouverte à Lamto (Cote d'Ivoire)

| Groupe | Auteur *1 | Technique d'extraction | Taille des échantillons | |
|---|---|---|---|---|
| | | | Surface ou volume | Profondeur |
| **MESOFAUNE** | | | | |
| Enchytréides | 23 | Lavage sur tamis de 0,8 mm | 400 cm2 | 0-50 cm |
| Symphyles | 1-2-3 | Berlèse-Tullgren | 20 cm2 | 0-15 cm |
| Pauropodes + Polyxénides | id | id | id | id |
| Acariens | id | id | id | id |
| Pseudoscorpions | id | id | id | id |
| Protoures | id | id | id | id |
| Diploures | id | id | id | id |
| Collemboles | id | id | id | id |
| Cochenilles (larves) | id | id | id | id |
| **MACROFAUNE** | 27 | | | |
| Vers de terre (Eudrilidae + Acanthodrilidae) | 17 a 22 | Tri manuel et lavage sur tamis de 0,8 mm | 1 m2 | 0-50 cm |
| Araignées *2 | 5 | Tri manuel | 1 m2 | 0-50 cm |
| Chilopodes | 23 | Tri manuel | 4 m2 | 0-50 cm |
| Diplopodes | id | id | id | id |
| Termites fourrageurs | 10 a 14 | Grand carrés Tri manuel | 2 500 m2 1 nid | 0 cm 0-100 cm |
| Termites champignonnistes | id | Tri manuel | 16-25 m2 1- 4 m2 1 m2 | 0- 25 cm 0-100 cm 0- 50 cm |
| Termites "humivores" | id + 7 | id | 1 m2 | 0-50 cm |
| Fourmis | 24 | id | 16-25 m2 | 0-25 cm |
| Larves Coléoptères | 9 | id | 4 m2 | 0-40 cm |

| Fréquence et dates d'échantillonnage | Densité nombre d'individus/m2 | Biomasse (poids frais) en mg/ m2 | Equivalent énergétique en cal/m2 |
|---|---|---|---|
| 10 (III-VI 72) | 0 - 400 | 0 - 10 | |
| 16/mois (XI 69-III 71) | 2 100 (90-7 000) | 50 | |
| id | 270 (0-3 500) | 5 | |
| id | 14 000 (1 600-68 000) | 140 | |
| id | 70 (0-600) | 50 | |
| id | 200 (0-1 500) | 4 | |
| id | 650 (0-2 000) | 130 | |
| id | 1 500 (150-5 400) | 38 | |
| id | 440 (0-3 800) | 7 | |
| 200 (63-67) 12/mois (XII 68-XII 72) 12/mois (VII 71-XII 72) | 230 (180-300) | 30 000 *3 (27 000-35 000) | 2 500 |
| 120 (III-V 71) | 4 | 25 | |
| 3/mois (XII 68-XII 69) | 0,2-1 | 10-40 | |
| id | 0,8-5 | 90-460 | |
| 20 (II-III 69-70) 14 (69-70) | 170 (30-400) | 850 (150-2 000) | 810 (145-1 900) |
| 80 (I 68-XII 70) 60 (XII 68-XII 70) 340 (X 70-V 71) | 450 (70-750) | 490 (79-750) | 570 (92-930) |
| 20/mois (I 71  ) | 400 (1-1 500) | 500 *3 (1 - 1 900) | 560 (1-2 100 |
| 12 series (1968) | 500 | 2 000 | |
| 20/mois (II 68-IV 69) | 2-11 | 100-1 000 | |

| Groupe | Régimes alimentaires principaux | Consommation anuelle/m2 | Production annuelle/m2 |
|---|---|---|---|
| **MESOFAUNE** | | | |
| Enchytréides | Géophages ? | | |
| Symphyles | Rhizophages | | |
| Pauropodes + Polyxênides | Saprophages | | |
| Acariens | Saproph.-fongivores : 52 % Predat. : 48 % | | |
| Pseudoscorpions | Prédateurs | | |
| Protoures | ? | | |
| Diploures | Rhizophages | | |
| Collemboles | Saprophages fongivores | | |
| Cochenilles (larves) | Rhizophages (piqueurs) | | |
| **MACROFAUNE** | | | |
| Vers de terre (Eudrilidae + Acanthodrilidae) | Saproph. : 10 % Géophages: 90 % | 10-20 g litière sèche 80-100 kg terre seche | 60-100 g = 30-50 kcal |
| Araignes *2 | Prédateurs | | |
| Chilopodes | Prédateurs | | |
| Diplopodes | Saprophages | | |
| Termites fourrageurs | Graminées + cannibalisme | 2,0-2,5 g mat. seche | 0,9-1 g = 1-1,2 kcal |
| Termites champignonnistes | Saprophages + cannibalisme | 100-120 g mat. sèche | 5 - 6 g = 6-8 kcal |
| Termites "humivores" | Géophages + cannibalisme | 1 kg terre sèche | |
| Fourmis | Granivores Omnivores Prédateurs | | |
| Larves Coléoptères | Surtout Rhizophages | | |

R e m a r q u e s :   Les animaux qui constituent la MICROFAUNE - Protozoaires, Tardigrades  et Rotifères - n'ont pas fait  l'objet de recherches; les résultats  des études  sur les Nématodes  ne sont pas
encore disponibles.

[*1]   Les numéros renvoient à la bibliographie.

[*2]   Il s'agit des Araignées à moeurs typiquement endogées.  La majorité
des Araignées, de même que les Blattes  et des Coléoptères tels que
les Carabiques  font partie  de la faune épigée  et ne figurent par
conséquent pas dans le tableau.
Les Mollusques, les Crustacés et les larves de Diptères ne figurent
pas dans le tableau  en raison  de leur rareté  dans les savanes de
Lamto.

[*3]   Biomasses  des animaux  dont  le tube digestif  a été préalablement
vidé.

systèmes africains, mais très diversifiés  (plus de 50 espèces).  Il en
va de même avec les Fourmis (plus de 150 espèces) qui exercent leur action à tous les niveaux (24).

Notons en revanche le rôle effacé  des *Chilopodes*,  des *Diplopodes*
et des larves de *Coléoptères*, ainsi que l'absence quasi totale des *Mol-*
*lusques*, des *Crustacés* terricoles et des larves de *Diptères*.

ELÉMENTS DU RÉSEAU TROPHIQUE ENDOGÉ

En se référant  au régime alimentaire,  le peuplement animal  endogé se
répartit en quatre grands groupes.  Leurs biomasses respectives ont été
estimées d'après le tableau 2; (voir aussi 27).

## Consommateurs de végétaux vivants (1200 mg/m$^2$, poids frais)

Les principaux consommateurs  de végétaux vivants  sont  les Fourmis et
les Termites fourrageurs (900 mg/m$^2$).  Il convient de remarquer  qu'ils
récoltent leur nourriture (exsudats, graines et organes aériens des Graminées) au-dessus du sol.  Les rhizophages-larvés de Coléoptères  (surtout des Scarabéides), *Protoures*, *Diploures*, *Symphyles* et Cochenilles -
s'avèrent peu abondants (300 mg/m$^2$), surtout par rapport à l'importante
masse racinaire disponible: 1,2 kg/m$^2$ (6).

SAPROPHAGES, CONSOMMATEURS DE VÉGÉTAUX ET D'ANIMAUX MORTS
PLUS OU MOINS DÉCOMPOSÉS (4900 mg/m$^2$, POIDS FRAIS)

Les principaux consommateurs de litière non décomposée sont les Termites champignonistes: malgré une modeste biomasse moyenne (490 mg/m$^2$),
ils n'utilisent pas moins de 120 g/m$^2$/an de litière sèche. Aidés par
leurs champignons et bactéries symbiotiques, ils conduisent la matière
organique à une minéralisation avancée.

   Les autres saprophages - 10 % de la biomasse des Vers de terre, les
Collemboles, les Pauropodes, les Diplopodes, 52 % des Acariens, une
partie des Fourmis et des larves de Coléoptères - forment une chaine de
dégradation de la matière organique au cours de laquelle les ingestions,
digestions partielles et excrétions se succèdent ou alternent avec
l'action de la microflore.

   Les gros Diplopodes et les Vers de terre constituent souvent le
premier maillon des chaines de saprophages en incorporant au sol la litière récoltée en surface.

   Peu de saprophages s'attaquent aux racines mortes, et seulement
après la dégradation de cette source de nourriture très abondante par
la microflore.

## Géophages (27500 mg/m$^2$, poids frais)

On entend par géophages, les animaux qui consomment de la terre et as-
similent des débris et la matière organique peu ou pas figurée, ainsi
que la microflore édaphique. Par leur biomasse (27.500 mg/m$^2$, tube digestif vidé), les géophages dominent la faune du sol; ils sont cependant
peu diversifiés et comprennent essentiellement les Enchytréides, des
Vers de terre et les Termites "humivores".

   Pendant leurs périodes d'activité (pF<3), les Vers de terre ingèrent journellement 5 à 30 fois leur poids de terre, selon l'espèce et
la taille des individus. On estime ainsi à 100 kg/m$^2$ le poids de terre
sèche remaniée chaque année par ces Oligochètes. Leur activité minéralisatrice n'est cependant évaluée qu'à 80-100 g/m$^2$/an de matière organique en raison de la pauvreté des sols en matière organique et de leur
faible capacité d'assimilation.

**Prédateurs (1600 mg/m$^2$, poids frais)**

Les principaux prédateurs dans le sol sont les *Chilopodes*, les *Araignées*, les *Pseudoscorpions*, 48 % des *Acariens*, une partie des Fourmis et une partie des larves de *Coléoptères*. La majorité d'entre eux, sauf certains *Acariens*, chassent surtout à la surface du sol et dans la litière. Un groupe écologiquement aussi important que les Vers de terre géophages est de ce fait peu affecté par la prédation; il en va de même pour les *Termites*, sauf lorsque les sexués ailés quittent le milieu endogé.

CONCLUSIONS

Le peuplement animal endogé des savanes de Lamto est largement dominé, au point de vue des biomasses, par un petit nombre d'espèces géophages. Leur activité minéralisatrice, par rapport à leur biomasse est relativement faible; en revanche, ils jouent un rôle important dans la structuration du sol.

Les consommateurs de végétaux vivants et les saprophages, parmi lesquels les Termites champignonistes jouent un rôle important, se caractérisent par le fait qu'ils récoltent la majeure partie de leur nourriture à la surface du sol. La faune endogée s'attaque donc peu aux racines, malgré une biomasse disponible très importante, sauf après une dégradation préalable par la microflore.

Les prédateurs récoltent également la majeure partie de leur nourriture à la surface du sol.

La représentation en figure 1 est une première approche d'un modèle de la vie dans les sols de Lamto.

GENERAL FEATURES OF SOIL COMMUNITIES IN THE LAMTO SAVANNA
(IVORY COAST)

Total mineralization in Lamto's tropical ferruginous soils is of 1 to 2 kg/m$^2$/year vegetal litter and 8 to 10 kg/m$^2$/year dead roots.

The soil meso- and macrofauna includes:

1900 mg/m$^2$ (fresh weight) primary consumers (300 mg of them being root-feeders);

4900 mg/m$^2$ saprophageous;

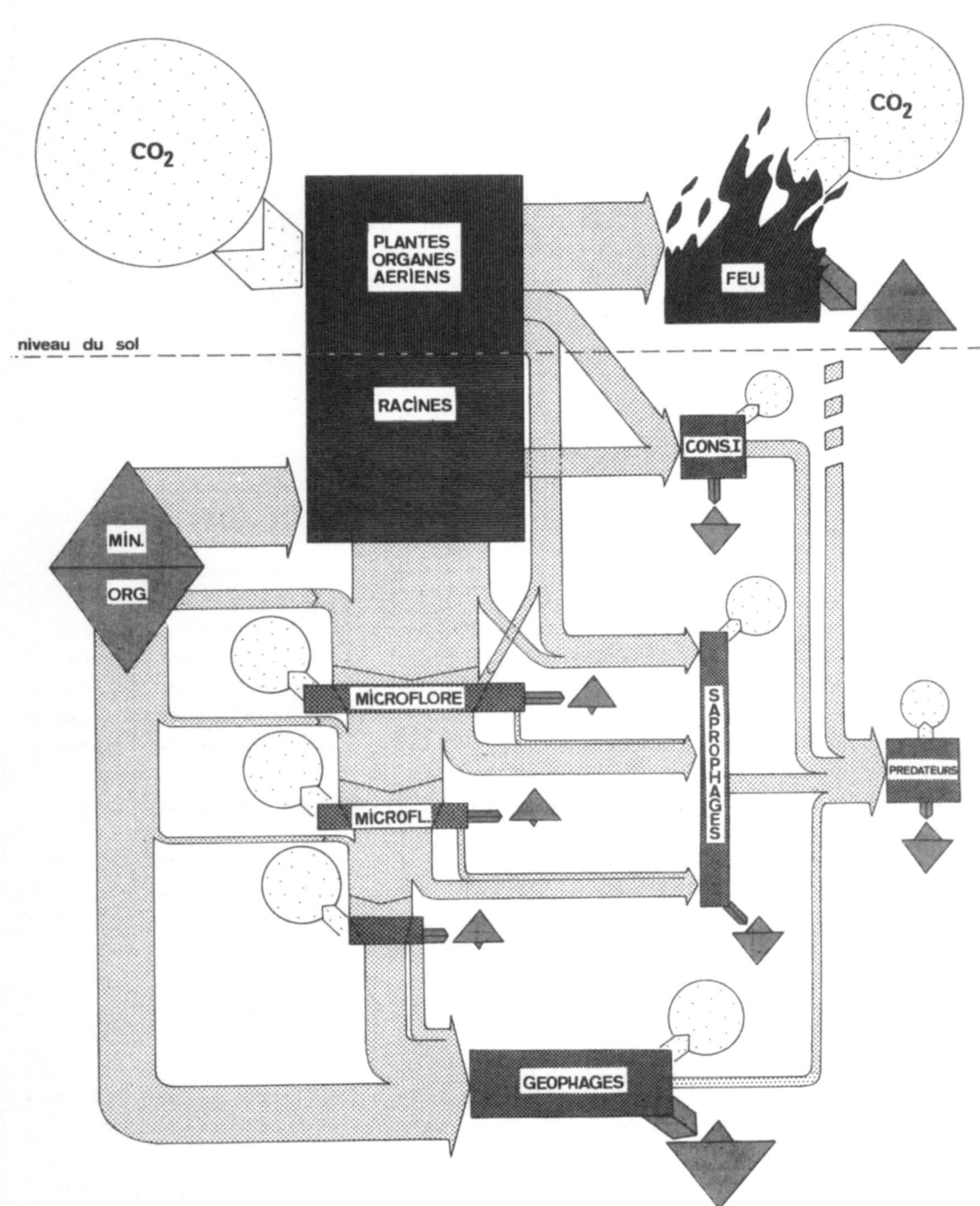

Fig.1.  Modele de la vie  dans le sol de la savane Lamto
(Cote d'Ivoire).  Les differents groupes  de producteurs
et consommateurs sont representes par des surfaces pro-
portionnelles aux  racines cubiques  de leurs biomasses
(CONS.I = consommateurs de vegetaux vivants).  Le niveau
de la microflore saprophytique  a ete divise symbolique-
ment en 3 pour sugere la multiplicite  de son action. Le
groupe de la microfaune devra etre ajoute ulterieurement
lorsque les donnees  qui s'y rapportent seront disponibles. Les cercles representent le $CO_2$ qui est utilise par
le systeme.  Les triangles  avec sommet vers le haut represent les substances minerales utilisees par le systeme.  Les triangles  avec sommet vers le bas representent
les substances organiques (cadavres, feces...) presentes
(et utilisables) par les système.

2750 mg/m$^2$ geophageous;
1600 mg/m$^2$ predators.
The most important group are ants, termites and earthworms.

Bibliographie

1 A t h i a s , F. (1971): Recherches  écologiques  dans  la savane  de
     Lamto (Côte d'Ivoire):  Étude quantitative préliminaire des Micro-
     arthropodes du sol. Terre et Vie, 3, 395-409.

2        (1973): Étude quantitative du peuplement  en Microarthropodes
     du sol  d'une savane  de Côte d'Ivoire.  Thèse de 3è cycle. Paris,
     103 pp.

3        (sous presse): Note préliminaire sur l'importance de certains
     facteurs mésologiques vis à de l'abondance des Acariens  d'une sa-
     vane de Côte d'Ivoire. Rev. Ecol. Biol. Sol.

4 A t h i a s , F., J o s e n s , G. and L a v e l l e , P. (1973):
     Influence du feu de brousse annuel  sur le peuplement endoge  dans
     les savanes de Lamto (Côte d'Ivoire)- ce fascicule.

5 B l a n d i n , P.: Communication personnelle.

6 C e s a r , J. (1971): Étude quantitative  de la strate herbacée de la
     savane de Lamto (Côte d'Ivoire). Thèse de 3è cycle, Paris, 125 pp.

7 C o r v e a u l e , D.: In litt.

8 D e l m a s , J. (      ):  Recherches  écologiques  dans  la savane de
     Lamto (Côte d'Ivoire):  Premiers aperçus sur les sols  et leur va-
     leur agronomique. Terre et Vie 21, 216-227.

9 G i r a r d , C.: Communication personnelle.

10 J o s e n s , G. (1971a): Variations thermiques  dans les nids de Tri-
     nervitermes geminatus, en relation  avec  le milieu extérieur dans
     la savane de Lamto (Côte d'Ivoire). Ins. Soc. 18/1/, 1-14.

11        (1971b): Le renouvellement  des meules à champignous constru-
ites par quatre *Macrotermitinae* (Isoptères)  des savanes de Lamto-
Pacobo (Côte d'Ivoire). C.R.Ac.Sc., Paris, 272, 3329-3332.

12 J o s e n s ,  G. (1971c): Recherches  écologiques  dans  la savane de
Lamto (Côte d'Ivoire): données préliminaires  sur le peuplement en
Termites. Terre et Vie 2-71, 255-272.

13        (1972): Etudes biologiques et écologiques  des *Termites (Isoptera)* de la savane de Lamto-Pacobo (Côte d'Ivoire). Thèse, Bruxelles, 262 pp.

14        (sous presse):  L'étude écologique des Termites à Lamto (Côte
d'Ivoire)  un moyen d'approcher les phénomènes  de symbiose  et de
parasitisme au point de vue bioenergetique.  Colloque Franco-Belge
d'Ecologie Animale, 26-29/X/1972b.

15 L a m o t t e ,  M. (1967):  Recherches écologiques  dans  la savane de
Lamto (Côte d'Ivoire). Présentation du milieu et programme de travail. Terre et Vie 21, 197-215.

16        (1970):  La participation  au  PBI  de la station  d'Ecologie
Tropicale de Lamto (Côte d'Ivoire). Bull. Soc. Ecol. 2, 58-65.

17 L a v e l l e ,  P. (1971a):  Etude préliminaire  de la nutrition  d'un
ver de terre africain: *Millsonia anomala* Omodeo (*Acanthodrilidae-
Oligochètes*).  IVèCIZS,  Dijon, sept. 70;  Ann. Zool., Ecol. Anim.
n°hors, série: 133-145.

18        (1971b):  Etude démographique  et  dynamique  des populations
de *Millsonia anomala* Omodeo  (*Acanthodrilidae*-Oligochètes). These
3ècycle,  Paris, 88 pp.

19        (1971c):  Production annuelle  d'un ver  de terre: *Millsonia
anomala (Acanthodrilidae*-Oligochètes) dans la savane de Lamto (Côte
d'Ivoire). Terre et Vie 2, 240-254.

20        (1971d):  Recherches  sur  la démographie  d'un ver  de terre
d'Afrique: *Millsonia anomala* Omodeo (*Acanthodrilidae*-Oligochètes).
Bull. Soc. Ecol. 2(4), 302-312.

21        (1973a): Peuplement et production des vers de terre  dans les
savanes de Lamto (Côte d'Ivoire). Ann. Univ. ABIDJAN.

22        (1973b): Consommation annuelle  de terre  par une population
naturelle de vers de terre (*Millsonia anomala* Omodeo  (*Acanthodrilidae*-Oligochètes) dans  la savane  de Lamto  (Côte d'Ivoire). Ie
Coll. Soc. Ecol.

23 L a v e l l e ,  P.: Non publié.

24 V e v i e u x ,  J. (1971): Données écologiques et biologiques  sur le
peuplement en Fourmis terricoles d'une savane préforestière de
Côte d'Ivoire. Thèse d'Etat, Paris, 300 pp.

25 P o c h o n ,  J. and B a c v a r o v ,  I. (1973): Données préli-

minaires sur l'activité microbiologiques des sols  de la savane de
Lamto (Côte d'Ivoire). Rev. Ecol. Biol. Sol. 10 (1), 35-45.

26 R a m b e l l i , A. and B a r t o l i , A. (1972): Recherches sur
la microflore fongique des sols de Lamto (Côte d'Ivoire). Rev.
Ecol. Biol. Sol. 9, 41-53.
27 V i n c e n t , J. P. (1970): Recherches écologiques sans la savane
de Lamto (Côte d'Ivoire): Observations préliminaires  sur les Oligochètes. Terre et Vie 2, 22-39.

D i s c u s s i o n

M. G ó r n y : For enchytraeid extraction you have used, samples of
$400 \text{ cm}^2$, and you have found extremely small numbers of them in the soil.
It seems to me the samples were too big and this may be the reason for
extracting the worms from a part of a sample only.

F. A t h i a s : La méthode utilisée était destinée à extraire les
petits vers de terre (Eudrilidae) et leurs cocons. Ces valeurs sont
données à titre indicatif et sont certainement sous-estimées. Toutefois,
il est certain que les Enchytraeides sont très peu importants dans ces
savanes.

J. N o s e k : How do you explain the relatively high density  of Pro-
tura in African soils?

F. A t h i a s : Les Protoures peuvent être abondants dans des sols
européens à certaines époques de l'année. Ces organismes sont sans
doute des biolytiques, et sont favorisés dans des sols resemblamt aux
sols cultivés européens, auxquels la savane de Lamto présente beaucoup
de caractères similaires. Il semble que la même remarque pourrait être
faite à propos des Symphyles.

M. S. G h i l a r o v : Quelles Diploures sont rhizophage? Dans les
pays tempérés ces insectes sont ravageurs ou saprophage! 100 $\text{kg/m}^2$ re-
mués par vers de terre (250 $\text{ex/m}^2$ est fantastiquement beaucoup. Quelle
est la méthode d'estimation?

F. A t h i a s : D'après Mr.Pajès qui a determiné les Diploures, ceux-
ci sont surtout réprésentés par des Parajapygidae, qui sont rhizophages

P. L a v e l l e : La méthode d'estimation est expliquée dans le même
volume (voir p.). La quantité de terre consommée par ces vers est effectivement très importante: en période d'activité chaque animal ingère

de 4 à 30 fais  son  propre  poids de terre  chaque jour.  Ce fait peut
s'expliquer  par la très faible teneur en matiere organique de la terre
(1 a 2 %)  et le faible taux  d'assimilation  de  cet  aliment (environ
10 %).

# Influence du feu de brousse annuel sur le peuplement endogé de la savane de Lamto (Côte d'Ivoire)*

F. ATHIAS, G. JOSENS, P. LAVELLE

Laboratoire de la Faune du Sol:
Chargé de Recherches du F. N. R. S.,

Université de Bruxelles;

Laboratoire de Zoologie, Ecole Normale Supérieure,
Paris

Le feu de brousse est un élément essentiel de la dynamique des écosystèmes herbacés tropicaux.

Dans une région de savanes préforestières comme celle de Lamto, le feu annuel maintient une formation herbacée alors que la végétation climax est une forêt semi-décidue (Adjanohoun, 1964). De fait, dans des parcelles protégées du feu depuis 9 ans, Vuattoux (1970) note un très fort embuissonement tandis que se développent de nombreuses essences forestieres.

Allumé chaque année, et ce depuis des siècles pour faciliter la chasse et les déplacements, le feu survient généralement à la fin du mois de janvier, au plus fort de la saison sèche. L'incendie détruit entièrement la couverture végétale herbacér, soit de 6 à 10 t/ha (poids sec) d'herbes sur pied et de litière au sol, ainsi que des jeunes pousses d'arbres (Cesar, 1971, Monnier, 1971).

Le front de flammes se déplace trop rapidement (50 à 600 m/h) pour provoquer un échauffement important du sol. En affet, si la température atteint alors 300 $^{\circ}$C au sommet des herbes, elle ne dépasse pas 60 $^{\circ}$C à la base des touffes de graminées (Gillon et Pernes, 1968, Menager, 1971). Dans le sol la température n'augmente que de 2 $^{\circ}$C à -5 cm dans les cinq minutes qui suivent le passage du feu, et à -10 cm, l'écart thermique est imperceptible.

L'incendie n'a ainsi aucun effet immédiat sur le peuplement du sol. Son effet à moyen terme, dans les six mois qui suivent son passage, et; l'évolution du peuplement dans une parcelle protégée du feu depuis 9 ans, font l'objet de cette étude.

* La présente étude entre dans le cadre des recherches poursuivies à la station d'Ecologie Tropicale de Lamto (Côte d'Ivoire), installée avec l'aide du Centre National de la Recherche Scientifique (R.C.P. n°60) et de l'Université d'Abidjan, dans le but d'analyser la structure et la vie d'une biocénose terrestre (PBI).

## ACTION DU FEU SUR LE MILIEU ENDOGÉ

Dans les mois qui suivent le passage du feu, l'absence de couvert végétal influence le microclimat du sol et les sources d'aliments disponibles pour la faune endogée.

Bien que la repousse soit rapide après le feu, le sol reste très
peu protégé pendant plusieurs mois: la structure cespiteuse de la végétation herbacée laisse à nu plus de 80 % de la surface du sol. Ce n'est
que vers le mois de juillet que le recouvrement est de nouveau total
et que la litière commence à se déposer.


## TEMPÉRATURE ET HUMIDITÉ DU SOL

Les principales modifications après le passage du feu concernent la
température: le sol, dénudé et recouvert d'une couche de cendres noi-
râtres, est alors bien plus chaud en moyenne, tandis que l'amplitude
thermique journalière s'accroît considérablement (fig. 1).

Les mois qui suivent le feu - février, mars et avril - sont les
plus chauds. Les maximums journaliers dépassent alors 37 $^{\circ}$C dans les 10
premiers centimètres; ce n'est qu'à partir du mois de juillet qu'ils
deviennent inférieurs aux maximums atmosphériques, la végétation recon-
stituée jouant le rôle d'écran. Après le mois de juiller, les températures sont peu différentes de celles enregistrées dans une parcelle de
savane non brûlée témoin: les maximums sont régulièrement inférieurs
aux maximums etmosphériques tandis que les minimums sont supérieurs aux
valeurs correspondantes de l'air.

L'influence très nette de l'absence de couvert herbacé sur la température du sol est également perceptible en ce qui concerne son humidité. C'est en effet dans les semaines qui suivent l'incendie qu'on
note les valeurs les plus basses de l'humidité du sol: le pF 4,2 est
fréquemment dépassé et on atteint quelquefois le pF 4,7 (entre 0 et
-10 cm), lorsque les premières pluies sont tardives. Les animaux su sol
ont alors à faire face à des conditions très dures de température et de
sécheresse.


## LES RESSOURCES ALIMENTAIRES DES ANIMAUX DU SOL

S'ajoutant aux conditions microclimatiques passagèrement arides, la suppression quasi totale de la litière va poser un problème d'alimentation
aux animaux saprophages.

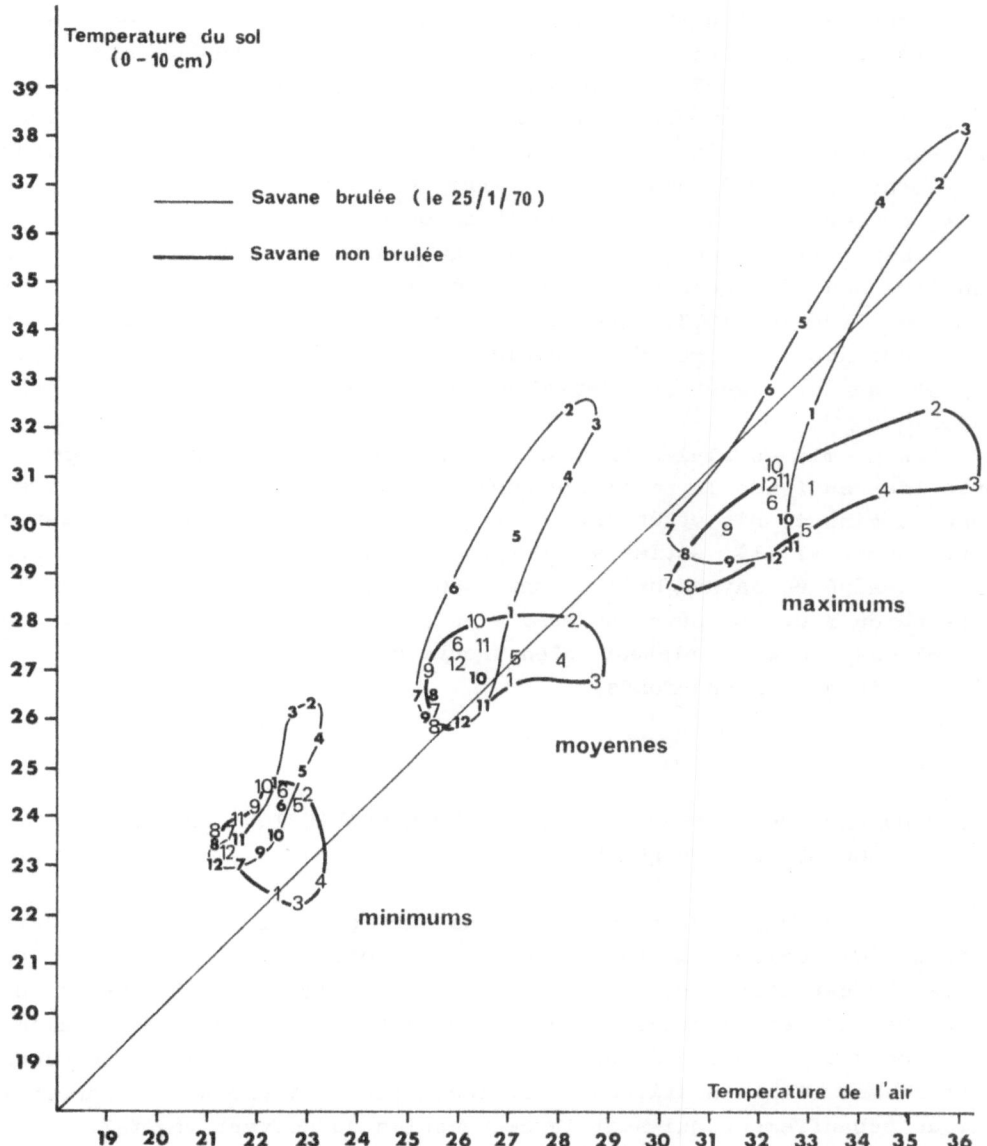

Fig.1. Températures moyennes relevées en 1970, dans l'air et dans le sol (0-10 cm), dans une savane régulièrement brûlée et dabs une savane protégée du feu depuis 1962. 1 = Janvier; 2 = Fevrier etc. ...

Leurs ressources alimentaires se limitent alors aux débris végétaux épargnés au centre des touffes de graminées et aux bases foliaires souvent enfoncées de plusieurs cm dans le sol.

Selon les années et selon l'importance de la strate arbustive, une quantité parfois importante de feuilles d'arbres (jusqu'à 600 kg/ha en poids sec, Menaut 1971) peut tomber juste après les feux, compensant partiellement la destruction de la litière. Des débris végétaux se trouvent également enfouis dans les couches superficielles du sol, consécutivement au travail des Vers de terre (turricules), des Termites (galeries couvertes en terre) ou par l'action de la pluie (alluvions).

Cesar (1971) pense que meurent au cours de la saison sèche des quantités non négligeables de racines de graminées, mais on a vu (Athias, Josens et Lavelle 1973) que les saprophages qui les consomment sont rares; cet apport de matière organique profitera donc surtout aux "géophages" après digestion partielle de ces racines par les microorganismes.

Les premières pluies érodent le sol dénudé et entraînent vers les zones de bas-fonds la pellicule superficielle, riche en matière organique et fins débris végétaux. Dans des savanes analogues à celles de Lamto, Roose (1971) estime à 150 kg/ha/an le poids de terre annuellement entraîné en savane brulée contre 50 kg seulement dans une parcelle protégée du feu. Le développement des animaux géophages établis dans les sols de pente et plateau s'en trouve défavorisé au profit de ceux qui colonisent les bas-fonds.


## ACTION DU FEU SUR LE PEUPLEMENT MÉSO- ET MACROFAUNIQUE DU SOL EFFET DU COUPLE CHALEUR-HUMIDITÉ

A la suite du feu, les effets de la chaleur et de l'humidité sont difficiles à dissocier. Le feu survenant au coeur de la saison sèche accentue le caractère sec et chaud du microclimat endogé. Cependant, l'assèchement et l'échauffement progressifs du sol qui se produisent dès le mois de décembre induisent une adaptation de la part de beaucoup d'organismes aux conditions du milieu; ils sont prêts à supporter le brusque échauffement qui suit la destruction du couvert végétal.

Dès le début de la saison sèche, on observe une migration en profondeur des Vers de terre suivie de leur entrée en quiescence lorsque le pF du sol dépasse la valeur 3. Après le feu, les températures maximales enregistrées près de la surface (> 35 $^{o}$C) sont létales pour la plupart d'entre eux. Chez *Millsonia anomala*, l'espèce la plus répandue (15 g/m$^2$ en poids frais, tube digestif vidé), les jeunes individus sont plus résistants aux températures élevées que les adultes; de plus, ils sont favorisés dès les premières pluies de février-mars, leur consommation étant maximale entre 28 et 30 °C alors que l'optimum pour les adultes se situe aux alentours de 25 °C. Le feu se montre ainsi, par ses conséquences médiates, un facteur important de la dynamique des populations de *M. anomala* (Lavelle 1971a, b, c et d; 1973a et b).

Pour les *Arthropodes*, qui ne présentent pas de quiescence et sont moins sensibles à la sécheresse, le feu est suivi d'une migration en profondeur relativement peu importante chez les Oribates, plus nette pour la plupart des autres groupes. Il produit en outre une moratlité brutale chez les Microarthropodes, mais la plupart des groupes et les *Collemboles* en particulier, recolonisent rapidement le milieu (Athias, 1971, 1973).

Les populations de Termites migrent également en profondeur, mais la période qui suit le feu semble leur être favorable: c'est en effet à ce moment que dans la majorité des espèces se développent les sexués ailés qui peuvent représenter (chez *Trinervitermes geminatus*) jusqu'à 50 % de la production annuelle (Josens 1971a,b et c; 1972a et b).


REACTION A LA SUPPRESSION DES RESERVES
ALIMENTAIRES SUPERFICIELLES

La destruction presque totale de la litière, suivie de la désorganisation et de l'érosion superficielles du sol lors des premières tornades de printemps. limitent la recolonisation du premier centimètre du sol et de sa surface par les microarthropodes et les *Diplopodes*. Au cours de cette période, et jusqu'en juillet où se reconstitue une litière, la compétition pour les débris végétaux est vive. Certains groupes de saprophages ont alors un effectif très bas: c'est le cas des Acariens Oribates et Tydeidés dont l'abondance est significativement liée à celle de la litière.

Les *Termites* champignonistes et fourrageurs capables d'accumuler des réserves dans leurs nids, semblent peu affectés par la pénurie passagère. Les Vers de terre saprophages exploitent la litière enfouie dans le premier centimètre du sol avant l'incendie et adoptent un régime à tendance géophage plus marquée. Il est intéressant de noter que les espèces de Vers de terre saprophages et les Termites champignonistes sont nettement plus abondantes dans les faciès savaniens les plus arborés, ce qui souligne l'importance des feuilles d'arbustes tombées après le feu.

La reproduction du Ver de terre saprophage *Millsonia lamtoiana*, qui débute en mars dans la savane protégée du feu ne se manifeste qu'en juin en savane brûlée.

On constate ainsi  que le peuplement des savanes de Lamto  se compose d'espèces adaptables aux conditions difficiles liées au passage du
feu de brousse,  soit par migration  en profondeur  et   modification de
leur régime alimentaire,  soit  par accumulation  de réserves  dans les
nids, soit par adaptation de leur cycle de reproduction.


EVOLUTION DU PEUPLEMENT ENDOGÉ APRÈS SUPPRESSION DU FEU

On a comparé le peuplement  d'une savane régulièrement brûlée  (Athias,
Josens et Lavelle, 1973)  à celui  d'une savane protégée  du feu depuis
9 ans.  Cette dernière  se caractérise  par une strate arbustive en expansion, par la présence d'une litière toujours abondante (de 200 à 500
g/m$^2$  en poids  sec selon  les saisons (Cesar 1971) - et par des températures moins élevées au cours de la saison sèche (fig. 1).

T a b l e a u   l.   Biomasses moeyenne  comparées  des animaux  endogés
et dans une savane arbustive ouverte régulièrèment brûlée  (B.)  et dans
une savane protégée du feu depuis neuf ans (N.B.) ā Lamto (Côte d'Ivoire)

| Consommateurs de végétaux vivants | | | Saprophages | | |
|---|---|---|---|---|---|
| Groupe | B. Biomasse en mg/m2 | N.B. Biomasse en mg/m2 | Groupe | B. Biomasse en mg/m2 | N.B. Biomasse en mg/m2 |
| Symphyles | 50 | 20 | Vers de terre | 3 000 | 7 600 |
| Protoures | 4 | 4 | Pauropodes + Polyxénides | 5 | 10 |
| Diploures | 130 | 190 | Diplopodes | 250 | 750 |
| Larves de Cochenilles | 7 | 10 | Acariens | 70 | 170 |
| Termites fourrageurs | 425 | 340 | Collemboles | 38 | 80 |
| Fourmis | 400 | 400 | Termites fourrageurs | 425 | 340 |
| Larves de Coléoptères | 200 | 80? | Termites champignonistes. | 490 | 490 |
| | | | Fourmis | 200 | 200 |
| | | | Larves de Coléoptères | 70 | 40? |
| Total (arrondi) | 1 2oo | 1 000 | Total (arrondi) | 4 900 | 9 700 |

Suite du Tableau 1.

| Géophages | | | Prédateurs | | |
|---|---|---|---|---|---|
| Groupe | B. Biomasse en mg/m2 | N.B. Biomasse en mg/m2 | Groupe | B. Biomasse en mg/m2 | N.B. Biomasse en mg/m2 |
| Enchytréides | 10 | 0 | Chilopodes | 25 | 75 |
| Vers de terre | 27 000 | 15 000 | Pseudoscorpions | 50 | 620 |
| Termites humivores | 500 | 500 | Acariens | 70 | 120 |
| | | | Araignées | 25 | 100? |
| | | | Fourmis | 1 400 | 1 600 |
| | | | Larves de Coléoptères | 20 | 10 |
| Total (arrondi) | 27 500 | 15 000 | Total (arrondi) | 1 600 | 2 500 |

Le tableau i donne une image du peuplement endogé en savane brûlée et non brûlée (voir l'article précédent pour les méthodes et la densité d'information).

Le groupe des **consommateurs primaires**, avec une biomasse moyenne de 1.0 g/m$^2$, a peu évolué. Certains groupes se développent, tels les *Diploures* et les *Cochenilles*, tandis que diminuent les *Symphiles* et les *Termites* fourrageurs, animaux héliophiles défavorisés par l'embuissonnement croissant du milieu.

Le groupe des **saprophages** en revanche est de beaucoup plus important qu'en savane brûlée. Sa biomasse passe de 4,9 à 9,7 g/m$^2$. Cette augmentation est essentiellement le fait des Vers de terre dont la biomasse passe de 3 à 7,6 g, des *Diplopodes* et des *Microarthropodes*. Les populations de *Termites* champignonistes restent stables.

Le groupe des **géophages** (15,5 g/m$^2$), qui est toujours le plus important, est toutefois en nette régression: les populations de Vers de terre qui le constituent en majorité diminuent avec l'augmentation du boisement; le terme ultime de cette évolution serait probablement l'extinction presque totale des vers géophages avec l'installation de la forêt semidécidue.

Les **predateurs** (2,5 g/m$^2$), plus importants en secteur non-brûlé semblent suivre en celà les saprophages qui constituent l'essentiel de leurs proies.

Après neuf ans  de protection,  le peuplement de la savane  a donc
évolué  de façon importante. Les saprophages  et  leurs prédateurs qui
vivent essentiellement dans la litière ou dans les premiers centimètres
du sol  se sont développés, favorisés  par une plus grande stabilité du
milieu; leur biomasse a doublé et le nombre des espèces a augmenté.

En revanche,  les géophages régressent,  de même que les consommateurs primaires. Le peuplement endogé évolue  ainsi  vers celui de la
forêt semi-décidue voisine, caractérisé par l'abondance et la diversité
des formes saprophages et par leur localisation  dans les horizons très
superficiels du sol.

B i b l i o g r a p h i e

A d j a n o h o u n , E. (1964): Végétation des savanes et des roches
     en Côte d'Ivoire centrale. Mem. n°7, O.R.S.T.O.M., Paris, 178 pp.
A t h i a s , F., J o s e n s , G. and L a v e l l e , P. (1973):
     Traits généraux du peuplement animal endogé de la savane  de Lamto
     (Côte d'Ivoire). Même fascicule.
C e s a r , J. (1971): Etude quantitative de la strate herbacée de la
     savane de Lamto (Moyenne Côte d'Ivoire).  Thèse 3ème cycle, Paris,
     111 pp.
G i l l o n , D. and P e r n e s , J. (1968):  Etude  de l'effet du
     feu de brousse sur certains groupes d'Arthropodes  dans une savane
     préforestière  de Côte d'Ivoire. Ann. Univ. Abidjan, sér. E,  I,
     101-112.
M e n a g e r , M. T. (1971): Etude méso - et microclimatique du con-
     tact forêtsavane en Côte d'Ivoire. D.E.S., Paris, 130 pp.
M e n a u t , J. C. (1971): Etude  de  quelques  peuplements  ligneux
     d'une savane quinéenne de Côte d'Ivoire.  Thèse 3ème cycle. Paris,
     153 pp.
M o n n i e r , Y. (1971): Les effets du feu de brousse sur une savane
     préforestière de Côte d'Ivoire. Etudes Eburnéennes, 9, 260 pp.
R o o s e , E. (1971): Influence de la modification  du  milieu  sur
     l'èrosion, le ruissellement, le bilan hydrique et chimique.  Suite
     à la mise en culture. Résultats sous pluie naturelle. Comité tech-
     nique O.R.S.T.O.M., Abidjan 16/11/71. Comm. pers. Dactylogramme,
     20 pp.
V u a t t o u x , R. (1970): Observations sur l'évolution des strates
     arborées et arbustives dans la savane de Lamto. Ann. Univ. Abidjan,
     sér. E, 3 (1), 285-315.

Discussion

H. F r a n z :  Wie lange  ist die Trockenzeit?  Welche der 4 genannten
Böden wurden untersucht?

F. A t h i a s :  La saison  sèche  dure  environ  4 mois,  de décembre
à mai.  La durée de la saison  des pluies et l'abondance des pluies varient considérablement d'une année vers l'autre. Nous avons échantillo-
nés en sol ferrugineux tropical, qui est  un sol sableux, pauvre en matière organique. Dans la savane de Lamto les argiles noires ne couvrent
qu'une partie du territoire de la station  Cependant quelques échantil-
lonages y ont été effectués,  mais  ces résultats  ne figurent pas dans
ce travail.

H. F r a n z :  On voit bien que les Géophages  ne sont pas touché  par
le feu, ils semblent être capable de migrer  dans les couches profondes
du sol. Il serrait très intéressant de mesurer la température  dans les
différentes couches  du sol.  Les animaux  qui vivent  dans la littière
doivent être très réduits par le feu. Par exemple les Oribates.

F. A t h i a s :  Les Oribates  sont effectivement très atteints par le
feu,  leurs populations  ne se reconstituent entièrement  qu'au mois de
décembre, époque où la littière est la plus abondante en savane brûlée.
En savane  non brûlée,  ils sont beaucoup  plus nombreux  et ils vivent
principalement dans la littière et sous les touffes de Graminées. D'une
manière générale,  les Microarthropodes, et  particulièrement  les Acariens, sont moins abondants  en savane brûlée, et ils vivent en profondeur (5-10 cm) une grande partie de l'année.  En savane protégée du feu
ils sont plus abondants,  particulièrement  les saprophages,  et vivent
en majorité  dans les strats superficielles (O-2,5 cm)  et la littière.

# The influence of cultivations
# on soil animal populations

C. A. EDWARDS AND J. R. LOFTY

Rothamsted Experimental Station,
Harpenden, England

Wild habitats such as woodlands and grassland usually support a diverse flora and fauna. Even these are used for agriculture, this flora is replaced, either by a few species of plants when grassland is reseeded or by a single species when an arable crop is sown; this, in turn, usually causes a decrease both in the total numbers and species diversity of the animals associated with the flora. Such changes have been thoroughly investigated for the insects that live on the aerial parts of plants, but there have been few experimental investigations of the effect of cultivation on populations of soil animals (Edwards and Lofty, 1969). Although it has been assumed that cultivation decreases numbers and diversity of soil animals, such assumptions have been based on relatively sparse data from surveys of the fauna of different habitats, and may be unjustified because most of them have been based on single population estimates and have taken no account of changes in populations with time.

The data reported here are from two complex long-term experiments at Rothamsted, designed to investigate how cultivation affects the soil fauna. Due to shortage of space, only some of the treatments can be considered and the data have been averaged over long periods. Full accounts of the experiments will be published elsewhere.

LONG-TERM CULTIVATION EXPERIMENT

This experiment was started in the spring of 1969 using old pasture on a silt clay loam soil containing flints. Four randomly – distributed plots, 24 x 21 feet (7.2 x 6.3 m) were left with grass and uncultivated, four were ploughed, rolled, disked twice, rolled again (fewest cultivations) and reseeded with a meadow fescue, timothy grass and clover mixture, and four were ploughed, rolled, disked five times, harrowed, rolled again (most cultivations) and also reseeded. From each plot,

four 2" diam x 6" deep (5 x 15 cm) soil cores  were taken prior to cultivation  and subsequently at 2-monthly intervals, and the soil animals
extracted  in  a modified Tullgren  high gradient funnel extraction apparatus.  Every spring and autumn, earthworm populations  were assessed
by watering three 2 x 2 foot quadrats  (60 x 60 cm)  in each plot  with
dilute formalin (55 parts in 1000) (Raw, 1959). Worms that were brought
to the surface were collected and identified.

    The effects of the cultivations on soil arthropods (for six months
after treatment) are summarised in Fig. 1.  Only the more common groups

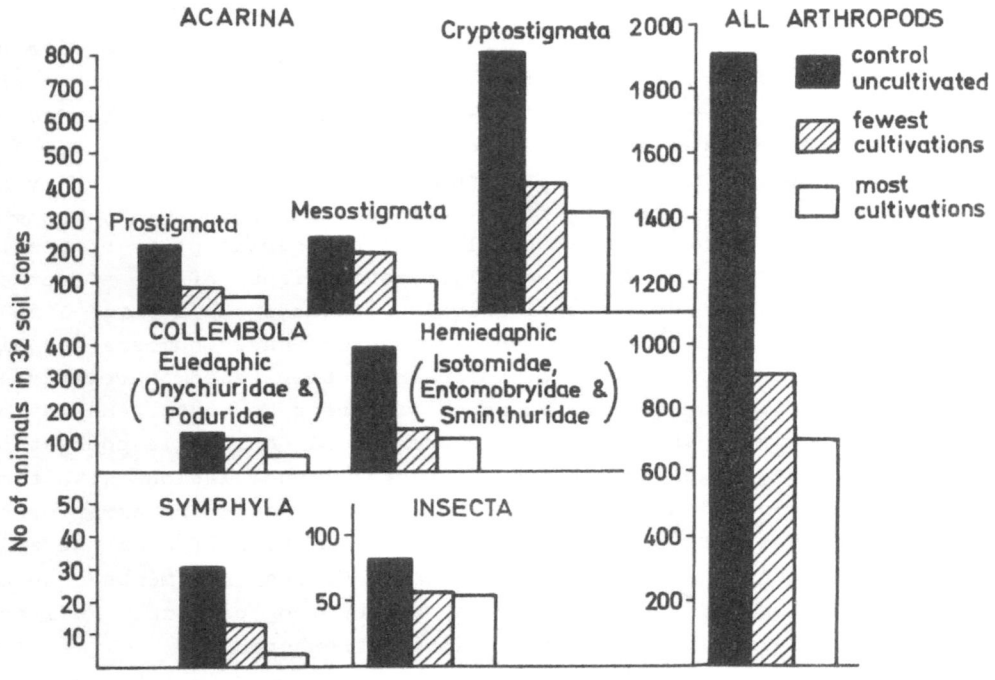

Fig.1. Effect of cultivating  and  reseeding old pasture
on soil arthropods.

of animals are represented in the histograms. The effects of the cultivations  were to decrease numbers  of all major groups  of soil animals
significantly, particularly the cryptostigmatid mites  and the surface-
dwelling (hemiedaphic) *Collembola,* the extent of the decrease depending
on the amount of cultivation.  The only exceptions  to this general decrease in numbers  were the larger mesostigmatid mites and coleopterous
larvae (particularly wireworms), neither of which were greatly affected.
    Although the effect  of cultivations  on earthworms  differed con-

siderably between species, all the cultivations increased the numbers of all species for two years after the treatment. For the two species *Lumbricus terrestris* and *Octolasium cyaneum,* the greater the amount of cultivation, the more the populations increased. For the other two common species, *Allolobophora longa* and *A. caliginosa,* there was a greater increase in numbers in plots with minimum cultivations than in those with most.

LONG-TERM WHEAT SLIT-SEEDING EXPERIMENT

The current trend in agriculture, to keep cultivation of arable land to a minimum, has culminated in England in the use of herbicides ('Paraquat') to kill all vegetation completely and then the subsequent sowing of crops in the uncultivated soil. Such techniques have been quite successful on lighter land, so it seemed of considerable interest to investigate the long-term effects of such practices on populations of soil invertebrates.

An experiment was laid out in 1965 on an arable site with sandy silt loam soil. Four plots, 21 x 65 feet (6.3 s 19.5 m), were treated with herbicide ('Paraquat' at 2 lb per acre i.e. 2.24 kg per ha) then sown with Capelle winter wheat (190 lb per acre i.e. 213 kg per ha), using a special I.C.I. drill that cut slits in the row, and dropped the seed into these slits. A further four plots, randomly-selected, were ploughed then sown at the same seed rate with the same drill. The treatments were repeated annually with wheat until 1971, when the incidence of the disease 'take-all' necessitated the ending of the experiment.

Soil cores similar to those in the first experiment were taken at two-monthly intervals throughout the six years and the animals extracted in the same way. Earthworm populations were also assessed as in the first experiments, three quadrats per plot being sampled in the spring and summer.

Space permits only a general consideration of the effects of the two treatments on soil arthropods, although data are available showing the gradual changes in populations throughout the course of the experiment. Fig. 3 summarises the differences in numbers of soil arthropods resulting from the two treatments over the whole period of the experiment. Although overall changes in populations were relatively small, the herbicide/slitseeding treatment favoured most groups of soil arthropods, the principal exceptions being the larger predatory mites, the deeper soil-dwelling (euedaphic) *Collembola* and dipterous larvae.

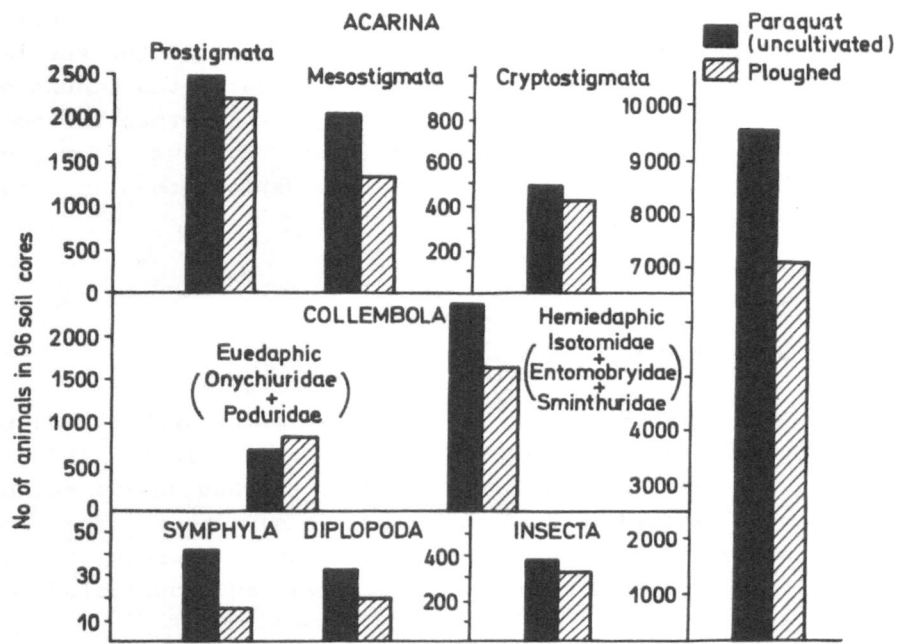

Fig.2. A comparison of the effects of ploughing and paraquat treatments on soil arthropods (total animals over 5 years).

The effects on the larvae of *Thysanoptera* varied, ploughing favouring them for the first two years of the experiments; thereafter more were found in the herbicide-treated plots.

The effects of the treatments differed between different groups of arthropods. Numbers of prostigmatid mites in the two treatments were similar until the last two years of the experiment, when numbers increased dramatically in the uncultivated plots, but the numbers of rhodacarid mites were consistently greater in the uncultivated plots throughout the experiment. There were no great differences in numbers of *Collembola* between treatments, except *Entomobryidae*, which were consistently more numerous in the uncultivated plots. There were consistently more *Symphyla* and *Diploda* in the uncultivated plots. In both experiments, the effects of the treatments on earthworm populations were much more variable than they were on those of arthropods, or than were the effects of cultivations on earthworms in the first experiment. This was particularly so for *Lumbricus terrestris*, which was much more numerous in the uncultivated plots at the end of the first year of the experiment but, thereafter, increased greatly in numbers in the ploughed plots, so that over the whole period of the experiment, ploughing greatly favoured this species.  The effects on numbers of *Allolobophora longa, A. chlorotica, A. rosea* and *O. cyaneum* were not great, but the changes  in numbers of *A. caliginosa,*  the dominant species,  were con- siderable The effects on *A. caliginosa* resembled the effects on *L. ter- restris* for the first two years but,  thereafter,  there were consider- ably more *A. caliginosa*  in the ploughed plots than in the uncultivated ones. Thus, although the numbers in ploughed and herbicide-treated plots differed little over the whole six years, the effects of the treatments differed greatly with time.


DISCUSSION


The effects of cultivations on arthropod populations in the two experi- ments were consistent, both when grassland  was cultivated and reseeded and when arable land  was cultivated  and planted with wheat.  Although soil cultivations  tended to decrease numbers  of most species  of soil arthropods,  there were indications in both experiments,  that cultiva- tion  tended  to favour  a few groups  of arthropods,  particularly the larger predatory mites and dipterous larvae.  Certainly,  the decreases in numbers of soil animals were not large, even in the long-term herbi- cide/ slit-seeding  experiment.  W. Wilkinson  (Jealott's Hill Research Station, I.C.I.)  commented, in the discussion after a paper by Edwards and Lofty  (1969),  that  he had also found  consistently higher micro- arthropod populations  in 'Paraquat'-treated, direct-drilled plots than in ploughed ones.

     It seems  likely  that,  when cultivations cease,  populations  of arthropods soon  return  to their previous levels;  Edwards  and  Lofty (1969) reported that,  after a single cultivation,  populations of most microarthropods  recovered  within  six months.  There is evidence that recolonisation of soil, even by truly edaphic microarthropods is mainly aerial, by animals carried  in the wind  (Buahin and Edwards, 1965), so that unless extremely large areas are planted to arable crops over very long periods,  it is  unlikely  that the soil microarthropod population would be seriously depleted of affected by cultivation.

     The effects of cultivations on earthworm populations are much more varied.  It has usually been reported that  earthworm populations  are much smaller in arable land than in old pasture or woodland  (Evans and Guild, 1948;  Graff, 1953;  Zicsi, 1958),  although the opposite effect has also been claimed (Hopp and Hopkins 1964; Zicsi, 1969). Most workers that have stated that cultivations  are deleterious to earthworms based

their  conclusions  on spot assessments  of numbers  in  arable fields,
taking little account of the past history  of the sites.  There is good
evidence  in the first of our experiments that cultivation favours most
species of earthworms, at least for the two years  that we recorded the
populations (see Fig.2). However, even during this time it became clear
that  for many species too much cultivation  was  less favourbale  than
a little.  There is evidence  from the same experiment, in data not reported here, that repeated annual cultivation eventually  had an effect
on adverse populations  of most species  of earthworms, particularly *L.
terrestris*. The species favoured most by cultivation was *A. caliginosa,*

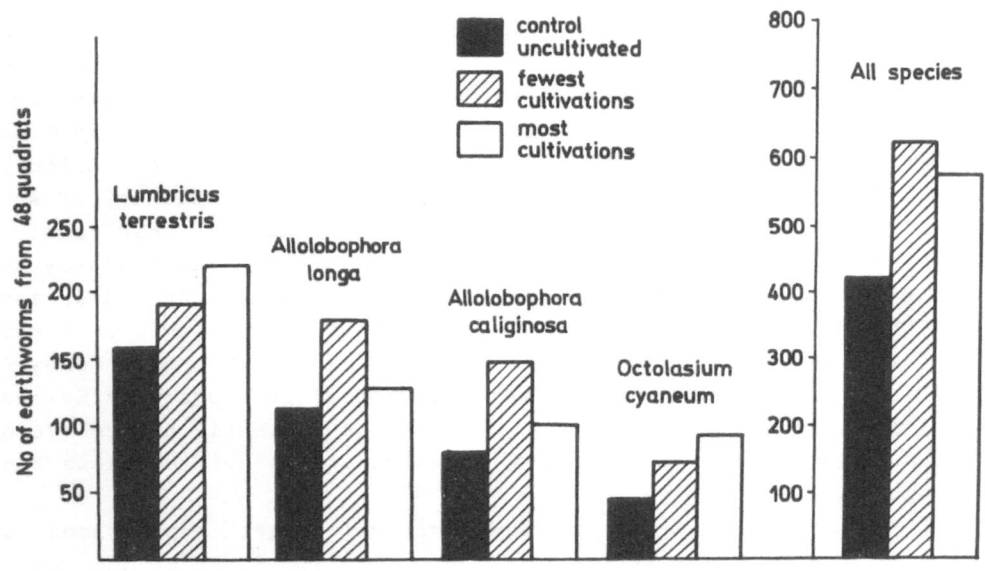

Fig.3. Effects of cultivating  and reseeding old pasture
on earthworms (totals for two years).

which was twice  as numerous  in the plots  with fewest cultivations as
in uncultivated plots,  even three years  after  the cultivations  were
completed. Presumably, for all species, a modest amount of cultivation,
by opening up the soil,  provides better soil conditions  for earthworm
burrowing and movement,  and moves surface organic matter  to the lower
soil layers,  where it becomes available to those species  that  do not
normally come to the soil surface.
    The changes in numbers of earthworms in the second experiment were

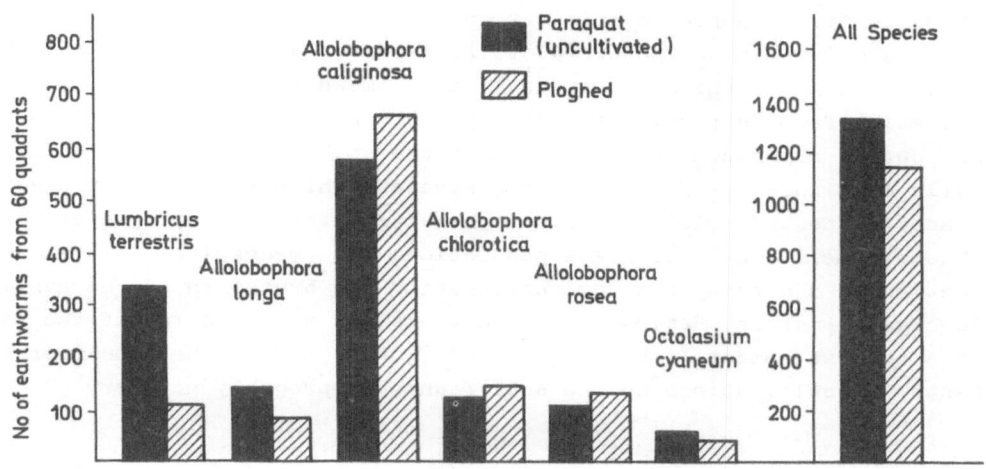

Fig.4. A comparison of the effect of ploughing and paraquat treatments on earthworms (total worms over 5 years).

more variable.    In contrast   to the first experiment,    there   were more
L. *terrestris*   in   the uncultivated plots than  in  the cultivated ones
throughout the six years of the experiment, but this was a much lighter
soil and would be easy to burrow through,   moreover,   the large amounts
of organic matter lying  on the soil surface   in  the herbicide-treated
plots provided a very suitable source of food for L. *terrestris,* a species that collects organic matter  from the soil surface  and  pulls it
down  into its burrows.  The effects  of the cultivations  on the other
species of earthworms  were much more similar to those in the first experiment.   There was a marked buildup in numbers of A. *caliginosa,* particularly   during  the later years  of the experiment, with up to eight
times  as many individuals  of this species  in the ploughed  as in the
unploughed plots in 1971. There were also fewer individuals of the species A. *chlorotica*  and  A. *rosea* in the uncultivated plots than in the
ploughed ones, particularly in the later years  of the experiment.  The
gradual depletion  of organic matter  in the sub-surface layer  of soil
in the uncultivated plots  was probably  as much responsible  for these
decreases as was the opening-up of soil  for earthworm activity by cultivation.

On the basis  of these results  it is possible  to  summarise  the
general effects  of cultivation  on earthworm populations  and  thereby
explain some of the anomalous results previously reported.  For L. *terrestris*  a modest degree of cultivation at infrequent intervals favours
population growth, not only by opening up the soil, but also by leaving
an adequate supply of organic matter at the surface, where this species

prefers to feed.  Repeated cultivation removes organic matter  from the
surface and is eventually harmful to *L. terrestris*. The overall effects
of cultivation  on the other species  is  somewhat different.  A single
cultivation favours them, not only  by opening up the soil, but also by
distributing organic matter  throughout the soil  where it is much more
readily available  as food for those species  which feed below the soil
surface.  Repeated cultivations to some extent favour most species, but
if too frequent, may  decrease populations  by  mechanical damage, destruction of habitats  and  by accelerating the breakdown of the available organic matter. Reseeding of old grassland seems to favour the increase of most species.  We can conclude,  therefore,  that the overall
effects  of cultivations on the soil fauna  is probably not very great.

R e f e r e n c e s

B u a h i n ,  G. K. A. and  E d w a r d s ,  C.A. (1965):  The reco-
        lonisation of sterilised soil by arthropods. Rep. Rothamsted Expfl
        Stn. 1964, 185-186.
E d w a r d s ,  C. A. and  L o f t y ,  J. R. (1969):  The  influence
        of agricultural practice  on soil microarthropod populations.  In:
        The Soil Ecosystem,  Systematics Assoc., Publ. 8., Ed. J.G. Sheals,
        237-47.
E v a n s ,  A. C.  and  G u i l d ,  W. J. McL. (1948): Studies on the
        relationships between earthworms and soil fertility.  V. Field po-
        pulations. Ann. Appl. Biol. 35(4) 485-93.
G r a f f ,  O. (1953): Investigations  in  soil zoology  with  special
        reference  to the terricole *Oligochaeta*. Z. Pfl. Ernähr. Düng. 61,
        72-7.
H o p p ,  H.  and  H o p k i n s ,  M. T. (1946):  The effect of crop-
        ping systems on the winter populations of earthworms. J.Soil Water
        Conserv. 1(1), 85-8.
R a w ,  F. (1959): Estimating earthworm populations by using formalin.
        Nature, Lond. 184, 1661.
T i s c h l e r ,  W. (1955): Synökologie der Landtiere. Gustav Fischer,
        Verlag, Stuttgart, 414 pp.
Z i c s i ,  A. (1958): Einfluss der Trockenheit und der Bodenarbeitung
        auf das Leben der Regenwürmer in Ackerböden. Acta Agron. 7, 67-74.
Z i c s i ,  A. (1969): Über die Auswirkung der Nachfrucht  und  Boden-
        arbeitung auf die Aktivität der Regenwürmer. Pedobiologie 9 (1-2),
        141-6.

# Discussion

**J. W. Reynolds :** The behavioural characteristocs of *Octolasium cyaneum* in North America is similar to that of *L. terrestris*. Do you think that this burrowing behavioural pattern may account for differences in the density of these two species when compared to the other earthworm species in your study? The burrowing habit of *Octolasium cyaneum* is less than *L. terrestris* and this trend is also reflected in our density data.

**C. A. Edwards :** Thank you for your comment which helps to explain these results.

**R. W. Sims :** Has Dr Edwards reached any tentative conclusions to account for the differences in the numbers of *Lumbricus terrestris* in ploughed land in comparision with most other earthworm species studied? Could it be that the differences may be attributable to the behaviour of *Lumbricus terrestris* which being a surface browser, seems to inhabit more or less permanent burrows whereas most other species do not? Thus regular ploughing may produce a greater effect by destroying burrows than by removing the surface food source.

**C. A. Edwards :** I agree that this would be an important factor in the effect of cultivation on *L. terrestris*.

**L. Vlym :** "Cultivation" might mean different things for different animals. Can something be said about the causal factors which are involved?

**C. A. Edwards :** Yes, for earthworms cultivation creates crevices in soil which enable them to burrow more easily and moves organic matter through the top layers of soil. The effect differs for *L.terrestris* which feeds on organic matter at the soil surface. The effect of cultivation on soil arthropods is to destroy their habitats and mechanically damage them.

# The effects of clearing and grazing on the termite fauna (Isoptera) of tropical savannas and woodlands

T. G. WOOD

Centre for Overseas Pest Research,
Termite Research Unit,
London

## INTRODUCTION

It has been recognised for a long time that termites are one of the dominant groups of invertebrates in the tropics. In spite of the fact that their geographical range (45-48°N to 45°S) includes two-thirds of the earth's land surface they have received little attention from soil biologists. For instance, only recently (Lee & Wood 1971a, 1971b) has it been shown that there is little truth in the widely held belief (probably initiated by Drummond, 1886) that in tropical soils they play a similar role to that of earthworms in temperate soils. There are many differences and few similarities between the effects of these two groups on soils and a further difference is that some species of termites are significant pests (Harris, 1969). Again, in contrast to the wealth of quantitative information on soil invertebrates in temperate regions the available data on termite populations is largely qualitative, quantitative data is available for only a few selected species or groups of species and there is no estimate of the total numbers of termites in any one habitat.

Agricultural and pastoral development in the tropics is very largely dependent upon clearing and/or burning of the native vegetation. In this paper I will discuss changes in the termite fauna resulting from clearing and grazing in relation to the significance of termites as pests of crops and pastures and as a beneficial or harmful component of the soil fauna.

## ABUNDANCE

The spatial distribution of termites within an ecosystem is such that their nests and foraging galleries may be **arboreal** (within, or attached

to the exterior of trees and shrubs), **epigeal** (i.e. mounds) or **subterranean.** Of the 60 species recorded by Kemp (1955) in the coastal plains
of Tanzania 14 were arboreal (including 10 nesting in dead trees and
logs), 32 were mound-builders and 14 were subterranean.

There are no available estimates of the abundance of arboreal
termites although some species are known to maintain large individual
colonies [632 000 in *Nasutitermes ripperti* (Rambur), Andrews (1911) and
1 250 000 in *Coptotermes acinaciformis* (Froggatt), Greaves (1957)].
Many arboreal termites maintain contact with the soil so that foraging
parties may be included in the subterranean category. Some of the available data on abundance of mound-building and subterranean termites is
shown in Table 1. These two categories are not mutually exclusive as

T a b l e   1.   Abundance of mound-building and subterranean termites
in tropical savannas and woodlands

| Species or groups | Numbers/m$^2$ | Country | Author |
|---|---|---|---|
| MOUND-BUILDERS | | | |
| *Cubitermes exiguus* Mathot[1] | 612-701 | Congo | Bouillon (1970) |
| *Trinervitermes geminatus*[2] (Wasmann) | 110-2,860 | Nigeria | Sands (1965a,b) |
| SUBTERRANEAN | | | |
| All species | 833-2,345 | Uganda/ Tanzania | Salt (1952) |
| All species | 1,800 | Congo | Harris (1963) |
| All species | 2,000 | Australia | Lee & Wood (1971b) |
| *Apicotermes* "gurguliflex"[3] | 71 | Congo | Bouillon (1972) |

[1]   9,409 (after swarming) - 10,783 (before swarming) individuals per
     colony and 650 occupied mounds per ha.
[2]   19,000 - 52,000 individuals per colony and 58 - 550 occupied mounds
     per ha.
[3]   14,517 individuals per colony and 1 living colony per 203.5/m$^2$.

foraging parties of mound-builders are likely to be included in soil
samples and subterranean species often colonise termite mounds. In the
savannas studied by Bouillon (1970) and Sands (1965a, b) there were,
in addition to the dominant species shown in Table 1, mounds built by
other species indicating that the abundance of mound-building species
may total several thousand per square metre. The available estimates

of subterranean populations, besides lacking in statistical accuracy, are almost certainly underestimates as their depth (15cm or less) would exclude up to 2/3 of the total nest chambers.

CHANGES IN POPULATIONS FOLLOWING CLEARING AND GRAZING

There have been no specific studies of this subject and for information I have had to rely on inadequate published data.

**Effects on numbers of species**

It is obvious that removal of trees and shrubs will have the greatest effect on those arboreal termites which depend entirely upon them for food and within which they maintain their entire nest-system. It is less obvious what happens to those arboreal-nesting species which maintain constant contact with the ground and which are able to build mounds or even subterranean nests: examples of such species are *Coptotermes acinaciformis* and *Microcerotermes turneri* (Froggatt) in Australia (Gay and Calaby, 1970) and *Amitermes evuncifer* Silvestri in West Africa (Sands, pers. comm.).

The available information indicates that removal of trees and shrubs has some impact on the number of species of mound-building and subterranean termites. In northern Nigeria Sands (1965a) recorded 9 mound-building species in climax woodland, 8 in partly cleared woodland and 6 in totally cleared areas. The corresponding figures for subterranean species (some "species" consisted of species-complexes) were 10, 9 and 6. At different localities in the Ivory Coast Bodot (1967) recorded 5 mound-building and 5 subterranean species in cleared areas in contrast to 7 mound-building and 14 subterranean species in wooded savanna.

**Effects on nest-building**

In the absence of quantitative data on populations the closest approach to assessing the effects of clearing and grazing is to consider the relative frequency od nest-building.

In the areas studied by Sands (1965a) in northern Nigeria, *Trinervitermes geminatus* had 63 mounds/ha (including 5 abandoned) in climax woodland, 440/ha (including 270 abandoned) in partly cleared woodland and 753/ha (including 248 abandoned) in a totally cleared area. In climax woodland 90 % of the mounds were built in the open (i.e. unshaded by trees) and the increase in abundance of mounds in partly cleared and fully cleared areas could be related to the increased availability of unshaded nesting sites. It is not known, however, if there was a corresponding increase in population density of the termites. In contrast the abundance of mounds of *Trinervitermes oeconomus* (Trägårdh) which prefers to build mounds in shaded sites, decreased from 2.5/ha in climax woodland to 0.5/ha in partly cleared woodland and was absent from the fully cleared area. The frequency of occurrence of chambers of subterranean species was $4.5/m^2$ in climax woodland, $4.2/m^2$ in partly cleared woodland and $5.6/m^2$ in the fully cleared area. The dominant subterranean termites were various species of Macrotermitinae some of which declined in cleared areas whereas others increased. Surprisingly the humus-feeding *Anoplotermes* spp. increased in the fully cleared area ($1.67$ chambers/$m^2$ as opposed to $0.34/m^2$ in partly cleared woodland and $0.31/m^2$ in climax woodland) where levels of soil organic matter could be expected to be considerably lower than in climax woodland. In other situations shade provided by vegetation may be favourable to subterranean species (Lee and Wood, 1971b, Table 1).

Grazing of cleared areas can be expected to affect termite populations by:

a) Consumption of grass and herbaceous vegetation affecting grass-harvesting and litter-feeding species directly; other species may be affected indirectly by reducing shade and exposing the soil surface. My own studies (Wood & Lee, 1971) of the grass- and litter-feeding, mound-building *Amitermes laurensis* Mjöberg in savanna woodland in northern Australia showed that numbers of mounds were least on two relatively undisturbed areas (28/ha and 50/ha), greater on two areas which were largely cleared and lightly grazed (142/ha and 154/ha) and greatest on two areas which had been completely cleared and heavily grazed (268/ha and 210/ha). Hartwig (1955) maintained that over-grazing in the South African veld resulted in increased areas of bare ground which provided favourable nesting sites for the grass-harvesting Hodotermitinae.

b) Hoof-action disturbing surface runways, compacting soil and adding to plant debris on the soil surface by physically breaking down grasses and shrubs. There is little information although there are reports from Tsavo National Park in Kenya (Anderson and Coe, pers. comm.) that vegetation trampled and broken down by elephants is rapidly attacked by termites.

c) Adding dung to the soil surface. Weir (1971) showed that termites rapidly consumed the large quantities of elephant dung deposited
around water holes in the Wankie Game Reserve, Rhodesia. Ferrar and
Watson (1970) recorded 46 species and sub-species of termites from dung
(largely bovine dung) in northern Australia although it is not known
whether or not dung is a preferred food for any particular species of
termite.


## Effects on vigour of colonies

The health or vigour of a colony can be related to its ability to maintain itself and to defend itself against disease organisms and predators. In the areas studied by Wood and Lee (1971) in northern Australia
and Sands (1965a) in northern Nigeria there is evidence that although
mounds of *Amitermes laurensis* and *Trinervitermes geminatus*, respectively, were more abundant in cleared and grazed areas than in natural savanna woodland, colonies in the disturbed areas were less able to maintain themselves than in the undisturbed areas. Firstly, in the cleared
areas (also grazed in northern Australia) there was a much greater proportion of abandoned mounds and secondly, the mounds that were occupied
by termites were more susceptible to invasions by ants and other species of termites (Table 2). In the Ivory Coast Bodot (1967) suggested
that clearing of trees from savanna modified the effects of predation
by ants and inter-specific competition among termites.


## DISCUSSION

One of the major problems in tropical soil science is the maintenance
of soil structure and fertility in agricultural land. Following the
clearing of forests, woodlands and savannas there is a general decrease
in stability of soil structure (Boyer, 1970), decrease in rainfall ac-
ceptance, permeability and percolation (Pereira, 1954) and decrease in
levels of organic matter and nutrients (Nye and Greenland, 1960). These
trends can be reserved by introducing forest or ·grass follows into the
rotation as is done in traditional methods of "shifting" cultivation
or, in some circumstances, by the addition of fertilisers and organic
residues. From Fig. 1 it is obvious that the activities of termites are
intimately linked with soil structure and nutrient cycling. However,

T a b l e   2.    Invasion  and  occupation  of termite mounds in savanna woodland by ants  and  other species of termites.  All figures refer to ants only,   except figures in parentheses  which refer to ants and ter- mites.

| Area | Termites present | Mounds abandoned | Termites present | Mounds abandoned |
|---|---|---|---|---|
| **Northern Nigeria[1]** | | | | |
| Climax vegetation | 58 | 5 | 0  (0) | 0  (0) |
| Partly cleared | 170 | 270 | 5.3(32.6) | 10.8(69.7) |
| Fully cleared | 505 | 248 | 2.0(19.3) | 13.7(32.5) |
| **Northern Australia (Townsville)[2]** | | | | |
| Relatively undisturbed | 21 | 7 | 38 | 71 |
| Partly cleared and lightly grazed | 111 | 31 | 46 | 72 |
| Fully cleared and heavily grazed | 105 | 105 | 70 | 82 |

[1]   Sands (1965a): Termites are *Trinervitermes geminatus*; ants are *Mega-ponera* (a known predator of termites), *Pheidole*, *Camponotus* and *Cre-matogaster*.
[2]   Wood and Lee (1971): Termites are *Amitermes laurensis*; ants are *Iri-domyrnex purpureus* F. Smith.

there  is no precise information  on whether  or not humus-feeders  and other groups  which do not damage crops  have an overall beneficial ef- fect  on soil structure  and  fertility or whether  they are "pests" in an indirect way.  Their high assimilation  efficiencies result in rapid removal of plant debris  and  its return to the ecosystem as faecal ma- terial which is low  in plant nutrients  and  easily-decomposed organic matter and, in some cases, inhibitory to micro-organisms (Lee and Wood, 1971a, b).  There is  a need for investigations into the advantages and disadvantages of the presence of various groups of termites in tropical soils  used for agriculture  to facilitate the wise application of con- trol measures against pest species.


SUMMARY

Studies in Australia and Africa have demonstrated that there are quali- tative  and  quantitative changes  in the termite fauna  when  tropical

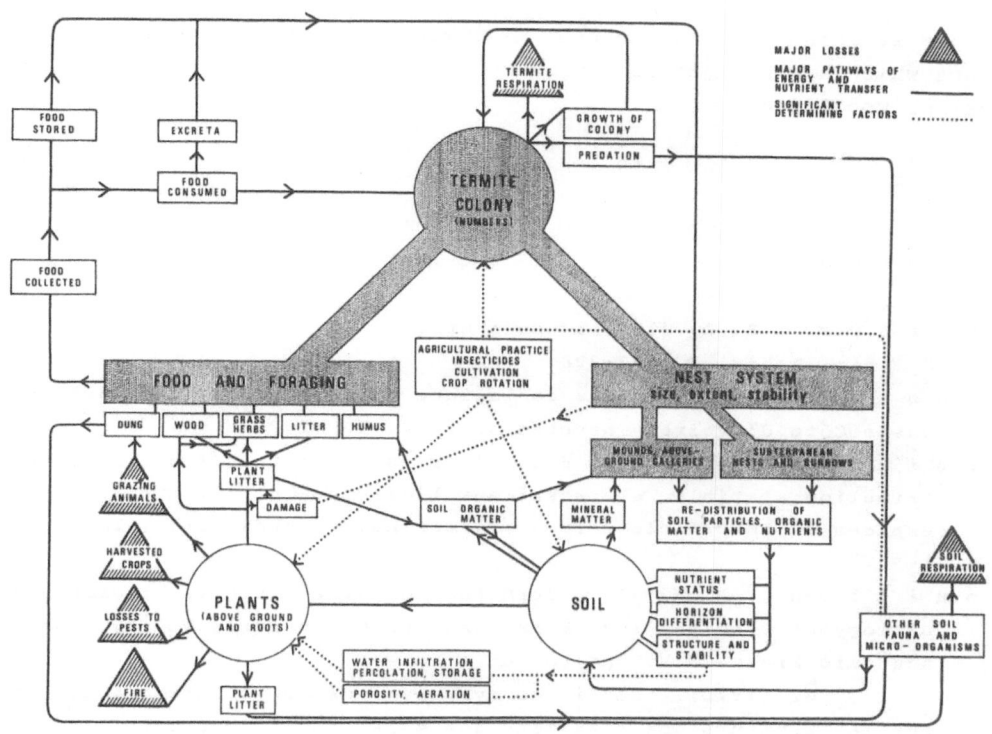

Fig.1. Simplified representation  of the impact  of termites in a tropical agricultural ecosystem.

woodlands and savannas are partially or completely cleared for pastoral
and agricultural purposes.  These changes have not been investigated in
detail and consequently we do not known the conditions under which certain species attain the status of pests.

Removal of trees and increase in herbaceous vegetation affects the
wood-litter- and grass-feeding species by changing the distribution and
amount of their food.  In addition certain species increase  and others
decrease due to reduction in shade  and increase in bare ground effecting the availability of sites for the foundation and maintenance of colonies.  Grazing animals affect the termite fauna by direct or indirect
competition for food  with grass-harvesting termites, the disruption of
surface runways by hoof-action  and  the provision of a new food source
in the form of dung.

Measures  to control the numbers  of the few species  which affect
economic damage to pastures and crops should be directed with knowledge

of the effects  of these control measures  on  the rest  of the termite
fauna which are a major component of the soil fauna  in tropical savannas and woodlands.

References

A n d r e w s ,  E. A. (1911): Observations  on  termites  in  Jamaica.
    J. Anim. Behav. 1, 193-228.
B o d o t ,  P. (1967):  Etude écologique  des termites  des savanes de
    Basse Côte d'Ivoire. Insectes soc. 14, 229-259.
B o u i l l o n ,  A. (1962): Etudes sur les termites Africains 1. Dis-
    tribution spatiale  et  essai sur l'origine  et  la dispersion des
    espèces du gênre Apicotermes (Termitinae). Studia Univ. lovan. 15,
    1-35.
B o u i l l o n ,  A. (1972):  Termites  of  the Ethiopean region.  In:
    Biology of Termites (ed. K.Krishna and M.Weesner), Vol.2, 153-280,
    Academic Press, N.Y. and London.
B o y e r ,  G. (1970):  Essai  de synthèse  des connaissances acquisés
    sur les facteurs  de fertilité des sols  en Afrique intertropicale
    francophone. Committee on Tropical Soils, London.
D r u m m o n d ,  H. (1886): On the termite  as  the tropical analogue
    of the earthworm. Proc. R. Soc. Edinb. 13, 137-146.
F e r r a r ,  P.  and  W a t s o n ,  J. A. L. (1970): Termites  as-
    sociated with dung in Australia. J. Aust. Entomol. Soc.9, 100-102.
G a y ,  F. J.  and  C a l a b y ,  G. H. (1970):  Termites  from the
    Australian region.  In: Biology of Termites  (ed. K.Krishna and M.
    Weesner), Vol. 2, 393-448, Academic Press, N.Y. and London.
G r e a v e s ,  T. (1957):  Experiments  to determine  the populations
    of tree-dwelling colonies  of termites  [Coptotermes acinaciformis
    (Froggatt) and C.frenchi (Hill)]. Tech.Pap.Div.Entomol. C.S.I.R.O.
    Aust. 7, 19-33.
H a r r i s ,  W. V.  (1963):  Exploration  du Parc National  de la Ga-
    ramba, Part 42. Isoptera, 43 pp. Imprim. Hayez, Brussels.
H a r r i s ,  W. V. (1969): Termites as pests of crops and trees. Com-
    monw. Inst. Entomol., London, 41 pp.
H a r t w i g ,  E. K.  (1955):  Control of snouted harvester termites.
    Fmg. S. Afr. 30, 361-366.
K e m p ,  P.B. (1955):  The termites of north-eastern Tanganyika: their
    distribution and biology. Bull. Entomol. Res. 46, 113-135.

L e e , K. E. and W o o d , T. G. (1971a): Physical and chemical effects on soils of some Australian termites, and their pedological significance. Pedobiologia 11, 376-409.

L e e , K. E. and W o o d , T. G. (1971b): Termites and Soils. Academic Press, N.Y. and London, 251 pp.

N y e , P. H. and G r e e n l a n d , D. J. (1960): The soil under shifting cultivation. Tech. Comm. 51 Commonw.Bur.Soils, Harpenden, 150 pp.

P e r e i r a , H. C. (1954): The assessment of structure in tropical soils. J. Agr. Sci. 45, 401-410.

S a l t , G. (1952): The arthropod population of the soil in some East African pastures. J. Anim. Ecol. 17, 180-201.

S a n d s , W. A. (1965a): Termite distribution in man-modified habitats in West Africa, with special reference to species-segregation in the genus Trinervitermes (Isoptera, Termitidae, Nasutitermitinae). J. Anim. Ecol. 34, 557-571.

S a n d s , W. A. (1965b): Mound population movements and fluctuations in Trinervitermes ebenerarius Sjöstedt (Isoptera, Termitidae, Nasutitermitinae). Insectes Soc. 12, 49-58.

W e i r , J. S. (1971): The effect of creating additional water supplies in a central African National Park. In: The Scientific Management of Animal and Plant Communities for Conservation (ed. E. Duffey and A.S. Watt), 367-385. Blackwells, London.

W o o d , T. G. and L e e , K. E. (1971): Abundance of mounds and competition among colonies of some Australian termite species. Pedobiologia 11, 341-366.

D i s c u s s i o n

G. J o s e n s : In some regions of Africa, the fungus-growing termites are very important: in a fire-climax savanna of the Ivory Coast, only 0.5 g/m$^2$ of Macrotermitinae eats more than 100 g/m$^2$/year. These species feed almost exclusively on litter in this natural savanna but they are known as being pests in agricultural regions.

T. G. W o o d : Macrotermitinae (i.e. Fungus-growers) do appear to be responsible for most termite damage to crops in Africa. There are no Macrotermitinae in Australia and most crop damage appears to be due to Mastotermes darwiniensis.

H. E i j s a c k e r s: Is there any influence of termite nests  (above or under surface level) on other soil inhabiting arthropods?

T. G. W o o d : Yes, but  there  have been no detailed studies.  Some soil arthropods (eg. ants, other termites, *Coleoptera*) colonise termite nests. It is doubtful if a "normal" soil fauna is maintained underneath a termite mound.

H. F r a n z : Zur Frage  des Einflusses  von Termiten  auf die übrige Bodenfauna  möchte ich  an die belgischen Untersuchungen im Kongo erin- nern. Sie zeigten, dass in gerodetem Gelände andere Termiten lebten als im Regenwald. Nach der Rodung beeinflussten die Termiten weitgehend die physikalischen Eigenschaften des Bodens  und dadurch haben sie auch die übrigen Bodenfauna beeinflusst.

# Effects of sewage effluent disposal
# on community structure of soil invertebrates

D. L. DINDAL, D. SCHWERT AND R. A. NORTON

State University, College of Environmental Science
and Forestry,
New York

## INTRODUCTION

In 1962, researchers at Pennsylvania State University  pursued the following objective: to determine the feasibility  and effects of disposal
of municipal sewage effluents  on a variety of vegetative sites.  Since
then, numerous facts  have been accumulated  which were compiled by the
project  directors  Sopper  and  Kardos (1973). Generally they have demonstrated for the first time in the U.S. a satisfactory method for renovation of wastewater  and sludge  and secondly, some very interesting
physical, chemical and biological relationships resulting  from irriga-
tion-recycling  of the materials.  All the relationships described  are
focused on what they call  the  "living filter concept".  That is, they
suggest that plants, litter, organic matter  plus associated microflora
and fauna on  and incorporated  within  upper  soil horizons constitute
this dynamic biological filtering system.

During 1972, we joined  these men  to help ellucidate  some intri-
cacies and the composition of their "biological filter" system.  Therefore, one of our objectives, on which  I present  this  preliminary report, was to determine and compare the microcommunity structure of soil
invertebrates in different vegetative communities irrigated with municipal wastewaters.

## RESEARCH SITES

In addition to site descriptions a brief summary of Sopper´s and Kardos´

The work  upon which this report  is based  was supported by funds
provided  by  the Northeastern Forest Experiment Station  of  the  USDA
Forest Service,  through the Pinchot Institute Consortium  for Environmental Studies.

(1973) results is presented for total ecological understanding.

   Sites range in size from 0.2 to 8 ha (0.5 to 2oA). All treated (T) areas have a comparable control or untreated (UT) site. The origin of the effluent is the small college town in central Pennsylvania, USA, University Park. Transport of previously chlorinated wastewater is via pipe line to the experimental areas. There it is applied periodically from rotating sprinkler nozzles located at least 2 m from the soil surface. Unless otherwise noted, application rate is 5 cm/wk.

## Reed canary grass monoculture

*Phalaris arundinacea* L. has been planted and the area we studied has been continuously irrigated year round for 3 yr. Preliminary results show that the amount of effluent nitrates added per year to this site equals the nitrate amount that can be harvested in grass per year.

## Mixed oak hardwood community

This is the second area which receives continuous year round irrigation, and it has for 7 yr. The forest is about 60 yr old dominated by the following plants: white oak (*Quercus alba* L.), chestnut oak (*Q. prinus* L.), black oak (*Q. velutina* L.), red oak (*Q. rubra* L.), scarlet oak (*Q. coccinea* Muench.), red maple (*Acer rubrum* L.) and hickory (*Carya* spp.). Principal species in the undergrowth include violet (*Viola* spp.), wild sarsaparilla (*Aralia nůdicalis* L.), poison ivy (*Toxicodendron radicans* L.) and blueberry (*Vaccinium* spp.).

   Treatment has caused a more lush undergrowth production. In addition, the diameter growth per year of the trees has been 3.0 cm/yr (UT) and 5.3 cm/yr (T).

## Red pine plantation

Two *Pinus resinosa* Ait. stands were studied for 10 yr, one receiving irrigation at 5 cm/wk and a second at 2.5 cm/wk. Reduced diameter growth 2.8 cm/yr (UT) versus 1.5 cm/yr (T), was observed on the former site. Also, this stand was subject to blow down and was leveled by a 1969 ice storm.

The 2.5 cm/wk site exhibited no significant differences between diameter growths, 2.5 cm/yr (*UT*) and 2.0 cm/yr (*T*). Undergrowth vegetation such as poison ivy, and pokeweed (*Phytolacca americana* L.) flourished in the irrigated sites and were absent in the *UT* area.


**Old field herbaceous community**

Prior to application, white spruce, *Picea glauca* (Moench) Voss, seedlings were sparsely planted in both *T* and *UT* fields. Growth in heighth has been significant, 28.0 cm/yr (*UT*) compared to 58.5 cm/yr (*T*).

The predominant *UT* old field ground cover was poverty grass (*Danthonia spicata* Beauv.), goldenrod (*Solidago juncea* Ait), dewberry (*Rubus flagellaris,* Willd.) and occasionally butter and eggs (*Linaria vulgaris* Hill). After treatment all but goldenrod were absent.

Dry matter production was affected dramatically with 1848 kg/ha/yr (*UT*) and 6196 kg/ha/yr (*T*) produced.


METHODS

Annelid collections were obtained using the formalin extraction method. Results from digging and hand sorting compared very closely to chemical extraction when we used 1 L (25 ml 10% formalin in 4 L of water) per a 710 aq cm sampling area. All worms responding and exiting the soil in 10 min were considered our sample.

Microarthropod assay and microbial respiration was determined on 600 soil cores (10 cores/site/sampling period). Arthropods from each core (54 mm diameter x 40 mm deep) were extracted in a modified Tullgren funnel for 7 da.

Microbial respiration was measured by $CO_2$ evolved in 24 hr from incubated soil in closed systems (Stotzky, 1965).

Species diversity ($\bar{H}$) calculations were made using the modified Shannon-Weiner formula using $\log_e$ after Lloyd, Zar and Karr (1968). Species richness (*d*) was determined employing the equation by Menhinick (1964).

RESULTS AND INTERPRETATIONS
## Microbial respiration

Within any of the four vegetative communities  there appeared  to be no
significant difference during any season between respiration values due
to treatment.  Shifts in microbial species composition may and probably
have occurred,  however,  this method used did not provide that type of
information.

## Annelida

P o p u l a t i o n s . With two exceptions (Reed canary grass, August;
Red pine, April), earthworm populations  were stimulated  by irrigation
(Fig.1). The increase in the mixed oak site was highly significant both
statistically (P≤.01) and biologically.  Normally oak litter  has a re-

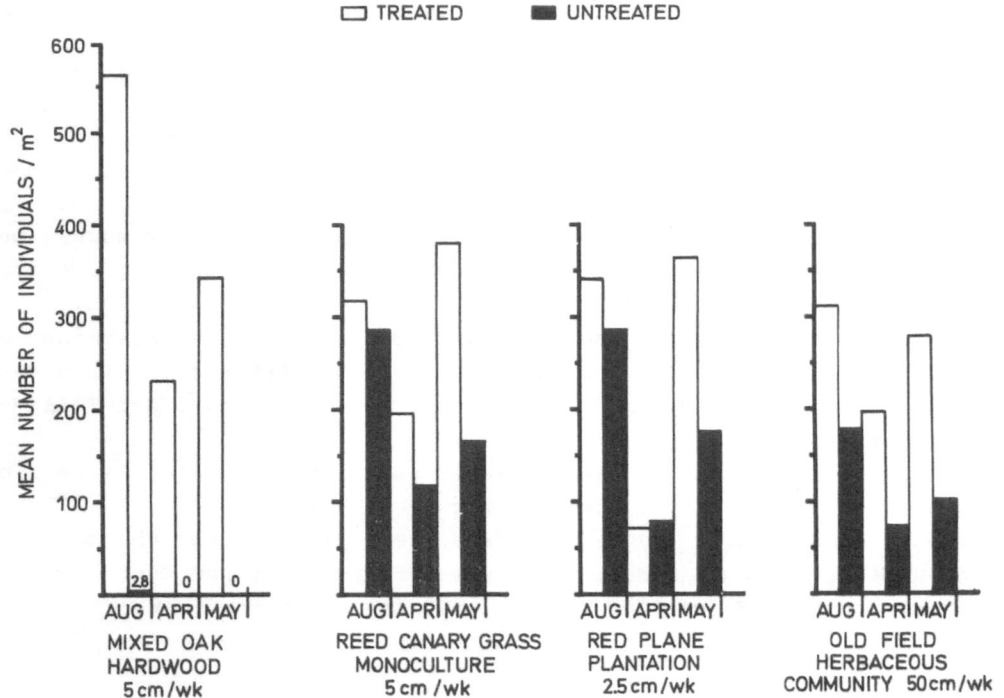

Fig.1. Comparative earthworm populations on sites treated with sewage effluent (University Park, Pennsylvania).

latively high C/N ratio  in the range of 40/1 to 50/1  (Kühnelt, 1950). Nitrates and organic nitrogen  in wastewater no doubt  alters the ratio thus inhancing utilization by earthworms  and other soil organisms. Recycling would be hastened,  and the longevity  of leaf litter  would be reduced.

        S p e c i e s   d i v e r s i t y . Figure 2  illustrates  the dynamic differences in earthworm diversity and richness in autumn samples.

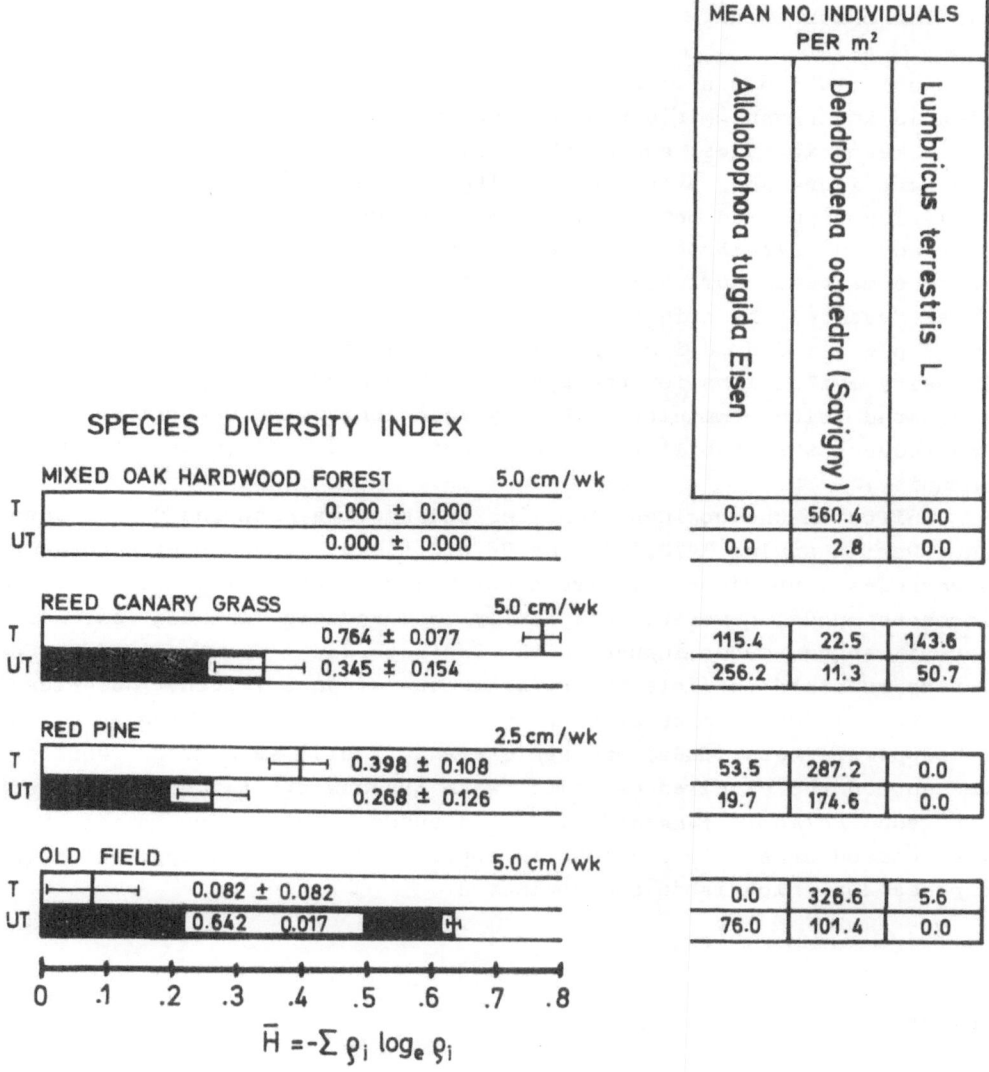

Increased biomass  is associated  with increased earthworm populations.  This implies an increased movement  and  aeration of soil along
with a greater incorporation of organic matter.  Also, the trend in oak
litter  would be away from a potential laminated mor toward a mull condition.


## Soil microarthropods

P o p u l a t i o n s .  On all sites  during spring and summer periods
oribatid and prostigmatid populations  were generally reduced by treatment.  Mesostigmatids  shared this trend except  in the old field where
they were increased  over  the *UT* site  in the summer 1972.  Astigmatid
populations appeared not to differ on the two sites. Seasonal variation
occurred  in *Collembola* with  enhancement of numbers  by irrigation at
least  some time  during the year.  Perhaps  collembolans  were filling
niches "vacated" by oribatids.
S p e c i e s  d i v e r s i t y .  Table 1  summarizes details of
diversity evaluations for the spring 1972 period. If high diversity can
be equated  with community stability than long range wastewater irrigation reduces the stability  of the acarine  and collembolan soil microcommunity.
Shifts in the dominant families and species occurred due to treatment. During spring 1972,  in the *UT* old field,  species of *Eupodes* and
*Cocceupodes* (Eupodidae)  showed numerical dominance over all other soil
microarthropods; however, this family  was reduced  in half  on *T* sites
where it shared subdominance  with  *Scutacaridae, Brachychtonidae* and
*Isotoma eunotabilis* (Folsom). *Sminthurinus elegans* (Fitch)  numerically
surpassed all other microarthropods on the *T* herbaceous area.
*Oppiella nova* (Oud.) and *Oppia minus* (Paoli) as a group  were predominant on the *UT* mixed oak site  with *Eupodes* sp. A, *Cocceupodes* spp.
A, B  and  *Folsomia fimetaria* (L.)  as subdominants.  Irrigation of oak
sites caused drastic reductions in oppiid  and *Folsomia* spp. while *Cocceupodes* spp. flourished and dominated.


## SUMMARY

Annelid populations  are generally increased  by wastewater irrigation.
Treatment selects for *Dendrobaena octaedra* (Savigny)  in all cases. Po-

T a b l e   1.    Diversity characteristics  of soil Microarthropods from sewage effluent treated (T)* and untreated (UT) sites.(University Park, Pennsylvania, U.S.A.) Spring samples, June 1972; N = 10/site

| Mixed Oak Forest Community | | | | | | | | |
| --- | --- | --- | --- | --- | --- | --- | --- | --- |
| Taxon | Mean (n) Total Individuals | | Mean (s) Total Species | | Mean (H̄) Species Diversity | | Mean (d) Species Richness | |
| | UT | T | UT | T | UT | T | UT | T |
| Prostigmata | 108 | 102 | 21 | 7 | 1.2524 | 1.0273 | 2.06 | 0.69 |
| Mesostigmata | 50 | 11 | 9 | 6 | 0.3465 | 0.2023 | 1.27 | 1.82 |
| Oribatei | 199 | 8 | 17 | 6 | 1.0180 | 0.0693 | 1.20 | 2.09 |
| Collembola | 236 | 74 | 10 | 17 | 1.0979 | 0.8743 | 0.65 | 1.98 |

| Old Field  Herbaceous Community | | | | | | | | |
| --- | --- | --- | --- | --- | --- | --- | --- | --- |
| Taxon | Mean (n) Total Individuals | | Mean (s) Total Species | | Mean (H̄) Species Diversity | | Mean (d) Species Richness | |
| | UT | T | UT | T | UT | T | UT | T |
| Prostigmata | 483 | 443 | 36 | 12 | 1.8397 | 1.0702 | 1.64 | 0.57 |
| Mesostigmata | 38 | 22 | 7 | 4 | 0.4028 | 0.3053 | 1.13 | 0.85 |
| Oribatei | 196 | 24 | 18 | 8 | 1.1094 | 0.2485 | 1.28 | 1.63 |
| Collembola | 105 | 232 | 17 | 11 | 1.1379 | 0.8421 | 1.65 | 0.72 |

* Summer irrigation at rate of 5 cm/wk for 10 yr.

pulation numbers and species diversity of *Acari* are suppressed by treatment, while collembolan populations  varied  and  diversity  decreased. Therefore, microcommunity stability of *Acari* and *Collembola* is decreased.  Niche replacement  among soil organisms probably occurs.  Finally, potential agents of litter-organic matter decomposition  are shifted by irrigation  from  many organisms  with low biomass (microarthropods) to those with greater biomass (annelids).

R e f e r e n c e s

K ü h n e l t , W. (1950): Soil Biology. Rodale Books, Inc. Emmaus, Pa.
     397 pp.
L l o y d , M., Z a r , J. H.  and  K a r r , J. R. (1968):  On the
     calculation of information-theoretical measures of diversity.  Am.
     Midl. Nat. 79(2), 257-272.

M e n h i n i c k , E. F. (1964): A comparison of some species diversity indices applied to samples of field insects. Ecology 35, 859-
861.
S o p p e r , W. E. and K a r d o s , K. T. eds. (1973): Recycling
treated municipal wastewater and sludge through forest and crop-
land. The Penn. State Univ. Press, University Park and London, 479
pp.
S t o t z k y , G. (1965): Methods in microbial respiration: $CO_2$ evolution. 1550-1558 pp. In: C. A. Black, methods of soil analysis.
Pt. 2, Am. Soc. Agron., Madison, Wisc., 1572 pp.

## D i s c u s s i o n

C. A. E d w a r d s : Could I ask which method of extraction of animals was used?

D. L. D i n d a l : A modified Tullgren funnel system was used permit-
ting samples to air dry: various lamps (25, 10, 7.5 Watt and different
colours) as heat sources in small plastic funnels were pretested with
soil cores under our lab conditions. From this we found greatest efficiency with the system we selected. Hand sorting of extracted cores
veryfied funnel extraction efficiency.

K. H. D o m s c h : It is surprizing that a complete recycling of nutrients was found after a year-round sewage application. In our experience the crucial point is that maximum requirement of plants does
not coincide with maximum mineralization during spring and fall.

D. L. D i n d a l : This phase of the work was completed and reported
by Sopper and Kardos (1973). It is my understanding that complete recycling occurred in reed canary grass sites after grass removal. However in fairness to the Penn State researchers reference to the original article should be made.

M. G o r n y : Were your control plots watered or not? Otherwise one
can talk about the influence of watering also, not only about the sewage
influence. We have in Poland some forest areas on which the influence
of watering on soil fauna is studied. Generally all the groups of soil
fauna increased in numbers, but some species dissapeared in the third
year of watering.

D. L. D i n d a l : Control sites were not watered since the major
objective of the Penn State workers was to determine effects of waste

water effluent in its entirety.   Therefore, the effects I described may
be due to either addition of water  or  nutrients  or  a complex interaction of both.   Carey, Leaf and I, from our college, have  some unpublished information regarding irrigation  only  of Red Pine plantations.
Effects on *Acari* are comparable to yours and correspond well to my data
from this study just described.

# Changes in the structure of animal populations in soil under the influence of farm crops

M. M. ALEINIKOVA AND N. M. UTROBINA

Soil Laboratory, Kazan Institute of Biology,
Academy of Sciences

The study of the animal population structure of agrobiocoenoses is of great practical value and considerable theoretical interest for the solution of general problems of biocoenology and the determination of the ways of purposeful landscape development. The study on the soil fauna is of particular interest, since its activity largely determines soil fertility.

A long-term investigation of soil fauna under farm crops has been carried out in various landscapes and soils of the Middle Volga Region, as well as in experimental crop rotations, monocultures and continuous fallow on the territory of one farm (Stolbishchensky State Farm, situated in the Tatar ASSR, 20 km from Kazan).

The population density of large invertebrates has been determined by digging in plots measuring 50 x 50 cm, and collecting the animals by hand.

Samples of 1 $dm^3$ and, in the experimental crop rotations and monocultures, of 125 $cm^3$ were taken for a quantitative estimation of micro-arthropods. The former were treated in Tullgran funnels, the latter in cardboard funnels without the use of electricity by Ballogh's method (1958). The volume of samples for the recording of *Enchytraeidae* population density was 500 $cm^3$. They were extracted in accordance with O'Connor's procedure (1955). Concurrently with soil sampling, beetles were collected in cylinders dug into trench bottoms.

It has been established that the total number of both large and small invertebrates depends on the type of soil, its structure and humus content, and increases in the Middle Volga Region from soddy-podzolic soils of the taiga towards ordinary chernozems of a typical forest steppe (Aleinikova, 1965, 1968).

Within the limits of one soil type, the population density, the ratio of soil animal groups and the trophic structure depend on the character of farm crops and their cultivation and, in particular, on soil tillage (Aleinikova and Utrobina, 1969).

T a b l e  1.  Average population density of the principal invertebrate groups (per $1m^2$) in some rotation fields in different soil types of the Middle Volga Region

| Soil type | Culture | Insecta | Geophilo-morpha | Lumbri-cidae | Collem-bola | Acarina |
|---|---|---|---|---|---|---|
| Sod-medium-podsolic | Maize | 14,3 | 0,5 | 8,8 | 891 | 1278 |
| | Wheat | 12,3 | 2,8 | 8,0 | 3816 | 1461 |
| | Second year of use | 32,5 | 1,3 | 9,2 | 2970 | 7353 |
| Gray wooded | Maize | 24,8 | 3,1 | 4,1 | 4265 | 24659 |
| | Wheat | 38,0 | 7,3 | 7,9 | 1208 | 15968 |
| | Third year of use clover + alfalfa | 53,5 | 2,2 | – | 36019 | 44018 |
| | Third year of use alfalfa | 51,0 | 4,5 | – | 5000 | 74150 |

Table 1 continued

| Soil type | Culture | Insecta | Geophilo-morpha | Lumbri-cidae | Collem-bola | Acarina |
|---|---|---|---|---|---|---|
| Leached cher-nozem and or-dinary cher-nozem | Maize | 39.0 | 8.1 | 6.8 | 9941 | 9616 |
| | Wheat | 42.4 | 11.2 | 5.0 | 15189 | 16093 |
| | Third year of use alfalfa | 65.4 | 12.8 | 6.7 | 7369 | 11123 |
| | Second year of use sainfoin | 39.0 | 0.5 | - | 32262 | 4162 |
| | Third year of use alfalfa + crestad wheatgrass+timothy | 137.3 | 14.1 | 3.8 | 14300 | 14883 |
| | Seventh year of use crested wheatgrass | 112.0 | 21.0 | - | 6187 | 8362 |
| Carbonated chernozem | Maize | 28.5 | 2.5 | - | no data avail-able | |
| | Wheat | 46.5 | 8.1 | 1.5 | 10937 | 15375 |
| | Third year of use alfalfa | 94.0 | 9.2 | r | 10562 | 47937 |

General regularitites can be observed in population density changes
in the principal soil fauna groups, namely its increase in the order
maize-wheat-perennial grass. This tendency is evident in all soil types,
although it is not equally pronounced (see table 1). The maximum accumulation of animals under perennial grass is due to the moist microclimate, abundance of microflora and decaying plants which provide
nutrition for small saprophages, and the absence of land cultivation
over a number of years.

The adverse effect of land cultivation and the absence of plant
cover on the soil fauna is evident in an 8-year-old continuous fallow.
Thus, large invertebrates are practically absent, the number of *Collembola* in the monocultures of maize, rye and wheat of the same age, in
comparison with the fallow, is from 3 to 7 times higher, the total number of *Acarina* from 2 to 5 times higher, that of *Enchytraeidae* 10 times
higher.

The variety of soil invertebrate species increases in the same way
as their density reaching a maximum under perennial grass and in field-
protecting tree plantations (see table 2).

T a b l e  2.    Species variety of some groups of soil invertebrates
under various cultures in the state farm Stolbishchensky, Tatar ASSR
(dark gray soil)

| Invertebrate groups \ Number of species | Fallow | Maize | Wheat | Alfalfa + clover perennials | 20-year old birch field protecting plantations |
|---|---|---|---|---|---|
| *Carabidae* | 23 | 38 | 29 | 42 | 57 |
| *Staphylinidae* | 17 | 21 | 29 | 21 | 31 |
| *Curculionidae* | 7 | 10 | 15 | 15 | 46 |
| *Elateridae* | 4 | 5 | 5 | 7 | 10 |
| *Histeridae* | 6 | 6 | 6 | 12 | 11 |
| Coleoptera total number | 57 | 80 | 84 | 97 | 155 |
| *Araneina* | 8 | 13 | 11 | 12 | 20 |
| *Lubricidae* | 1 | 4 | 3 | 4 | 7 |
| Large invertebrate total number | 66 | 97 | 98 | 113 | 182 |
| *Collembola* | 16 | 15 | 15 | 17 | no data available |
| *Oribatei* | 11 | 14 | 21 | 20 | - |

Alongside with the variety of species  and the presence of certain
species the percantage ratio of dominants in the individual soil animal
groups  is also characteristic of different cultures.  This can be seen

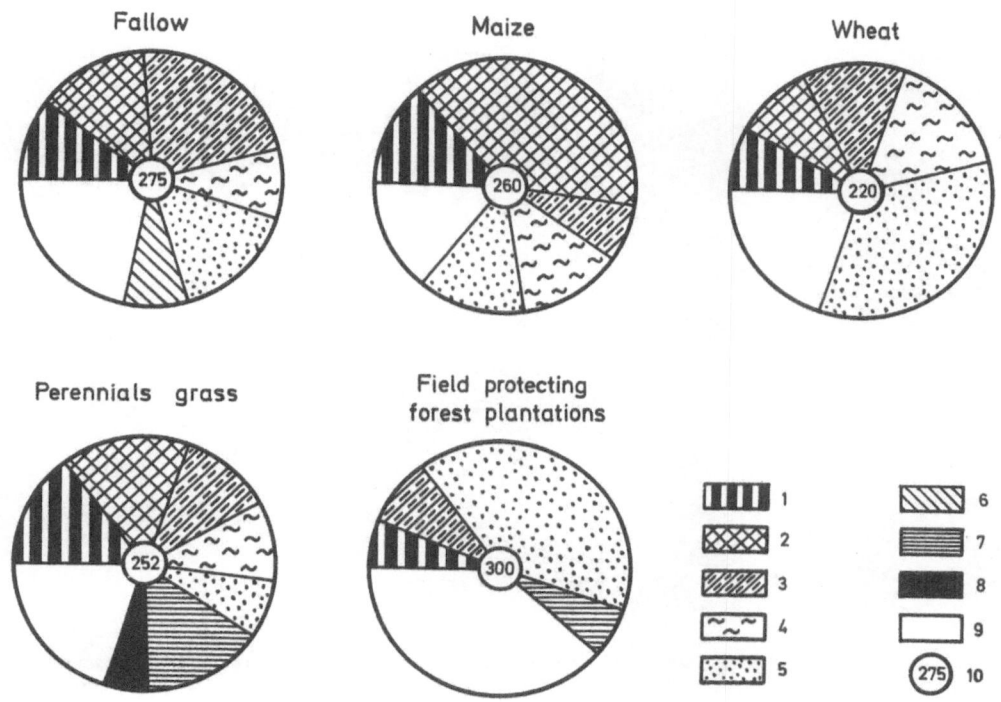

Fig.1. *Carabidae* species ratio   and   their numbers under
different cultures: 1 - *Bembidion lampron*, 2 - *Bembidion
quadrimaculatum*, 3 - *Pterostichus punctulatus*, 4 - *Pterostichus versicolor*, 5 - *Ophonus rufipes*, 6 - *Ophonus calceatus*,   7 - *Carabus cancellatus*,   8 - *Calosoma investigator*, 9 - Other species, 10 - Average number of beetles
collected during 10 days.

on the example of *Carabidae* (fig.1) which are represented by 64 species
(Aleinikova  and  Utrobina, 1969)  in the fields of the Stolbishchensky
State Farm.   Thus, in field-protecting forest plantations, the relative
importance  of *Ophonus rufipes* Dej.  increases sharply,  whereas in the
most  contrasting  agrobiocenoses,  i.e.,  in the fallow  and  perennial
grass, the dominant species (in maize and wheat) do not constitute more
than 2 - 3 % of the carabid fauna;  in the fallow  this is *O. calceatus*
Duft, in grassland *Carabus cancellatus* L. and *Calosoma investigator* Ill.
     The effect of plant cover  is most striking  in the case of *Micro-
arthropoda* and *Acarina,* in particular, when comparing  their population

density and group composition in four-crop rotations after two rotations, and, respectively, in eight-year-old monocultures and an eight-
year-old continuous fallow (fig. 2).

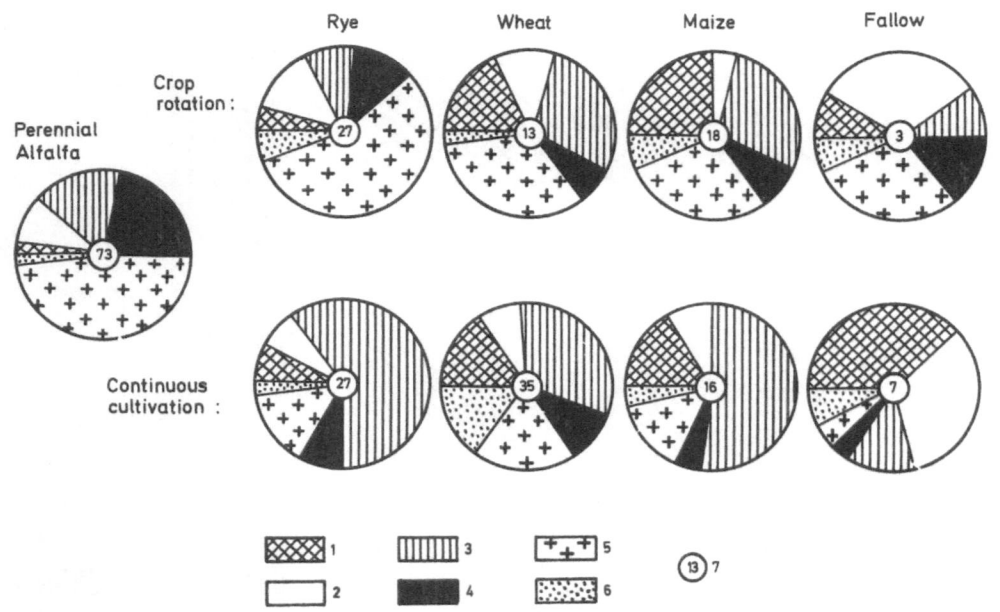

Fig.2. Population density and group ratio of *Acarina* (in
%) in crop rotations under continuously cultivated cultures and in continuous fallow:
1 - *Endeostigmata*,    2 - *Prostigmata*,    3 - *Tarsonemina*,
4 - *Gamasoidea*,    5 - *Oribatei*,    6 - *Acaroides*,    7 - in
thousands per 1 m$^2$.

The method of plant cultivation has no appreciable effect on the
total number of acarofauna, at the same time it affects considerably
the ratio of its individual groups. The following general regularity is
noticeable. *Oribatei* have a greater dominance under all the crops and
in the fallow of crop rotations, whereas *Trombidiformes* clearly domi-
nate under monocultures, in particular, the cohorts of *Tarsonemina* (of
the *Pyemotidae* family). Under individual cultures various *Acarina* groups
predominate. Thus, under rye and maize *Tarsonemina* constitute 50 - 75 %,
whereas under wheat theit dominance is reduced to 35 % and in the continuous fallow they represent only 10 % of the acarofauna. At the same
time under wheat versus the other monocultures, *Gamasoidea* and *Acaro-
idea* become more numerous.
    This phenomenon is caused by different soil regimes under dif-

ferent cultures, by the accumulation  of root fragments  and after-reap
plant remnants  and the specific effect of lizosphere which selectively
accumulates  soil bacteria  and  *Protozoa*.  All these factors  act upon
*Acarina*.  Under extreme conditions  of long-term continuous fallow  the
*Oribatei* numbers decrease considerably while *Endeostigmata* with preference for well-aerated soils  become disproportionately numerous (40 %).

Depending on the culture, also the variety  of species  in the individual *Acarina* groups  and  composition  of dominants change,  but to
a lesser extent  than  their group composition.  Thus, in monocultures,
as compared with crop rotations, *Oribatei* species  become somewhat less
varied, the dominance of certain species tends to increase: e.g., under
rye and wheat, *Oppia minus* Paoli, in the continuous fallow *Tectocepheus
velatus* Mich.  which constitutes 68 %  of *Oribatei*.  Among *Tarsonemina*,
the dominance of *Pygmephorus gracilis* Krczal under wheat and *P. sellnicki* Krczal  under maize monoculture  and  in the continuous fallow tends
to increase.

Depending on the plant cover, the dominance of the individual *Collembola* species also changes, thus *Collembola* of the *Onychiurus* species
have the highest dominance index  under  perennial grass, the lowest in
the continuous fallow.

It follows  from  our investigations  that the plant cover affects
greatly the structure of animal population  within one region  and soil
type.  Hence,  by changing farm crop composition,  man can control soil
animals and soil fertility.

R e f e r e n c e s

A l e i n i k o v a , M. M. (1965): Die Bodenfauna des Mittleren Wolga-
landes und ihre regionalen Besonderheiten. Pedobiologia 5, 17-49.
A l e i n i k o v a ,  M. M. (1968):  The regularities of the landscape
distribution  of soil invertebrates  in  the  Middle Volga Region.
Zool. journ., v. XLII, No. 7, 1022-1034 (in Russian).
A l e i n i k o v a ,  M. M. and  U t r o b i n a ,  N. M. (1969): Soil
animal population  in  the agrobiogeocenoses  of  the Middle Volga
Region.  In: Soil animal population of agrobiogeocenoses  and  its
changes  under  the influence  of  agriculture,  Kazan, University
Press, Kazan, 3-61 (in Russian).
B a l o g h ,  J. (1958): Lebensgemeinschaften der Landtiere.  Akademie
Verlag, Berlin, 560 S.
O'C o n n o r ,  F. B.  (1955):  Extraction  of  enchytraeid  worm from
a coniferous forest soil. Nature, Vol. 175, N 4462, 815-816.

# The deterioration of Fagus sylvatica stands and the development of ectophytic mycorrhizas

V. K. MEJSTŘÍK

**Czechoslovak Academy of Sciences,
Institute of Landscape Ecology,
Průhonice near Prague**

INTRODUCTION

Studies on mycorrhizas in the world have been in progress over half a century. During the last twenty years, research efforts have become particularly intense.

Studies on mycorrhizal ecology are still scarce. Some publications by Dominik and Boullard (1957, 1958, 1959, 1961), Singer (1960, 1965), Mejstřík (1969, 1970) and Selivanov (1968, 1969, 1970) have shown that it is very important to study plant- and soil fungi associations together.

Mejstřík (1969) pointed out the high homogeneity of the mycorrhizal spectrum in old stands (virgin forests) of *Fagus sylvatica* and its high diversity in stands deteriorated through man-made activities. In this paper I should like to present some results of my research concerning the dependency of mycorrhizal frequency, mycorrhizal intensity and the productivity of short mycorrhizal roots on the extent of stand deterioration.

MATERIALS AND METHODS

Four *Fagus sylvatica* stands of different quality were selected for a one year investigation of mycorrhizal frequency and short mycorrhizal root production. All these stands are located in Northeast Bohemia in the Broumov district at the Broumovské stěny (Broumov Rocks) and in the Javoří hory (Javoří Mts.). Site conditions are described in Table 1. Climatic conditions are favourable for the growth of beech (Fig. 1).

An old beech forest (age 80 - 120 years) growing on shallow sandy soil. The soil surface was covered with litter, debris and moor. Soil

Table 1.  Stand conditions

| Stand number | Elevation (m) | Aspect | Slope gradient (°) | Soil texture class | Parent rock | Age of trees | Site deteriorization |
|---|---|---|---|---|---|---|---|
| 1. Broumov Rocks | 668 | - | - | loamy sand | sandstone | 100 | medium |
| 2. Broumov Rocks | 680 | SW | 20 | loamy sand | sandstone | 100 | light |
| 3. Broumov Rocks | 580 | NE | 25 | loamy sand | sandstone | 120 | none |
| 4. Javoří Mts. | 500 | E | 10 | loamy sand | porphyry | 150 | high |

was acid, but still favourable for good growth of beech (Table 2). The best stand of *Fagus sylvatica* was stand No.3, which was not influenced by unfavourable man-made activities. The second stand, No.2, with few trees of *Picea abies* and with good microclimatic conditions was still very good for beech growth. Stand No.1 was slightly deteriorated by the planting of *Picea abies* and by recreation utilization. Stand No.4 was highly deteriorated; it was situated on the edge of a wood with many weed plant species and a high content of nitrogen (Table 3).

These four stands were not very rich in plant species (Table 4). The stands belong to the association *Abieti-Fagetum hercynicum*.

Samples of roots were taken in June and September from a depth of 0 - 15 cm from all stands; on stand No.2, other samples were taken from a depth of 0 - 4 cm, 4 - 8 cm, 8 - 15 cm, 15 - 20 cm, 20 - 25 cm and 25 - 30 cm for investigation of the vertical distribution of short mycorrhizal roots in the soil profile. Five beech trees of about the same age were selected in each stand. Ten samples were taken from a radius of 3 m diameter around each beech tree. A special corer (Marks et al. 1967) was used to remove soil samples, each measuring 100 ccm. All ten samples were homogenised; all stones and large roots were removed and subsamples of a fixed volume (100 ccm) were used for extraction. The extraction of short mycorrhizal roots, the counting and calculation

of short root productivity  were performed with the method described by
Marks et al. (1967).  For calculating the surface area  of short mycorrhizal roots  the average short mycorrhizal roots, 1.3 mm  in  diameter
and 3 mm in length was estimated. For the classification of mycorrhizal
subtypes we used Dominik´s classification (Dominik, 1969).

RESULTS AND DISCUSSION

This paper  sets out to provide a comprehensive description of subtypes
of  ectophytic mycorrhizas,  their ecology,  distribution  in  the soil
profile  and productivity  and distribution  of short mycorrhizal roots
in three different stands  of *Fagus sylvatica,*  each of them  at a different stage of deterioration.

T a b l e  2.    Chemical analysis of the soil

| Stand number | Depth cm | pH/KCl | Available nutrients mg/1g of soil + | | | Note |
|---|---|---|---|---|---|---|
| | | | $P_2O_5$ | $K_2O$ | MgO g | |
| 1 | 0 - 15 | 3.0 | 14 | 96 | 71 | + in 1 % citric acid |
| | 15 - 30 | 3.1 | 5 | 48 | 76 | |
| 2 | 0 - 15 | 2.9 | 16 | 170 | 90 | |
| | 15 - 30 | 3.0 | 4 | 48 | 61 | |
| 3 | 0 - 15 | 3.0 | 9 | 48 | 85 | |
| | 15 - 30 | 3.8 | 3 | 21 | 71 | |
| 4 | 0 - 15 | 3.0 | 27 | 130 | 110 | |

Fig. 2  illustrates the humus horizon and upper soil layer starti-
graphy, which is important for the growth  and development of short mycorrhizal roots and mycorrhizal associations. The distribution of short
mycorrhizal roots  in the soil profile  may provide information on conditions and processes occurring in different depths of the soil. Fig. 3
shows  the general trend  in root distribution,  namely the decrease in
the number of short mycorrhizal roots  with  increasing depth  of soil.
The maximum of short mycorrhizal roots  was found  in the F layer, between 4 - 6 cm  under the litter layer (L) (Fig. 3), where the number of
dead short mycorrhizal roots was minimal. There was a decreasing number
of short mycorrhizal roots in the H layer; this increased again in layer A;  in the deeper soil layers  the number of short mycorrhizal roots

Broumov  (410 m a. s. l.)    7,5 °C      685 mm
1901 — 1950

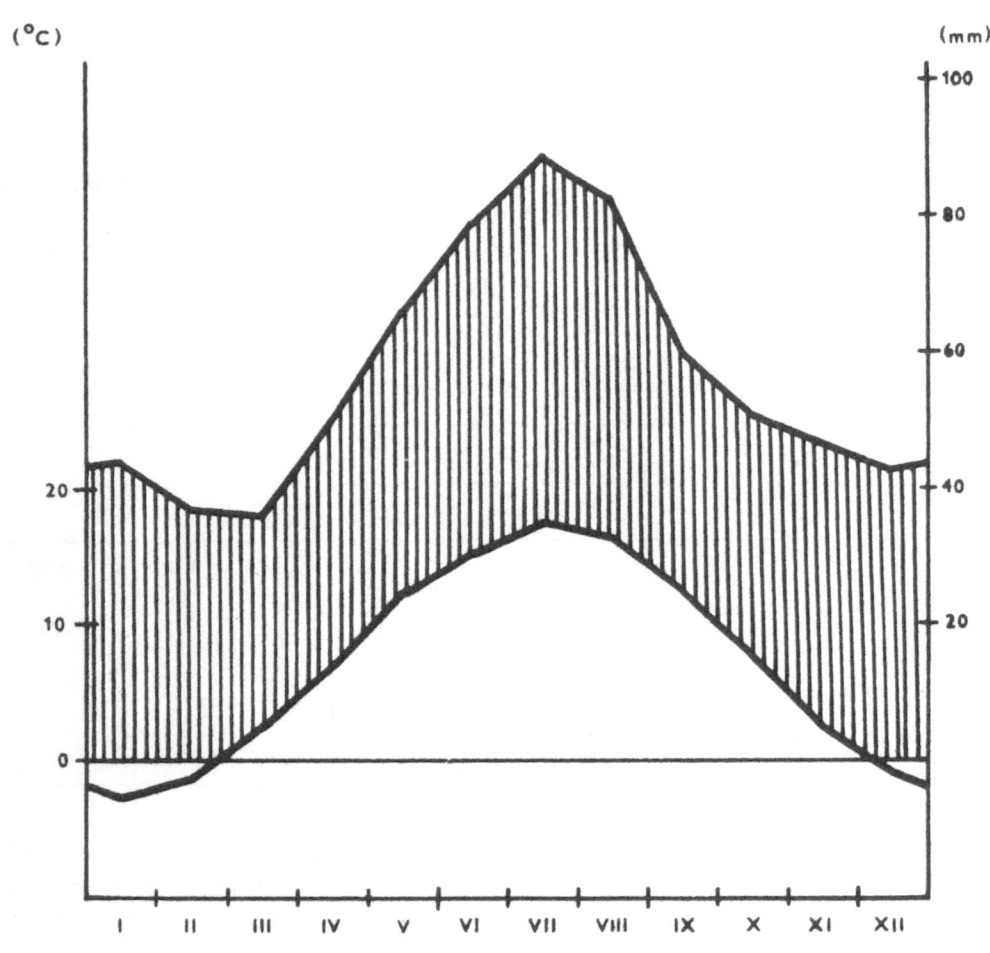

Fig.1. Climadiagram.

was sharply decreasing.  The soil layer  was investigated  to the depth
of 30 cm only; the highest number of short mycorrhizal roots  was found
in a depth of 16 cm.  Lyr (1961)  described  mycorrhizal roots of *Fagus
sylvatica* from the depth of 260 cm.  The observed distribution of short
mycorrhizal roots in humus horizons is influenced by favourable and unfavourable conditions  in these layers.  According to Lyr´s experiments

T a b l e   3.    Physical analysis of the soil

| Stand | Depth | Mineral particles size in mm | | | | | Cation exchange capacity values | | | | | Note |
|---|---|---|---|---|---|---|---|---|---|---|---|---|
| | | 0.25 % | 0.25-0.05 % | 0.05-0.01 % | 0.01-0.001 % | 0.001 % | H mequiv | S mequiv | T mequiv | V % | C ox. % | |
| 1 | 0-15 | 53.7 | 6.4 | 16.5 | 14.3 | 9.1 | 42.0 | 25.4 | 67.4 | 36.2 | 19.3 | S=T-H. V=S/T |
| | 15-30 | 73.8 | 7.9 | 7.4 | 7.0 | 3.9 | 21.0 | 7.6 | 28.6 | 26.6 | 7.6 | |
| 2 | 0-15 | 48.2 | 9.1 | 14.1 | 16.5 | 12.1 | 61.5 | 41.5 | 103.0 | 40.3 | 16.9 | |
| | 15-30 | 66.8 | 5.6 | 5.8 | 16.1 | 6.7 | 21.5 | 14.5 | 36.0 | 40.3 | 10.0 | |
| 3 | 0-15 | 66.8 | 9.4 | 11.7 | 3.7 | 8.4 | 2.5 | 22.5 | 25.0 | 90.0 | 9.4 | |
| | 15-30 | 60.9 | 11.7 | 16.5 | 8.8 | 2.1 | 6.0 | 3.3 | 9.3 | 35.5 | 1.2 | |
| 4 | 0-15 | 15.3 | 11.0 | 40.6 | 18.7 | 14.4 | 56.0 | 44.0 | 100.0 | 44.0 | 15.0 | |

T a b l e   4.    List of plant species

| Stand number | 1 | 2 | 3 | 4 |
|---|---|---|---|---|
| Species<br>Layer - $E_3$ | Presence+ | | | |
| *Abies alba* Miller | - | - | - | + |
| *Acer pseudoplatanus* L. | - | - | + | - |
| *Fagus sylvatica* L. | + | +  . | + | + |
| *Picea abies* (L.) Karsten | + | + | + | + |
| *Sorbus aucuparia* L. | + | - | - | - |
| Layer - $E_2$ | | | | |
| *Betula pendula* Roth. | + | - | - | + |
| *Fagus sylvatica* L. | + | - | + | - |
| *Lonicera xylosteum* L. | - | - | + | - |
| *Picea abies* (L.) Karsten | - | + | + | - |
| *Sambucus racemosa* L. | - | - | - | + |
| *Sorbus aucuparia* L. | + | + | + | + |
| Layer - $E_1$ | | | | |
| *Acer pseudoplatanus* L. juveniles | - | - | + | - |
| *Actea spicata* L. | - | - | + | - |
| *Angelica silvestris* L. | - | - | - | + |
| *Anemone nemorosa* L. | - | - | - | + |
| *Calamagrostis arundinacea* (L.) Roth. | - | - | - | + |
| *Calamagrostis villosa* (Chaix.) J.F.Gmel | + | - | - | - |
| *Carex sylvatica* Huds. | - | - | - | + |

(1961)  no correlation  was observed  between mycorrhizal frequency and
chemical properties of soil composition  with increasing depth,  a more
important factor  might be the content  of special growth factors, content of $O_2$ and $CO_2$ there.

According to earlier studies, ectophytic mycorrhizas  develop better in species growing in poor forest soil than in rich soil. Later, it
was maintained that  ectophytic mycorrhizas  may well develop  in poor
sandy soils  and  that a high concentration  of nutrients  in the soil,
especially nitrogen, is the factor  that reduces development  of mycorrhizas.  This indicates that the host plant possesses an effective pro-
tective mechanism  which  keeps the fungal symbiont  under control.  If
nutritional  conditions  change, this applies, particularly,  to certain
differences in nitrogen, the fungus gains entrance  into  the host tissues, because the host plant  is unable to uptake a sufficient quantity

Table 4 continued

| Stand number | 1 | 2 | 3 | 4 |
|---|---|---|---|---|
| Species Layer - $E_1$ | | Presence+ | | |
| Deschampsia flexuosa (L.) Trin. | + | + | - | + |
| Dryopteris filix-mas (L.) H.W.Schott | - | - | - | + |
| Dryopteris dilatata (Hoffm.) A.Gray | + | + | + | - |
| Fagus sylvatica L. juveniles | - | - | + | + |
| Festuca altissima All. | - | - | + | - |
| Gymnocarpium dryopteris (L.) Newm. | - | - | + | - |
| Chamaenerion angustifolium (L.) Scop. | - | - | - | + |
| Lamium galeobdolon ssp. montanum (Pers.) Hayek | - | - | + | + |
| Maianthemum bifolium (L.) F.W.Schmidt | + | - | - | + |
| Mercurialis perennis L. | - | - | + | - |
| Mycelis muralis (L.) Dum. | - | - | + | + |
| Moehringia trinervia (L.) Clairv. | - | - | + | + |
| Oxalis acetosella L. | + | + | + | + |
| Poa nemoralis L. | - | - | - | + |
| Phegopteris connectilis (Michx.) Watt | - | - | + | - |
| Rubus idaeus L. | - | - | + | + |
| Sorbus aucuparia L. juveniles | - | - | - | + |
| Senecio nemorensis subsp. fuchsii (Gmel.) Celak. | - | - | + | + |
| Vaccinium myrtillus L. | + | + | + | - |
| Viola raichenbachiana Jord. ex Bor. | - | - | - | + |

Nomenclature according to Rothmaler, 1966.

+ present    - missing

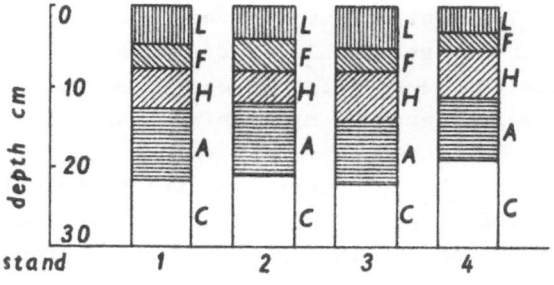

Fig.2. Profile of humus layers.

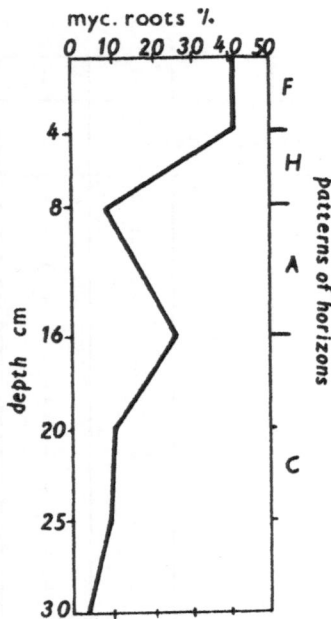

Fig.3.    Distribution of short mycorrhizal roots   in soil
profile, stand No. 2.

of nutrients without the help of the fungus.   Whereas   under conditions
of a high content   of available nutrients the roots    are able to absorb
enough nutrients   for the plant   without requiring the presence   of the
symbiont fungus.   But, as Slankis (1967)   pointed out, nothing is known
about the metabolism   by which nitrogen concentration   in the roots in-
terferes   with the stability   of the specific   physiological   state   in
ectophytic mycorrhiza.  With better management   in forestry   and   agriculture, with intensive use   of fertilizers, higher   and higher quantities of nutrients enter the soil.   The equilibrium   between mycorrhizal
association   and   host plants   in these artificially enriched soils   is
destroyed.   We can easily recognize the degree of soil deterioration of
these stands   by using the relative frequency of mycorrhizas   as a bio-
indicator of soil conditions.

        The most frequent subtype   of endophytic mycorrhiza   was F   in all
sites (Fig. 4).   In a not deteriorated stand (No. 3)   the relative frequency   of   subtype F   reached   55 %,   in   a lightly deteriorated stand
(No. 2) only 45 %, in a medium deteriorated stand   (No. 1) 40 %, and in
a heavily deteriorated stand (No. 4) only 35 %. In stand No. 3 we found
four mycorrhizal subtypes: B, D, F and G; in stand No. 2 four subtypes:
B, D, F and G; in stand No. 1  7 subtypes:  A, B, D, E, F, G and H, and

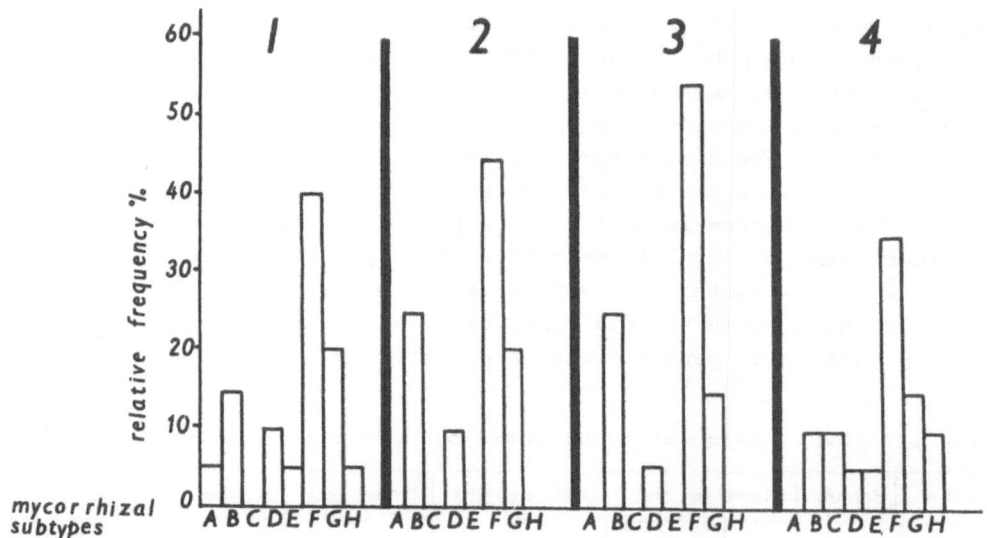

Fig.4.   Relative   frequency   of mycorrhizal subtypes   in different stands.

in the heavily deteriorated stand No. 4   seven subtypes: B, C, D, E, F, G and H.   A large diversity in the mycorrhizal spectrum occurred in the deteriorated stands No. 1 and 4.

The following eight subtypes  of endophytic mycorrhizas  were distinguished on *Fagus sylvatica* in native stands in the district of Broumov: A, B, C, D, E, F, G and H.  In our article  (Mejstřík and Dominik, 1969) we described subtypes  of ectophytic mycorrhizas  from  different stands of *F. sylvatica* from Czechoslovakia in comparison with the presence of mycorrhizal subtypes in Poland and France. The F subtype is dominant  in native stands of beech  in Czechoslovakia.  The intensity of mycorrhizal infection depends upon conditions of the habitat, but beech has a particularly wide ambit of tolerance of soil conditions, and this is the main reason why variation  in the development of the mycorrhizal intensity may be observed. The differences in the composition of mycorrhizal subtypes in different stands  or under different conditions  are developed in different associations of symbiotic fungi.  The variations in the mycoflora in different soil horizons, especially in layer H, and under changing soil moisture conditions  are quite large and quick. And competition between different fungal species  is the reason  for  variability in the mycorrhizal spectrum. Mycorrhizal fungi have a restricted existence except in associations  with their host growing  under natural conditions.  The mycorrhizal state (Harley, 1969) may be the only normal state  of both hosts  and  fungi under natural conditions, i.e.,

t'ey may both be ecologically obligate symbionts. With changes of soil
conditions in the habitat due to man-made activities, the natural equilibrium between hosts and fungi is destroyed; consequently, the mycorrhizal spectrum is wider and new subtypes might occur. The period for
establishing a new equilibrum may be either long or relatively short
dependent on the extent of changes in the environment. However, there
is room for much experimental work on this subject.

There were no significant differences in the number of short mycorrhizal roots during different seasons of the year (Table 5). Parkinson and Balasooria (1969) in their study on seasonal variation in
fungal populations reported that the degree of fungal colonization of

T a b l e  5.  The quantity of short mycorrhizal roots in samples

| Stand number | Date | Number sht. myc. roots | Wet weight g | Oven dry weight g | Number dead myc. roots | % dead myc. roots |
|---|---|---|---|---|---|---|
| 1. | 1.6 | 1.573 | 0.4920 | 0.0872 | 280 | 17.7 |
|    | 10.10 | 1.719 | 0.5283 | 0.0891 | 210 | 12.2 |
| 2. | 1.6 | 2.576 | 0.8500 | 0.1122 | 386 | 14.9 |
|    | 10.10 | 3.253 | 1.2432 | 0.1565 | 315 | 9.6 |
| 3. | 1.6 | 3.934 | 1.5863 | 0.2200 | 412 | 10.5 |
|    | 10.10 | 4.605 | 2.7381 | 0.2516 | 370 | 8.0 |
| 4. | 1.6 | 515 | 0.1820 | 0.0309 | 163 | 31.6 |
|    | 10.10 | 528 | 0.1677 | 0.0316 | 131 | 24.8 |

soil microhabitats was governed more by soil moisture content than by
any seasonal factor.

Harley (1969) pointed out that a relative increase of the root-
surface area is exhibited by trees which become mycorrhizal as compared
with their uninfected counterparts. In our experiments we found large
differences in quantity of short mycorrhizal roots in different beech
stands. The results about biomass productivity of short mycorrhizal
roots in various stands at different stages of deterioration are presented in Table 6. There are extremely large differences in the quantity of short mycorrhizal roots; from these data we can conclude on development - biological succesion - or degradation of the ecosystem, and
this can be expressed in several ways (Bormann, Likens, 1967). To know
nutrient cycling in the ecosystem in order to manage it for a well-balanced utilization is very important for the existence of ecological
systems and life itself. Uptake and recycling of phosphorus is very im-

Table 6.    Biomass of short mycorrhizal roots in different stands

| Stand number | Date | Total number sht. myc. roots per 1 ha x $10^8$ | Wet weight sht. myc. roots kg/ha | Oven dry weight sht. myc. roots kg/ha | Active surface area sht. myc. roots $km^2$/ha |
|---|---|---|---|---|---|
| 1. | 1.6 | 235.95 | 7.380 | 1.308 | 26.8 |
|    | 10.10 | 257.85 | 7.924 | 1.359 | 29.3 |
| 2. | 1.6 | 386.40 | 12.750 | 1.683 | 44.0 |
|    | 10.10 | 487.95 | 18.641 | 2.344 | 58.6 |
| 3. | 1.6 | 590.10 | 23.548 | 3.300 | 67.2 |
|    | 10.10 | 690.75 | 31.071 | 3.774 | 78.7 |
| 4 | 1.6 | 77.25 | 2.740 | 444 | 8.8 |
|    | 10.10 | 79.20 | 2.515 | 477 | 9.0 |

portant for trees with ectophytic mycorrhizas. There are large differences in the uptake and translocation of phosphorus by different species of mycorrhizal fungi (Mejstřík, Krause, 1973) and by different subtypes of ectophytic mycorrhizas (Mejstřík, 1970). This information is very valuable for recycling of phosphorus. In a well-balanced and not deteriorated stand the volume of active short mycorrhizal roots with only few different mycorrhizal subtypes is high and so is the occurrence of subtypes with a large ability for phosphorus uptake. In deteriorated stands the quantity of active short mycorrhizal roots is smaller, the mycorrhizal spectrum is wide and there is a large competition between symbiotic fungal species. The influence of man-made activities on the whole ecosystem is strong. In a balanced ecosystem nutrient recycling is optimal.

Large differences in production of biomass short mycorrhizal roots in stands with different values of deterioration (Table 6) demonstrated that changes in mycorrhizal quantity and in quality of mycorrhizal spectrum can be used as sensitive bioindicators of the value of soil deterioration.

It should be noted that in heavily deteriorated stands (No.4) ectoendophytous mycorrhizas increased significantly. The fraction of moribund short mycorrhizal roots was higher in deteriorated stands.

Acknowledgements

I should like to express my appreciation to Prof. E. Hadač corresponding member of the Czechoslovak Academy of Sciences for his interest and
discussion,   and   to Dr. H. Nekvasilová   and   M. Dvorská for their help
with the root samples.

References

B o r m a n n ,  F. H.,  L i k e n s ,  G. E. (1967):   Nutrient cycling.
    Science, 155, 424-429.
B o u l l a r d ,  B.,  D o m i n i k ,   T. (1958): Badania porównawcze
    nad mikoryzami  Pinus strobus L.  z róznych stanowisk  we Franciji
    i w Polsce. Prace Inst. Badaw. Lesn., Nr. 177-182, 45-85.
B o u l l a r d ,  B.,  D o m i n i k ,   T. (1959): Recherches compara-
    tives entre le mycotrophisme  du Fagetum carpaticum  de Babia Gora
    et celui d'autres Fageta precedemment etudies.  Zesz. Nauk. Wyszej
    Szkoly Roln., Szczecin, 3: 1-20.
D o m i n i k ,   T. (1957): Badania mykotrofizmu zespolow buka nad Bal-
    tykiem. Ekol. Polska, Ser. A, Nr. 7, 213-256.
D o m i n i k ,  T.,  B o u l l a r d ,  B.  (1961):  Les  associations
    mycorrhoziennes  dans  les hetraies Francaises. I. Recherches pre-
    liminaries, Prace Inst. Badaw, Lesn., Nr. 207, 3-30.
D o m i n i k ,   T. (1969): Key to ectotrophic mycorrhizae.  Folia For.
    Polon., Ser. A, 15, 309-321.
H a r l e y ,   J. L. (1969):   The Biology  of Mycorrhiza.  Leonard Hill
    (Books), London, 334 pp.
L y r ,  H. (1961):  Über  die Abnahme  der Mykorrhiza - und Knöllchen-
    frequenz mit zunehmender Bodentiefe.  In: Mykorrhiza Internationa-
    les Mykorrhizasymposium, Weimar 1960, 303-313, VEB G. Fischer Ver-
    lag, Jena.
M a r k s ,  G. G.,  D i t c h b u r n e ,  N.,  F o s t e r ,  R. C.
    (1967):  A technique  for  making quantitative estimates of Mycor-
    rhiza populations in Pinus radiata forests  and its application in
    Victoria. XIV. IUFRO Kongress, Sect. 24, 67-83, München.
M e j s t ř i k ,  V.,  B e n e c k e ,  U, (1969): The ectotrophic my-
    corrhizas of Alnus viridis (Chaix) D.C.  and their significance in
    respect to phosphorus uptake. New Phytol. 68, 141-150.
M e j s t ř i k ,  V.,  D o m i n i k ,  T. (1969): The ecological dis-
    tribution of the mycorrhizas of the beech. New Phytol.68, 689-698.

M e j s t ř i k ,   V. (1970): The uptake   of $^{32}$P  by different kinds of
ectotrophic mycorrhiza of *Pinus*. New Phytol. 69, 295-298.

M e j s t ř i k ,   V.,   K r a u s e ,   H. H. (1973): Uptake  of $^{32}$P  by
*Pinus radiata* roots inoculated with *Suillus luteus*  and *Cenococcum
graniforme* from different sources of available phosphate. New Phytol. 72, 137-140.

P a r k i n s o n ,   D.,   B a l a s o o r i j a ,   I.   (1969):  Studies
on fungi  in pinewood soil IV.  Seasonal and spatial variations in
the fungal populations. Rev. Écol. Biol. Sol. 4, 147-153.

S e l i v a n o v ,   J. A.,   E l u s e n o v a ,   N. G.,   L u r a n ,
A.V. (1968): Materiali k charakteristike mikosimbiotrofnych svazej
v nekotorych fitocenosach Shaz-Kzumov.  Utsch. Zap. Permskogo gos.
pedagog. instituta 64, 326-332.

S e l i v a n o v ,   J. A.,   S k a r a b a ,   E. M. (1969): Mikosymbio-
trofnye svazy v elovych lesach Severo-Zapadnovo Predurala.  Utsch.
Zap. Perm. Gos. pedagog. instituta 68: 5-55.

S e l i v a n o v ,   J. A.,   U t e m o v a ,   L. D. (1970):  Charakteristika mikotrofnosti zlakov  v zavisimosti  ot ich  biomorfologi-
tsheskych ossobenostej i uslovij proizrostanija. Utsch. Zap. Perm.
gos. pedagog. instituta 80, 65-75.

S l a n k i s ,  V.,   L i s t e r ,  G. R.,   K r o t k o v . K.,   N e l -
s o n ,   C. D.: Physiology of *Pinus strobus* L.  seedlings  grown
under high or low soil moisture conditions, Ann. Bot. 31, 121-132.

Discussion

K. P u g h :   Was there any correlation  between   the mycorrhizal subtypes and higher *Basidiomycetes?*

V. M e j s t ř i k :   Yes, different species  of *Basidiomycetes create*
different subtypes of ectophylous mycorrhizae  with a different ability
of nutrient uptake.

K. H. D o m s c h :   Would you please give a brief characterisation of
mycorrhizal subtypes?

V. M e j s t ř i k :  The subtypes  were described  according to anato-
mical and morphological structure,  structure of fungal mantel (see Dominik, 1969).

J. S a t c h e l l :  Would you please  explain  the way  in  which the quality of the four sites was defined and tell us what effect you think the subtypes of mycorrhiza have on the growth of *Fagus?*

V. M e j s t ř í k :  The quality of the four localities  was evaluated by using chemical, physical  and  vegetation analyses.  There are  different influence by different fungal species on the uptake of nutrients especially of *Fagus.*

# Copper contamination effects on earthworms by disposal of pig waste in pastures

J. A. VAN RHEE
Research Institute for Nature Management,
Arnhem

## INTRODUCTION

The addition  of large amounts  of copper salts  to pig rations  to increase weight gain  in the fattening pig  has led  to the production of
excreta with high copper concentrations. Regulations governing the discharge of, among others, agricultural effluents  into watercourses, the
high cost of transporting pig waste slurries to waste treatment plants,
and the difficulty involved in the treatment of this material  have resulted in the application of pig waste on adjacent pastures.

Copper  can have  toxic effects  on  earthworm populations.  It is
known that in orchard soils a copper level  of 85 ppm  is not tolerated
by worms  (van Rhee, 1963, 1969), and  in grassland high copper concentrations (260-360 ppm)  proved sufficiently toxic to eradicate worm populations almost entirely (Nielson, 1951).  Laboratory experiments provide confirmation of the view that this element checks the reproductive
capacity, albeit  at higher concentrations  than under field conditions
(van Rhee, 1969). Extermination of worms will inhibit the biodegradation
of organic matter.  In pastures such effects result in the formation of
a thick mat  of undecomposed dead plant  parts generally carrying loose
and poor sods.

A similar problem  arises  from  the disposal  of  harbour sludge.
Regular  dredging  of the Rotterdam harbour  is  necessary  to maintain
adequate depth especially  for the large ships  touching  at this port.
Because of technical difficulties  associated with the discharge of the
mud in the North Sea, an ideal disposal process  was found in land-fill
of areas in the vicinity of the harbour.  On such deposits a favourable
development of natural vegetation was observed.  However, it would also
be desirable  to investigate  the development  of shortlived organisms,
i.e. such decomposers  as  earthworms, which contribute  to a rapid recycling of mineral components in plant wastes back into the plants.  It

seemed possible that earthworms would be excluded by the influence of deleterious substances occurring in the harbour mud as a result of, for instance, the high level of industrial wastes in the Rhine water. In this case complete recycling would not take place, and disturbances would gradually appear in the system.

The following is a brief report of recent studies on both problems.


DISPOSAL OF PIG WASTE IN PASTURES

For the evaluation of the effects of the application of pig waste in pastures it was of primary interest to know what level copper accumulation had reached already in such soils and to study the worm populations.

The sites chosen were mainly on poor-class sandy soils and a few on peaty soils, all located in the Veluwe (Province of Gelderland, The Netherlands). From a large number of pastures, which received pig waste over a period of 10 years, samples of surface soil (0-5 cm) were analysed for their content of copper and organic matter. The organic matter content serves as an indication for the rate of decomposition of plant residues. Worm sampling was done in three places (0.25 m$^2$) per pasture, the worms being dug out and removed by hand.

As for the effect of copper accumulation on worm populations it should be mentioned that the populations of the pastures under study were very similar in species composition. *Lumbricus rubellus* was most abundant numerically, followed by *Allolobophora caliginosa*. *A. rosea* was found in very small numbers. Therefore, in Table 1 worm densities are expressed only in total numbers. It is seen in this Table 1 that not all of the pasture soils were comparable, the variations being due mainly to recent ploughing. But a toxic effect is nevertheless evident, because in pastures with high copper levels worm numbers were on average lowest. More evidence is expected from a planned comparison between pastures with and without application of pig waste.

In Fig. 1 a positive correlation between the copper and organic matter content is shown. This suggests a poor development of earthworm populations in pasture soils where high copper concentrations are reached, leading to inhibition of the biodegradation of organic matter. The low value of *r* is in correspondence with the variations in the pasture soils such as caused by ploughing, which must have interfered with the relationship.

## DISPOSAL OF HARBOUR MUD

To investigate the chances for penetrating earthworms to survive in deposits of harbour mud, worms were introduced to such areas. The inoculated worms, predominantly *A. caliginosa*, failed to establish. After a period of two months most of them were in very bad condition and some had even died. Breeding experiments with worms in soil samples takem from three deposits showed the same effects (see Table 2) Bad physical conditions is manifested in a decrease of body weight as well as complete disappearance of the clitellum.

Further research on the extent of contamination showed that the mud soils are polluted with large amounts of various rather toxic substances. To collect more data about toxic effects on worms, laboratory experiments were made in polder soil mixed with these substances up to concentrations corresponding with those found in the mud. Soil of the Oostelijk Flevoland polder is known to be very suitable for earthworms.

The results (Table 3) showed that oil only had no effect at all. Both DDT and NaCl checked cocoon production, and mixtures of these substances had even more noxious effects, as shown by a sharp decrease of body weight and complete disappearence of clitellum. With respect of NaCl, it should be kept in mind that high concentrations in soil are usually rapidly reduced as a result of the action of rain.

Table 1. Comparison between pastures with high and low copper contents

| Farm | Pastures with high copper content | | | Pastures with low copper content | | | Remarks |
|---|---|---|---|---|---|---|---|
| | org.matter (%) | copper content (.%) | worms per m² | org.matter (%) | copper content (%) | worms per m² | |
| 1 | 7.4 | 64.6 | 16 | 8.5 | 7.4 | 143 | |
| 2 | 12.3 | 36.0 | 10 | 6.3 | 7.1 | 176 | |
| 3 | 6.4 | 45.8 | 169 | 5.7 | 6.0 | 215 | |
| 4 | 12.2 | 31.4 | 138 | 9.6 | 11.8 | 182 | |
| 5 | 7.1 | 33.0 | 69 | 4.7 | 5.6 | 0 | low numbers because the top 10 cm layer was yellow sand as a result of recent ploughing |
| 6 | 12.1 | 34.1 | 153 | 4.3 | 8.2 | 42 | |

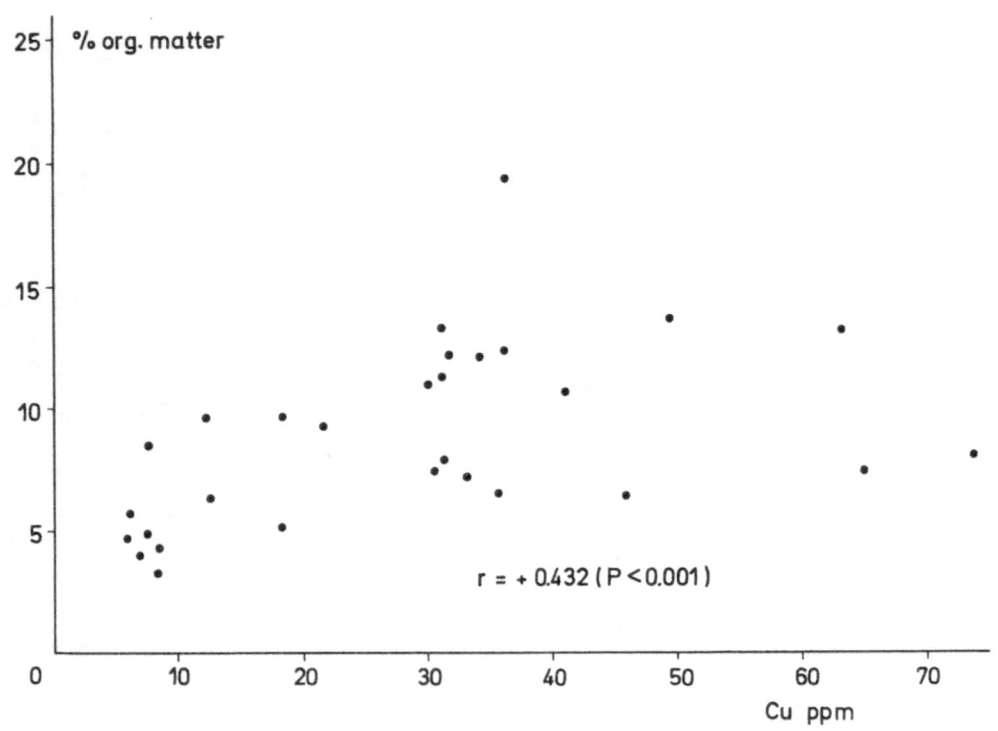

Fig.1. Organic matter content in relation to copper content.

T a b l e   2.    Breeding experiments  with  *Allolobophora caliginosa* in deposit soils (duration of experiment 13 weeks)

| Soil | Body weight (% initial weight) | Cocoons/worm week | Worm with clitellum (%) | Mortality (%) |
|---|---|---|---|---|
| Control (polder soil) | 100 | 0.5 | 100 | 0 |
| Deposit A | - | 0 | 0 | 77 |
| B | 57 | 0 | 0 | 0 |
| C | 80 | 0 | 0 | 0 |

Table 4 shows the effect of non-ferrous metals. A significant loss of body weight  was observed only in soils mixed with zinc and in those containing a mixture of metals.  Some of the animals died, and the sur- vivors lost their clitellum.  The effect of other metals was manifested in a reduction  of cocoon production,  and copper also reduced the via- bility of the cocoons.

T a b l e   3.    Breeding experiments  in polder soil   mixed   with   oil,
DDT, and NaCl (duration of experiment 4-7 weeks).  Polder soil containing negligible quantities of soil. DDT, and NaCl

| Treatment | Body weight (% initial weight) | Cocoons/worm week | Worms with clitellum (%) | Species (adult specimens) |
|---|---|---|---|---|
| nil | 137 | 0.75 | 65 | |
| Oil 500 mg/kg | 140 | 0.4 | 65 | *Allolobophora* |
| DDT 75 ppm | 140 | O | 30 | *longa* |
| NaCl 2.5 g/kg | 100 | O | 20 | |
| nil | 137 | 1.3 | 100 | |
| DDT+oil+NaCl | 53 | O | O | *Allolobophora* |
| Oil+NaCl | 66 | O | O | *caliginosa* |

T a b l e   4.    Breeding experiments   with   *Allolobophora caliginosa* in
polder soil   mixed   with   non-ferrous metals (duration of experiment $8\frac{1}{2}$
weeks). Polder soil containing in ppm:   0.05-0.2 Hg, 2 Cu, and 16-60 Zn

| Treatment (metals in ppm) | Body weight (% of initial weight) | Cocoons/worm week juveniles ( ) | Mortality (%) | Worms with clitellum (%) |
|---|---|---|---|---|
| nil | 154 | 1.95 (0.54) | O | 100 |
| 10Hg + 20Co | 127 | 0.68 (-) | O | 80 |
| 110 Cu | 170 | 1.43 (0.17) | O | 90 |
| 1100 Zn | 73 | O | 22 | O |
| Hg+Co+Cu+Zn | 52 | O | 22 | O |

CONCLUSION

The subjects  discussed form  only two aspects  of the problem  of soil
pollution.  Although the work  is not yet complete, it is somewhat disturbing to note that in both cases the contamination of soil is largely
irreversible.  Little more can be done at the moment  than point to the
existence of a particular hazard such as ligh levels of especially zinc
and copper where maintenance of worms becomes very difficult.  This can
have consequences for natural life in the soil, and disturbances of the
system are to be expected in due time.

R e f e r e n c e s

N i e l s o n , R. L. (1951): Effect of soil minerals  on earthworms, N.Z.J. Agric. 83, 433-435.

R h e e , J. A. van (1963): Earthworm  activities  and  the breakdown of organic matter  in agricultural soils.  In: Soil organisms, (J. Doeksen and J.van der Drift, eds). North-Holland Pub.Co.Amsterdam, 55-59.

R h e e , J. A. van (1969): Effects of biocides  and their residues on earthworms. Ned. Rijksfaculteit Landbouwwetenschappen Gent,  1969, XXXIV, (3), 682-689.

D i s k u s s i o n

C. A. E d w a r d s :  Might I ask if you investigated the influence of heavy metal contamination on pH in your first experiment?

J. A. van  R h e e :  I know that,  on mud deposits,  a pH  of 7.5  was found.

K. H. D o m s c h :  I like to point out, that data  on heavy metals of waste products are generally based on total content.  The amount available to organisms  is  certainly  much lower (1-2 orders of magnitude). The concentrations used in rearing experiments - using soluble salts of heavy metals - are therefore much more toxic and not directly comparable to field observations.

J. A. van  R h e e :  In my experiments salts  of heavy metals  (chloride) were added.  This means a large increase of the chloride content. First, these high concentrations of chloride have to be reduced to much lower concentrations.

J. S a t c h e l l :  There seemed to be a very wide variability in the relationships you showed us  between  organic matter, earthworm populations and copper content.  The relationship  is presumably very complex involving the microflora on which the earthworms feed. Would the microbiologists present like to comment  on microbial toxicity of heavy metals which have been mentioned?

J. A. Van  R h e e :  It is known  to the author  that the influence of copper - contaminated  pig waste  has been studied.  An increase  of COD was found as a result of high copper levels.

A. J. R e i n e c k e :  You  showed  an approximately direct  relation between organic matter content  and  copper content.  How is this to be explained?  If we can expect organic matter stratification can't we expect Cu stratification as well in a certain soil and a resulting vertical migration of earthworms as a result of this stratification.

J. A. van  R h e e :  No analyses  were made  for the copper content in different layers.  The 0-5 cm top layer only was analysed.  Earth worms were found in the 0-15 cm layer.

# Comparative analysis of faunal changes in soils of the south-east part of Aserbaidjan as result of man's economic activity

N. G. SAMEDOV AND L. A. BABABEKOVA

Institute of Zoology, Acad. Sci. USSR,
Baku

Along with soil conditions, an important part in changing the structure
and abundance of soil organisms and their role in agrobiocoenoses belongs to anthropogenic factors. In this connection, a definition of the
distribution patterns of soil insects and other invertebrates in different types of agrocoenoses as compared with natural ones is sure to
be of great interest for solving a number of practical and theoretical
problems. With this in view, in 1968 - 1972, we studied changes in the
structure of species and abundance of the soil mesofauna in the process
of developing of virgin lands for crop cultivation in brown, meadow-
brown, yellow-podsol gley and mountain-forest brown soils of the Lenkoran Region of Aserbaidjan; our studies were different from previous
studies by Samedov et al. (1970); Samedov (1970); Samedov, Bababekova
(1970); Samedov et al. (1972).

Each soil type was presented by a plot with natural vegetation and
genetically equal arable soil: tea plantation on irrigated yellow-podsol gley and rain-fed grain crops on brown, meadow brown and mountain-
forest-brown soils.

The soil fauna was recorded by the usual method of taking soil
samples in the spring, summer and autumn. It was established that the
abundance and structure of soil-inhabiting invertebrates have been
changed. Some species typical of virgin plots disappeared from cultivated plots or their abundance was heavily reduced. On the other hand,
certain species increased their abundance and some new species accompanying the cultivated crop appeared in cultivated soils (Table 1).

In yellow-podsol gley soil on virgin plots earthworms dominated
(59.8 %): *Dendrobaena byblica* Rosa, *Eophila asiatica* Mal. and *Allolobophora jassyensis*. On the tea plantation we found *Allolobophora caliginosa* Sav. f. *trapezoides* associated with arable lands and insects
prevailed (67.2 %); these were mainly *Carabidae (Ophonus, Harpalus)*,
and of the *Elateridae Agriotes sputator* L., *A. brevis* Cand. and *Draste-*

*rius bimaculatus* Rossi,  while on virgin plots  there  was  only *Athous* ,
*mingrelicus* Rtt. Of the mollusks we found *Parmacella ibera* Eichw. known
as a tea pest. *Diplopoda* disappeared mostly from arable soils, on virgin
plots we found 9 species of *Diplopoda*,  but of these only *Schizophillum
caspium* Lohm., apparently an eurytopic species in tea plantations.

T a b l e   1.   Population  density  of soil invertebrates  (mesofauna)
in different soil types (average density per m$^2$)

| Soil types | Coenoses | | | | | |
|---|---|---|---|---|---|---|
| | Natural | | Arable | | | |
| | | | Tea | | Grain | |
| | number of species | population density | number of species | population density | number of species | population density |
| Yellow-pod-sol gley soil | 49 | 109.3 | 20 | 18.0 | – | – |
| Brown soil | 30 | 84.4 | – | – | 22 | 30.3 |
| Meadow-brown soil | 43 | 50.0 | – | – | 21 | 14.7 |
| Mountain-forest brown soil | 47 | 48.8 | – | – | 13 | 20.2 |

     In rather dry-brown  and  meadow-brown soils insects dominated (50
and 36 % respectively).  In brown soils on arable plots we observed the
absence of *Tanyproctus ovatus* Motsch. and  *Rhizotrogus serrifunis* Mars.
*(Scarabaeidae)*  which usually inhabit virgin brown soils; also a reduction of population density of *Elateridae* and *Curculionidae* was observed.
*Agriotes sputator* L., *Melanotus fisciceps* Gyll., *Athous mingrelicus* Rtt.
were present on grain crop in brown  and  meadow-brown soils.  Contrary
to the virgin lands, *Synandromorphus etruscus* Quens., *Zabrus tenebri-
oides elongatus* Mén., *Ditomus obscurus* Dej. appeared  in  association
with grain crops.  Of the *Diplopoda*, also *Strongilosoma lencoranum*  and
some species of *Polydesmidae* were found.
     *Myriapoda* dominated in virgin plots in mountain-forest-brown soils
(47.7 %).  Out of 13 species of *Diplopoda*  inhabiting virgin lands, we
found also *Chromatoiulus fagorum* Att.,  *Chr. brachyurus brachyurus* Att.
*Leptoiulus tanymorphus* Att.  After soil cultivation, *Melolontha kraatzi*
Reitt *(Scarabaeidae)*  and  *Athous mingrelicus* Rtt., *A. rosinae, A. cir-
cumductus* Mén. *(Elateridae)* disappeared.  On the grain crop, we found
also *Agriotes brevis* Cand. of *Elateridae*  and  *Acupalpus meridianus* L.,
*Harpalus* sp. of *Carabidae*.

The activity of man influences the structure and character of distribution of the mesofauna in different soil types and plays a very important role in the regulation  of biological productivity  of agrobiocoenoses.

R e f e r e n c e s

S a m e d o v ,   N. G. (1970): Soil zoology  as  a biocoenotic  element
    of landscapes  in Aserbaidjan.  Ser. Biol. nauk 3, Baku, 53-60 (in
    Russian).
S a m e d o v ,   N. G.,  B a b a b e k o v a ,   L. A.,  R a s u l o v a ,
    Z. K. and  R a c h m a n o v ,  R. R. (1970):  Soil-inhabiting in-
    sects and their importance in wood biogeocoenoses  of the Lenkoran
    zone in Aserbaidjan. Inf. Mat. MBP 1969, Baku, 39-40 (in Russian).
S a m e d o v ,   N. G. and B a b a b e k o v a ,   L. A. (1970): On eco-
    logical groups  and  structure of insect population  and other in-
    vertebrates in some soil types.  Annotacii dokladov VI Sjezda VEO,
    Voronez, 159 (in Russian).
S a m e d o v ,   N. G.,  B a b a b e k o v a ,   L. A. and  R a c h m a-
    n o v ,  R. R. (1972): Distribution of *Myriapoda* in different soil
    types of Lenkoran.  Zool. zurn. 51, 8, Moskva, 1244 - 1247 (in Rus-
    sian).
S a m e d o v ,   N. G.,  B a b a b e k o v a ,   L. A. and  R a c h m a-
    n o v ,  R. R. (1972):  Complexes of soil inhabiting invertebrates
    in different soil types. Materialy Iu Vsesojuzn.sovescania po pro-
    blemam pocvennoi zoologii, Baku, 121-122 (in Russian).

# Soil microfauna changes under the influence of various fertilizers

T. I. ARTEMJEVA, F. G. GATILOVA

Soil Zoology Laboratory at the Kazan Institute
of Biology, USSR Academy of Sciences,
Kazan

The extensive use of chemical fertilizers requires a long-term forecast of their effect on all the components of agrobiocoenoses, in particular, on the soil layer inhabited by animals. This paper makes use of the data collected over many years by the Soil Zoology Laboratory of the Kazan Institute of Biology, USSR Academy of Sciences, together with the Chair of Agrochemistry, Kazan Agricultural Institute.

We have studied the influence of two systems of fertilizers: manure-mineral and mineral (NPK) on *Marcoarthropoda* under wheat and maize cultivated in four-plot rotations after one or two rotations, as well as in four- and eight-year-old maize monocultures and in a continuous fallow. Over a four-year period within the manure-mineral system, 36 tons of manure, 80 kg of N, 150 kg of $P_2O_5$, 100 kg of $K_2O$ were applied; of the mineral fertilizers, 260 kg of N, 249 kg of $P_2O_5$ and 316 kg of $K_2O$ were applied per hectare of primary nutrient. (In the mineral system, the total N, P, K content was equalized with that of the manure--mineral system).

Mineral fertilizers were applied in the form of ammonium sulphate, powdered superphosphate and potassium chloride. Quantitative estimations of *Microarthropoda* were made after the application of the fertilizers for three- and eight-year-periods.

Samples were taken in May, July and September. Fifteen 125 $cm^2$ samples were taken simultaneously from each variant of the experiment from four different depths, i.e., 0 - 5 cm, 5 - 10 cm, 10 - 15 cm and 15 - 20 cm.

*Microarthropoda* were extracted in cardboard funnels according to Balogh's method (1958) without electrical heating for seven days. *Microarthropoda* were differentiated with varying degrees of accuracy. *Collembola. Tarsonemina. Acaridiae, Oribatei* were determined down to the species, *Endeostigmata* to the family, *Gamasina* and *Prostigmata* to the group. *Collembola* were identified by E.F. Martynova and E.L. Solntzeva, *Tarsonemina, Acaridiae* by V.D. Sevost'janov.

The numbers of *Microarthropoda* on the plots ranged from 7,000 to 23,000 individuals per 1 m$^2$. The lowest numbers were recorded for an eight-year-old continuous fallow due to a prolonged absence of plant cover and frequent land cultivation. *Collembola* and *Oribatei* numbers in the fallow were one-half of those under maize and wheat. At the same time, the mites of the *Trombidiformes* group became disproportionately abundant. On the whole, mites were predominant in the experimental plots with a predominance of *Trombidiformes*, whereas approximately less than 50 % of the acrofauna is constituted by *Oribatei*. The proportion of *Collembola* among *Microarthropoda* is relatively small (5 - 10 %).

The fauna of all groups of *Microarthropoda* was similar at all the stages of the experiment and was represented by wide-spread species, characteristic of arable lands of the Middle Volga Region.

Thirty *Collembola* species have been found; among these *Onychiurus cancellatus* Gisin and *Ceratophysella succinea* Gisin predominated (34 - 55 % and 14.5 - 21.0 %, respectively), while under maize and in the fallow *Mesaphorura krauslaueri* Börn dominated alongside with the former (13.6 - 16.0 %). *Tarsonemina* (15 species in all) were represented by *Pyemotidae, Scutacaridae, Tarsonemina*. *Pyemotidae* mites were the most numerous (8 species) and *Pygmephorus sellnicki* Krozal and *P. gracilis* Krozal were the most abundant among them. Ten *Oribatei* species have been found; the most prominent in the fallow was *Tectocepheus velatus* Michael. followed by mites of the genus *Oppia* Grandj.; under maize and wheat, *Oppia* mites dominated. Of the *Endeostigmata*, the dominant mite families were *Nanorchestidae* Grandj. and *Lordalychidae* Grandj.

It has been established that the character of effect of a mineral fertilizer is determined by the plant cover and the specific reaction of the individual *Microarthropoda* groups and species.

In crop rotations under maize and wheat as well as under the mono-culture of wheat, the application of NPK resulted in a considerable in-crease of *Microarthropoda* numbers which occurred mainly in view of an increase of *Tarsonemina, Acaridae* and *Oribatei* after a four-year and eight-year application of the fertilizers (figs. 1, 3). It should be pointed out that the effect of NPK was most favourable in the autumn, which may be due to an accumulation of debris after harvesting provid-ing food for soil animals. Not all *Microarthropoda* groups reacted equal-ly to NPK. *Collembola* as well as *Prostigmata* and *Endeostigmata* changed insignificcantly in numbers in comparison with the control plot (figs. 2, 3c, d). In the continuous fallow, where no nutrients are extracted by plants microarthropod numbers tended to decrease under the influence of NPK and, in some cases, there was a definite decrease *(Tarsonemina, Oribatei, Prostigmata)*. *Collembola* in the fallow as well as under maize and wheat responded but weakly to the application of NPK.

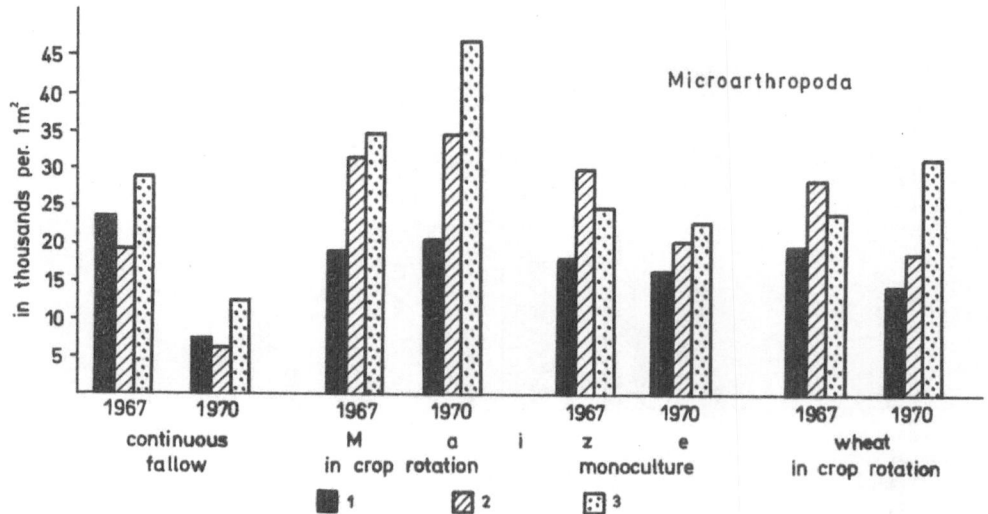

Fig.1. The effect of different fertilizers  on the total
numbers of *Microarthropoda*:  1 - control plot,  2 - NPK,
3 - NPK + mature.

The reaction of the individual invertebrate groups  was determined
by the reaction of the dominant species. Thus, the increase of *Tarsone-*
*mina* numbers under maize was determined by an increase in the abundance
of *P. sellnicki*  and  *P. gracilis*.  The decrease of *Oribatei* numbers in
the fallow was connected with a decrease in *T. velatus* and *Oppia* mites,
while under maize and wheat the numbers of *Oppia* mites  increased under
the influence of NPK.

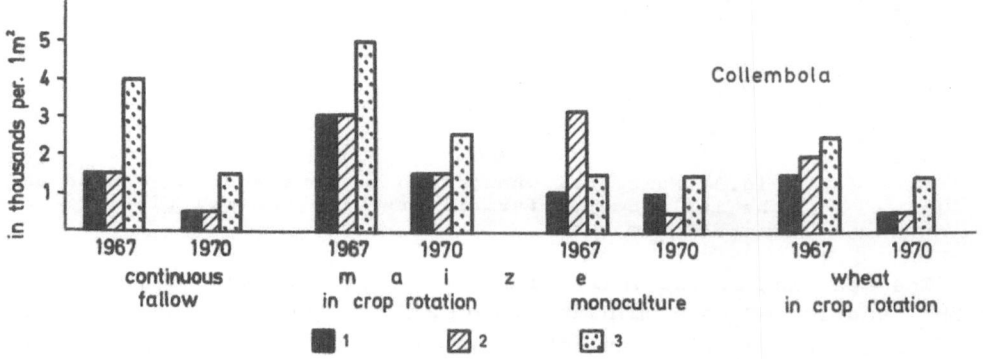

Fig.2. Numerical changes  in  *Collembola*  under  the influence  of  fertilizers: 1 - control plot, 2 - NPK, 3 -
NPK + manure.

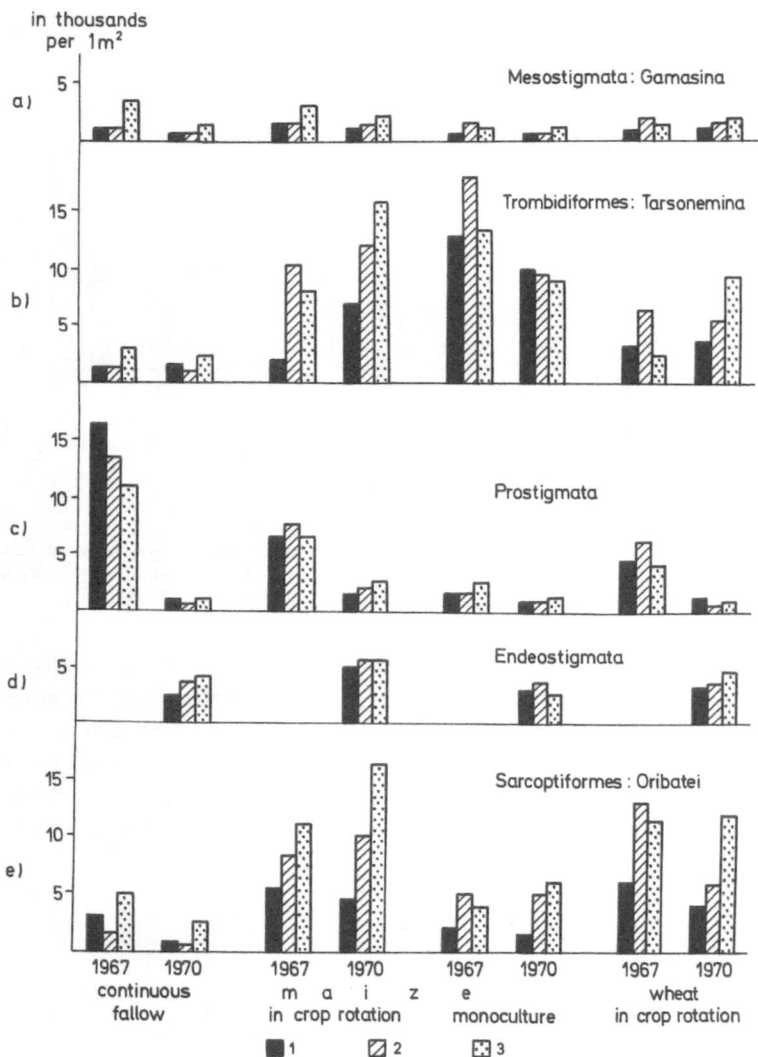

Fig.3. Numerical changes in various *Acarina* groups under
the influence of fertilizers: 1 - control plot, 2 - NPK,
3 - NPK + manure.

     The reaction of the individual invertebrate groups  was determined
by the reaction of the dominant species. Thus, the increase of *Tarsone-*
*mina* numbers under maize was determined by an increase in the abundance
of *P. sellnicki*  and  *P. gracilis*.  The decrease of *Oribatei* numbers in
the fallow was connected with a decrease in *T. velatus* and *Oppia* mites,
while under maize and wheat the numbers of *Oppia* mites  increased under
the influence of NPK.

The majority of investigators came to the conclusion that mineral fertilizers affect the soil fauna indirectly, by an increase of microflora and plant productivity. However, a direct negative effect of large doses of fertilizer on soil animals may also be possible. We believe that the reduction of the individual *Microarthropoda* groups in the fallow may also have been due to the negative effect of NPK.

A fairly certain increase in the numbers of most soil invertebrate groups occurred in the fallow, and under maize and wheat under the influence of manure and mineral fertilizers (figs. 1, 2, 3). It should be noted that after an 8-year-application of manure-mineral fertilizers the reaction was mostly stronger than after a 4-year-application. At the same time, *Endeostigmata* did, practically, not react to manure-mineral fertilizers (fig. 3d).

A favourable effect of manure and mineral fertilizers was brought about mainly by the presence of manure in these mixtures. It has been established that an increase in the various soil invertebrate groups occurred at different times. Our investigations have shown that the populations of *Tarsonemina* and *Acaridiae* increased after 10 months following the application of half-rotted manure, and the same is true of *Collembola* and *Oribatei* (after 12 months). This is due to an association of the individual invertebrate groups with certain stages of manure decomposition. An increase of typical soil species also confirmed the favourable effect of organo-mineral fertilizers. Thus, in the fallow and under maize, such collembolan species as *M. krausbaueri* and *O. cancellatus* increased considerably in abundance.

Agrochemical investigations indicated that manure-mineral fertilizers in the fallow and under maize and wheat increased the humus content more than mineral fertilizers. At the same time, in a continuous fallow, its certain decrease under the influence of NPK has been noted which is in agreement with the observed tendency of a decrease in *Microarthropoda* numbers as compared with the control plot.

Thus, quantitative changes in the total number and the number of individual *Microarthropoda* groups and species caused by various fertilizers reflect, to a certain extent, the changes occurring in the soil. These data can be used in the elaboration of fertilizer systems.

References

B a l o g h , J. (1958): Lebensgemeinschaften der Landtiere. Akademie-
Verlag, Berlin, 560 S.

Discussion

C. A. Edwards : Do you believe the increases in numbers of micro-
arthropods are due to direct effects  or to the influence  of the crop?

T. J. Artemjeva : Fertilizers influence *Microarthropoda*  predominantly in an indirect way, i.e., through the crop.

# The effect of the herbicides 2, 4 - D and Simazin on the coenosis of Collembola and Acari in arable soil

J. PRASSE

Pedological and Microbiological Institute
of the University,
Halle

INTRODUCTION

Although problems of the effect of 2,4-D and Simazin on the autochtho-
nal soil fauna have been solved in general, several questions of considerable soil-biological significance are still unclear. This requires
further investigations of a physiologically-biochemical and faunisti-
cally-ecological character. It might be advisable to choose very different sites and ecosystems of a very unstable character, i.e., high-
productive systems, for the arrangement of further field experiments.

In our field experiment started at Etzdorf/Saalkreis, G.D.R. in
1970 and conducted for the purpose of establishing the effect of perennial, continuous herbicide application on the ecosystem, we used six
plots for soil-biological analyses, each measuring 20 m x 10 m. Two of
the plots were treated constantly each year with a single dose of herbicide (variant S), two plots received a double dose p.a. (D), the remaining two plots were kept untreated and served as controls (variant
C). The type of herbicide to be applied depended on the specific plant
cover. Populations of *Collembola* and mites were analysed both as regards their quantity and quality, and separately for the individual
soil layers. The soil was loess-black earth, of a granular structure
down to a depth of 20 cm, composed of clay (22.6%), silt (21.2%) and
sand (6.2%). The average humus content was 2.9%, the mean annual rate
of precipitation 486 cm, the mean annual temperature 8.6 °C. The results
of our research work during the first two years are presented in the
text.

METHODS

On May 7 1971, 2,4-D sodium salt ("Spritzhormit") was applied to winter
wheat in a dose of 1.5 and 3 kg/ha respectively; on April 24 1972,
Simazin ("W 6658") was applied to maize in a dose of 4 and 8 kg/ha respectively. Four weeks preceding the application of the herbicides,
soils of all four plots were tested for the abundance of soil animals.
After the application of 2,4-D, soil samples were taken with a semi-
cylindrical, tubular borer on May 20, July 8, September 2, November 8
1971, and on April 10 1972 and, similarly, after the application of
Simazin, on May 5, June 16, August 4, September 21, November 7 1972 and
April 13 1973. We obtained 64 soil samples from each plot measuring
3 cm$^2$ across the surface and 10 cm in depth (in the first year of our
investigation), 20 cm in depth during the second year. Each sample was
divided into 2 and 4 partial samples respectively (depth 5 cm), and 8
partial samples of every soil layer were combined to obtain one average
sample. The animals were extracted in a Tullgren funnel within 4 days
with a gradual increase in temperature to a maximum of 26 °C. For a statistical evaluation of the material we used the parameter-free method,
the X-test by Van der Waerden (1957) for the independent samples. Differences confirmed statistically with the bilateral test and a probable
error of 5 and 1 & respectively, were marked + for feebly significant
and ++ for highly significant. All absolute figure values corresponded
to a volume of 120 cm$^3$ of soil. Weed was not controlled mechanically,
it was insignificant in variant C (winter wheat), but highly significant
in maize overgrowing the maize during the vegetation period.

RESULTS

Tables 1- 4 show the total effect of herbicide treatment for the period
of our investigation, i.e., from May 1971 to April 1972 and from May
1972 to April 1973. Fig. 1 shows the effect at the individual times of
application. The effect of 2,4-D on *Collembola* and *Acari* populations
disclosed no statistically significant differences, the general trend
pointing toward overpopulation (Table 1). At the individual times of
application, populations incrased in soil layer 5-10 cm two weeks
after herbicidal treatment; an overpopulation of mites occurred in November in the uppermost soil layer (fig. 1). A decrease in the population starting for the springtails in autumn, for the mites evidently
later, had not been redressed in April of the following year. Simazin,

T a b l e  1.  Abundances of the Collembola and of the Acari

| Group | Variant | 2,4-D (May 1971 - April 1972) | | Simazin (May 1972 - April 1973) | | | |
|---|---|---|---|---|---|---|---|
| | | 0 - 5 cm | 5 - 10 cm | 0 - 5 cm | 5 - 10 cm | 10 - 15 cm | 15 - 20 cm |
| Collembola | C | 26,3 | 29,7 | 14,1 | 6,9 | 7,8 | 5,2 |
| | S | 28,1 | 38,4 | 4,4++ | 3,9+ | 2,9++ | 2,5+ |
| | D | 26,9 | 34,3 | 3,8++ | 3,6+ | 2,6++ | 2,3+ |
| Acari | C | 34,0 | 14,8 | 46,6 | 16,8 | 10,2 | 7,7 |
| | S | 42,4 | 16,9 | 35,5 | 10,9 | 6,7 | 7,5 |
| | D | 40,1 | 15,8 | 31,4 | 9,4+ | 6,6 | 6,6 |

C = control,  S = single dose,  D = double dose.

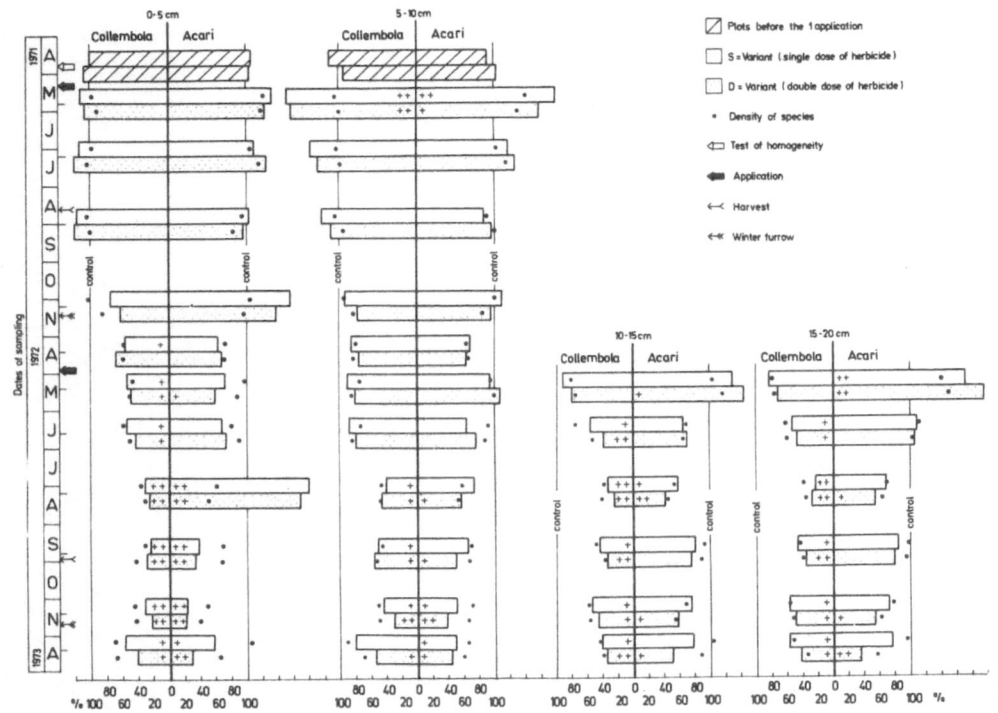

Fig.1. Percentile deviations of the abundance and of the
density of species of the S-  and D- varians of the *Collembola* and *Acari* in comparison to the control (C=100 %).
The significance is related to the abundance.

however, decreased  significantly the abundance of *Collembola* in almost
all soil layers  throughout  the year of our investigation.  Its effect
was greatest in the uppermost soil layer. The same applied to the *Acari*
immediately  after  treatment, with an increase  in abundance  in April
apart from a significant overpopulation in the deeper soil layers. Data
obtained in April  of the following year  suggested that the depression
of the populations had persisted during the winter.

    With regard to the abundance  of important prevailing species (Ta-
ble 2) 2,4-D  stimulated  particularly  the development  of populations
of *Tullbergia krausbaueri, Folsomia* sp., *Isotoma notabilis*  and  *Tyrophagus longior* It decreased the abundance of *Ceratophysella denticulata*
and *Tectocepheus velatus*.  A significant overpopulation occurred in May
with the species *Folsomia* sp.,  *I. notabilis, T. longior, Histiostoma
feroniarum*  and  *Bakerdania exigua,* in November almost exclusively with
the species  *T. longior*.  The species  affected most  by  a decrease in
abundance in the spring of 1972 were *I. notabilis, C. denticulata, Per-*

T a b l e   2.   Abundances of important prevailing Collembola- and mite species

| Species | 2,4-D (May 1971 - April 1972) | | | | | | Simazin (May 1972 - April 1973) | | |
|---|---|---|---|---|---|---|---|---|---|
| | 0 - 5 cm | | | 5 - 10 cm | | | 0 - 5 cm | | |
| | C | S | D | C | S | D | C | S | D |
| Onychiurus jubilarius | 1,5 | 1,6 | 1,7 | 2,1 | 2,3 | 2,4 | | | |
| Tullbergia krausbaueri | 8,0 | 10,6 | 10,6 | 10,9 | 14,5 | 13,4 | 5,9 | $3,0^{+}$ | $2,3^{++}$ |
| Folsomia sp. | 5,4 | 5,2 | 5,8 | 7,2 | 9,9 | 9,7 | 2,0 | $0,3^{++}$ | $0,3^{++}$ |
| Isotoma notabilis | 4,1 | 5,5 | 5,6 | 3,9 | $6,7^{+}$ | 5,0 | 2,1 | $0,7^{++}$ | $0,8^{++}$ |
| Ceratophysella denticulata | 4,0 | $1,0^{++}$ | $0,7^{++}$ | 2,0 | $0,8^{+}$ | $0,7^{+}$ | | | |
| Rhodacarellus silesiacus | 1,8 | 1,9 | 1,6 | 2,7 | 2,6 | 2,5 | 4,0 | $1,9^{+}$ | $1,9^{+}$ |
| Speleorchestes formicorum | | | | | | | 7,7 | $24,1^{++}$ | $22,2^{++}$ |
| Tyrophagus longior | 12,6 | $22,7^{+}$ | $20,9^{+}$ | 0,8 | 1,2 | 0,9 | 16,0 | $2,6^{++}$ | $1,5^{++}$ |
| Tectocepheus velatus | 7,5 | 4,7 | 5,1 | 1,1 | 0,7 | 0,8 | 16,3 | $5,6^{++}$ | $6,1^{++}$ |

gamasus suecicus, T. velatus and Eupodes acuminatus Willmann. Simazin caused variation in the abundance of populations of all dominant species in all soil layers except for the species Speleorchestes formicorum. Significant overpopulations of mites in the lowest soil layers occurred particularly with the species T. longior and H. feroniarum, while in the uppermost soil layer in August, with the species S. formicorum only.

The values referring to the dominance of the individuals were similar to those obtained for absolute abundance. Increases or decreases after herbicide application were in accord with a coincidental shifting of dominance values for the individual species. In Table 3 dominant and influent species of the individual variants have been arranged in three groups according to the degree of dominance variation after application of the herbicide. Changes in dominance have been examined with a method suggested by Moritz (1965). After application of 2,4-D, a positive

Table 2 continued

| Species | Simazin (May 1972 - April 1973) | | | | | | | | |
|---|---|---|---|---|---|---|---|---|---|
| | 5 - 10 cm | | | 10 - 15 cm | | | 15 - 20 cm | | |
| | C | S | D | C | S | D | C | S | D |
| Onychiurus jubilarius | | | | | | | | | |
| Tullbergia krausbaueri | 3,9 | 3,2 | 2,9 | 4,6 | 2,2[+] | 1,7[+] | 2,9 | 2,1 | 1,9 |
| Folsomia sp. | 1,4 | 0,6[+] | 0,5[+] | 2,0 | 0,7[+] | 0,8[+] | 1,7 | 0,6[+] | 0,6[+] |
| Isotoma notabilis | 0,9 | 0,3 | 0,6 | 0,5 | 0,2 | 0,3 | 0,2 | 0,1 | 0,2 |
| Ceratophysella denticulata | | | | | | | | | |
| Rhodacarellus silesiacus | 2,1 | 2,0 | 2,1 | 3,8 | 1,6[+] | 1,9[+] | 2,4 | 1,3[+] | 1,3[+] |
| Speleorchestes formicorum | 1,2 | 2,9[++] | 2,3[+] | 0,2 | 0,9 | 0,7 | 0,2 | 0,7 | 0,7 |
| Tyrophagus longior | 3,5 | 0,7[++] | 0,7[++] | 1,0 | 0,6 | 0,5 | 1,3 | 1,3 | 1,1 |
| Tectocepheus velatus | 4,3 | 2,0[+] | 2,0[+] | 0,7 | 0,4 | 0,4 | 0,6 | 1,0 | 1,0 |

shifting of dominance has been recorded for 3 species in both soil
layers and for 4 species in the lowest soil layer, a negative shifting
of dominance for two species in both soil layers and for one species
in the lowest soil layer. However, no appreciable variation occurred
in the gradation of the species. A different situation occurred after
Simazin application - a positive shifting of dominance was observed in
a few species only, while a negative shifting of dominance occurred
with a relatively large number of species, whereby the relative abundance of S. formicorum increased in all soil layers, that of Folsomia
sp. decreased in all soil layers; several species were positive in
certain soil layers only, other negative or shifting their dominance,
others again changed from negative to positive. Relatively significant
variations in the rank order of species suggested an important regroup-
ing in the microarthropod coenosis after the application of Simazin.
    The spectrum of species was not greatly affected by the herbicides.

T a b l e   3.    Dominances of individuals of dominant (more than 5 %) and influent (2-5 %) species

| | 2,4-D (May 1971 - April 1972) | | | | | | Simazin (May 1972 - April 1973) | | |
| | 0 - 5 cm | | | 5 - 10 cm | | | 0 - 5 cm | | |
| | C | S | D | C | S | D | C | S | D |
|---|---|---|---|---|---|---|---|---|---|
| I | T.lo.20,5<br>T.kr.14,0<br>I.no. 6,8 | T.lo.32,8<br>T.kr.15,2<br>I.no. 7,9 | T.lo.31,1<br>T.kr.15,9<br>I.no 8,3 | T.kr.24,4<br>Fols.16,2<br>I.no. 8,7 | T.kr.26,3<br>Fols.18,0<br>I.no.12,1<br>P.al. 2,4<br>T.lo. 2,1<br>P.is. 2,1 | T.kr.26,8<br>Fols.19,3<br>I.no. 9,9<br><br>P.is. 2,0<br>A.fr. 2,0 | S.fo.10,6 | S.fo.50,5<br>N.ar. 4,8 | S.fo.52,7<br>N.ar. 3,2 |
| II | T.ve.12,3<br>C.de. 5,1 | T.ve. 6,7 | T.ve. 7,7 | C.de. 4,8<br>T.ve. 2,3<br>T.qu. 2,0 | | | T.lo.29,5<br>T.ve.22,3<br>T.kr. 8,1<br>R.si. 5,4<br>Tars. 3,0<br>I.no. 2,9<br>Fols. 2,8<br>(O.ar. 2,2) | T.lo. 5,3<br>T.ve.11,7<br>T.kr. 6,3<br>R.si. 3,9 | T.lo. 3,5<br>T.ve.14,5<br>T.kr. 5,3<br>R.si. 4,3 |
| III | Fols. 9,0<br>R.si. 3,0<br>O.ju. 2,5 | Fols. 7,3<br>R.si. 2,8<br>O.ju. 2,2 | Fols. 8,6<br>R.si. 2,4<br>O.ju. 2,5 | R.si. 6,0<br>O.ju. 4,7 | R.si. 4,7<br>O.ju. 4,2 | R.si. 5,0<br>O.ju. 4,8 | | | |
| | 73,2 | 74,9 | 76,5 | 69,1 | 71,9 | 69,8 | 86,8 | 82,5 | 82,5 |

Table 3 continued

**Simazin (May 1972 – April 1973)**

| | 5 – 10 cm | | | 10 – 15 cm | | | 15 – 20 cm | | |
|---|---|---|---|---|---|---|---|---|---|
| | C | S | D | C | S | D | C | S | D |
| I | S.fo. 4,6<br>S.va. 4,2 | S.fo.21,8 | S.fo.14,5<br>B.ex. 2,8<br>S.va. 2,0 | | S.fo. 7,4<br>S.va. 5,0 | S.fo. 6,8<br>S.va. 2,9 | T.ve. 3,7 | T.ve. 8,4<br>S.fo. 5,5<br>Tars. 3,8 | T.ve. 9,6<br>S.fo. 6,4<br>Tars. 3,5 |
| II | T.ve.15,2<br>T.lo.12,4<br>Fols. 4,9<br>I.no. 3,0<br>P.su. 2,1 | T.ve.11,2<br>T.lo. 4,1<br>Fols. 3,1 | T.ve.12,3<br>T.lo. 4,0<br>Fols. 3,1<br>I.no. 3,5 | T.kr.21,4<br>R.si.17,3<br>Fols. 9,1<br>D.mo. 4,3<br>O.ju. 3,1<br>(O.ar. 3,0) | T.kr.18,7<br>R.si.13,7<br>Fols. 6,3<br>O.ju. 2,2 | T.kr.16,4<br>R.si.17,9<br>Fols. 7,7<br>D.mo. 2,9 | T.kr.19,0<br>R.si.15,7<br>Fols.11,2<br>D.mo. 3,6<br>O.ju. 2,9<br>(O.ar. 2,0) | T.kr.17,6<br>R.si.11,1<br>Fols. 4,6<br>D.mo. 2,3 | T.kr.17,6<br>R.si.12,4<br>Fols. 5,9 |
| III | T.kr.13,9<br>R.si.16,3 | T.kr.17,8<br>R.si.11,2<br>N.ar. 2,4 | T.kr.17,9<br>R.si.12,9<br>Tars. 2,5<br>N.ar. 2,0 | T.lo. 4,4<br>T.ve. 3,2<br>P.su. 2,7<br>A.pl. 2,3<br>I.no. 2,3 | T.lo. 5,2<br>T.ve. 3,7<br>Tars. 3,7<br>A.pl. 3,3<br>H.fe. 3,3 | T.lo. 4,8<br>T.ve. 3,9<br>Tars. 3,4<br>P.su. 3,4<br>H.fe. 3,6<br>I.no. 2,4 | T.lo. 8,6<br>A.pl. 2,3 | T.lo.10,9<br>H.fe. 3,4<br>A.pl. 2,5<br>N.ar. 2,2 | T.lo.14,6<br>H.fe. 4,0 |
| | 72,4 | 75,8 | 77,5 | 73,1 | 72,5 | 76,1 | 69,0 | 72,3 | 74,0 |

Percentages of the total populations of *Acari* and *Collembola*. I = species with increased relative abundance (positive shifting of dominance) in S and D in comparison to C; II = decreased relative abundance (negative shifting of dominance) in S and D in comparison to C; III = the relative abundances in C, S and D are to be considered as equivalent. (A.fr. = *Alicorhagia fragilis* BERLESE; A.pl. = *Ameroseius plumigerus* (OUDEMANS); B.ex. = *Bakerdania exigua* (MAHUNKA); C.de. = *Ceratophysella denticulata* BAGNALL; D.mo. = *Discourella modesta* (LEONARDI); Fols. = = *Folsomia (fimetaria* and n.sp.); H.fe. = *Histiostoma feroniarum* (DUFOUR); I.no. = *Isotoma notabilis* SCHÄFFER; N.ar. = *Nanorchestes arboriger* BERLESE; O.ar. = *Onychiurus armatus* (TULLBERG); O.ju. = *Onychiurus jubilarius* GISIN; P.al. = *Pseudosinella alba* PACKARD; P.is. = *Pseudanurophorus isotoma* BÖRNER; P.su. = *Pergamasus suecicus* (TRÄGHARDH); R.si. = *Rhodacarellus silesiacus* WILLMANN; S.fo. = *Speleorchestes formicorum* TRÄGHARD; S.va. = *Scutacarus valentini* BALOGH and MAHUNKA; Tars. = *Tarsonemus* sp.; T.ve. = *Tectocepheus velatus* (MICHAEL); T.kr. = = *Tullbergia krausbaueri* BÖRNER; T.lo. = *Tyrophagus longior* (GERVAIS); T.qu. = *Tullbergia quadrispina* (BÖRNER).

The species present in a soil layer of variant C only, were species of insignificant steadiness (= constancy) and dominance. After treatment with 2,4-D, the species *Uropoda virgata* Hull and *Parasitus niveus* (Wankel) disappeared completely (their average dominance in variant C had been 0.07 and 0.04 respectively). After Simazin application, we recorded the presence of the species *Isotomodes productus* (Axelson) - 0.17, *Pseudonurophorus isotoma* - 0.61, *Lepidocyrtus cyaneus* Tullberg - 0.13, *Onychiurus armatus* - 2.0, *Scutacarus eucomus* (Berlese) - 0.22, *Hypochthonius luteus* Oudemans - 0.36, *Liochthonius piluliferus* Forsslund - 0.07, and again *U. virgata* - 0.13.

As regard the species density, no differences were observed after treatment with 2,4-D, while significant differences were found in the abundance of springtails and mites in all soil layers after the application of Simazin (Table 4). Of the mites, the decreases were highest with the *Mesostigmata*. Apart from the species referred to under "Species spectrum", a decrease in abundance occurred particularly with the species *Arctoseius cetratus* (Sellnick), *Alliphis siculus* (Oudemans) and *Discourella modesta*. Of the springtails, a reduction of population density occurred with the species *Tullbergia quadrispina* and *Pseudosinella alba*. Interesting results were obtained from examinations following the individual applications of herbicides. These showed an important increase of mites in layers 5 - 10 cm and 15 - 20 cm. In April, the density of *Collembola* and *Acari* species was remarkably decreased in soil layers 0 - 5 cm and 5 - 10 cm in both variants, while in April 1973, the density of mites in the S variant decreased in soil layer 5 - 10 cm only.

T a b l e  4.    Density of species of the main groups

| Group | 2,4-D (May 1971 - April 1972) | | | | | | Simazin (May 1972 - April 1973) | | | | | | | | | | | |
| | 0 - 5 cm | | | 5 - 10 cm | | | 0 - 5 cm | | | 5 - 10 cm | | | 10 - 15 cm | | | 15 - 20 cm | | |
| | C | S | D | C | S | D | C | S | D | C | S | D | C | S | D | C | S | D |
|---|---|---|---|---|---|---|---|---|---|---|---|---|---|---|---|---|---|---|
| *Mesostigmata* | 2,8 | 3,0 | 2,7 | 3,6 | 3,6 | 3,4 | 2,7 | 1,6 | 1,0 | 2,6 | 1,4 | 1,6 | 2,8 | 1,5 | 1,7 | 2,5 | 1,5 | 1,5 |
| *Tarsonemina* | 2,6 | 2,4 | 2,4 | 1,8 | 1,8 | 1,9 | 1,6 | 1,6 | 1,3 | 1,2 | 1,3 | 1,2 | 0,9 | 1,2 | 0,9 | 0,6 | 1,0 | 0,9 |
| *Prostigmata* (rest) | 3,0 | 3,1 | 3,1 | 2,7 | 2,7 | 2,7 | 2,8 | 2,0 | 2,0 | 1,8 | 1,4 | 1,3 | 1,1 | 0,8 | 0,7 | 0,7 | 0,7 | 0,7 |
| *Acaridiae* | 1,1 | 1,3 | 1,3 | 0,8 | 1,2 | 1,0 | 1,2 | 0,8 | 0,5 | 0,8 | 0,6 | 0,5 | 0,6 | 0,6 | 0,5 | 0,7 | 0,8 | 0,5 |
| *Oribatei* | 1,3 | 1,4 | 1,3 | 1,0 | 1,1 | 0,9 | 1,3 | 1,2 | 1,1 | 1,2 | 1,0 | 0,8 | 0,6 | 0,6 | 0,4 | 0,6 | 0,8 | 0,7 |
| Total Acari | 10,8 | 11,2 | 10,8 | 9,9 | 10,4 | 9,9 | 9,6 | 7,2[+] | 6,3[+] | 7,6 | 5,7[+] | 5,4[+] | 6,0 | 4,7[+] | 4,1[+] | 5,1 | 4,8 | 4,3 |
| Collembola | 6,2 | 5,8 | 5,6 | 6,1 | 6,0 | 5,7 | 5,6 | 2,3[++] | 2,2[++] | 3,3 | 1,9[+] | 1,9[+] | 3,2 | 1,8[+] | 1,6[++] | 2,5 | 1,4[+] | 1,3[+] |

DISCUSSION

The results of our investigations  are consistent in general with those
obtained in earlier studies. Rappoport and Cangioli (1963), Fox (1964)
and Edwards (1965) observed after a single application of 2,4-D, Davies
(1965)  and Bieringer (1969)  after repeated application of this herbicide for a period of 10  and  11 years respectively,  that it had none
or  very little influence  on collembolan- and mite populations  in the
soil.  The population  increase  following  2,4-D  treatment at Etzdorf
might be ascribed  to an increased microbiological  and,  particularly,
bacteriological activity in the soil as a direct response to the herbicide. The same observation was made by Hickisch.  A population decrease
following 11 months after herbicide treatment might evidently be due to
a lower ratio of plant remnants in the soil in comparison with the control variant.  As regards Simazin, Edwards (1965)  inferred that it reduced populations of soil animals  to 1/3rd or even 1/2 of those in untreated soil. "The predatory mites and hemiedaphic *Collembola,* particularly the *Isotomidae,* were most affected  by simazine,  but the earthworms, enchytraeid worms, dipterous  and coleopterous larvae, and populations of other mites and springtails  were all decreased, and significant differences between the numbers of these in treated and untreated
soil  were still  obvious  three to four months  after the chemical had
been applied".  On the other hand, Steinbrenner at al. (1960)  observed
on sandy soil  that  *Collembola* increased  significantly in abundance,
*Acari* slightly only, 6 - 7 weeks after Simazin application in a dose of
2 kg/ha and more.  It is evident that, at Etzdorf, considerable changes
in the composition  of the microarthropod coenosis persisting  for more
than a year should be attributed to weed control  worsening living conditions of most of the soil-inhabiting species.


R e f e r e n c e s

B i e r i n g e r ,  H. (1969):  Untersuchungen über den Einfluss lang-
    jähriger Herbizidanwendungen  auf  Boden-Collembolen. Diss. Hohen-
    heim.
D a v i s ,  B. N. K. (1965):  The immediate  and  long-term effects of
    the herbicide  MCPA  on  soil arthropods.· Bull. Entomol. Res. 56,
    357-366.
E d w a r d s ,  C. A. (1965): Effect of pesticide residues on soil in-
    vertebrates and plants. Fifth Symp. Brit. Ecol. Soc. 239-261.
F o x , C. J. S. (1964): The effects of five herbicides  on the numbers

of certain invertebrate animals in grassland soil. Canad. J. Plant Sci. 44, 405-409.

M o r i t z , M. (1965): Untersuchungen über den Einfluss von Kahl-schlagmassnahmen auf die Zusammensetzung von Hornmilbengesell-schaften *(Acari: Oribatei)* norddeutscher Laub- und Kiefernmisch-wälder. Pedobiologia 5, 65-101.

R a p p o b r t , E. H. and C a n g i o l i , G. (1963): Herbi-cides and the soil fauna. Pedobiologia 2, 235-238.

S t e i n b r e n n e r , K. F., N a g l i t s c h , F. und S c h l i c h t , I. (1960): Der Einfluss der Herbizide Simazin und W 6658 auf die Bodenmikroorganismen und die Bodenfauna. Al-brecht-Thaer-Arch. 8, 611-631.

W a e r d e n , B. L. van der (1957): Mathematische Statistik. Berlin--Göttingen-Heidelberg.

## D i s c u s s i o n

K. H. D o m s c h : Für bodenzoologische und mikrobiologische Untersu-chungen werden Kontrollen gefordert, die es ermöglichen, direkte von undirekten Effekten zu trennen. Bei Aussagen über herbizide Wirkstoffe ist es ratsam, durch Anwendung von Leerformulierungen die Wirkung von Formulierungshilfstoffen von der aktiven Wirkstoffe abzutrennen.

J. P r a s s e : Um eben Wirkstoffeffekt erfassen zu können, sind spe-zielle Untersuchungen erforderlich. Hier sollten nur die durch Herbi-zideinsatz verursachten Veränderungen der Mikroarthropodenzönose er-fasst werden.

T. G. W o o d : Could you tell us something about the relative pH of the two soils. I would suspect that the surface accumulation of unde-composed plant material would lower pH in the "green" soil and produce a change towards a dominantly fungal, as opposed to bacterial, decom-position.

J. P r a s s e : The greens were originally very acid but they had been limed some years before we began this work. The pH of the greens and fairways is now similar, at about pH 6.

C. A. E d w a r d s : In similar experiments 2-3 years ago we obtained similar results to those of Dr Prasse but attributed the effects of simazine to direct toxicity by keeping all plots weed-free.

# Effects of the herbicide 2, 4, 5 - T on Onychiurus quadriocellatus Gisin (Coll.)

H. EIJSACKERS
**Research Institute for Nature Management,
Arnhem**

INTRODUCTION

The herbicide 2,4,5 trichloro phenoxy acetic acid (2,4,5-T), a representative of the hormone herbicides, is used in coniferous forests in
The Netherlands to control the black cherry *Prunus serotina* Ehrh. This
species, which was imported from Nothern America and planted in our
forests mainly because it was thought to improve soil conditions, gradually became a serious problem. Because of its vigorous growth and competitive powers, it became dominant in many coniferous forests and in
addition a number of deciduous stands with value from the point of view
of nature conservancy were also invaded, interesting undergrowth being
replaced by a uniform black cherry vegetation.

Therefore control became necessary. This can be accomplished mechanically or chemically. Mechanical control, i.e., cutting of the
shrubs, is inefficient, because the rapid and vigorous regrowth makes
regular repetition necessary and this is too expensive. Chemical control can be accomplished by spraying the leaves with a 3 % aqueous solution of 2,4,5-T or by smearing the stems with a mixture of the herbicide and fuel oil (the herbicide contained 40 % a.i.). Especially with
a waterly spray, a considerable amount of herbicide reaches the forest
floor. We estimated that under field conditions between three and ten
litres of 2,4,5-T per hectare can drip from the leaves onto the forest
floor. This contamination caused some concern because of the possible
side-effects of 2,4,5-T on the soil fauna, particularly in view of the
alarming information about the teratogenic effect of 2,4,5-T.

EXPERIMENTAL SET-UP

To study the effects of 2,4,5-T on the soil fauna, experiments were set

up with three common representatives  of the soil fauna,  which  can be
arranged in a food chain.

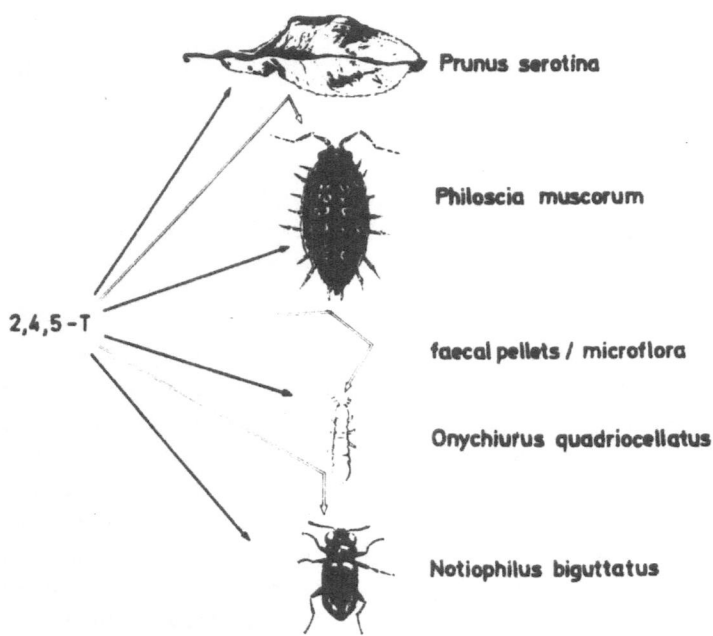

Fig.1. Experimental set-up  in a food chain  with direct
(black) and indirect (white) influences of 2, 4, 5-T .

Figure 1 gives an idea of this food chain and of the possible ways
of 2,4,5-T interference. The basic element in the food chain is the litter of sprayed black cherry shrubs.  This is consumed  by the woodlouse
*Philoscia muscorum* (Scopoli), which is a representative of the group of
saprophagous macro-arthropods.  In the faecal pellets  of the woodlouse
the leaf litter  is found fragmented  but  mostly chemically unaltered.
This fragmentation promotes the fungal attack  on the organic material.
The fungus  on the pellets  is grazed upon by the springtail *Onychiurus
quadriocellatus* Gisin,  representing  the meso-arthropods.  In its turn
the springtail is eaten by the carabid beetle *Notiophilus biguttatus* F.,
a representative of the predacious macro-arthropods, which feeds largely
on springtails.  It should be kept  in mind that this food chain  could
occur under field conditions, but is only one of the many chains in the
complex food web in the soil.
The herbicide, can either have a direct influence  on a species or
exert an indirect influence through its food.  For instance the spring-

tail  could be affected  by a change  in the composition  of the faecal
pellets of the woodlouse and the microflora on these.

This report presents some experimental results  concerning the direct effects of 2,4,5-T  on the springtail  *Onychiurus quadriocellatus*.
For these experiments  an agar-soil substrate  was  used  (adapted from
Sanocka-Woloszyn and Woloszyn, 1970).  The herbicide was mixed with the
substrate and the springtails were directly placed on it.

During the experiments  the temperature  was kept at 15 $^{\circ}$C and the
relative humidity at 90 %.

Observations  were made on: a) the mortality  under  constant exposure; b) the influence of the herbicide on the activity level; c) the
repellent  action  of the herbicide, and  d) the total effect  of these
factors under conditions enabling the springtails to avoid contaminated
parts of the substrate.


INFLUENCE ON MORTALITY

To determine whether 2,4,5-T has a lethal effect on the springtail, 100
individuals  were exposed for 60 days to herbicide  in amounts of 1.25,
2.5, and 5cc/m$^2$, corresponding  to doses of 12.5, 25  and  50 litre/ha.
Each day the dead specimens  were counted and removed to prevent necro-
phagy.  The results  are shown  in fig. 2, in which the cumulative percentages of dead springtails after 60 days exposure to different amounts
of herbicide are plotted against time.

It is clear from these results  that an increasing dose  of herbicide is correlated with an increased mortality.  After 60 days the mortality in the controls was low (8 %).  At a dose of 1.25 cc/m$^2$ over the
same period  the mortality  was 39 %, and at 2.5 cc/m$^2$ it reached 65 %.
With a dose of 5 cc/m$^2$ the 100 % mortality level was reached in 34 days
Especially  during  the first ten days  the mortality rose sharply.  In
this interval the mortality in the control group  was nil, and  for the
three amounts of herbicide, 2 %, 16 %, and 63 %, respectively.

After this period too, mortality was higher  the larger the amount
of herbicide.  Because  the main effect  was obtained  in the first ten
days,  subsequent experiments were restricted to this period.

These results  clearly  show  that  under  continuous exposure the
herbicide can have a pronounced effect on the springtail.  However, the
mobility  of the springtails  and the patchy distribution  of the spray
under field condition  (dripping from the leaves onto the forest floor)
may enable the animal to avoid contaminated areas.

Percentage mortality

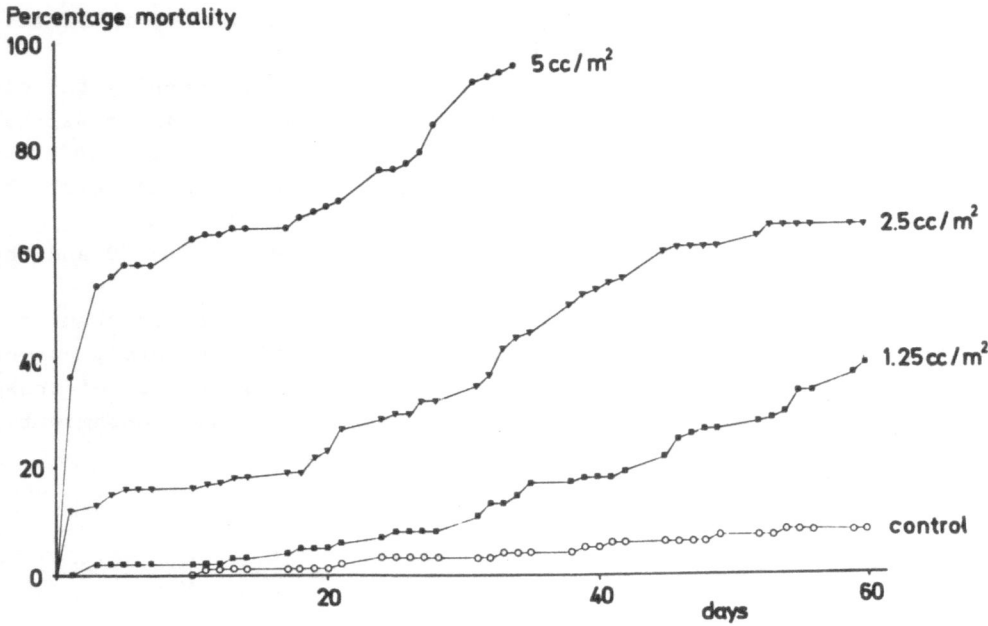

Fig.2.Cumulative mortality of springtails under constant exposure.


Therefore the effect of the herbicide on the activity level of the springtail had to be studied next.


INFLUENCE ON ACTIVITY

To get a rough impression of the influence of the herbicide on their activity, springtails were kept under direct observation for 12 hours after being placed on the substrate. Five types of activity could be distinguished: normal resting, normal walking, walking with irregular movements of the legs, immobility with tremour like contractions of the body and complete immobility. Immobility can easily be distinguished from resting because immobile springtails keep their legs and abdomen stretched. The observations indicated that at high dose the herbicide had a paralysing and at low dose an activating action on the springtails.  In petri dishes  without herbicide  there was first a period of "normal walking", due to placement on the substrate, but after six hours all the springtails were resting.  In the presence of a small amount of herbicide  (0.3 cc/m$^2$ corresponding to three litres  per hectare or the lowest amount we observed  under field conditions)  a higher percentage of the population "walked normally" and for a much longer period.  Some of the animals showed "irregular movements of the legs",  but after  11 hours all moved normally.  At the rather high dose  of  2.5 cc/m$^2$ there were more springtails  with  irregular movements  of the legs and a no- table percentage  showing tremour-like contractions;  only a few showed normal walking.

At a dose of 5 cc/m$^2$ the great majority of the animals became com- pletely immobile, 27 % of them after showing tremour-like contractions.


REPELLENT ACTION

These activating  and paralysing effects of low  and high doses of her- bicide  respectively,  would have importance  only if springtails could distinguish  between areas  with and without herbicide.  To investigate this point, experiments  were  performed  in petri dished  divided into four equal quadrants and provided with herbicide  in two opposite quad- rants.  Springtails were given the choise  between the treated  and un- treated quadrants  by placing them individually  at the intersection of the two diameters forming the quadrants. The distribution of the spring- tails was noted every hour and daily for 10 days. The results are shown in fig.3.

The distribution  is plotted here as the percentage of the spring- tails in the control and herbicide quadrants against time.

After one day,  most of the springtails  were in the control quad- rants of the dishes.  The results  of more frequent observation  during the first twelve hours of exposure (to be published in detail elsewhere) showed that an uneven distribution appeared as early as the first hour. In the dishes  of the control group  the springtails  were  distributed evenly over the quadrants.  Differences  between the various amounts of herbicide were small.  Only during the first three days was the unequal distribution  somewhat more distinct  with  higher amounts of herbicide. It is concluded that the repellent effect  is correlated only to a lim- ited extent with the amount of herbicide. Gradually, this unequality of the distribution diminished, although after ten days  about 60 % of the springtails were still in the non-treated quadrants.

Continued observation  showed that this situation persisted for at least the next 50 days.

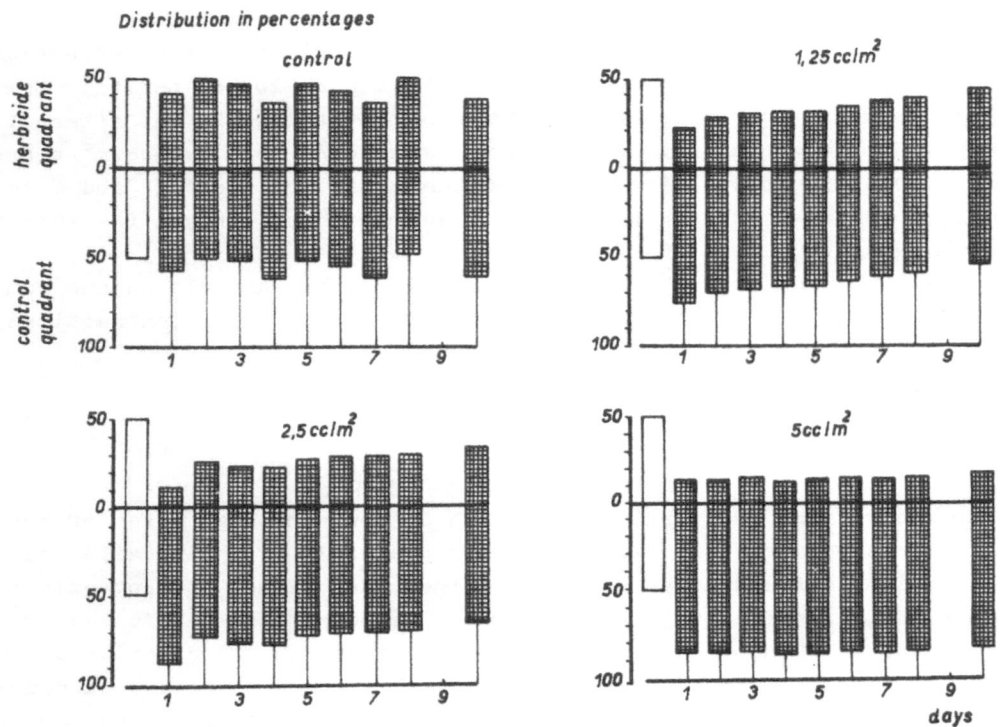

Fig.3. Distribution of springtails in herbicide and control quadrants  after given a choice  between  the quadrants.

From the results  of this experiment it is clear that 2,4,5-T  has a repellent  effect, but not whether  this  effect  is strong enough to force the springtail to withdraw from a contaminated quadrant.


IMPLICATIONS FOR ESCAPE AND MORTALITY

The experiment  was  therefore  repeated, but now the springtails  were placed directly  in the contaminated quadrants.  The results  are plot-ted in  fig. 4.  At the start  of  the experiment  all  the springtails (100 %)  were in the herbicide quadrants.  In the control dishes  there was a slight displacement during the first day, probably resulting from

Distribution in percentages

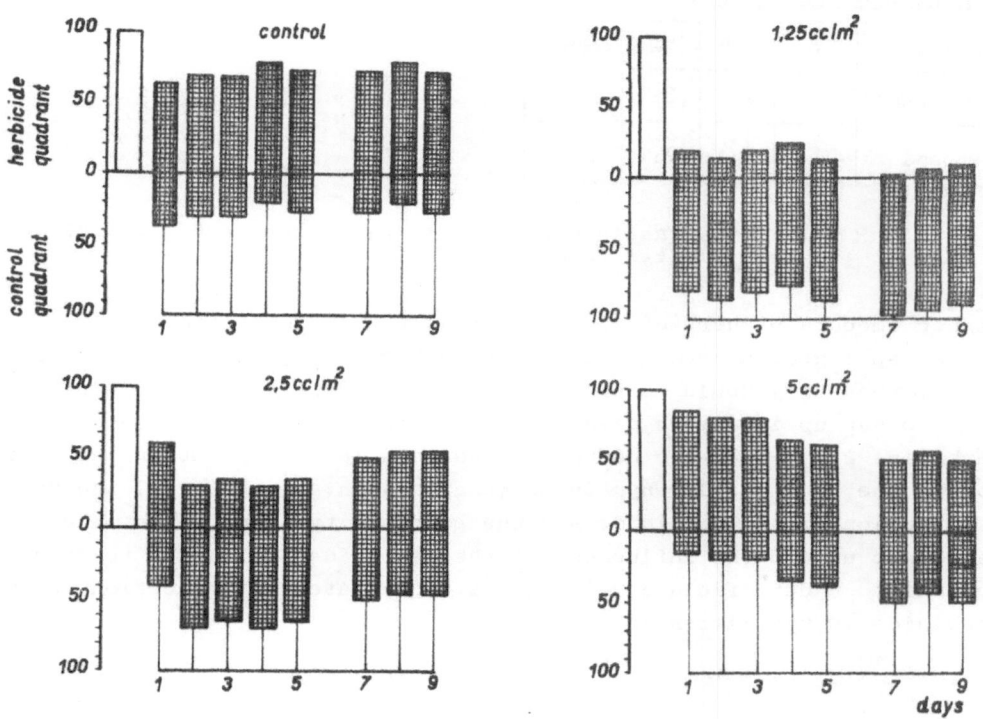

Fig.4. Distribution of springtails in herbicide and control   quadrants   after   placing   the springtails   in the
herbicide quadrant.

the disturbance activity   caused   by the transfer to the medium.   After
the second day the distribution did hardly change.

In the quadrants with herbicide two phenomena were observed: a) with
a low amount of herbicide   there was very rapid decrease   of numbers in
the herbicide quadrants   and   b) with a high amount a delayed   and slow
decrease.   These two phenomena agree with the activating/paralysing effect the herbicide was found to have on the activity level. To show the
influence on mortality in the two experiments   and for different doses,
fig. 5   gives the ratios found in the herbicide   and control quadrants.
In the repellency experiment   there was no distinct difference   in mortality between these quadrants and a relationship   with the presence of
herbicide is not clear. In the escape experiments on the contrary there
was   a very distinct difference:   the mortality in the herbicide quadrants far exceeded   that in the control quadrants, particularly for the

| RATIO OF MORTALITY HERBICIDE : CONTROL | | | | |
|---|---|---|---|---|
| % | 5 | 2,5 | 1,25 | control |
| repellency | 1:1 | 2:1 | 3:1 | 1:7 |
| escape | 40:1 | 5:1 | 17:1 | 1:1 |

Fig.5. Ratio mortality in herbicide  and  control quadrants after 10 days.

higher amounts of herbicide.  These results  indicate that introduction
of the springtails into a quadrant with herbicide, increased the probability that they would be affected  by the herbicide, possibly fatally.
     To sum up it may be said that there is a definite influence of the
herbicide 2,4,5-T on the mortality and to a lesser extent on the activity of the springtail *Onychiurus quadriocellatus*.  However, the repellent action of the herbicide and the possibility  for part of the population to escape the influence  of the herbicide  will tend to decrease
mortality  under  field conditions  i.e.  a patchy distribution  of the
herbicide in the litter layer.


R e f e r e n c e

S a n o c k a - W o l o s z y n ,  E.,  W o l o s z y n ,  B.W. (1970):
     Med. Fac. Landb. wet. Gent, 35, 731-738.


D i s c u s s i o n

C. F. v. d. B u n d :  We observed  the same reaction of *Onychiurus bi-*
*compatur* Gisin  when they were placed in small vials partly filled with
soil  with a high content of DDT and in uncontaminated soil.  With high
contamination  there was  a strong preference for clean soil.  With low
concentration there was no preference.

# The effects of fungicides on microbial activities in the soil

G. J. F. PUGH, J. I. WILLIAMS, and M. WAINWRIGHT
Department of Botany,
University of Nottingham, U. K.

INTRODUCTION

Fungicides and other pesticides are considered to be indispensable aids in agricultural and horticultural practice and a vast array of biocides are applied directly to soil or enter soil as run-off from treated aerial systems or from drifting sprays. Soil biologists have long been aware of the importance od the non-pathogenic soil microflora in main- taining soil fertility and soil structure and it is also well-known that few pesticides are sufficiently specific to affect pathogens alone. With the rapid escalation in biocide usage it is becoming increasingly urgent that we evaluate the results of repeated applications of bio- cides on the soil flora and fauna and the consequent effects upon soil fertility. The long-term, repeated usage of fungicides containing heavy metals such as mercury is of special interest as the heavy metal is chelated by the mineral and humus fractions so that a gradual accumula- tion of the element occurs in treated soils. Changes in soil structure on golf greens have been associated with the long-term prophylactic usage of an organo-mercury fungicide (Pugh and Williams, 1971). Such effects are mediated by changes in the soil microflora. Saprophytic fungi are among the more important decomposer organisms particularly in acid soils so that any long term inhibition of fungal activity in such soils caused by repeated pesticide usage may be expected to inter- rupt the flow of energy through the soil ecosystem. Nitrogen is a major nutrient for higher plants and soil micro-organisms and nitrogen levels are probably the most limiting of factors in the development and growth of the soil biota. It is therefore of importance that we examine the circulation of nitrogen in soil following pesticidal treatment.

## MATERIALS AND METHODS
### a) Saprophytic fungi

The fungal potentials of a green  and  an adjacent fairway  at Wollaton
Golf Course, Nottingham, were compared  over a period of 12 months following  the appearence  of a "mat" of poorly decayed organic matter beneath the sward of the greens.  No such "mat" was apparent on the fairway although the major grass species of both green  and  fairway swards
was *Agrostis tenuis* Sibth.  An organomercury fungicide, (2.5 % mercury,
organically combined), was applied to the greens at least twice annually over a period  of about twenty years prior to the development of the
"mat" in order to control fungal pathogens of the turf,  especially *Fusarium nivale* (the cause of Fusarium patch disease).  Replicate samples
from two depths of both green and fairway were used for dilution counts
and for soil plates, using two media. The techniques are fully described by Pugh and Williams (1971).  The effect of these treatments  on the
bacterial components of the soil microflora was also investigated using
the dilution method with yeast extract agar as the culture medium.

### b) Circulation of Nitrogen

Air dried, sieved soil was used for the estimation of nitrate and ammonium N after the application of the fungicide.  The techniques used are
fully described  by Wainwright and Pugh (1973).  Similarly, the methods
used in the extraction of amino acids have been described by Wainwright
and Pugh (1974).

## RESULTS
### 1. Saprophytic fungi

On both media  the fairway samples  were seen  to possess  considerably
larger fungal potentials at the two sampling depths throughout the year.
Peaks in isolations  from fairway samples  were observed  during autumn
and early summer, low numbers being associated  with high soil temperature and low soil moisture  in summer or low soil temperature  and high
soil moisture in  winter. The lower fungal potentials of the green samples fluctuated quite markedly, with notable declines following the applications of the fungicide (after Oct. samples and before April samples

were collected). Comparable trends were noted on soil plates where, following the spring application, there was a marked reversal of the upward trend in numbers of isolates as seen in the fairway sample. Not only was there a significant dimunition in fungal potential in samples from the green compared with the fairway, but there was also a change of the spectrum of common species of fungi (Table 1). Whereas a large group of cellulolytic species were frequently isolated from fairway samples (group 1), four were regularly isolated from green samples (group 3), while a few species were frequently obtained from both green and fairway samples (group 2). With non-cellulolytic species these trends were reversed, with a larger group of species frequently isolated from the green (group 6) and a few species confined solely to the fairway (group 4).

Following the addition of the fungicide to soil there was an immediate and dramatic reduction in numbers of viable fungal propagules at both concentrations of the fungicide. This was followed by a recovery which reached its peak three weeks after the fungicidal treatment. However, even six weeks after the applications the fungal potentials were still not comparable with those of untreated soil. Cellulolytic fungi were less tolerant of the treatment than non-cellulolytic forms. Although the fungicide brought about an immediate reduction in numbers of bacteria, this was followed by a rapid proliferation which reached its peak three weeks after the fungicidal treatment. There was then a decline so that by the 42nd day the bacterial population had returned more or less to normality in terms of numbers.

## 2. Nitrogen circulation

Two general affects on nitrification were apparent following the addition of the fungicide. At very low concentrations of the fungicide there was a slight increase in the amount of $NO_3^-$-N formed. At higher concentrations however, there was a progressive decrease in $NO_3^-$N production; so that the process was completely inhibited at 10 µg fungicide per g of soil. Addition of the fungicide at levels above 1 µg/g of soil led to a marked increase in the level of $NH_4^+$-N present in the soil. The lowest level of the fungicide resulted in most nitrification and least ammonification. The effect of the fungicide on the free amino acid content of soil was also examined.

Following addition of the fungicides aspartic acid was extracted with only minor differences from the control. All other amino acids

T a b l e  1.  Common cellulolytic and non-cellulolytic fungi isolated from green and fairway

| Cellulolytic species | 2-3 cm F | 2-3 cm G | 7-9 cm F | 7-9 cm G |
|---|---|---|---|---|
| **Group 1** | | | | |
| Botryotrichum piluliferum | + | - | + | - |
| Cephalosporium sp. | + | - | + | - |
| Chaetomium funicolum | + | - | + | - |
| Fusarium culmorum | + | - | + | - |
| F. equiseti | + | - | - | - |
| Gliomastix murorum v. felina | + | - | + | - |
| Myrothecium striatisporum | + | - | + | - |
| Paecilomyces percicinus | + | - | + | - |
| Penicillium pulvillorum | + | - | + | - |
| Trichoderma koningi | - | - | + | - |
| Verticillium ? terrestre | + | - | + | - |
| **Group 2** | | | | |
| Chrysosporium pannorum | - | + | + | + |
| Cladosporium herbarum | + | + | + | - |
| Paecilomyces carneus | + | + | + | - |
| Trichoderma hamatum | + | + | + | + |
| **Gp 3** | | | | |
| Tricellula inequalis | - | + | - | - |
| Sterile forms | - | + | - | + |
| No. of common species | 13 | 6 | 13 | 3 |

| Non-cellulolytic species | (G) F | (G) G | (F) F | (F) G |
|---|---|---|---|---|
| **Gp 4** | | | | |
| Absidia spinosa | + | - | + | - |
| + Penicillium lilacinum | + | - | + | - |
| Sterile stromatous form | + | - | + | - |
| **Gp 5** | | | | |
| Mortierella minutissima | + | + | + | - |
| Penicillium piscarium | + | + | + | - |
| * P. spinulosum | + | + | + | - |
| Sterile dematiaceous forms | + | + | + | + |
| "     hyaline | + | + | + | + |
| **Group 6** | | | | |
| Mortierella elongata | - | + | + | + |
| M. spinosa | - | - | - | + |
| Mucor heimalis | - | + | - | - |
| Penicillium crustosum | - | + | - | - |
| Yeasts | - | + | - | + |
| Zygorhynchus heterogamus | - | - | - | + |
| Z. moelleri | - | + | - | + |
| No. of common species | 8 | 10 | 8 | 7 |

* these species were sometimes weakly cellulolytic,
+ recorded with a frequency of > 20 % by soilplate or dilution method,
- recorded with a frequency of < 20 % by soilplate or dilution method.

showed greater or lesser degrees of variation from the control. Trypto-
phan was less frequently isolated following application of 50 µg fungi-
cide/g of soil, and at this level only glutanic acid could be extracted
after 28 days.  The amino acid glycine  was not extracted from the controls, but was frequently extracted from the two treated soils. Quanti-
tatively the amount  of free amino acid N  extracted fell to less  than
20 % of the control, 28 days  after  treatment  at 50 µg/g of soil. The
lower  concentration  of the fungicide  resulted  in greater amounts of
amino acid -N,  but in both cases  the amounts extracted  was less than
the control.

## DISCUSSION
### a) Saprophytic fungi

Nowadays it is convenient to regard soil as an open ecosystem  in which
the energy output  is exactly balanced  by energy input  (Brock, 1967).
The role of decomposer organisms in the flow of energy through soil has
been largely ignored but MacFadyen (1961)  drew attention to the importance of decomposers  in grasslands  by stating  that over 80 %  of the
energy entering soil is utilised by bacteria and fungi.  Fungi are considered  to be  the major group  of organisms responsible  for  organic
decay in acid soils (Alexander, 1960; Brock, 1966) and thus any factors
which are inhibitory to fungal growth  and  activity will interrupt the
flow of energy  through  the soils.  In the absence  of suitable methods
for estimating fungal activity  in soils the dilution technique  is one
of the few alternatives that gives quantitative results (Montégut, 1960)
and can be justified in following the response of the fungal population
to environmental conditions (Griffiths and Siddiqi 1961; Garrett 1963).
The absence  of a "mat" of undecayed organic matter  on the fairway allowed this site  to be used  as a control against  which the fungal potential of the green could be compared.  In this context  the continued
smaller  fungal  potential  of the samples  from the green must reflect
reduced fungal activity.  The scarcity of viable propagules of cellulolytic fungi in particular indicated that the flow of energy through the
greens  must have occurred  at a slower rate  than through the fairways
and the "bottleneck" in the flow of energy  was responsible for the accumulation of undecayed organic matter in the form of the "mat".

Microbial populations  in any particular habitat reflect the phys-
ico-chemical conditions  of the environment  as well  as  the result of
competition between different organisms (Garrett, 1963). Of the environ-

mental parameters likely to inhibit fungal growth  and  activity in the
greens, the use of the organomercury fungicide  is the most potent. The
regular use of this pesticide  over a protracted period  has caused the
accumulation of up to 10 µg/g of mercury in the "mat"  and this remains
a potential source of continued toxicity apart  from the fungitoxic effects of fresh applications.  The immediate devastating effect  of this
fungicide  on the fungal potential  of soil  is seen following applications at concentrations likely to occur in the field (250 and 500 µg/g).
Organomercury fungicides are therefore persistent environmental poisons
to soil fungi.  Although the fungicide  is the primary cause  of fungal
inhibition the gradual accumulation of organic matter in the greens has
had secondary effects  on  the fungal population.  The increased  water
holding capacity  of the "mat"  has resulted  in the intermittent existence of waterlogged conditions  which  further reduce fungal activity.


## b) Circulation of Nitrogen

Application of less  than 1 µg Verdasan  (active ingredient) /g of soil
resulted  in  inhibition of nitrification,  while the amount of $NH_4$ - N
increased dramatically.  The levels of $NO_2$ - N  remained low throughout
the incubation period.  It appears therefore that Verdasan inhibits the
$NH_4^+$- N- oxidising bacteria, while the heterotrophic ammonyfying micro-
-organisms are differentially affected.  Some, mainly fungi, are killed
by  applications  of the fungicide  while others,  mainly heterotrophic
bacteria increase  and are able to mineralize the dead biomass to $NH_4^+$.
     Following addition  of Verdasan  to soil the levels  of amino acid
-N found were lower than the control.  This is consistent with the view
that increased dead microbial biomass made available  after addition of
fungicides is quickly mineralized to $NH_4^+$.  The exception appears to be
glycine, an amino acid more frequently extracted  from the treated than
from the control soils. This may be due to the resistance of glycine to
degradation or to it being excreted in copious amounts  by the increasing heterotrophic bacteria.
     The application of fungicides to soil can lead to increased growth
response of certain crops. This response is thought to be due mainly to
increased mobilization of N.  The inhibition of nitrification by fungicides  was once thought to be detrimental  to soil fertility.  However,
the conversion of the nitrogen  in soils to the $NH_4^+$ form  can be beneficial, since this ion  is held  by cation exchange  and  is not easily
leached from soil.  Such benefits have to be weighed against the pollution of soil caused by retention of heavy metals such as mercury.

R e f e r e n c e s

A l e x a n d e r ,  M. (1961):  Introduction  to soil microbiology. J.
    Wiley and Sons, Inc. N. York, London and Sydney.
B r o c k ,  T. D. (1967):  The ecosystem  and  the steady state.  Bio-
    science 17, 166-169.
B r o c k ,  T. D. (1966):  Principles of microbial ecology.  Prentice-
    Hall, New Jersey.
G a r r e t t ,  S. D. (1963):  Soil fungi and Soil Fertility. Pergamon
    Press Ltd.
G r i f f i t h s ,  E.,  S i d d i q i ,  M. A. (1961):  Some  factors
    affecting the occurrence  of Fusarium culmorum in the soil. Trans.
    Br. Mycol. Soc. 44, 343-353.
M a c F a d y e n ,  A. (196 ): Metabolism of soil invertebrates in re-
    lation to soil fertility. Ann. Appl. Biol. 49, 215-218.
M o n t e g u t ,  J. (1960): Value of the dilution method in The Ecol-
    ogy of Soil Fungi. Liverpool Univ. Press.
P u g h ,  G. J. F.,  W i l l i a m s ,  J. I. (1971): Effect of an or-
    ganomercury fungicide  on saprophytic fungi  and  on litter decom-
    position. Trans. Br. Mycol. Soc. 57, 164-166.
W a i n w r i g h t ,  M.,  P u g h ,  G. J. F. (1973):  The effect of
    three fungicides on nitrification and ammonification in soil. Soil
    Biol. Biochem.
W a i n w r i g h t ,  M.,  P u g h ,  G. J. F. (1974): Changes in free
    soil amino acids following fungicide treatment. Soil Biol. Biochem.

D i s c u s s i o n

D. E. R e i c h l e :  My question has two parts: 1. Have you found any
evidence  of fungal detoxication  of mercurials or accumulation of elemental mercury?  2. Do you anticipate performing tests on mercurial inhibition of nitrifying bacteria?

G. J. F. P u g h :  Yes, Chrysosporium  is able to detoxify this fungicide in pure culture. We have not yet looked very closely at the bacteria and their activities.

J. S a t c h e l l :  Is there any evidence that the few species  which
are not susceptible  to the mercury fungicide might develop  in time to
utilize the available substrate more completely?

G. J. F. P u g h :  The fungi which we find commonly in the greens  may
be 1. stimulated, 2. tolerant  or  3. intolerant but able to recolonise
rapidly  when  the fungicide level falls to a tolerable level. *Chryso-*
*sporium* can grow  on the fungicide, while *Mortierella*  and  *Trichoderma*
for example  are  not  tolerant,  but are able  to recolonise partially
sterilised soil quickly.

C. A. E d w a r d s :   To supplement  Dr.Satchell's question:  Are you
sure that no chemical  has been used  to keep the greens free of worms?
Most greenkeepers use such chemicals in my experience.

G. J. F. P u g h :  I cannot be sure of this, but I do not know of such
applications.

# Variations in meadow associations of earthworms caused by the influence of nitrogen fertilizers and liquid-manure irrigation

I. ZAJONC,
College of Agriculture,
Nitra

## INTRODUCTION

The decomposition of the produced plant mass which becomes again a source of plant nutrition represents an important part of the circulation of substances in ecosystem characteristics. Of the agrobiocoenoses it is the intervention of man just into this phase of the natural metabolism of the total system. Man harvests the crop and translocates it to another system so that no decomposition and natural renovation of soil fertility can occur here. Uptaken substances are then replaced mostly by mineral manures and these influence the life of soil organism providing a single phase of decomposition together with another chemical intervention. Although it is obious that the present intensive agriculture depends, in the domain of nutrition and plant protection, first on chemical means we must focus our attention on the effects of agrotechniques on the living components of the soil which remain a natural renewer of its fertility.

In our study we aimed to follow the influence of the application of several synthetic and natural manures on the structure of earthworm associations *(Lumbricidae)* in meadows representing relatively the simplest system in agriculturally exploited areas and thus a suitable model.

In the literature information on this problem is scarce. The influence of application of dung and several mineral substances is discussed by Wilcke (1956, 1962), Rohde (1956), van Rhee and Nathans (1961). Doerell (1950) observed the influence of ammonium nitrogen on lumbricids, and Slater (1954) the action of superphosphate. A valuable knowledge of the effects of natural dungs is offered by Voisin (1961).

## METHODS AND RESEARCH OBJECTS

The action of the fertilizers applied was followed by comparing the material obtained from fertilized  and  control areas,, whereby we always
started  from a series of ten parallel samples  from an area of 0.25 m$^2$
each. On several occasions, we sampled only once, on others, we compared the results  obtained from several series of samples  collected during different seasons of the year.  For the collection of earthworm material we used a combined method based on the manual picking up the individual samples from soil excavated to a depth of 30 cm from which the
worms were expelled with a weak formaldehyde solution.

The investigations were made in the following areas:

a) Nitra - a strip of land  of the Permanent State Agriculture Ex-
position  on the left bank of the  Nitra River,  altitude 143 m,  sandy
loam soil, annual precipitation 650 mm, mean temperature 9 °C.

b) Chyzerovce (Nitra district) - the area of the local cooperative
farm  near the  Žitava River, altitude 147 m, sandy loam soil, precipitation 650 mm, mean temperature 9 °C.

c) Becherov (Bardejov district) - area of the State Farm, altitude
550 m, clay  and clay-loam soil, precipitation 750 mm, average temperature 6.5 °C.

## INFLUENCE OF N-FERTILIZERS

The influence of high doses of N-fertilizer was observed first on meadow areas at Chyzerovce. Lovosice niter containing 25 % nitrogen, superphosphate (18 % $P_2O_5$)  and potassium salt (40 % $K_2O$)  were applied. The
doses  were  split  so that 50 %  of the total amount  was used  in the
spring and 25 % after each cut.  *Allolobophora antipai antipai* predominated  with 47 %, *Allolobophora rosea*  with 33.4 %  and  *Lumbricus terrestris* with 5.4 %. In addition we found another eight earthworm forms.
The mean abundance was 202.8 specimens/m$^2$, maximum values were recorded
in August 1967,  i.e. 576.0 specimens/m$^2$  in an average of ten samples.
The influence  of the fertilizer  applied in a high dose  was indicated
first by a decrease in the number of species. In the area treated abun-
dantly with nitrogenous fertilizers (300 kg/ha) three species occurring
in another case were lacking. Of these two forms  belong to *Allolobophora antipai;*  in two forms variation  in their abundance  is shown in
Table 1.

T a b l e   1.    Number of specimens   per 0.6 m$^2$   in plots treated   with
different doses of N-fertilizers

| Species | O | PK | N-100 kg + PK | N-200 kg + PK | N-300 kg + PK |
|---|---|---|---|---|---|
| Allolobophora antipai antipai | 87 | 76 | 51 | 42 | 13 |
| Allolobophora antipai tuberculata | 12 | 10 | 3 | 1 | O |

By contrast abundance   increased   with several species   *(A. rosea,*
*L. terrestris, Dendrobaena platyura)*.    These are mostly species   living
in the depth. While the abundance of juveniles   of the genus   *Allolobo-*
*phora* decreases with increasing doses of dung, that of juveniles of the
genus *Lumbricus* increases.

The highest abundance (Fig. 1) was found in areas receiving 100 kg
of nitrogenous fertilizers, the average   was 250.6 specimens per m$^2$. As

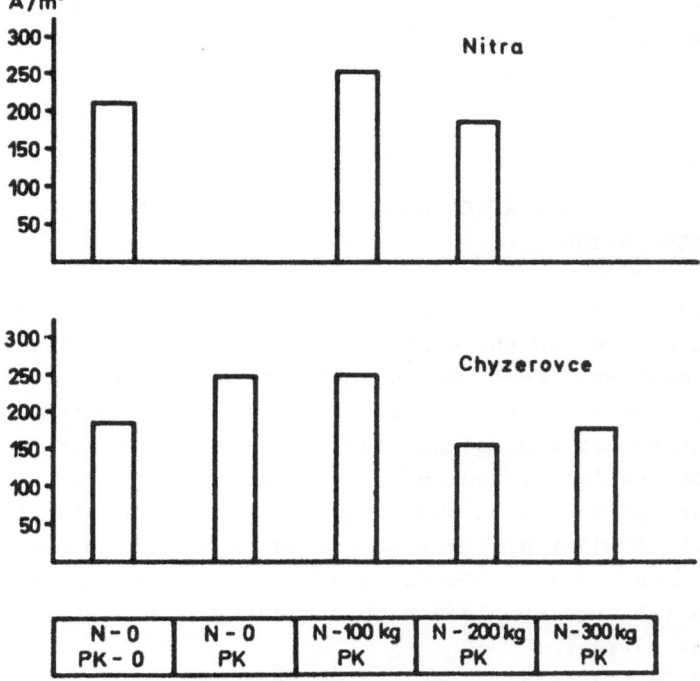

Fig.1. The abundance of earthworms on meadow areas treated with different fertilizers.

regards the abundance, the control area (untreated)  was third  in succession, while abundance  in the area  fertilized with N-fertilizers at
an amount of 200 and 300 kg  per ha  was lower than that of the control
area.  However, here the biomass  did not decrease but increase in view
of an increase  in  the proportion  of the representation  of the major
species.

The influence of high doses of nitrogenous fertilizers was further
observed in localities  near Nitra  where four species of *Lumbricidae* -
*A. antipai tuberculata, A. chlorotica, A. rosea* and *A. caliginosa* - were
found.  The most frequent was *A. chlorotica,* followed by *A. rosea;* less
frequent were *A. antipai*  and  *A. caliginosa*.  The average abundance of
these specimens was 198.6 specimens per $m^2$.

The highest abundance (225.1 specimens)  was observed  in variants
with 100 kg  (Fig. 1)  doses of N-fertilizer  per ha.  Similar, but 2 %
lower, was the abundance in the control area while in the areas treated
with 200 kg  of nitrogenous fertilizer, it was 176.8 specimens  per $m^2$,
i.e., 22 % less.  These variations  disclosed  that the dominance of *A.
chlorotica* increased  with higher N - doses.  A contrary phenomenon was
observed  in  *A. rosea*  with a dominance decreasing from 42.1 %, in the
control area to 27.3 % (100 kg doses per ha) and 26.8 % (200 kg per ha)
in an area with increased fertilizer application.  *A. caliginosa* showed
an increasing tendency, *A. antipai,* a decreasing tendency.


## THE INFLUENCE OF LIQUID MANURE IRRIGATION AND SEMILIQUID MANURE FARMING ON EARTHWORMS ASSOCIATIONS

This problem  was studied  on a locality near Nitra  in which we traced
the influence  of liquid manure  on the earthworm association of permanent meadows and sowing mixtures. We found seven forms of earthworms in
the irrigated areas  in comparison  with five forms only in the non-irrigated area.  The structure  of the association  appeared to different
in the representation  of species typical of agricultural land - *A. caliginosa*  with a dominance of 21.5 % in the irrigated area agains 6.5 %
in the non-irrigated area. Similar differences existed also in the dominance of the species *A. rosea* and *Octolasium lacteum*.  A low dominance
in the irrigated area to 1/3 when compared with the control  was stated
for *A. antipai* and *A. chlorotica*.

The abundance in irrigated meadows (Fig. 2) represented 48.0 specimens  per $m^2$ against 12.2 specimens per $m^2$ in non-irrigated areas. The
abundance on sowed areas  was  lower  by  6 to 50 % in comparison  with
meadow areas.

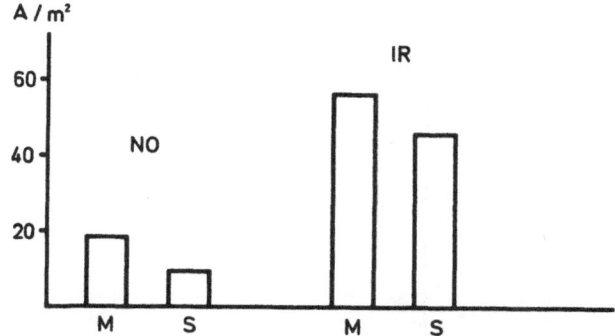

Fig.2. The abundance of earthworms in irrigated (IR) and
non-irrigated (NO) plots near Nitra; M - meadow, S - sowing mixture.

On the Becherov meadows  we   studied   the  influence   of semiliquid
manure irrigation on the earthworm association.  Eight species  of *Lumbricidae* with dominant species *A. rosea, A. caliginosa* and *O. lacteum* −
were found  in  the non-irrigated plots.  In the irrigated plots   there
were  only six species  present  whereby  the order of dominance of the
first three species  corresponded  to that of the previous case.   There
were no signicant differences either from the point of species composition or from that of the juvenile  and  adult individuals in the structure of both associations.

However, marked differences appeared in the quantitative representation of earthworms (Fig. 3).  The mean abundance  was 122.0 specimens
per $m^2$   at a biomass 36.4 g  in the irrigated plots. and 69.6 specimens
at a biomass 20.3 g  per $m^2$  in the non-irrigated plots- Thus, it was by
73 % higher in irrigated than on the control areas with biomass growth
up to 80 %. By means of a deeper analysis we found an increasing amount
of individuals especially those  of the species *A. rosea. A. caliginosa*
and *O. lacteum*  in irrigated areas.  Other species remained at approximately the same multitude.

SUMMARY

From interference  of man  through intensive farming of meadows  we observed first the influence of N-fertilizers  applied in amounts of 100,
200 and 300 kg  per hectar. The influence of manuring  varied with the
individual species;  the representation of the surface species  becomes
lower, while it remains stable or (increases moderately) in forms living
in the depth. Manuring of this type is best tolerated by *A. chlorotica,*

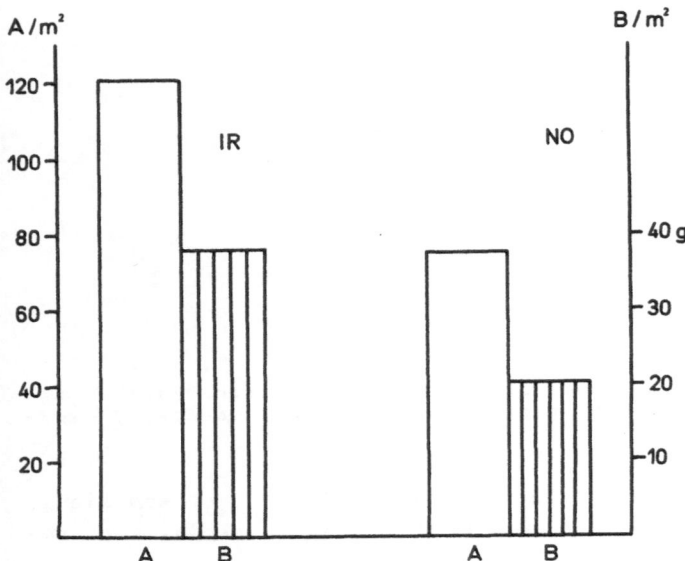

Fig.3. The abundance (A) and biomass (B) in irrigated
(IR) and non-irrigated (NO) areas near Becherov.

while *A. antipai* is mostly badly affected. The highest abundance was
established with the variant 100 kg/ha of N-fertilizer and that 250.6
specimens/m$^2$. The areas fertilized with P- and K-fertilizers only in
a standard dose were second in succession as far as abundance was concerned, and third in the non-fertilized control area. Higher doses of
nitrogenous fertilizers decrease the abundance.

In comparison with non-irrigated plots, liquid dung irrigation increased dominance by 75 % on the average. In meadows fertilized with
liquid manure we found also a marked increase of abundance as compared
with that in the control areas, i.e., from 68.1 to 122.2 specimens/m$^2$.

R e f e r e n c e s

D o e r e l , E. C. (1950): Was sagen die Regenwürmer zur Mineraldün-
gung. Deutsche landw. Presse, 4, 1.
R h e e van, J. A. and N a t h a n s , S. (1961): Observations on
earthworm population in orchard soils. Netherl. J. Agric. Sci. 9,
94-100.
R o h d e , G. (1956): Stalldünger und Bodenfruchtbarkeit. Berlin.
S l a t e r , C. S. (1954): Earthworms in relation to agriculture. US
Dept. of Agric. Res. Service Circular.

W i l c k e ,  D. E. (1956): Der Einfluss  der Düngung  auf die Regen-
    würmer, Kali-Briefe, Fachgebiet 8, Mineral und Wirtschaftsdüngung,
    1-7.

            (1962): Untersuchungen  über die Einwirkung von Stallmist und
    Mineraldüngung auf den Besatz  und  die Leistungen der Regenwürmer
    im Ackerboden. Monogr. Angew. Entomol. 18, 121-167.
V o i s i n ,  A. (1961): Lebendige Grasnarbe. BLV, München.

## D i s c u s s i o n

K. H. D o m s c h :  May I have some additional information on the type
of manure used, on the application rate, and the equivalents  of available nitrogen in the liquid manure?

I. Z a j o n c :  Semiliquid manure from thirty cows  was used for regular irrigation  of 7 ha  of pasture. This material  was not subjected
more specifically to chemical analysis.

J. S a t c h e l l :  What do you think  are the mechanisms causing the
population changes you observed?

I. Z a j o n c :  For the influence of applied chemical fertilizers, we
suppose  higher doses influence  directly  the body surface  of species
living  in the upper soil layers. Organic matters and respective water
volume contained in liquid  and  semiliquid manure  have posivively influenced the vital environment  of earthworms. This has been known for
a long time  from literary  sources.  In our case the study of the complex action of both factors was concerned. In relatively dry conditions
of our experiment  the environment  seems to have been first influenced
by a higher humidity  and then by better nutritive conditions of earthworms supplied with organic substances.

J. L. A u c a m p :  In which chemical form  was the nitrogen  applied?

I. Z a j o n c :  Lovosice niter - a mixture  of ammonium  and  calcium
nitrate with 25% N content - was used as nitrogenous fertilizer.

# Effects of DDT on community structure of soil microarthropods in an old field

D. L. DINDAL, D. FOLTS AND R. A. NORTON

State University of Environmental Science
and Forestry,
New York

INTRODUCTION

In June 1969 we initiated a three year study to determine the effect of a single DDT application on the community structure of soil micro-arthropods. Since data from the last year are still being analyzed, this paper constitutes a progress report of trends observed.

RESEARCH LOCATION

Research was conducted on a 4 ha (10 A) treated (T) and adjacent 4 ha untreated (UT) old field herbaceous site located within a 207 ha (518 A) abandoned state wildlife farm in west central Ohio, USA. The common soil type was an alfisol, aeric ochraqualf known previously as Crosby-silt loam. The site was undisturbed for 30 yr and dominated by the following vegetative species: orchard grass (*Dactylis glomerata* L.), blue grass (*Poa pratensis* L.), wild parsnip (*Pastinaca sativa* L.), Canada thistle (*Cirsium arvense* L.), white aster (*Aster ericoides* L.), wild carrot (*Daucus carota* L.) and early goldenrod (*Solidago juncea* Ait). Finally there was no previous record of direct pesticide application to the area.

METHODS

A single granular application of formulated technical DDT at 1.12 kg/ha (1 lb/A) was made in 1969.

Research supported by US AEC Grant Contract No. AT-(11-1)-3474.

Forty soil samples  were extracted  and  analyzed monthly for 3 yr
from the T and UT sites.  Each core was 5.4 cm in diameter x 4 cm deep.
After collection, the core  was divided  into an upper 3 cm section and
a lower 1 cm segment.  To date,  no significant  differences  have been
observed between T  and  UT samples in the lower 1 cm sample so, therefore, no further mention will be made.

A total  of 1,500 sample cores containing ·about  450,000 invertebrates have been studied. Microarthropods were extracted for 7 da using
modified Tullgren funnels. Then microbial respiration was assayed using
a $CO_2$ evolution method (24 hr incubation, closed system)  after Stotzky
(1965).  In addition  to arthropod and microbial respiration determinations, various physical factors were assayed and efforts are being made
to evaluate the relationships of all factors.

RESULTS AND DISCUSSION
## Total microarthropods

More than 200 species have been collected.  Of these, acari and collembolans were dominant.  When considered totally, arthropods in the upper
3 cm T samples  exhibited an initial suppression for 2 mo  (June, July)
following application.  During the first year  this was followed by a 2
mo (Aug., Sept.) lag period  which  is non-significantly different from
the UT site;  then the T populations  were  stimulated  for 3 mo (Oct.,
Nov., Dec.) before returning to comparable levels with the UT area.

## Microbial respiration

Since many microorganisms  comprise the food web base  for soil arthropods, microbial respiration was assayed. Respiration was initially suppressed for 1 mo  after treatment  followed by 17 mo period of stimulation (Fig. 1). The interruption noted during March and May 1970  cannot
be explained at this time.

Perhaps  microbial stimulation  was due to the following cause and
effect relationships:  reduction  in phytophagous insects,  causing increased primary productivity, in turn causing increased litter  and organic matter deposition,  ultimately providing increased microbial substrate.  Malone (1969)  and  Shure (1971)  reported  similar  increased
primary productivity  in an old field resulting  from diazinon applica-

tions.  When our organic matter assays are completed, those data should
help clarify the respiration trend.

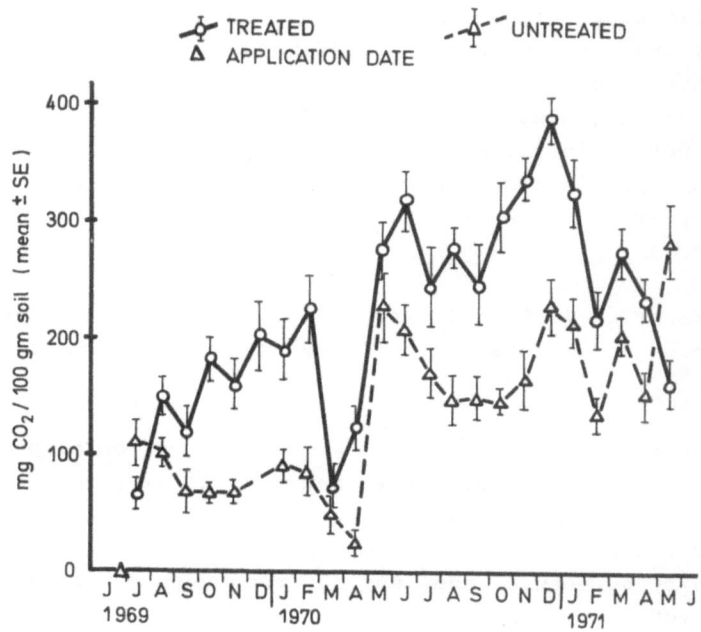

Fig.1. Soil microbial respiration of DDT treated and untreated old field community.


**Predator-prey relationships**

Two common soil groups, *Mesostigmata* and *Collembola*, illustrate possible predator-prey interactions.  Similar relationships  have been reported by Edwards (1969) and others, and our data support these earlier
observations.

*C o l l e m b o l a*
This group  comprising the prey  was represented  by 30 species in this
study.  The following species  were dominant: *Tullbergia clavata* Mills,
*Onychiurus subtenius* Folsom  and  *Folsoma fimetaria* L.  In  T samples,
collembolans  showed  an initial 3 mo  (June, July, Aug.)  suppression,
a 3 mo lag period  (Sept., Oct., Nov.)  followed by a 1 mo (Dec.) stimulation period.  Then populations  gradually  decreased to zero  during
June, July, Aug. 1970.

*M e s o s t i g m a t a*

Fifteen predaceous species were identified including laelapids, ascids,
veigaiids and parasitids. June, July and Aug. appeared to be months of
normal peak mesostigmatid populations on UT sites. In 1969 this peak
was reduced one half by treatment, whereas in 1970 mesostigmatid numbers
on the T area doubled over those UT.

Combining the two trends, the reduction period in predators was
followed in 3 mo by a twofold increase in prey species. During the following year mesostigmatid populations doubled with a corresponding complete reduction of the prey.

*A s t i g m a t a*

Adult *Tyrophagus putrescentiae* (Schrank) and *Histiosoma* sp. and *Calo-
glyphus* sp. hypopi represented this suborder. They occurred uncommonly.

*P r o s t i g m a t a*

Ninety species were identified thus comprising the largest arthropod
group. Dominant families and species were *Scutacaridae (Scutacarus
quadrangularis* Paoli), *Pyemotidae (Bakerdania* sp.), *Eupodidae (Eupodes
sp.* A) and *Tarsonemidae (Tarsonemus* sp.) in decreasing order.

Analysis of community structure and the diverse trophic roles of
prostigmatids have not yet been completed.

*O r i b a t e i*

This is the second most dominant taxon with 48 species collected.
*Oppiella nova* (Oud.) was the most abundant oribatid on both sites. Its
numbers were two times greater on T sites during Oct. and Nov. of 1969
and 1970. Voronova (1968) reports a ninefold increase in this species
due to application of the insecticide carbaryl.

S p e c i e s   d i v e r s i t y

Two methods of stating diversity were employed and both indicate similar trends. Figure 2 shows results using the modified Shannon-Weiner
formula ($\log_{10}$) after Lloyd, Zar and Karr (1968). Oribatid diversity
was significantly ($P \leq .05$) increased by treatment during sporadic intervals totaling 8 mo. When plotting total species only, (a simple indicator of species richness) the same array of significant increases
due to treatment were noted (Fig. 3). These results generally contradict
findings by Edwards (1965, 1969), Menhinick (1962) and others.

I n t e r s p e c i f i c   a s s o c i a t i o n s

Applying formulae of Cole (1949) to our data, the number of positive
associations among oribatids were increased during 15 out of 18 mo b,

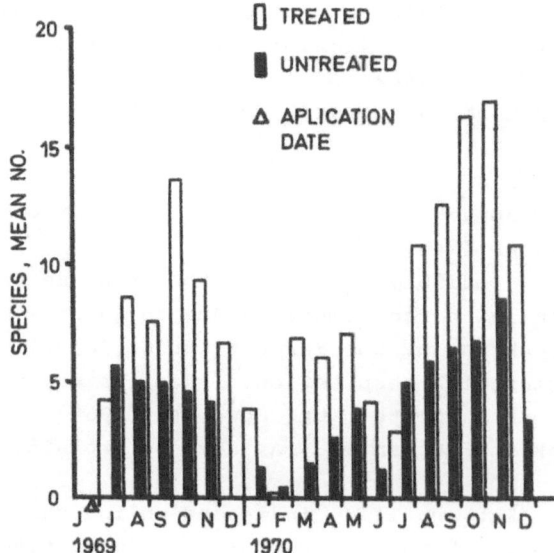

Fig.2. Total species of Oribatids in DDT treated and untreated old fields.

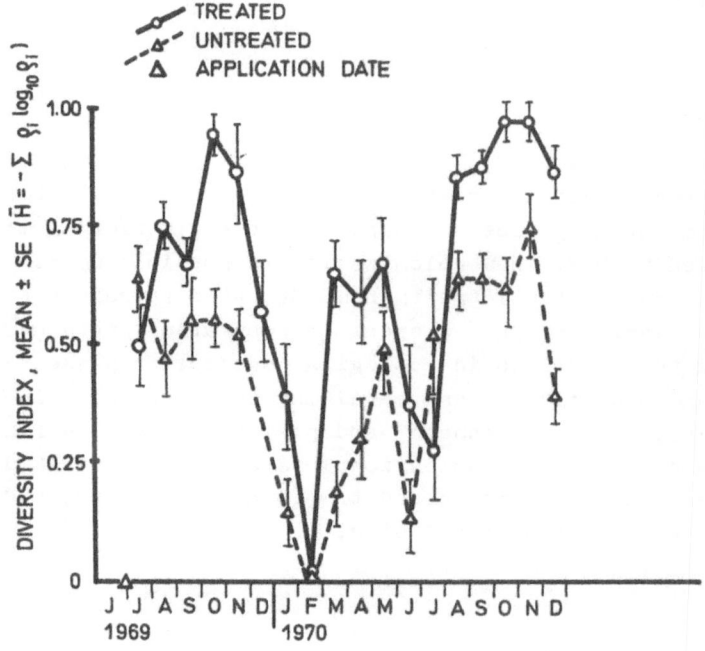

Fig.3.   Species   diversity   of Oribatid mites   from   DDT
treated and untreated old field communities.

treatment. Eight species make up the bulk of positive associations: *Oppiella nova* (Oud.), *Oppia minus* (Paoli), *Anoribatella* sp., *Schelo-ribates* sp. nr. *laminata* (Ewing), *Ceratozetes* sp. nr. *mediocris* Berlese, *Tectocepheus velatus* (Michael), *Eremobelba* sp. and *Brachychthonius* sp.

## Population patterns

Six species specific patterns were apparent relating the effects of DDT application to individual populations (Table 1).

In Pattern I, the first peak possibly reflected the response to reduction in oribatid competition with collembolans due to increased microbial activity. The second peak, being the largest, coincided with the relatively high increase in microbial respiration. Also, it may be a result of trophic level shift as suggested by Edwards (1969) and Menhinick (1962). The single peak in Pattern II may also be explained by the above.

Presence on the T site only (Pattern III) indicated an increased colonization potential caused by DDT. Perhaps predators were destroyed which normally kill propagules of the species found after treatment.

Pattern IV shows species that are not directly or indirectly affected by the application whereas, Pattern VI indicates such effects reducing existing species.

## CONCLUSIONS

Faunal simplification (decreased diversity) resulting from pesticide application as reported by many does not apply to the old field oribatid community subjected to DDT. New colonization is possible by some species of *Acari* as a result of DDT treatment. Oribatid responses to DDT treatment exhibit variable species specific population patterns. Finally, when referring to the trends in ecological succession presented by Margalef (1968) and Odum (1971), and recalling our results of increased species diversity, species richness and positive interspecific associations, it appears that the successional rate of the oribatid decomposer community is temporarily increased toward a more mature, and possibly, a more stable condition on the T site.

T a b l e 1. Response patterns of oribatid mite species to single application (15 kg/ha) of DDT

| Pattern | Characteristics | Number species exhibiting pattern | Dominant representative species |
|---|---|---|---|
| I | Species are commonly (and similarly) present on T and UT sites, but increasing more abundantly on T sites; two population peaks/yr. | 11 | Oppiella nova (Oud.) Oppia minus (Paoli) Quadroppia ferrumeguina Jacot Tectocepheus velatus (Michael) Galumna lanceatum (Ewing) Ceratozetes sp. nr. medocris Berlese Eremobelba sp. |
| II | Species are sporadically (and similarly) present on T and UT sites; more abundant on T sites; one population peak/yr. | 5 | Brachychthonius semiornatus Evans Belba olitor Jacot |
| III | Species present on T site only. | 8 | Liochthonius evansi (Forsslund) Brachychthonius erosus Jacot Anoribatella sp. |
| IV | Species present on T and UT site with no significant differences between sites. | 4 | Nothrus biciliatus C.L. Koch Cultroribula juncta (Michael) Rhysotritia ardua (C.L. Koch) |
| V | Species increased during first year on UT only; T and UT sites non-significantly different the second year | 5 | Scheloribates milleri Jacot Epilomannia elongata Banks Brachychthonius italicus Berlese Brachychthorius rostratus Jacot |
| VI | Population numbers on UT site greater than on T site | 2 | Ceratozetes gracilis (Michael) Xylobates sp. nr. lophotricus (Berlese) |

Old field herbaceous community, central Ohio, USA. T = treated, UT = untreated.

R e f e r e n c e s

C o l e , L. C. (1949): The measurement of interspecific association.
    Ecology 30(4), 411-424.

E d w a r d s , C. A. (1965): Effects of pesticide residue on soil in-
    vertebrates and plants. 239-261 pp. In: Goodman, G.T., Edwards,
    R.W., Lambert, J.M. (eds.), Ecology and the industrial society,
    Wiley and Son, New York.

E d w a r d s , C. A. (1969): Soil pollutants and soil animals. Sci.
    Am. 220(4), 88-99.

L l o y d , M., Z a r , J. H. and K a r r , J. R. (1968): On the
    calculation of information-theoretical measures of diversity. Am.
    Midl. Nat. 79(2), 257-272.

M a l o n e , C. (1969): Effects of Diazinon contamination on an old-
    field ecosystem. Am. Midl. Nat. 82(11), 1-27.

M a r g a l e f , R. (1968): Perspectives in ecological theory. Univ.
    Chicago Press, Chicago, 111 pp.

M e n h i n i c k , E. F. (1962): Comparison of invertebrate popula-
    tions of soil and litter in mowed grasslands in areas treated and
    untreated with pesticides. Ecology 43(3), 556-561.

O d u m , E. P. (1971): Fundamentals of ecology. W. B. Saunders Co.,
    3rd ed., Philadelphia, London, Toronto, 514 pp.

S h u r e , D. J. (1971): Insecticide effects on early succession in
    an old-field ecosystem. Ecology 52(2), 271-279.

S t o t z k y , G. (1965): Methods in microbial respiration: $CO_2$ evo-
    lution. 1550-1558 pp. In: Black, C.A., Methods of soil analysis,
    Pt. 2, Am. Soc. Agron., Madison, Wisc., 1572 pp.

V o r o n o v a , L. D. (1968): The effect of some pesticides on the
    soil invertebrate fauna in the South Taiga Zone in the Perm Region
    (USSR). Pedobiol. 8, 507-525.

D i s c u s s i o n

C. F. v a n  d e  B u n d : Have you evidence of the content of DDT in
your experimental field? I would like to make some remarks: I observed
in my experiments with DDT in arable soil that the recovered amount of
DDT after 15 years of annual applications with an amount of 0.3 grams
per square meter was 3.5 ppm in the topsoil; with an amount of 0.6 grams
per square meter it was 9 ppm; and with 2 gram per square meter it was
60 ppm.

With the lowest concentration the collembolans  were very abundant and the gamasids  decreased by about 50 to 70 %.  With the highest concentration  the increase  of the *Collembola*  was considerably less than the lowest dosage,  the gamasids  disappeared completely after 3 years.

Finally, the complete absence  of *Rhodacarus* spp.  connected  with a very high density of *Folsomia candida* or *Folsomia fimentaria* and *Onychiurus armatus*  may be used  as an indicator of the occurrence of high contamination of DDT in sandy arable soil.

D. L. D i n d a l :  Theoretically  if the DDT  was evenly  distributed there should have been approximately 2 ppm  in the upper 1 cm  of soil. However,  soil assays indicate,  that the initial distribution  was not that uniform. Perhaps the DDT concentrations at application ranged from less than 1 ppm to about 9 ppm.

Your remarks regarding *Collembola*  and  *Mesostigmata*  are very interesting to me since our preliminary results generally appear to agree with trends in your low DDT concentration fields.  I will be curious to see  if the specific patterns you indicate  will also hold true  in our field  since  we have representatives  of all the species you just mentioned.

A n d e r s o n : I have  several  questions.  First,  were  there  any major changes  in rank  of the oribatid species  in your treated sites? Secondly, were the changes in dominance in any way predictable based on the feeding habits of the oribatids  (that is,  any differences between microbial feeders versus higher plant material  feeders)?  Finally, have you been able to observe relationships  between dominance  and maturity of the field ecosystem?

D. L. D i n d a l :  In answer to your first question,  yes,  there appears to be a change  in rank order of oribatid species  in response to the treatment.  However, our quantitative evaluations of dominance have not yet proceeded to the point  where we can see possible relationships to feeding habitats and community maturity.  All species data  will ultimately be subjected  to a variety of formulae  to quantify dominance.

# The effect of chemical substances on the activity of Lumbricidae in the process of straw disintegration

O. ATLAVINYTE

Institute of Zoology and Parasitology,
Academy of Sciences of the Lithuanian SSR,
Vilnus

## INTRODUCTION

The effect  of chemical substances  such as herbicides and insecticides on soil organisms is but little investigated. Martin et al. 1959;  Klostermeyer, 1959; Van der Lan, Aspöck, 1962; Bauer, 1964; Edwards, 1965; Witkamp, Crossley, 1969; Ilin, 1969; Kurcheva,1971 show that the majority of widely  used  herbicides  and  insecticides have a more  or less negative effect on useful invertebrate organisms of the soil. Under the influence of these groups of chemical substances  the amount of biological agents in the soil represses,  very often,  all processes of their life activity which have a tendency to slow down.

The aim of this work  was to investigate the effect of some herbicides and insecticides on *Lumbicidae* under straw disintegration.

## METHODS

In 1970 - 1972 experiments were carried out under vegetative conditions in vegetative pots (5 l in volume).  For each variant 4 - 5 kg of soil, 50 g of straw and a definite number of earthworms (*Allolobophora caliginosa* Sav. f. typica) (10; 20; 30)  were placed in the 5 litre volume vegetative pots. All the variants of the experiments had 3 replications. In our experiments, we used these types of soil: soddy podzolized loamy sand (Vr), soddy podzolic sandy loam (Sd) and soddy gley sandy loam (K).

With  the purpose  of establishing  the effect  of the  individual chemical substances  on the biological processes  of the soil, the following herbicides and insecticides were used: sevin (1; 5; 10 and 20 mg). naphthalene (30 g), dipterex (5 mg), sodium trichloracetate - STA (2 g). Chemical substances, in the above mentioned amounts, were scatter-

ed on the surface  of the soil.  Intensity of straw disintegration  was
determined by weighing the remaining straw.


RESULTS

**The effect of naphthalene on the process
of straw disintegration**

Under the effect of naphthalene,  introduced in the amount of 30 g into
the straw layer, straw disintegration  was  considerably slowed down in
all the variants of the experiment. For example, in the control variant
with soil only, 60.5 per cent of straw  did not disintegrated, while in
the pots with the earthworms, the percentage of not disintegrated straw
was 29.4 only. Comparing the amount of the remaining straw in the variants  without naphthalene  with  that of the corresponding straw in the
variants with naphthalene  we found  that in soil K, naphthalene slowed
down  the disintegration  of straw  in the variants  without earthworms
(control)  by 24.7 per cent, while in the variants  with  earthworms it
decreased the corresponding process by 15.0 per cent.

A similar effect of naphthalene  on the disintegration of soil was
observed also in soil Sd (Table 1).


T a b l e  1.   The effect of naphthalene on the activity of earthworms
in the process of straw disintegration

| Date | 5.10.1970-2.3.1971 | | |
| Soil and variants | Not disintegrated straw | | Casts |
| | g | % | g |
|---|---|---|---|
| K | | | |
| Control | 21,5 | 35.8 | 63.7 |
| 10 earthworms | 12.3 | 20.6 | 149.3 |
| 20        " | 11.7 | 19.1 | 163.0 |
| 30        " | 8.7 | 14.4 | 108.0 |
| K | | | |
| Control + naphthalene | 36.3 | 60.5 | 24.3 |
| 30 earthworms + maphthalene | 17.7 | 29.4 | 42.3 |
| Sd | | | |
| Control | 24.7 | 41.1 | 48.0 |
| Control + naphthalene | 38.7 | 64.4 | 81.0 |
| 30 earthworms + naphthalene | 31.6 | 52.7 | 109.0 |

**The effect of herbicides and insecticides**

Results of investigations of the effect of sodium trichloracetate (STA), sevin and dipterex  on the activity of earthworms  in the zone of straw disintegration  disclosed  that  on day 71  on the experiment, 60.8 per cent of straw  remained  in the control variants  while in the variants with earthworms (20 of them) the corresponding figure  was 48.2. In the variants with 20 mg  and  1 g of sevin straw disintegration slowed down in the following manner:  in  control variants  by  5.0 per cent,  with earthworms  and 20 mg of sevin by 7.6 per cent  and with earthworms and 1 g of sevin  by 17.6 per cent.  A negative effect  was not observed in the variants  with  2 g of STA.  In these variants straw disintegration proceeded  with the same intensity  as in the variants without chemical substances (Table 2).

T a b l e   2.    The effect of herbicides and insecticides on the activity of earthworms in the process of straw disintegration in the soil Vr

| Date and variants | Straw | | Casts |
|---|---|---|---|
| | g | % | g |
| **27.4.1972-7-7-1972 (71 days)** | | | |
| Control | 24.3 ± 0.9 | 60.8 | 26.0 ±  3.5 |
| 20 earthworms | 19.3 ± 1.8 | 48.2 | 78.0 ±  6.7 |
| Control + 20 mg of sevin | 26.0 ± 1.1 | 65.0 | 39.3 ±  9.6 |
| 20 earthworms + 20 mg of sevin | 22.3 ± 2.4 | 55.8 | 81.7 ± 12.2 |
| Control + 1 g of sevin | 26.3 ± 1.2 | 65.8 | 43.0 ± 15.5 |
| 20 earthworms + 1 g of sevin | 26.3 ± 1.3 | 65.8 | 36.0 ± 10.0 |
| Control + 2 g of STA | 21.3 ± 2.2 | 53.2 | 26.7 ±  3.8 |
| 20 earthworms + 2 g of STA | 19.3 ± 2.2 | 48.2 | 103.3 ±  4.4 |
| | | | |
| **16.5.1972-28-8-1972 (104 days)** | | | |
| Control | 14.3 ± 0.1 | 35.7 | 21.0 ±  2.65 |
| 20 earthworms | 8.7 ± 1.7 | 21.7 | 105.7 ± 13.79 |
| 20 " + 1 mg of sevin | 11.3 ± 1.7 | 28.2 | 92.0 ± 25.15 |
| 20 " + 5 mg " | 11.7 ± 4.0 | 29.2 | 80.7 ± 12.02 |
| 20 " + 5 mg of dipterex | 8.2 ± 1.3 | 20,5 | 103.7 ± 17.35 |
| 20 earthworms + 2 g of STA | 8.5 ± 1.3 | 21.2 | 101.7 ±  9.35 |

A check-up of the intensity of straw disintegration  on day 104 of the experiment  demonstrated  36.6 per cent  of not disintegrated straw in the control variants while 21.7 % in the variants with earthworms. Sevin (1 mg; 5 mg; 20 mg) slowed down straw disintegration by 6.5 - 7.5 per cent. Dipterex (5 g) and STA (2 g) had no negative effect on both straw disintegration and the activity of earthworms. In these variants straw disintegration proceeded with the same intensity as in the control variants.

It is also assumed that the physiological activity of earthworms has a great impact on the intensity of straw disintegration. Under experimental conditions activity of earthworms was determined by the amount of casts excreted by these invertebrates. Results of investigations showed that the highest amount of casts was excreted by earthworms in soil fertilized by straw. In the straw layer, up to 163.0 - 193.0 g of casts were detected (Atlavinyte, 1971). When sevin and naphthalene were added, earthworms excreted muchı less casts (Tables 1 and 2).

A toxic effect of sevin on earthworms was noticed also by other authors (Van der Lan, Aspöck, 1962; Ilin, 1969). These authors assert that sevin paralyzes earthworms. This fact may also decrease earthworm activity in the process of straw humification.


CONCLUSIONS

Naphthalene and sevin slow down markedly earthworm activity. Straw disintegration is retarded under the influence of chemical substances. Under these circumstances a positive effect of earthworms on the processes of disintegration is very insignificant. Naphthalene decreased straw disintegration by 15.0 - 29.4 per cent, sevin by 6.5 - 17.6 per cent. STA had no negative effect on the processes of straw disintegration.


SUMMARY

Investigations carried out in vegetative pots have shown that following the introduction of different chemical substances to the soil, earthworm activity decreased at different rates. Sodium trichloracetate (STA) and dipterex slowed down moderately earthworms activity during straw humification. Naphthalene and minute amounts of sevin decreased considerably earthworm activity. Earthworms excreted less casts, and straw humification proceeded more slowly.

# References

V a n   d e r   L a n , H. and   A s p ö k , H. (1962):   Zur   Wirkung   von
    Sevin auf Regenwürmer. Anz. Schädlkunde 35, 7, 180-182.

A t l a v i n y t e , O. (1971):   The activity of *Lumbricidae, Acarina*
    and *Collembola* in the straw humification process. Pedobiologia 11,
    2, 104-115.

B a u e r , K. (1964): Studien über Nebenwirkungen von Pflanzenschutz-
    mitteln auf die Bodenfauna. Mitt. Biolog. Bundesanst. Land. Forst-
    wirtsch., Berlin-Dahlem, Heft. 112, 1-42.

E d w a r d s , C. A. (1965):   Effects of pesticide   residues   on soil
    invertebrates   and   plants.   In: Ecol. and the Industrial Soc. (V.
    Symp. Brit. Ecol. Soc.), Oxford, Blackwell Sci. Publ., 239-261.

I l i n , A. M. (1969):   On the toxic effect of herbicides on ants and
    earthworms. Zool. Ž. XLVIII, 1 (In Russian).

K l o s t e r m e y e r , E. C. (1959): Insecticide-induced population
    changes in four mite species on Alfalfa.   J. Econ. Entomol. 52, 5,
    991-994.

K u r c h e v a , C. F.(1971): The role of soil animals in the disinte-
    gration   and   humification of vegetative remnants.   Moskva, Nauka,
    1-155 (In Russian).

M a r t i n , J. P.,   H a r d i n g , R. B.,   C a n n e l l , G. H.
    and   A n d e r s o n , L. D. (1959):   Influence   of   five annual
    field applications of organic insecticides   on soil biological and
    physical properties. Soil Sci. 87, 6, 334-338.

W i t k a m p , M. and   C r o s s l e y , D. A., Jr. (1966): The role
    of arthropods   and   microflora   in breakdown   of White Oak litter.
    Pedobiologia 6, H. 3/4, 293-303.

# Discussion

T. W.   R e y n o l d s : I am interested to know your reason for selecting *Allolobophora caliginosa*   as the species   for your experiments. As
you know   I do not use a species complex   in my research and   I will be
interested to examine some of your specimens in the future.

O. A t l a v i n y t e : *Allolobophora caliginosá*   represents   80-90 %
of the species found in our region.

# Veränderungen der Moosmilbenzönosen (Acarina, Oribatoidea) im Verlaufe von 45 Jahren nach Bodensterilisation durch unterirdischen Kohlenflözbrand

J. VANĚK

Institut für Landschaftsökologie der Tschechoslowakischen Akademie der Wissenschaften, Průhonice, ČSSR

Das Institut für Landschaftsökologie und seine Arbeitsgruppe für Bodenbiologie befasst sich mit der Suche nach solchen Struktur- und Funktionscharakteristiken der natürlichen Landschaftskomponenten, die entweder durch ihre Werte den natürlichen Zustand derselben charakterisieren, oder durch Veränderung ihrer Werte Störungen der Komponenten infolge natürlicher oder künstlicher Einflüsse signalisieren. Solche Auswertung und Ausnützung der Indikationseigenschaften der Zönosencharakteristiken bildet die Grundlage der zönologischen Bioindikation.

Beim Studium der Bioindikationseigenschaften von Zönosen der Bodentiere und bei Versuchen um die Prüfung ihrer Empfindlichkeit untersuchten wir eine in diesem Sinne interessante Lokalität und die gewonnenen Ergebnisse sind dazu angetan, im Rahmen des Themenkreises dieses Kolloquiums vorgelegt zu werden.

Im Becken des Erzgebirgsvorlandes (NW-Böhmen) liegt in der Nähe der Stadt Teplice die Gemeinde Tuchomyšl. Vor 45 Jahren brach in der Nähe dieser Ortschaft ein unterirdischer Kohlenflözbrand aus. Der Brand pflanzte sich in Richtung gegen den Abhang des Rovný-Hügels fort. Am Brandherd brannte das Kohlenflöz ungefähr 0 - 5 m unter der Erdoberfläche. Im Jahre 1967, als die Proben entnommen wurden, brannte das Flöz 500 m vom Ort des Brandausbrechens entfernt und 30 m unter der Erdoberfläche.

Beim Fortschreiten der Brandfront konnten auf der Erdoberfläche folgende Begleiterscheinungen beobachtet werden:

a) Überhitzung der Erdoberfläche bis 75 - 80 °C;

b) Entweichen von Dampf- und Verbrennungsprodukten;

c) Ausbrennen des Lössbodens der ganzen Länge der Hangendscholle nach und Bildung von Ziegelstein;

d) Aufbrechen von bis 3,5 m tiefen und 1,5 m breiten Kluften;

e) langfristige Veränderungen des Mikroklimas  mit der Tendenz zur all-
   gemeinen Temperatur- und örtlichen Feuchtigkeitserhöhung;
f) Rutscherscheinungen und Bodensenkungen.

In Folge der ad a - c angeführten Begleiterscheinungen  wurde  der
Boden  an der Brandfront  fast vollkommen sterilisiert,  wobei sich die
sterilisierte Zone  der Brandfront  mit einer Geschwindigkeit von 500 m
in 45 Jahren  gegen den Abhang des Rovný-Hügels fortpflanzte.  Nach dem
Abklingen der akuten Erscheinungen des unterirdischen Brandes hielt die
primäre Rekolonisierung  und  die  mit ihr verbundene Regeneration  der
Phyto- und Zoozönosen ihren Einzug,  wobei die Sukzession der Rekolini-
sierung und Regeneration  durch die Geschwindigkeit  des Fortschreitens
der Brandfront bestimmt wurde.

Auf der untersuchten Lokalität  bestimmten wir einzelne Zonen nach
ihrem gegenwärtigen Habitat.  So wurde  die Zone A  für den Raum  abge-
grenzt, wohin  die Brandfront  bisher  noch nicht eingedrungen war  und
diese Zone  stellt den Ausgangs- und Vergleichszustand dar.  Die Zone B
schliesst  die aktuelle Brandfrontzone  mit allen ihren Begleiterscheinungen ein, die durch Bodensterilization  und Zerstörung der Bodenober-
flächenzönosen hervorgerufen wurden.  Die Zone C,  jene der primären
pflanzlichen  und  tierischen Rekolonisierung,  weist noch eine erhöhte
Oberflächentemperatur auf,  ermöglicht jedoch  die Entwicklung der Pio-
niervegetation  und  das Aufkommen der Tiere.  Flächenmässig überwiegt
hier die Gras- und Krautvegetation,  wenn auch noch nicht in vollkommen
geschlossener Form.  In der Zone D sind  die Begleiterscheinungen des
Brandes schon abgeklungen, die Temperatur der Bodenoberfläche  ist normal, es hat sich eine geschlossene Gras- sowie Krautvegetation ausge-
bildet und Sträucher wie auch Bäume haben Fuss gefasst.  In der Zone E
brach der Brand vor ungefähr 45 Jahren aus  und es ist bei den hiesigen
Zönosen  die längste Rekolonisierungs-, Regenerations- und Stabilisie-
rungsdauer vorauszusetzen.

Durch die Brandfortpflanzungsachse  wurde ein Transekt gelegt, aus
dem Probenreihen der Oberflächenbodenschicht entnommen und gemäss dem
geläufigen Selektionsverfahren mittels Tullgreen-Geräten analysiert
wurden.

Das gewonnene Material  diente der Bewertung  des Mezozooedaphons
als Ganzes, auf der Ebene der höheren Taxone. Dabei wurden die Veränderungen in der Vertretung  und in der Individuenhäufigkeit der typischen
Komponenten des Mezozooedaphons untersucht.

Die Milben der *Oribatoidea*-Gruppe wurden bis zu den Arten bestimmt
und daraufhin  eine eingehende Analyse  der *Oribatoidea*-Zönosen  in den
Zonen A – E dem Transekt entlang durchgeführt. Die gewonnenen Ergebnisse

werden in vollem Umfang in der Sammelschrift Quaestiones geobiologicae, Nr. 13 veröffentlicht. Für dieses Kolloquium wurden zumindest die gewichtigsten Erkenntnisse zusammengefasst und graphisch dargestellt.

Durch Untersuchung der ganzen Lokalität wurden 4.201 Tiere der Mezozooedaphon-Kategorie gewonnen, davon 2.570 Exemplare der *Oribatoidea*, die 69 Arten angehörten. Von diesen waren 7 Arten neu für die Fauna der ČSSR und eine Art (*Suctobelba italica* Mah., 1966) wurde als zweiter Fund in der Welt ermittelt.

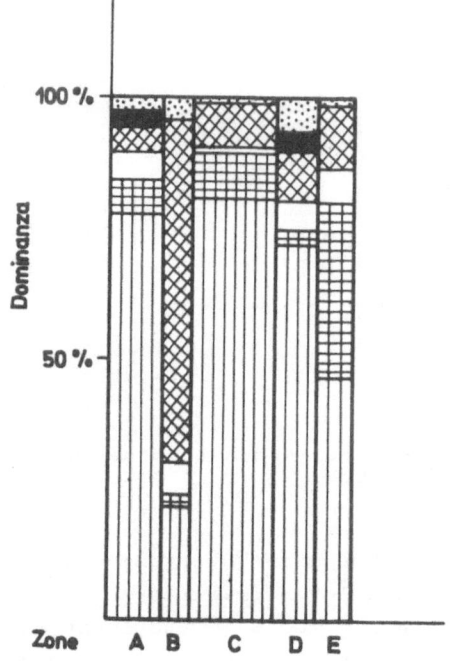

Abb.1. **Abundanz und Dominanz der Tiere in den Zonen A-E.**

Die Abbildung 1 zeigt eine kombinierte Darstellung der Abundanz und Dominanz der ermittelten Taxone des Mezozooedaphons in den Zonen A - E.

Die errechnete Besiedlungsdichte der Zonen A - E betrug per 1 m$^2$ Bodenfläche und bis 5 cm Bodentiefe 119.850 Tierexemplare in der Zone C (Maximum) und 23.950 Exemplare in der Zone E (Minimum). Der untersuchten Brandlokalität würden nach geobotanischen Rekonstruktionskarten Eichen-Hainbuchenwälder entsprechen, wobei der Stand des Mezozooedaphons in unseren Verhältnissen 3 bis 4mal höher sein sollte als die von uns ermittelten Werte. Auch die Stände des Mezozooedaphons der Wiesenersatz-

gesellschaften weisen unter entsprechenden Verhältnissen 1,5 bis 3mal
höhere Werte auf, als wir auf der untersuchten Lokalität ermittelten.
Dieser Sachverhalt ist durch den Hintergrund der auf der untersuchten
Lokalität bestehenden Verhältnisse bedingt. Das breitere Gebiet von
Teplice ist nämlich schon lange dem ständigen Einfluss von Emissionen
ausgesetzt. Zu diesen Umweltsbedingungen kommt noch die Wirkung des
unterirdischen Brandes hinzu, der Gegenstand unserer Untersuchung ist.
    Die in den *Oribatei*-Zönosen hervorgerufenen Veränderungen sind aus
den Abbildung 2 - 5 zu ersehen.

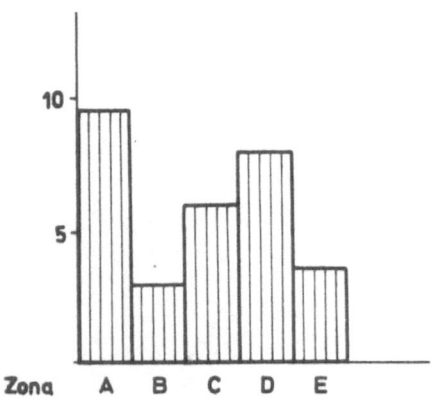

Abb.2. Mittlere Anzahl der Oribatei-Arten in einer Teil-
probe (100 cm$^3$) in den Zonen A-E.

    In der Zone der akuten Einwirkung des Brandfrontüberganges (Zone
B) kam es zur Erniedrigung der mittleren Anzahl von *Oribatei*-Arten per
100 cm$^3$ gegenüber dem Ausgangszustand (Zone A = 100 %), und letzterer
wurde im Transekt im Verlaufe von 45 Jahren der Zönosenrekolonisierung
und -regeneration nicht wieder erreicht. Interessanterweise zeigte der
aufsteigende Regenerationstrend dieses Indikators in der Zone E einen
steilen Abfall (Abb. 2).
    Die in den Zonen A - E ermittelten Arten wiesen eine unterschied-
liche Individuenanzahl auf. Die Abbildung 3 zeigt die Werte des Indikators "Individuenhäufigkeit der Arten" so wie er für die einzelnen
Zonen errechnet wurde. Wir finden, dass die höchsten Werte der Individuenhäufigkeit der Arten in der Zone C bei gleichzeitig ausgeprägt
niedrigerer Artenanzahl erreicht wurden (siehe auch Abb. 1). Beim An-
wachsen der Artenanzahl (Abb. 1) gleicht sich die Individuenhäufigkeit
der Arten mit dem Ausgangszustand aus (Abb. 2), wobei aber die Artenanzahl der Zone A nicht erreicht wird (Abb. 1).

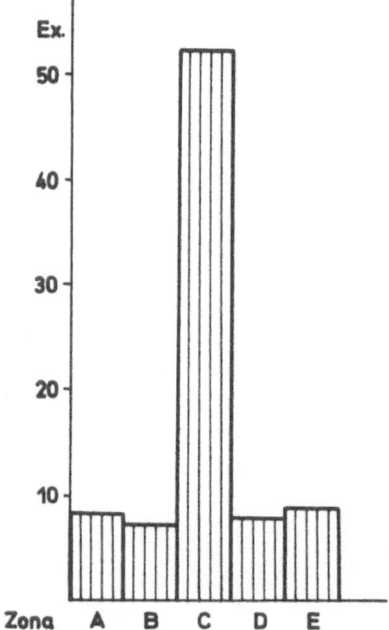

**Abb.3. Mittlere Individuenhäufigkeit der Milbenarten in den Zonen A-E.**

Zur Beurteilung der durch den Brand hervorgerufenen Veränderungen sowie des Regenerationstrends in der Sukzession der Zonenrekolonisierung benutzten wir den Indikator "Minimales Zönosenareal". Dieser Indikator ist die kleinste Fläche bzw. der kleinste Raum, wo theoretisch alle, zönotisch bedeutenderen Arten vorkommen, und er wird aus der Frequenzdominante mit niedrigster Abundanz errechnet. Diese Charakteristik bewährte sich bisher stets als besonders empfindlicher Indikator bei der Ermittlung der Veränderungen in den Zönosen des Mezozooedaphons. Auch in unseren diesbezüglichen Untersuchungen signalisierte sie durch ausgeprägte Erhöhung ihrer Werte eine Störung in den Zonen B und E, wobei sie in der Zone D fast die Werte der Vergleichszone (A = 100 %) erreichte.

Die Berechnungen der qualitativen Homogenität sind auf der Abb. 5 zusammengefasst. Bei ihrer Einschätzung ist in Betracht zu ziehen, dass die mittleren Werte der Homogenität in den Zonen C, D und E bei einer beträchtlich niedrigeren Artenanzahl (siehe auch Abb.2) gewonnen wurden.

Im oberen Teil der Abb. 6 ist das durch die Brandfortpflanzungs- achse, also kongruent mit dem untersuchten Transekt gelegte Profil des Brandfeldes eingezeichnet. In diesem Profil, das zweimal überhöht ist, sind die Zonen A - E bezeichnet. Ferner sind hier angeführt: von den

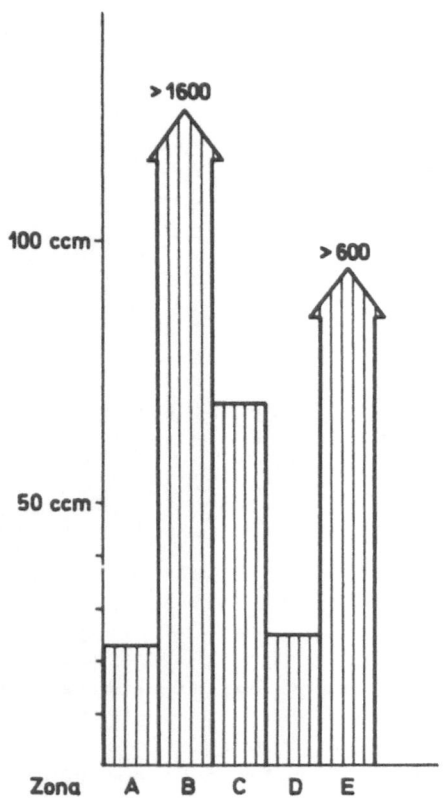

100 ccm

50 ccm

Zona   A   B   C   D   E

Abb.4. Minimale Areale· der *Oribatei*-Zönosen in den Zonen A-E.

ausgeprägsten  qualitativen  Charakteristiken  die mittlere Artenanzahl
(Abb. 6b); von den quantitativen Charakteristiken die Abundanz der *Oribatei* (Abb. 6c) und als letzte, ganz unten, graphisch dargestellt die
synthetische Charakteristik der Zönosenstruktur der *Oribatei*.  In der
letzteren  sind  4 berechnete Zönosenstruktur-Charakteristiken der *Oribatei* zusammengefasst und als Komplex bewertet: qualitative Homogenität,
minimales Areal, Anzahl der typischen  und Anzahl der frequenten Arten.
Sie drückt die Formationsstufe der *Oribatei*-Zönosen aus.  Im Vergleich
mit dem Wert dieser Charakteristik  für die Zone A (= 100 %) lassen die
für die Zonen B, C, D und E  errechneten Werte den Trend · der Zönosen-
rekonstruktion und -regeneration nach dem deteriorierenden Einfluss der
Brandfront erkennen.

        In allen angeführten Diagrammen  wie auch in weiteren  (hier nicht
vorgelegten) numerischen Analysen der *Oribatei*-Zönosen ergab sich wider
Erwarten,  stets ein niedrigerer qualitativer,  quantitativer  wie auch

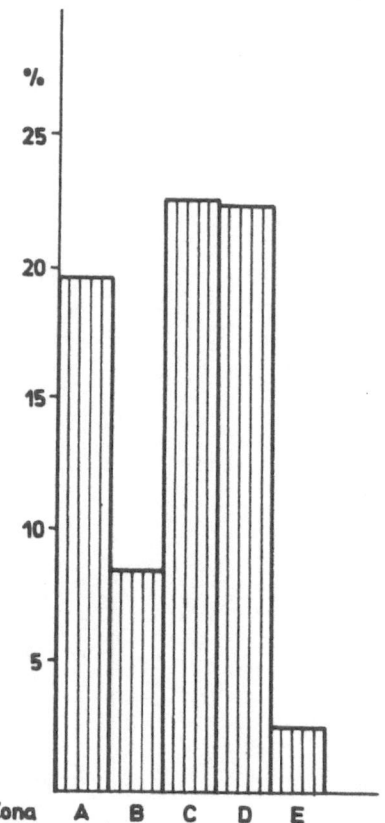

Abb.5. Qualitative Homogenität der *Oribatei*-Zönosen in den Zonen A-E.

struktureller Wert der *Oribatei*-Zönosen für die Zone E. Darüberhinaus wurden in dieser Zone einige Arten gefunden, die das Optimum ihrer Lebensbedingungen in feuchten Biotopen finden. Es wurde deshalb eine ausführliche Untersuchung der Entwicklungsgeschichte der Lokalität in der Zone E vorgenommen und es wurde ermittelt, dass hier früher zwei kleine Seen waren, die vor zwei Jahren entwässert wurden. Die Entwässerung hatte hier eine ausgeprägt ganzflächige Veränderung zur Folge, während sich in der Zone B die akute Auswirkung der Brandfront nur ungleichmässig durchsetzte, so dass die Bodensterilisation flächenmässig nicht hundertprozentig war. Dieser Unterschied im Charakter des Deteriorationseinflusses kam in den Werten der Zönosencharakteristiken der *Oribatei* mit hinreichender Indikationsempfindlichkeit zum Ausdruck. Dadurch konnte die Verwendbarkeit der zönologischen Bioindikation für die Untersuchung der Präsenz und der biologischen Wirkungskraft der deterio-

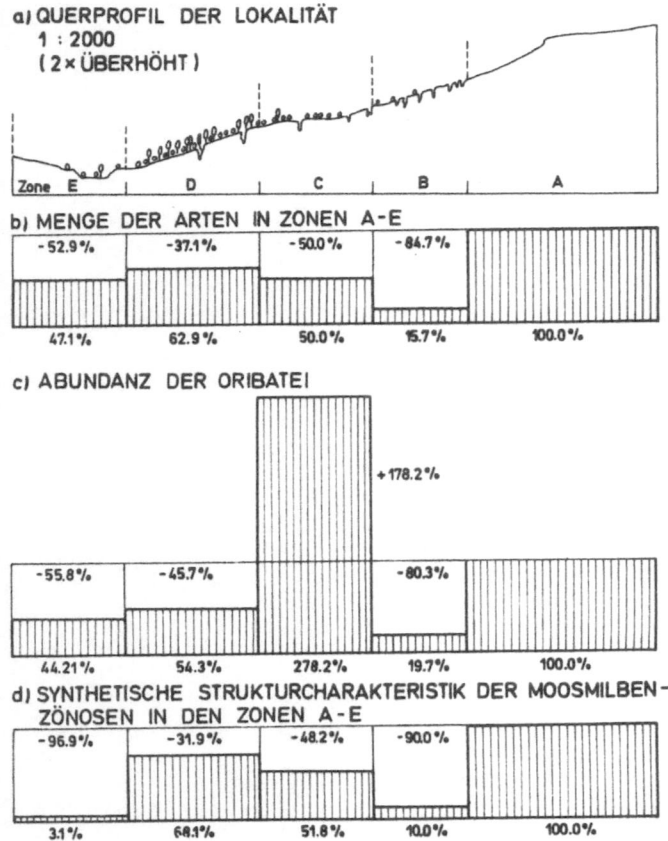

a) QUERPROFIL DER LOKALITÄT
   1 : 2000
   ( 2× ÜBERHÖHT )

Zone  E        D         C         B            A

b) MENGE DER ARTEN IN ZONEN A-E

- 52.9%    - 37.1%     - 50.0%     - 84.7%

47.1%      62.9%       50.0%       15.7%       100.0%

c) ABUNDANZ DER ORIBATEI

+ 178.2%

- 55.8%    - 45.7%                  - 80.3%

44.21%     54.3%       278.2%      19.7%       100.0%

d) SYNTHETISCHE STRUKTURCHARAKTERISTIK DER MOOSMILBEN-
   ZÖNOSEN IN DEN ZONEN A-E

- 96.9%    - 31.9%     - 48.2%     - 90.0%

3.1%       68.1%       51.8%       10.0%       100.0%

**Abb.6. Synthetische Bewertung der Hauptcharakteristiken der *Oribatei*-Zönosen in den Zonen A-E.**

risierenden - besonders anthropischen und anthropogenen - Einflüsse erneut bestätigt werden.

Aus der Untersuchung der Lokalität des unterirdischen Kohlenflözbrandes bei Tuchomyšl können die nachfolgenden Schlussfolgerungen gezogen werden. Die Zönosencharakteristiken brachten mit hinreichender indikatorischer Empfindlichkeit zum Ausdruck, dass

1. durch den Einfluss des unterirdischen Brandes in der Zone seiner aktiven Auswirkung weitgehende qualitative und quantitative Störungen des Mezozooedaphons verursacht wurden, besonders der Teilzönosen der *Oribatei*, deren Struktur zerstört wurde;

2. die durch den Brand hervorgerufenen Veränderungen langfristiger Natur sind und dass die *Oribatei*-Zönosen noch nach 45 Jahren nach dem Brandfrontübergang ihren Zustand weder qualitativ (Artenanzahl), noch

quantitativ (Anzahl der Individuen von *Oribatei* sowie anderer Gruppen des Mezozooedaphons), und nicht einmal strukturell (Zeichen der Formationsstufe und Stabilität der *Oribatei*-Zönosen; siehe Abb. 6d) erneuert haben;

3. der Unterschied im Charakter, im Zeitabstand und in den Folgen des Deteriorationseinflusses in den Zonen B (Brand) und E (Entwässerung vor zwei Jahren) durch die Werte der Zönosencharakteristiken mit bemerkenswerter Empfindlichkeit signalisiert wurde;

4. der qualitative Entwicklungstrend der Rekolonisierung und Regeneration der durch den Brand geschädigten Zönosen des Mezozooedaphons nicht zum Ausgangszustand (Zönosen der Kulturwiesen), sondern gleichlaufend mit der Entwicklung der Pflanzen zu einem andersartigen Zustand (Bodenzönosen der Gehölzformationen) gerichtet ist;

5. in der Zone C im Lichte der Zönosencharakteristiken der Zustand einer ökologischen Ausgeprägtheit festgestellt werden konnte, mit anderen Worten ein Zustand, der aus der determierenden Auswirkung eines ökologisch aktiven Faktors (in unserem Falle die erhöhten Bodenremperaturen) resultiert und der eine Übervermehrung der Vertreter einiger hier vorkommenden Arten nach sich zog. Ein solches Bild ist uns auch von anderen ökologisch ausgeprägten Orten bekannt, wie z.B. von Algen-, Flechten- und Moosablagerungen auf Steinen, Mauern und Dächern, oder auf Baumstämmen u.a.

Die Untersuchung der Lokalität des unterirdischen Brandes bei Tuchomyšl brachte ausser einer Vielzahl von Ergebnissen, die zur Klärung des Charakters und des Umfanges der Bodensterilisation infolge des brandbedingtes Deteriorationseffektes beitrugen, auch eine Reihe von Erkenntnissen über die Sukzession der Zönosenrekolonisierung und -regeneration des Mezozooedaphons nach dem Abklingen des Brandeinflusses, und ermöglichte die Einsicht in die Veränderungen der Zönosencharakteristiken der *Oribatei*-Gesellschaft, die alle diese Vorgänge und Erscheinungen begleiteten. Es konnte damit im wesentlichen der Nachweis erbracht werden, dass es in landschaftlich-ökologischen Untersuchungen unter Anwendung eines repräsentativen Verfahrens der Komplexbewertung der Bodenbiologie, mit Hilfe der Teilzönosen der *Oribatei*, möglich ist, die Methode der zönologischen Bioindikation erfolgreich zur Anwendung zu bringen.

## L i t e r a t u r

B a l o g h , J. (1953): Grundzüge der Zoozönologie. Budapest, 248 S.

B a l o g h , J. (1958): Lebensgemeinschaften der Landtiere. Budapest, 386 S.

V a n ě k , J. (1959): Použitelnost půdní zoologie pro lesnickou typologii. Sbor. les. fak. ČVUT, Praha, II, 25-40.

V a n ě k , J. (1966): Zoocenologická indikace biologického stavu půd. Sbor. VLÚ při VŠZ v Praze, 8, 253-267.

V a n ě k , J. (1967): Industriexhalate und Moosmilbengemeinschaften in Nordböhmen. Progress in Soil Biology, Braunschweig, 331-339 S.

V a n ě k , J. (1974): Mezoedafon nadloží požářiště sloje Barbora. Quaestiones geobiologicae, 13, Academia, Praha.

## D i s k u s s i o n

R. C o v a r r u b i a s : Did you calculate also diversity indexes in order to make comparisons, and not only percentage indexes?

J. V a n ě k : Die Diversitätsindexe wurden bisher nicht errechnet. Alle für diese Berechnung erforderliche Daten stehen zur Verfügung, so dass es möglich ist, diese später durchzuführen.

M. M o r i t z : Ist die Abundanzsteigerung in der Zone C neben interspezifischer Konkurrenz nicht auch auf eine Verbesserung (Optimierung) der abiotischen Faktoren zurückzuführen?

J. V a n ě k : Gewiss ja. Die zoozönologische Bioindikation ist eine Indikation des Komplexes der Lebensbedingungen. Die ermittelten Zönoseneigenschaften sind immer das Ergebnis der Entwicklung eines Ortes, an dem die Untersuchung durchgeführt wird, und die Wiederspiegelung des Einflusses des ganzen abiotisch-biotischen Komplexes der ökologischen Faktoren. Auf der Fläche der Zone C hatte diese Entwicklung des Ortes, desgleichen wie auch des Faktorenkomplexes, den von uns ermittelten Zustand zur Folge: die Oribatei-Zönosen enthalten wenige Arten, jedoch weisen die hier vorkommenden Arten grosse Individuenzahlen auf. Da uns ein solcher Zustand von Orten mit ausgeprägter Auswirkung eines oder einiger weniger Faktoren bekannt ist, setzen wir ähnliche Lebensbedingungen auch in der Zone C voraus. Dabei zählt zu den die Übervermehrung der hier vorkommenden Arten bewirkenden Einflüssen zweifelsohne jener der ganzjährig günstigen, entwicklungserförderlichen Lebensbedingungen.

A. P a l i s s a :  Die instabilsten Verhältnisse ergaben die Zonen
B und E.  Die verschiedene Entwicklung  beider Stellen  müsste in einem
verschiedenartigen Artenspektrum seinen Ausdruck finden?

J. V a n ě k :  Selstverständlich,  auf der Fläche  der Zone E  wurden
einige Arten gefunden, die normalerweise Teich- und Sumpfufer besiedeln.
Ihr Vorkommen  war nicht häufig, eher vereinzelt, aber sie signalisier-
ten eine grundsätzliche Unterschiedlichkeit der beiden Zonen  und waren
einer der Gründe die dem Streben nach Klärung der Entwicklungsgeschichte der Zone E zugrundelagen.

R. C o v a r r u b i a s :  Was the original field  homogeneous  enough
to justify the comparisons  between zones A until D, that is, was there
not also a superposition of the zonation problem?

J. V a n ě k :  In 45 Jahren pflanzte sich die Front des unterirdischen
Brandes um 500 m fort.  Die Geländeneigung  war  dabei  nicht so steil,
dass es nötig gewesen wäre,  mit einer Zonation  und  ihrem Einfluss zu
rechnen, der sich auf die gewonnenen Ergebnisse  in bedeutenderem Masse
hätte geltend machen können.

S. K. M i t r a :  Did you make  a qualitative  analysis  of  different
groups  of animals you have dealt  with  to ascertain species diversity
in relation to different zones of soil?

J. V a n ě k :  Eine qualitative Analyse verschiedener Tiergruppen wurde nur in groben Umrissen, orientierungshalber auf der Ebene  der höheren Taxone durchgeführt.  Nur die Milben der *Oribatoidei*-Gruppe  wurden
bis zu den Arten bearbeitet  und nur in dieser Gruppe  wurden eingehen-
dere Analysen der Gesellschaften durchgeführt.

     Was die Abhängigkeit von den Bodenzonen anbelangt,  so wurde diese
nicht verfolgt.  Untersucht wurde nur das Mezoedaphon  der Oberflächenbodenschicht  bis in die Tiefe von 5 cm.  Unter dem Begriff  "die Zone"
sind in diesem Beitrag  die Flächen der untersuchten Lokalität  zu ver-
stehen, die zum Zeitpunkt der Probenentnahme gleiche Beschädigungs- bzw.
Regenerationserscheinungen aufwiesen.

E. v. T ö r n e :  Wäre es für die Untersuchung  Ihres Problems  vielleicht zweckmässiger, vor allem die Diversität  des Prädatoren-Besatzes
zu ermitteln?

J. V a n ě k :  Möglicherweise ja,  doch meine taxonomischen Kenntnisse
reichen für die Untersuchung der Prädatoren nicht aus.

E. v. T ö r n e :  Hat  in der Zone B  eine hangabwärts gerichtete Ver-
schiebung des Oberbodens  stattgefunden und  könnten auch Unterschiede
in der Mächtigkeit des teilweise abgetragenen und teilweise akkumulier-

ten  Oberbodenmaterials  die Ursache für  die beobachteten  Besatzungs-
unterschiede gewesen sein?

J. V a n ě k :  Nein, der Boden  hat sich örtlich  nur gesenkt  und  es
sind stellenweise grosse Spalten  entstanden.  Eine seitliche Abtragung
des Bodenmaterials hat nicht stattgefunden.

G. W a u t h y :  Wieviele identische Arten  haben Sie im Transekt, den
Sie untersucht haben, gefunden? Von welchen Gattungen?

J. V a n ě k :  Es wurden nur sehr wenige Arten gefunden,  die in allen
Zonen der untersuchten Lokalität vorkamen.  Von den Moosmilben waren es
eigentlich  nur  drei Arten:  *Oppia nova* (Oud., 1902) sensu v.d.Hammen,
1952,  ferner  *Punctoribates  punctum*  (C.L. Koch – Berl., 1839)  sensu
Sellnick, 1928, und *Tectocepheus velatus*  (Mich., 1880)  sensu Knülle,
1954.  Die übrigen Arten waren in verschiedener Frequenz in den einzelnen Zonen vorhanden, jedoch kam  ausser den  drei erwähnten Arten keine
weitere Art in allen Zonen vor.

# The effect of clearing
# on the soil arthropods
# of a Nigerian rain forest

B. A. LASEBIKAN

Department of Biological Sciences,
Zoology Division,
University of Ife, Nigeria

INTRODUCTION

The influence of agricultural practices on soil microarthropod populations has been shown by many soil ecologists such as Sheals (1955),
Huhta et al. (1967), Olivier and Ryke (1969) and Edwards and Lofty
(1969). Apart from the work of Olivier and Ryke, most of these works
are concentrated in Europe. There is, however, no such investigation on
the soil arthropod fauna of tropical Africa.

The present investigation is an attempt to present the effect of
clearing of forests for purposes of farming, building, construction,
beautification, etc. on soil arthropod populations of tropical rain
forests. The work was initially aimed to last for several years but the
taking over of part of the investigation area by the University for her
building projects made this impossible. Although the work is far from
complete, the results so far obtained are very interesting and conclusive enough to warrant presentation.

SAMPLE SITE

The area selected for this study was part of the forest yet undeveloped
in the University of Ife Campus. Two sites each 50 metre square, were
chosen. The first site (A), forms part of the University of Ife Forest
Reserve while the second site (B) was a temporarily reserved forest
adjoining the Faculty of Agriculture buildings, this second plot has
now been taken over by the University Parks and Garden Unit, and has
been designated as Parks. The site was cleared in July 1972; the under-
bush as well as the small and medium sized trees were cut, leaving the
big trees on the site unfelled. These large trees cast a light patchy
shade over the soil.

The forest   in Ile-Ife forms part   of the lowland rain forest zone (Keay, 1965) or moist semideciduous forest (Charter, 1970) found in the Southern part of Nigeria.  The nature of the vegetation  of the site as well  as the physical  and  chemical composition  of the soil  has been given by Lasebikan (in preparation).

The University  of Ife Campus site  is about 86 km east of Ibadan, Western Nigeria  and  is located within latitude $7^{\circ}$ 29´N  and longitude $4^{\circ}$ 34´E.  The climate is humid-tropical (or hot and humid).  During the year 1972 the area received 109.2 cm of rainfall  during the wet season (April-October inclusive). Precipitation in the area dropped to 15.2 cm during the dry season (November to March inclusive). The yearly maximum mean temperature for the year was 30.6 $^{\circ}$C while the yearly minimum mean temperature was 22.2 $^{\circ}$C.

## METHODS

Six sample units each 12.4 cm  in diameter,  7.5 cm deep (120.7 sq.cm), were taken  in May 1972  from each of the sites A and B just before the latter was cleared. The samples were extracted in a Burhard type of the Berlese-Tullgren funnel.  During each  of the subsequent  sampling  oc- casions: October 1972, March 1973 and  May 1973,  10 sample units each 5.16 cm in diameter x 10 cm deep (24.6 sq.cm), were taken randomly from both sites A and B. The samples were taken with a split corer fashioned after O´Connor (1957).  This devise was employed because of the friable nature of the soil under investigation and the desire to avoid compres- sing the soil core by forcing it out of the corer  and  to preserve the natural structure  of the soil.  In order to keep the natural structure of the soil intact,  the samples  were taken  into bakelite cylindrical tubes each 5.16 cm internal diameter x 5 cm deep inserted  in the corer (see Lasebikan,  in preparation for further details).  Each 10 cm deep sample unit  was subdivided  into 5 cm depths before being extracted in the modified  Macfadyen  High Gradient Canister  extractor  (Macfadyen, 1961, 1962, and Lasebikan, in preparation). The total number of animals extracted from each subsample unit was ascertained but the data for the analysis  were based on the total animals recovered  from 0-10 cm depth since the preliminary investigations  have shown  that  the majority of the animals  are found  in this zone.  The *Acarina, Collembola* and many of the other groups were stored for future texonomic studies.

Except for the samples taken at the beginning of the investigation (i.e.  May 1972 samples)  t-test analysis  was carried out  on the data

obtained   during   the subsequent sampling occasions.   The analysis  was
however confined to the group of animals that were found common to both
sites on each occasion and also in fairly reasonable numbers.


RESULTS

The animal groups, together with their means  per unit sample, found in
sites A and B  before the clearing of site B  are presented in Table 1.
Tables 2, 3 and 4  give the t-test analysis, on the  mean  of the animal
groups   that   were found common   to both sites   on each of the sampling
occasions.   Although other animal groups were found in both sites, some
of these were recorded, especially in site B, in very low numbers as to
warrant any statistical analysis.   The subsequent changes   with regards
to the presence or absence of various animal groups   over the period of
sampling are summarised in Table 5.


DISCUSSION

There are a number of points worth to be taken into account in discussing the results presented in this work. First that the results presented here  may be considered as of preliminary nature to a much more comprehensive work  which  has now commenced  on a site purposely acquired
for soil ecological work.   Secondly although efforts  have been made to
identify the animal groups, particularly the *Acarina* and  *Collembola* to
family level, this taxonomic category  is still considered broad enough
as to mask any differences  that  may be detected  in the soil fauna of
cleared and uncleared forests.   Lastly because of lack  of knowledge of
the species content of soil fauna  in Nigerian soils and absence of any
information  on their biology, behaviour, feeding preferences etc. certain of the conclusions arrived at here may be regarded as speculative.
     From Table 1, it can be seen  that initially there is little or no
difference  in the composition of the fauna in sites A and B.   Although
site A  tends to have higher populations of animals  than site B  these
differences  do not appear significant.   The differences  in both sites
may, however, be due to differences  in  the intensity  of past disturbance as well as the stages of succession etc. of both sites.   Although
farming  was reported  to have stopped  in both sites about forty years
ago, site B  which is at the valley of an inselberg appeared to be much

T a b l e   1.   Mean   number   of individuals   per unit sample (24.6 cm$^2$) obtained from the uncleared forest (A)   and before the forest (B)   was   cleared.   (Data based on six replicates)

| Animal Group | Mean per Unit Sample | |
|---|---|---|
| | A | B |
| Belbidae | 3.00 | 3.50 |
| Carabodidae | 8.00 | 7.33 |
| Galumnidae | 6.17 | 5.33 |
| Haplozetidae | 14.16 | 14.66 |
| Liodidae | 1.16 | 0.00 |
| Lohnmaniidae | 2.00 | 1.00 |
| Machadobelbidae | 1.33 | 1.66 |
| Nothridae | 0.50 | 0.33 |
| Oppiidae | 8.50 | 4.83 |
| Oribatelidae | 2.00 | 0.66 |
| Phthiracaridae | 1.16 | 0.50 |
| Scheloribatidae | 7.50 | 5.00 |
| Suctobelbidae | 1.16 | 0.83 |
| Cryptostigmata (Adults) | 57.00 | 45.66 |
| Cryptostigmata (Juveniles) | 27.50 | 21.66 |
| Rhodavaridae | 3.00 | 2.33 |
| Gamasina (Adults) | 7.83 | 7.50 |
| Gamasina (Juveniles) | 7.66 | |
| Uropodidae | 5.83 | 2.16 |
| Bdellidae | 2.16 | 1.00 |
| Cheyletidae | 2.16 | 2.50 |
| Cunaxidae | 7.00 | 4.83 |
| Eupodidae | 19.30 | 8.16 |
| Rhagidiidae | 1.16 | 0.50 |
| Tenuipalpidae | 2.50 | 5.83 |
| Trombidiidae | 1.66 | 1.33 |
| Tydeidae | 4.83 | 4.16 |
| Scutacaridae | 0.50 | 0.66 |
| Other Prostigmata | 1.33 | 5.83 |
| Total Prostigmata | 42.53 | 34.80 |
| | | |
| Araneida | 0.33 | 0.16 |
| Pseudoscorpionidea | 1.83 | 0.83 |
| Schizopeltida | 0.66 | 0.50 |
| Chilopoda | 1.33 | 0.83 |
| Diplopoda | 1.00 | 0.66 |
| Pauropoda | 1.16 | 1.00 |
| Symphyla | 1.66 | 1.16 |
| Isopoda | 0.50 | 0.66 |
| Coleoptera (Adults) | 2.00 | 1.16 |
| Coleoptera (Larvae) | 1.33 | 2.50 |
| Diplura | 0.83 | 0.66 |
| Hemiptera (Coccidae) | 5.00 | 3.33 |
| Hymenoptera (Formicidae) | 4.00 | 5.33 |
| Psocoptera (Liposcelidae) | 5.16 | 4.00 |
| Thysanoptera | 1.83 | 0.00 |
| Total Insecta | 20.00 | 17.00 |
| | | |
| Entomobryidae | 39.8 | 32.66 |
| Isotomidae | 52.00 | 40.16 |
| Poduridae | 1.50 | 1.66 |
| Smithuridae | 2.33 | 1.66 |

Table 2.    Comparison  of the mean numbers  of arthropods obtained
from the uncleared forest (A) and cleared forest (B).  Data based on 10
replicates   were   obtained 3 months (October, 1972)   after the clearing
of B

| Animal Group | Mean per unit Sample | | Mean difference | Standard Error of mean difference (s.e.) | | Level of Significance |
|---|---|---|---|---|---|---|
| | A | B | d | | t | |
| Carabodidae | 1.40 | 0.70 | 0.70 | 0.79 | 0.89 | |
| Galumnidae | 0.50 | 0.10 | 0.40 | 0.31 | 1.31 | |
| Haplozetidae | 3.00 | 1.90 | 1.10 | 1.42 | 0.78 | |
| Lohmaniidae | 1.00 | 0.50 | 0.50 | 0.67 | 0.75 | |
| Machadobelbidae | 0.40 | 0.40 | 0.00 | 0.37 | 0.00 | |
| Oppiidae | 1.00 | 0.40 | 0.60 | 0.27 | 2.25 | |
| Scheloribatidae | 1.90 | 0.70 | 1.20 | 0.68 | 1.77 | |
| Cryptostigmata (Adults) | 9.40 | 4.80 | 4.60 | 1.89 | 2.43 | |
| Craptostigmata (Juveniles) | 2.80 | 0.90 | 1.90 | 1.21 | 1.58 | * |
| Gamasina (Adults) | 0.30 | 0.70 | 0.40 | 0.45 | -0.88 | |
| Gamasina (Juveniles) | 1.30 | 1.00 | 0.30 | 0.96 | 0.31 | |
| Uropodina | 0.50 | 0.60 | -0.10 | 0.31 | 0.32 | |
| Prostigmata | 1.20 | 0.60 | 0.60 | 0.31 | 1.96 | |
| Entomobryidae | 0.70 | 0.40 | 0.30 | 0.50 | 0.61 | |
| Hemiptera (Coccidae) | 0.30 | 0.40 | -0.10 | 0.35 | -0.29 | |
| Hymenoptera (Formicidae) | 0.70 | 0.30 | 0.40 | 0.31 | 1.31 | |
| Psocoptera (Liposcelidae) | 0.30 | 0.10 | 0.20 | 0.20 | 1.00 | |
| Insecta | 2.40 | 1.00 | 1.40 | 0.69 | 2.04 | |

Notes:    *Significant at 5% probability level.   Other t not significant.

more farmed than site A which is situated at the foot of the inselberg.
From  Table 2  which presents the results  of samples  taken in October
1972, it can be seen  that  with the exception of Coccidae (Hemiptera),
Uropodidae  and  the adults of Gamasina, which were recorded  in higher
numbers  in site B, site A  still supports a higher population  of soil
fauna.  It is, however, only one mean difference (that of adult Cryptostigmata)  that is significantly different.  The results  for Coccidae,
Uropodidae and adult Gamasina  represent an increase in population over
their initial population levels in site B, although this trend  was not
maintained during the subsequent samplings in this site.
    Tables 3 and 4 show  that many of the animal groups  were signifi-
cantly more abundant  in site A  than  in site B.  At the last sampling
occasion (May, 1973, Table 4)  more Prostigmata and Poduridae  were recorded in site B than in site A although the mean differences  were not

T a b l e   3.    Comparison of the mean numbers  of arthropods  obtained
from the uncleared forest (A) and the cleared forest (B). Data obtained
8 month (March, 1973)  after the clearing of B were based on  10 replicates

| Animal Group | Mean per Unit Sample | | Mean difference | Standard Error of mean difference (s.e.) | t | Level of Significance |
|---|---|---|---|---|---|---|
| | A | B | d, | | | |
| Carabodidae | 1.80 | 0.40 | 1.40 | 0.50 | 2.81 | * |
| Galumnidae | 1.30 | 0.20 | 1.10 | 0.31 | 3.50 | ** |
| Haplozetidae | 2.70 | 0.30 | 2.40 | 0.60 | 4.00 | *** |
| Lohmaniidae | 0.80 | 0.20 | 0.60 | 0.27 | 2.25 | n.s. |
| Phthiracaridae | 0.75 | 0.20 | 0.50 | 0.17 | 3.00 | * |
| Scheloribatidae | 1.40 | 0.20 | 1.20 | 0.25 | ˙4.81 | **** |
| Cryptostigmata (Adults) | 17.40 | 1.50 | 14.90 | 2.42 | 6.16 | **** |
| Cryptostigmata (Juveniles) | 5.70 | 0.60 | 5.10 | 0.62 | 8.19 | **** |
| Rhodacaridae | 0.80 | 0.60 | 0.20 | 0.25 | 0.80 | n.s. |
| Gamasina (Adults) | 2.40 | 1.70 | 0.70 | 0.33 | 2.09 | n.s. |
| Gamasina (Juveniles) | 1.60 | 0.50 | 1.10 | 0.43 | 2.54 | * |
| Cunaxidae | 1.60 | 0.80 | 0.80 | 0.42 | 1.92 | n.s. |
| Europidae | 3.60 | 0.20 | 3.40 | 0.90 | 3.79 | *** |
| Tydeidae | 1.40 | 0.60 | 0.80 | 0.33 | 2.45 | * |
| Prostigmata | 10.40 | 5.40 | 5.00 | 1.55 | 3.23 | * |
| Entomobryidae | 8.00 | 0.40 | 7.60 | 1.86 | 4.08 | *** |
| Sminthuridae | 0.70 | 0.50 | 0.20 | 0.25 | 0.80 | n.s. |
| Coleoptera (Adults) | 1.20 | 0.60 | 0.60 | 0.40 | 1.50 | n.s. |
| Coleoptera (Larvae) | 0.80 | 0.80 | 0.0 | 0.63 | 0.0 | n.s. |
| Psocoptera (Liposcelidae) | 1.40 | 1.10 | 0.30 | 0.30 | 1.00 | n.s. |
| Insecta | 6.60 | 1.80 | 4.80 | 0.57 | 8.37 | **** |

Notes: n.s.  = not significant,
       *  = significant at 5% probability level,
      ** = significant at 1% probability level,
     *** = significant at 0.05% probability level,
    **** = significant at 0.01% probability level.

statistically  significant.  This will suggest, if the trend continued,
that the clearing  did not seriously affect the density of certain species of Prostigmata and Poduridae.

    In Table 5 it can be seen that from the first sampling after clear-
ing, i.e., in October 1972, certain groups of animals  were observed to
have disappeared from site A and much more from site B. This disappearance is most likely due to the effect  of the dry season on the animals
so that  the phenomenon observed is due to normal seasonal fluctuations
of the groups involved.  Such seasonal fluctuations may mark the effect

of clearing  on soil arthropods, so that  in order to assess the effect much more accurately  normal  seasonal  fluctuations  of  the different groups  must be taken into consideration.  Unfortunately, such data are not yet available in Nigeria.   It is, however, interesting to see  that while all the groups not recorded  in site A  in October  were recorded again by March  and  May 1973, most of these groups, with the exception of *Prostigmata:  Cheyletidae, Eupodidae, Raphignathidae, Tenuipalpidae,* the *Collembola: Isotomidae, Poduridae* and *Sminthuridae*  and  the larvae of *Coleoptera,* had completely disappeared in the May 1973  samples from site B.   Certain groups, e.g. *Galumnidae, Lohmaniidae, Scheloribatidae, Psocoptera  (Liposcelidae), Rhodacaridae* and  *Entomobryidae*  occurred throughout the period of the investigation  in site B.  Although *Galum- nidae*  and  *Scheloribatidae*  were recorded throughout the period of in- vestigation in site B the two families  were represented by one species each as against four species  which were found in both sites at the be-

T a b l e   4.     Comparison  of the mean numbers  of arthropods obtained from the uncleared forest (A)  and  the cleared forest (B).  Data based on 10 replicates were obtained 10 months (May, 1973) after the clearing of B

| Animal Group | Mean per Unit Sample | | Mean dif- ference | Standard Error of mean difference | t | Level of Signif- icance |
|---|---|---|---|---|---|---|
| | A | B | d | (s.e.) | | |
| *Galumnidae* | 3.70 | 1.00 | 2.70 | 0.94 | 2.86 | * |
| *Oppiidae* | 4.20 | 0.10 | 4.10 | 0.82 | 4.98 | **** |
| *Scheloribatidae* | 6.10 | 1.30 | 4.80 | 1.02 | 4.71 | *** |
| *Cryptostigmata* (Adults) | 29.60 | 2.80 | 26.80 | 5.75 | 4.66 | *** |
| *Cryptostigmata* (Juveniles) | 7.00 | 1.80 | 5.20 | 2.27 | 2.29 | * |
| *Rhodacaridae* | 6.20 | 1.40 | 4.80 | 1.83 | 2.62 | * |
| *Gamasina* (Adults) | 16.70 | 3.80 | 12.90 | 4.00 | 3.23 | * |
| *Gamasina* (Juveniles) | 4.40 | 2.30 | 2.10 | 1.13 | 1.86 | n.s. |
| *Uropodidae* | 7.50 | 0.30 | 7.20 | 2.02 | 3.57 | ** |
| *Prostigmata* | 6.10 | 7.10 | 1.00 | 2.31 | 0.43 | n.s. |
| *Entomobryidae* | 1.70 | 0.30 | 1.40 | 0.31 | 4.58 | *** |
| *Isotomidae* | 17.00 | 6.60 | 10.40 | 5.96 | 1.75 | n.s. |
| *Poduridae* | 0.70 | 4.80 | 4.10 | 3.93 | 1.04 | n.s. |
| *Sminthuridae* | 1.00 | 0.90 | 0.10 | 0.66 | 0.15 | n.s. |
| *Hymenpptera* (Formicidae) | 14.10 | 6.20 | 7.90 | 11.68 | 0.68 | n.s. |
| *Psocoptera* (Liposcelidae) | 1.90 | 0.60 | 1.30 | 0.45 | 2.90 | * |

Notes: See Table 3.

T a b l e   5.   The composition of the arthropod groups in sites  A and B  during   each sampling occasion

| Animal Group | Before Clearing B | | After Clearing of Site B | | | | | |
| --- | --- | --- | --- | --- | --- | --- | --- | --- |
| | | | Experiment 1 (October, 1972 Samples) | | Experiment 2 (March, 1973 Samples) | | Experiment 3 (May, 1973 Samples) | |
| | A | B | A | B | A | B | A | B |
| *Belbidae* | + | + | + | − | + | − | + | − |
| *Carabodidae* | + | + | + | + | + | + | + | − |
| *Galumnidae* | + | + | + | + | + | + | + | + |
| *Haplozetidae* | + | + | + | + | + | − | − | − |
| *Liodidae* | + | + | − | − | + | + | − | − |
| *Lohmaniidae* | + | + | + | + | + | − | + | + |
| *Machadobelbidae* | + | + | + | + | + | + | + | − |
| *Nothridae* | + | + | + | − | + | − | − | − |
| *Oppiidae* | + | + | + | + | + | − | + | + |
| *Oribatelidae* | + | + | + | − | + | − | + | − |
| *Phthiracaridae* | + | + | − | − | + | − | + | − |
| *Scheloribatidae* | + | + | + | − | + | + | + | + |
| *Suctobelbidae* | + | + | + | + | + | − | + | + |
| *Rhodacaridae* | + | + | + | + | + | + | + | + |
| *Uropodidae* | + | + | + | + | + | − | + | + |
| *Bdellidae* | + | + | + | + | + | + | − | + |
| *Cheyletidae* | + | + | + | + | + | − | + | + |
| *Cunaxidae* | + | + | − | − | + | − | + | + |
| *Eupodidae* | + | + | − | − | + | − | + | + |
| *Raphignathidae* | − | + | − | − | − | + | + | + |
| *Rhagidiidae* | + | + | + | + | + | + | − | + |
| *Tenuipalpidae* | + | + | − | − | + | + | + | + |
| *Trombidiidae* | + | + | − | − | + | + | − | − |

Table 5 continued

| Animal Group | Before Clearing B | | After Clearing of Site B | | | | | |
| --- | --- | --- | --- | --- | --- | --- | --- | --- |
| | | | Experiment 1 (October, 1972 Samples) | | Experiment 2 (March, 1973 Samples) | | Experiment 3 (May, 1973 Samples) | |
| | A | B | A | B | A | B | A | B |
| Tydeidae | + | + | - | - | + | + | + | - |
| Scutacaridae | + | + | + | - | + | + | + | - |
| Araneidae | + | + | - | - | + | + | + | + |
| Pseudoscorpionidae | + | + | + | + | + | - | + | - |
| Schizopeltida | + | + | - | - | + | - | - | - |
| Chilopoda | + | + | - | - | + | - | + | - |
| Diplopoda | + | + | + | + | + | - | + | - |
| Pauropoda | + | + | - | - | + | - | + | - |
| Symphyla | + | + | + | - | + | - | + | - |
| Isopoda | + | + | - | - | + | - | - | - |
| Coleoptera (Adults) | + | + | + | + | + | + | + | - |
| Coleoptera (Larvae) | + | + | + | - | + | + | + | + |
| Diplura | + | + | + | + | + | - | + | - |
| Hemiptera (Coccidae) | + | + | + | + | - | + | - | + |
| Hymenoptera | + | + | + | + | + | + | + | + |
| Psocoptera | + | + | + | + | + | + | + | + |
| Entomobryidae | + | + | - | - | + | + | + | + |
| Isotomidae | + | + | - | - | + | - | + | + |
| Poduridae | + | + | - | - | + | - | + | + |
| Sminthuridae | + | + | - | - | + | + | + | + |

Notes: + = present; - = absent.

ginning of the investigation  and in site A throughout.  Furthermore at
the last sampling occasion  (May 1973)  the entomobryid fauna of site B
was  predominantly  *Lepidocyrtus* sp. while that of A  comprised *Lepidocyrtus* sp., *Cyphoderus* sp., *Seira* sp. and Paronellinae.  Similarly *Folsomides americanus* Denis and  *Isotomina thermophila* (Axelson)  were the
only *Collembola* recorded in site B  in May 1973  samples while in addition to these two species  *Isotomiella minor*  (Schäffer)  was  recorded
from site A.  These observations are not surprising in view of the fact
that site A still, except for normal seasonal variations, preserves its
environmental attributes while many of these have virtually been removed from B.  Furthermore, the observations  show  clearly  that clearing
reduced both the number  of species  and  the number  of individuals of
certain soil faunal groups.

Clearing may exert its influence in a number of ways which include:

1. Exposing the surface  of the soil.  This will cause temperature
conditions to be more extreme  and also aggravate daily as well as seasonal fluctuations  in temperature.  In addition,  moisture  conditions
change as the evaporation from the soil will now be greater.

2. Increasing the impact of rain drops  on the bare surface;  this
may destroy the soil crumb structure,  the pores as well as aggravating
the intensity of leaching.

3. The reduction  or elimination  of litter cover  which serves as
the micro-habitat  of some soil animals,  as food for others  and as an
insulating cover to the mineral soil below.

4. The loss of humus and its effect on the microbiological population of the soil.

The overall effect  of clearing,  therefore,  on the soil fauna is
its changing of climatic  as well  as biotic factors  operating in the
cleared site so that a number  of animal groups  unable to tolerate the
new microclimate  disappear  entirely either  as a result of drought or
lack of food. Such groups that possess a remarkable ecological plasticity  which  fits them  to varying habitats are,  however,  likely to be
little influenced by the clearing.

## SUMMARY

Results  of an investigation  of the effect  of clearing  of a tropical
rain forest in Nigeria on the soil micro-arthropods  are presented. Six

sample units were taken in May, 1972 from each of the two sites  (A and B) before site B was cleared.   Three series of samples, each consisting of ten sample units were taken from each of the sites 3 months (October, 1972), 8 months (March, 1973) and 10 months (May, 1973) after the clearing of site B.

The results obtained from the samples  taken in both sites A and B before  the clearing  of the latter  showed that there was little or no difference  in the composition  and densities  of the animals recorded, although site A appeared to support higher populations. While this trend was maintained  in site A  throughout  the period of investigation, the faunal composition  of B  became poorer and poorer both in species composition as well as in density.

Although different animal groups react differently  to the changed conditions  caused  by the clearing,  the overall results  suggest that clearing reduced both the number of species  and the number of individuals of certain soil faunal groups.

## A c k n o w l e d g m e n t s

The work is supported by the University of Ife Research Grant No. F/EC/ /ET.2/0960.

## R e f e r e n c e s

C h a r t e r ,  J. R. (1970):  Vegetation Ecological Zones. Fed. Dept. Forest Res., Ibadan, Nigeria.

E d w a r d s ,  C. A. &  L o f t y ,  J. R.  (1969):  The influence of agricultural  practice  on  soil micro-arthropod populations.  The Soil Ecosystem: Systematic Association Publication, No.8 (Ed. J.G. Sheals), pp. 237-247.

H u h t a ,  V.,  H a r p p i n e n ,  E.,  N u r m i n e n ,  M.    and V a l p a s ,  A. (1967): Effect  of silvicultural practices  upon arthropod, annelid  and nematode populations  in coniferous forest soil. Ann. Zool. Fenn. 4, 87-143.

K e a y ,  R. W. J. (1965): An outline of Nigerian Vegetation.  Federal Ministry of Information, Lagos, Nigeria, 46 pp.

L a s e b i k a n ,  B.  A d e b a y o  (in preparation):  Studies  on micro-arthropods  from  a tropical rain forest in Nigeria. I. Description of the habitat.

L a s e b i k a n ,  B.  A d e b a y o   (in preparation):   A modified
    Macfadyen  High Gradient Canister apparatus  for  soil arthropods.
M a c f a d y e n , A. (1961): Improved funnel-type extractors for soil
    arthropods. J. Anim. Ecol., 30, 171-184.
M a c f a d y e n , A. (1962): Control of humidity in three funnel-type
    extractors for soil arthropods. Progress in Soil Zoology (Ed. P.W.
    Murphy), Butterworths, London, 158-168.
O'C o n n o r ,  F. B. (1957):  An ecological study of the  enchytraeid
    worm population  of  a coniferous forest soil.  Oikos, 8, 162-199.
O l i v i e r ,  P. G. &  R y k e ,  P. A. J. (1969):  The influence of
    citricultural practices on the composition of soil  *Acari* and *Col-
    lembola* populations. Pedobiologia 9, 277-281.
S h e a l s ,  J. G. (1955): The effects of DDT and BHC on soil *Collem-
    bola* and *Acarina*. Soil Zoology (Ed. D.K.McE. Kevan), 241-252.

# SECTION D
## MODERN METODS OF INVESTIGATING
## SOIL ORGANISMS
## AND THEIR INFLUENCE
## UPON SOIL PROPERTIES

# Scope of the gelatin-embedding technique for studying soil arthropod fauna, with particular reference to Oribatei

Y. D. PANDE

Department of Agricultural Zoology and Entomology,
University of Udaipur,
Rajasthan, India

One of the main problems encountered in studying soil arthropods is that of suitable extraction. The principal methods commonly used for extraction are mainly of two kinds: dynamic and mechanical, each with its devotees and its critics. Although the suitability of each method greatly depends on the aims of the investigation, one conclusion which is common is that no method yet used successfully extracts all the invertebrate animals from all types of soils. Edwards and Fletcher (1971), on the basis of a survey, reported that at present most workers use same form of Tullgren funnel.

On the other hand, there has, from the very beginning, been a great need to know about their activities in their natural habitats. This has led to the development of a range of methods for the preparation of undisturbed soil sections which have been reviewed in detail by Nicholas and Parkinson (1967). These techniques suffer from one or other serious drawback and, therefore, their application has been very restricted, sometimes only with the inventor. Some improvements in the gelatin-embedding technique were reported by Anderson and Healey (1970). In an ecological survey of a black pine forest floor this technique was intensively used by the author and its efficiency compared with that of the most commonly used Tullgren funnel apparatus. An attempt was also made to try this technique on a pasture soil. This article is intended to communicate the utility and limitations of the gelatin-embedding technique. It must be emphasized that the survey discussed here, which was confined to a very few and distinctive localities, goes only a little way towards utilizing this technique in other biotopes.

MATERIALS AND METHODS

The methods involved in the preparation of soil sections (collection of
soil samples, embedding, curing and sectioning) were similar to those
proposed by Anderson and Healey (1970). Two mm thick serial sections
were used in the present study. Both faces of the sections were scanned
under intense illumination with a binocular microscope (60 x) fitted
with horizontal and vertical scales. Needles and decaying wood present
in the sections were routinely dissected in order to locate hiding
species. For comparison, samples were also obtained by a tubular corer
and extracted in a Tullgren funnel extraction apparatus. Details about
methods of sampling, extraction, locality and soil have been given
elsewhere (Pande and Berthet, 1973b).

OBSERVATIONS

The success of this technique depends very much on obtaining thin sections of good quality. This is possible only if the soil is rich in organic matter for gelatin is unable to bind mineral particles effectively. It was possible to obtain 2 mm thick serial sections of 10 x 6
cm size. One mm thick soil sections were also obtained but proved to be
difficult to handle because they break very easily. It seems evident
that a compromise has to be found between the thickness and size of the
sections. The maximum size and depth of the sample for obtaining good
sections will depend upon the amount of organic material present. In
coniferous forest soils, which provide almost uniform set of conditions
throughout the year, 10 x 10 x 6 cm seems to be the optimum sample size.
In deciduous forest soils where the decomposition process is usually
faster than in coniferous soils and the depth of organic matter layer
is shallow and varies considerably from one season to another, 3 to 5 cm
deep samples according to the season can be used for obtaining good
sections. In pasture soils sample depth has to be restricted to 2 to
3 cm in view of their shallow organic matter layer. On the basis of the
present investigation it is reasonable to suggest that this technique
would be very difficult to apply in the study of animal populations in
agricultural or greenhouse soils which contain little organic material.

    The frequent presence of undecomposed twigs more than 4 mm in dia-
meter and gravels creates difficulty in sectioning. Such soils severely
limit the minimum thickness of sections which can be obtained. The possibility of using an electrically operated rotary slicer were examined

but good sections  were not obtained possibly due to high  and constant
r.p.m. The hand operated domestic rotary slicer served the purpose very
well.  The magnification  to be used  for examining soil sections  also
requires  a compromise  since  the intensity  of illumination decreases
with  the increase  of magnification.  For studying mesofauna  the most
satisfactory magnification was found to be 60 x.

   The major advantage  of this technique  is that  the procedure involved  in the preparation  of soil  for  visual examination  allows no
marked disturbance of the soil  and  thus permits observation of animal
populations in situ under high magnification.  The sampling disturbance
is usually restricted  on the sides only.  The disturbance on two sides
can very well  be eliminated by discarding the first  and  the last few
sections.  Soil sections obtained in this way make it possible to study
the fabric of the soil  and  to recognize  the morphological results of
many biological processes going on in it.

   If the soil sectioning technique  is to be used to its best advantage  it should be applied  to problems  requiring data  on the precise
vertical and horizontal distribution, index of affinity between various
species, number and size of clumps and aggregation patterns at a micro-
scale, food and feeding behaviour, association of fauna  and  flora in
microhabitats, etc. Populations of less mobile forms and their immobile
stages, such as eggs, larvae  and  pupae of insects, some big and wood-
-boring species of mites, pseudoscorpians, etc. which are able to resist
adverse conditions  created  by  the common dynamic methods, can be estimated more accurately by this method.  It is also possible to use the
animals  obtained from soil sections  for gut content analyses.  Bigger
size samples, which are more representative  of the natural conditions,
can be used  in this technique and thus cutting effect and the pressure
exerted in obtaining samples  are considerably reduced in comparison to
the commonly used core method.  This leads to a better recovery of soil
animals. This technique also offers an opportunity to examine the fauna
in situ  at a later date  since frozen samples  can be stored  for  any
duration in a deep freeze and sections can be preserved in formalin for
several months.

   Although this technique opens new opportunities  for studying soil
fauna, the practical difficulties  and inconvenience involved limit its
utilization.  Perhaps, one of the greatest disadvantages of this technique is the considerable amount of time consumed in examining sections.
It requires 2 to 3 and sometimes 4 hours to examine thoroughly one section  (12 cm$^3$ of edaphic material).  In a biotope  such as forest soil,
where faunas  are unevenly distributed, it is necessary to take a large
number of samples for sound estimates. Thus this technique is not suit-

able for studying population dynamics in such ecosystems even for these
species which are extracted more efficiently by this method.

Again, the minute species of animals and larval stages of mesofauna are poorly estimated by this method. The technique also does not
appear to be ideal for soft-bodied animals such as collemboles and enchytraeids, which are damaged as a result of vacuum impregnation of
gelatin so that difficulty is encountered in their identification. The
compact mineral layers and tender vegetation in pasture soils also
create difficulty in examining the fauna. This technique does not allow
the determination of the viability of microorganisms and it is not possible to identify them. Like many other commonly used methods, the
studies cannot be conducted in vivo.

From the density and number of species points of view, oribatid
fauna is the most important in temperate forest soils. A summary of the
results obtained by studying the common species of oribatids in a black
pine (*Pinus nigra* Arnold) forest floor is reported here.


## 1. Relative efficiency

A comparison with the Tullgren funnel extraction method (Table 1) showed that this method is more efficient for big and inside wood and needle feeding species. Relative efficiency percentage for individual
species estimated by the two methods was calculated by $\frac{B-C}{B+C}$ x 100, where
B and C are the number of individuals found by sections and cores respectively in the same volume of soil. The larval and nymphal stages of
the various species were difficult to identify in soil sections and,
therefore, observations were confined to the adult stage. Detailed information on this account has been published (Pande & Berthet, 1973b).


## 2. Food and feeding habits

Although it was not possible to study this aspect in vivo, the presence
of animals and their faecal pellets in situ provided sufficient indication about their feeding preferences and feeding habits. It was also
not possible to estimate the exact amount of food consumed by individuals of each species, but feeding marks and the presence of individuals and their faecal pellets gave some indication. It was noted
that the duration of decomposition of faecal pellets varies from one

T a b l e   1.     Relative efficiency of the two techniques

| Species | Relative efficiency percentage |
|---|---|
| *Microtritia minima* Berl. | + 4.16 |
| *Rhysotritia duplicata* Grandjean | +16.98 |
| *Nanhermannia nanus* Nic. | - 6.97 |
| *Nothrus silvestris* Nic. | +17.71 |
| *Platynothrus peltifer* Koch | + 6.63 |
| *Hypochthonius rufulus* Koch | + 6.98 |
| *Eniochthonius minutissimus* Berl. | - 2.83 |
| *Tectocepheus velatus* Mich. | - 8.18 |
| *Chamobates incisus*  V.d. Ham. | - 3.78 |
| *Phthiracarus* sp. | +72.35 |
| *Carabodes marginatus* Koch | +15.33 |
| *Adoristes ovatus* Koch | +20.28 |
| Average | + 4.07 |

N.B. + and - signs indicate superiority of the section  and core method
respectively.

species to another according to the type of food ingested. Congregation
of faecal pellets were mainly found in the case of wood-feeding species
suggesting that these stay at one place for a long time  and the decomposition  of their faecal pellets  occurs very slowly.  The presence of
only a few faecal pellets  around microphytophagous species  and  well-
decomposed plant material feeders suggests that either these species do
not stay at one place  for long or their faecal pellets  are decomposed
quickly.  In any case it is reasonable to believe that the main role of
oribatid mites seems to lie  in the fragmentation of plant residues and
production of humus  and accumulation of plant nutrients in soil. Wallwork (1967)  has reported  that microarthropods  assist in the vertical
transportation  or organic material  to deeper layers  of soil profile.
According to the present study  they are not of much importance in this
aspect.  The food  and feeding habits  of individual species  and their
possible role in the decomposition of plant residues have been reported
in detail elsewhere (Pande & Berthet, 1973a).

## 3. Vertical repartition

It was possible  to measure the exact location  of each individual present in the sections with the help of two sliding scales, one horizontal
and the other vertical, and in this way  the depth distribution of each
species was precisely estimated. The percentages occurring in each half

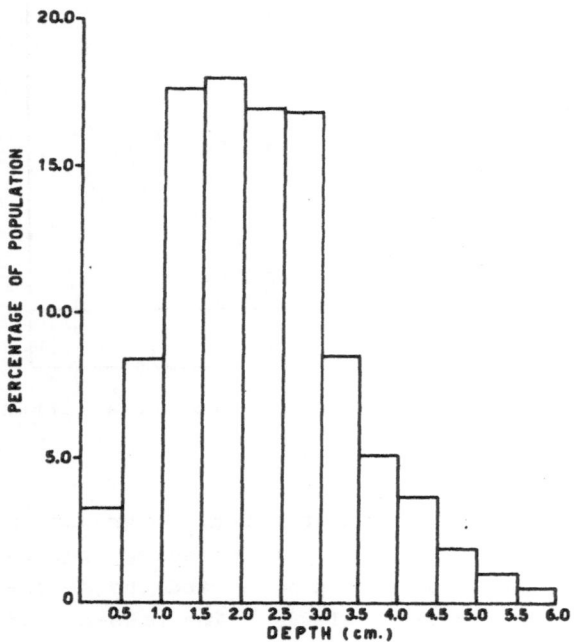

Fig.1. Mean annual depth distribution  of adult Oribatei
consisting of 12 abundant species.

cm depth  of the upper 6 cm profile  of the sampled populations  of the
12 species as a whole were calculated and are illustrated in Fig.1. The
present investigation  confirmed the earlier workers' observations that
fauna  is usually concentrated  in the upper layers of soil  and in the
overlying litter. Individual species showed marked differences in depth
distribution. Fig. 2 shows the depth distribution of the 4 species. The
individual blocks presented in the figure contained 40 serial sections.
It clearly shows that even within the same block the vertical distribution of *R. duplicata* varies considerably  and the distribution  depends
very much upon the distribution  of suitable material.  *M. minima* shows
a more wide depth distribution whereas *T. velatus* and *C. incisus* prefer
mainly fermentation layer.  No significant correlations were found bet-

ween the size of species  and  the depth of distribution.  Detailed observations on this aspect will be published elsewhere (Pande & Berthet,
in prep.).

## 4. Horizontal repartition

A number of workers  have shown that the sample counts  are distributed
asymmetrically about their mean,  suggesting a non-random or aggregated
repartition. The results obtained in this way are approximative indeed.
As in the case  of vertical distribution,  it is possible  to study the
precise  horizontal  distribution  of each species  by  this technique.
A computer may be profitably utilized in analysing the data and detect-
ing the number and size of clumps.  This aspect  will form the basis of
another paper  in the near future.  Fig. 3 indicates the horizontal repartition  of the 4 species  in individual blocks, each covering 80 $cm^2$
surface of soil 6 cm deep.

## Acknowledgements

I am greatly indebted  to my Director of Studies, Professor P. Berthet,
not only for the facilities which he made available to me, but also for
his continued advice  and encouragement.  Both Prof. Berthet  and Dr E.
Feytmans have read and critized this manuscript.
        Thanks are due to the U.C.L. from whom I was in receipt of a schol-
arship during the course of this work.

R e f e r e n c e s

A n d e r s o n ,  J. M. &  H e a l e y ,  I. N.  (1970):  Improvements
    in the gelatin-embedding technique  for  woodland soil  and litter
    samples. Pedobiologia 8, 108-120.
E d w a r d s ,  C. A.  and  F l e t c h e r ,  K. E. (1971): A compar-
    ison of the extraction methods for terrestrial arthropods. 150-185
    pp. In: Methods of study in quantitative soil zoology: population,
    production and energy flow,  ed. J. Phillipson, Blackwell, Oxford.

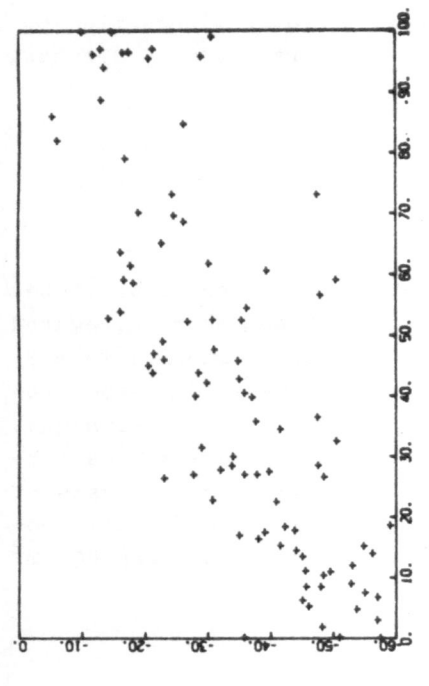

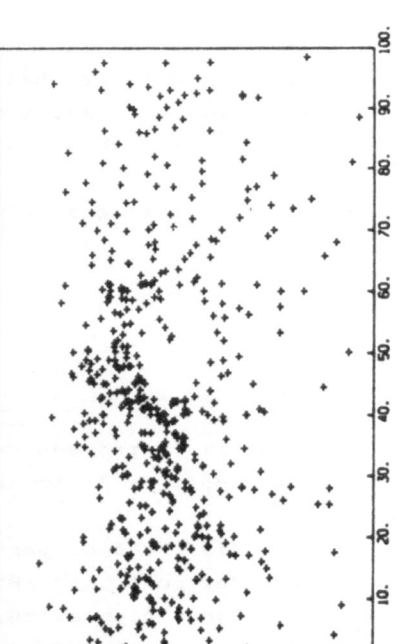

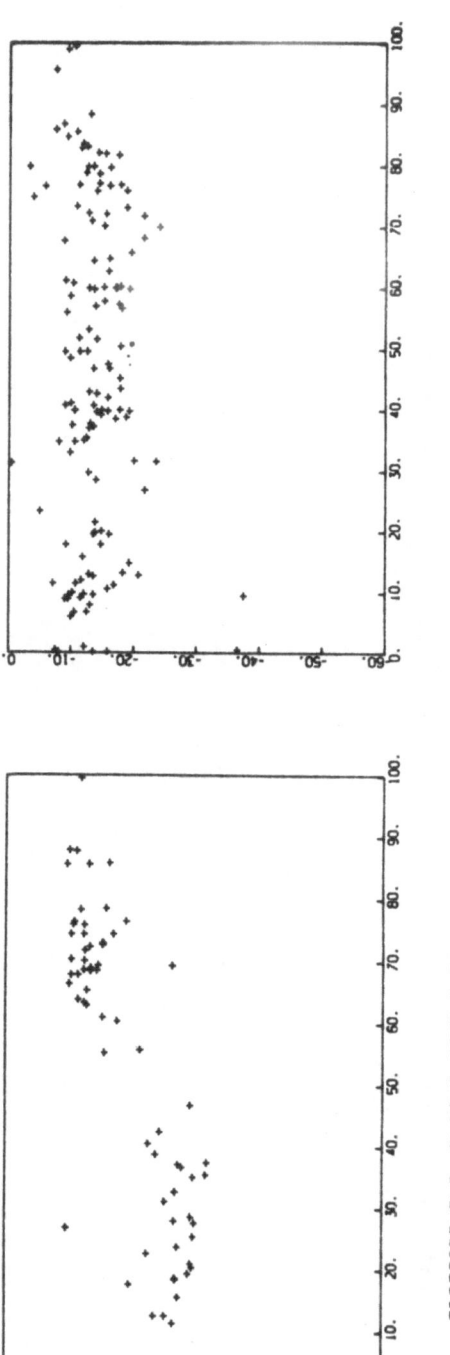

CHAMOBATES INCISUS SAMPLE 017

TECTOCEPHEUS VELATUS SAMPLE 021

Fig.2. Lateral projection of the location of the individuals of 4 species observed in the 40 serial sections
of different blocks (the scale in mm and zero of the
ordinate corresponds to surface level.

RHYSOTRITIA DUPLICATA SAMPLE 021

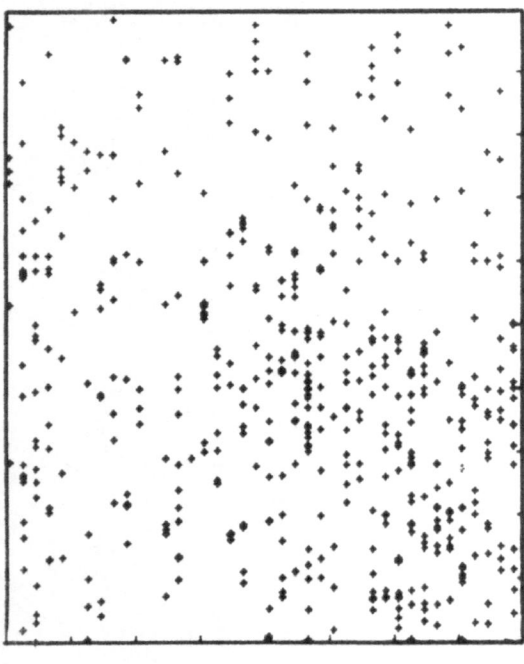

MICROTRITIA MINIMA SAMPLE 010

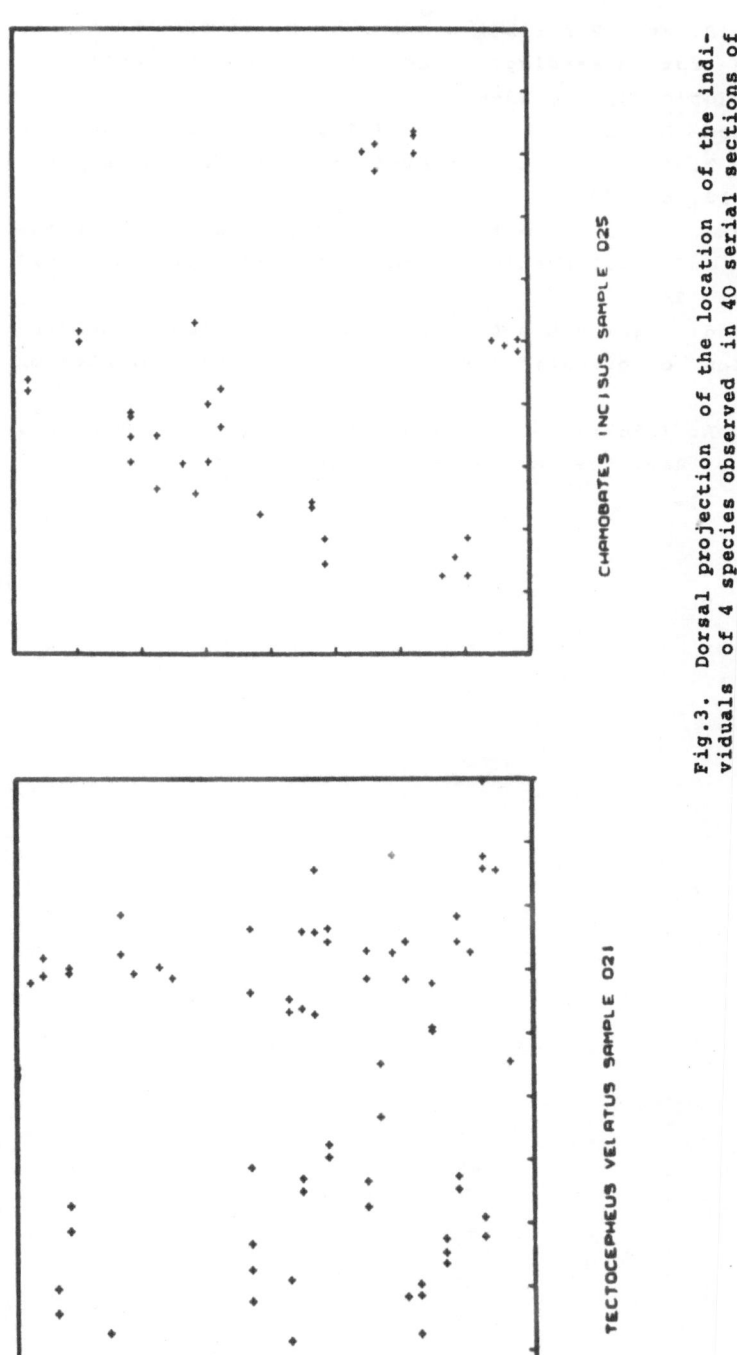

CHAMOBATES INCISUS SAMPLE 025

TECTOCEPHEUS VELATUS SAMPLE 021

Fig.3. Dorsal projection of the location of the individuals of 4 species observed in 40 serial sections of
4 different blocks (for scale see legend beneath Fig.2).

N i c h o l a s , D. P. and P a r k i n s o n , D. (1967): A comparison of methods for assessing the amount of fungal mycelium in
soil samples. Pedobiologia 7, 23-41.

P a n d e , Y. D. and B e r t h e t , P. (1973a): Studies on the
food and feeding habits of soil *Oribatei* in a black pine plantation. Oecologia 12, 413-426.

P a n d e , Y. D. and B e r t h e t , P. (197.3b): Comparison of the
Tullgren funnel and soil section methods for surveying oribatid
population, Oikos, 24.

P a n d e , Y. D. and B e r t h e t , P. (in preparation): The vertical distribution of oribatid mites in a black pine plantation
soil.

W a l l w o r k , J. A. (1967): *Acari*. 363 395 p. In: Soil Biology,
ed. A. Burges & F. Raw, Academic Press, London.

# Specific haemoglobin content as an ecological characteristic in earthworms (Lumbricidae)

J. B. BYZOVA

Laboratory of Soil Zoology, Institute of
Evolutionary Morphology and Ecology of Animals,
Acad. Sci. USSR,
Moscow

The presence of haemoglobin in the blood is believed to permit earth-worms, as skin-breathing animals, to attain a large body size, and to inhabit different substrates and biotopes.

Haemoglobin is dissolved in the blood fluid. Its concentration is the highest so far recorded in invertebrates - up to 13.5 g in *Octola-sium lacteum Oerley* in October (Byzova 1972). Ecological differences in earthworm habitats, especially variations in respiratory conditions suggest adaptive characters of the systems involved in the gaseous ex-change. In fact, it has been revealed that haemoglobin has a higher oxygen affinity in *Allolobophora terrestris longa* Ude, dwelling in the deep soil layers, than in *Lumbricus terrestris* L., regularly appearing on the soil surface and in the litter (Haughton, Kerkut, Munday, 1958).

The present paper deals with the results of a comparative study on total haemoglobin content in an ecological series of earthworms, and the seasonal variations of these characteristics. The haemoglobin con-tent was quantified by the pyramidon micromethod (Korzhuev, Radzinskaja, 1957) in homogenates of the whole animal. The following species were studied: *Lumbricus rubellus* Hoffmeister, *Allolobophora caliginosa* (Savigny), *Eisenia foetida* (Savigny), *Octolasium lacteum* Oerley and *Ei-seniella tetraedra* (Savigny). Only mature active worms have been taken for analyses. In the Moscow region these worms inhabit the top soil layers, mineral soil layers, composts, moist habitats, wet soils and flooded grounds, respectively. This series of habitats may be charac-terized by the deterioration of respiratory conditions.

The total haemoglobin content per weight unit increases in the ecological series of lumbricid species from the surface-dwellers to the compost-dwellers and amphibiotic species (Table 1).

The measurements were carried out in the period of high activity of all these species. At least 10 measurements were made for each spe-cies to establish the content of both haemoglobin and water. No intra-

T a b l e   1.   Haemoglobin and water content in *Lumbricidae*   (in June)

| Species | Hb content mg/g wet weight | Water content % wet w. | Hb content mg/g dry matter |
|---|---|---|---|
| *Lumbricus rubellus* Hoffm. | 3.1 ± 0.2 | 78.8 ± 0.7 | 14.4 |
| *Allolobophora caliginosa* (Sav.) | 3.6 ± 0.5 | 81.5 ± 1.0 | 19.5 |
| *Eisenia foetida* (Sav.) | 4.2 ± 0.4 | 81.6 ± 0.7 | 2?.8 |
| *Octolasium lacteum* Oerl. | 5.0 ± 0.3 | 82.4 ± 0.7 | 28.4 |
| *Eiseniella tetraedra* (Sav.) | 7.8 ± 0.8 | 80.1 ± 0.4 | 39.2 |

specific differences  were found in haemoglobin  content in relation to
body size.

The estimation  of total haemoglobin  content  during  the  seasons
without snow revealed a correlation between the stability of this index
and ecological characteristics  of every species  (Table 2 and Fig. 1).

T a b l e   2.   Annual changes  of haemoglobin  content  in the body of
earthworms (mg/g dry matter)

| Species | Months | | | | | | | |
|---|---|---|---|---|---|---|---|---|
| | IV | V | VI | VII | IX | X | XI | XII |
| *Lumbricus rubellus* Hoffm. | 15.5 | 14.3 | 14.4 | 16.0 | 15.1 | 19.1 | 12.6 | 15.2 |
| *Allolobophora caliginosa* (Sav.) | 9.6 | 10.6 | 19.8 | 22.8 | 18.0 | 12.0 | | 25.2 |
| *Octolasium lacteum* Oerl. | 33.3 | | 24.8 | 48.7 | | | | 76.6 |

The haemoglobin content per 1 mg dry matter is almost constant throughout the year  in  *Lumbricus rubellus*  inhabiting top soil layers, where
this species hibernates.  By contrast the content of haemoglobin varies
in  *Octolasium lacteum*  according  to  the habitat.  *Octolasium lacteum*
lives  in moist soil  and migrates to more humid places when the previously inhabited sites become dry.  These worms were found at the water-
-level of a brook in hot and dry periods of the summer. *Octolasium lacteum* is known also to be resistant to the flooding of the soil.

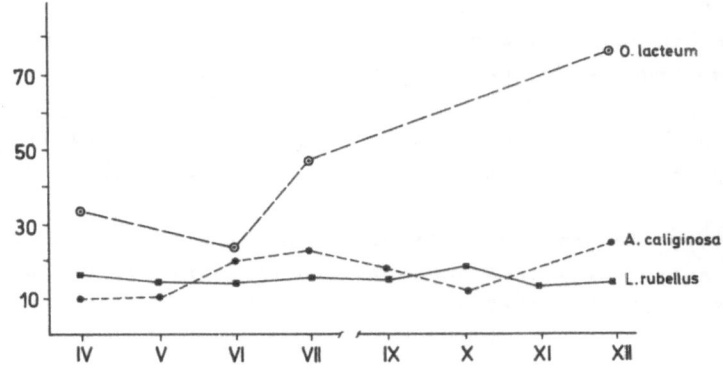

Fig.1.   Changes   of   haemoglobin content   in   earthworms
during the year (mg/g of dry matter).

The haemoglobin   content decreased nearly three times in *Allolobo-*
*phora caliginosa* and in *Octolasium lacteum* during hibernation. Probably
such a decrease is due not only to the weaker haemoglobin synthesis but
also to the utilization   of haemoglobin proteins   as a source of energy
in the winter. The utilization of haemocyanin proteins during starvation
was observed in   woodlice and other crustaceans (Wieser, 1965; Alikhan,
1970; Cuzon, Ceccaldi, 1971/1972).   The autumnal variation of the haemoglobin   content in *Allolobophora caliginosa*   may be of the same nature.
*Allolobophora caliginosa*   undergoes   a facultative   aestivation   in the
Moscow region.   Many specimens   of *Allolobophora caliginosa*   were found
with a low specific haemoglobin content in the early autumn. Such worms
perhaps aestivated.

Thus, the haemoglobin   content per unit dry matter correlates with
ecological characteristics of earthworms   and   is of an adaptive importance.

R e f e r e n c e s

A l i k h a n ,   M. A. (1971): Blood proteins in the woodlouse *Porcellio*
    *laevis* Latreille with special reference to haemocyanin. Comp. Bio-
    chem. Physiol., A39, 4, 735-741.
B y z o v a ,   J. B. (1972): Haemoglobin content of the blood in earth-
    worms. J. Evol. Biochem. Physiol. (Leningrad), 8, 5, 548-549.
C u z o n ,   G.   and   C e c c a l d i ,   H. J. (1971/1972): Evolution des
    proteines de l'hemolymphe de *Penaeus kerathurus*   durant   le jeune.
    The Tethys, 3, 2, 247-250.

H a u g h t o n , T. M. , K e r k u t , G. A. and M u n d a y , K.A.
    (1958): The oxygen dissociation and alkaline denaturation of haemo-
    globins from two species of earthworms. J. Exptl. Biol. 35, 2,
    360-368.

K o r z h u e v , P. A. and R a d z i n s k a j a , L. I. (1957):
    O mikrometode opredeleniya gemoglobina. Voprosy ichtiologii 9,
    192-196.

W i e s e r , W. (1965): Electrophoretic studies on the blood proteins
    in an ecological series of isopod and amphipod species. J. Marine
    Biol. Assoc. U. K., 452, 2, 507-523.

# D i s c u s s i o n

G. M a r c u z z i : Wishes to know  whether in *Lumbricidae*  there are
any modifications  in haemoglobin  content  as adaptations to different
oxygen contents in soil.

B. B. B y z o v a :  Yes, I spoke about it in my contribution.

M. B o u c h é : Votre  *"Allolobophora caliginosa"*  est-il un animal
pigmenté  ou  apigmenté, parce que ce nom  est la source de confusions.

J. B. B y z o v a :  *Allolobophora caliginosa*  is an unpigmented earthworm.

# Microrespirometry of oribatid mites (Acari: Cryptostigmata) using gas chromatography

**M. J. MITCHELL**

Environmental Sciences Centre, Seebe,
Alberta, Canada

ABSTRACT

A microrespirometry technique using gas chromatography employing a helium-ionization detector was described. The method was used to determine the respiration rates of several oribatid species. The respiration data were applied to an adult *Ceratozetes* sp. to determine its metabolic role in an aspen (*Populus tremuloides* Mich.) woodland soil. For 1972 it was estimated that the *Ceratozetes* sp. adult population consumed 36.66 ml of $O_2/m^2$.

In studying the role of oribatid mites in energy flow through the soil system, it is important to measure their respiratory metabolism which may constitute 96 percent of their assimilated energy (Engelmann, 1961). Since mites have low respiration rates, the Cartesian diver, which can detect oxygen consumption rates of $10^{-6}$ µl/h (Petrusewicz and Macfadyen, 1970), has usually been employed. However, the diver has some important disadvantages such as the confinement of the animal in a small space, necessity for a constant temperature environment, absence of $CO_2$ in the atmosphere, and the great skill needed for operation. To overcome some of these problems, a method using gas chromatography has been developed (Mitchell, 1973). Using this technique respiration rates are determined by measuring the increase of $CO_2$ within a vessel of known volume. $CO_2$ volume is analysed by removing small samples of the vessel's atmosphere and determining the $CO_2$ concentration by standard gas chromatographic techniques, using a He ionization detector which is highly sensitive to permanent gases. The respiration data given in this paper were derived by this method.

## STUDY AREA

The mites  used in this study  came from an aspen woodland soil located in  the Canadian Rockies of Alberta. The dominant overstory vegetation consists of aspen (*Populus tremuloides* Mich.) and balsam poplar (*P. balsamifera* L.). There is also  a luxuriant ground cover  of herbs and grasses. The soil organic layer is about 7 cm thick and is sharply differentiated from the underlying inorganic horizons. A more detailed description of the site is given by Dash and Cragg (1972).

## RESPIRATORY METABOLISM

Mites, which were to be used in weighing  and respiration studies, were extracted by a Tullgren funnel the night before experimentation. A Cahn Gram Electrobalance was used  for determining the weights of live non-gravid adults (Table 1). For respiration determinations,  it was necessary to use groups of mites owing to the sensitivity limitations of the present gas chromatographic technique. For the smaller species (i.e. *Ceratozetes* sp. A.) 20 individuals  were used,  while  for the larger species (i.e. *Eupterotegaeus rostratus*) 5 individuals were found to be sufficient. Respiration rates  are given  in terms of both $CO_2$  output and $O_2$ consumption  since the latter formulation  is most commonly used (Table 1).

For the genera studied, the only data available for comparison is for  *Ceratozetes gracilis*  (Mich.), for which  Wood and Lawton  (1973) using  a Cartesian diver  found a mean weight-specific respiration rate of 274.93 ± 499.54 µl/g/h (±95% confidence limits) at 10 °C  which is within the range of calculations  for *Ceratozetes* sp. A. at 10 °C. Also all of the respiration rates in Table 1  are within the range  of those determinations made by Berthet (1964)  using a Cartesian diver for respiration  of 16 species  of adult oribatids.  Thus similar results  are obtained utilizing Cartesian diver and gas chromatographic respirometry.

For *Ceratozetes* sp. A. a $Q_{10}$  of 3.46  over  the temperature range of 5-15 °C  was calculated  by a regression  of temperature against the log weight specific respiration. This value compares with Webb's (1969) $Q_{10}$ value of 2.56  for *Nothrus silvestris* Nicolet  for between 5-20 °C. Both of these values  are substantially lower  than  those presented by Berthet (1964), who determined a mean $Q_{10}$ of 4.0  for the oribatids he studied.

To extrapolate respiration rates to those temperatures not actual-

T a b l e 1. Respiratory metabolism of oribatid mites

| Species | Temperature °C | Live Weight μg | Respiration rate | | |
|---|---|---|---|---|---|
| | | | $10^{-3}$ μl $CO_2$/h/indiv. | $10^{-3}$ μl $O_2$/h/indiv.* | μl $O_2$/g/h |
| | | mean ± 95% confidence limits (number of samples) | | | |
| Ceratozetes sp. A.** | 15 | 8.1 ± 0.7 (12) | 1.47 ± 0.17 (10) | 1.80 | 222.2 |
| Ceratozetes sp. A | 10 | 8.1 ± 0.7 (12) | 0.74 ± 0.24 ( 3) | 0.91 | 112.3 |
| Ceratozetes sp. A | 5 | 8.1 ± 0.7 (12) | 0.42 ± 0.44 ( 3) | 0.51 | 62.9 |
| Ceratozetes sp. B.** | 15 | 28.2 ± 1.4 (18) | 4.42 ± 0.85 ( 7) | 5.39 | 190.8 |
| Scheloribates spp. | 15 | 9.8 ± 0.5 ( 3) | 2.07 ± 0.48 (14) | 2.53 | 257.3 |
| Eupterotegaeus rostratus Higgins and Wooley | 15 | 34.7 ± 2.6 ( 8) | 4.28 ± 1.51 ( 4) | 5.22 | 150.4 |

* Assuming an R.Q. of 0.82
** Both Ceratozetes sp. A. and sp. B. are new species on which taxonomic descriptions will be published.

ly experimentally determined, the data were fitted to Krogh's Curve by
the method of Nielsen and Evans (1960) as suggested by Webb (1969). For
*Ceratozetes* sp. A. the following relationships were derived:

$$v_1 = 14.67 \times 1.17^t + 18.35, \quad v_1 = 10^{-5} \text{ µl } O_2/\text{h/individual},$$
$$v_2 = 18.14 \times 1.17^t + 22.56, \quad v_2 = \text{µl } O_2/\text{h/g},$$
$$°C.$$

## POPULATION METABOLISM

To calculate population metabolism, laboratory respiration determinations must be related to field temperatures and population densities.
On each sampling date 9 core samples of 22.9 cm$^2$ were taken to a depth
of 10 cm divided into four equal depths, and were extracted over 5 days
in a Macfadyen (1962) high gradient unit. After extraction the animals
were cleared in lactic acid and counted. The population data was nor-
malized using a log x + 1 transformation (Gérard and Berthet, 1966)
and the means with ± 95 percent confidence limits were calculated.
Population estimates for adult *Ceratozetes* sp. A. for 1972 are given
(Figure 1).

Soil temperature data were collected by continuous recording using
a distance thermograph. Mean weekly temperatures are given for 4 cm,
the depth at which *Ceratozetes* sp. A. is abundant (Figure 1). These mean
temperature values were applied to Krogh's Curve to estimate respiration rates of individual mites in the field for a given time period.
The population metabolism was then derived by multiplying the temperature corrected respiration rate by the population density (Figure 1).
By a summation of the population metabolism throughout the year, the
respiration for the adult *Ceratozetes* sp. A. was estimated at 36.66 ml
of $O_2/\text{m}^2$ for 1972.

## DISCUSSION

To relate mite respiration to energy flow in the soil an estimate of
energy input is necessary. In the autumn of 1972, litter trap collections were made to determine overstory leaf input, and plot clippings
of herbaceous vegetation and grass were made to determine above ground
standing crop, most of which enters the soil decomposition process. The

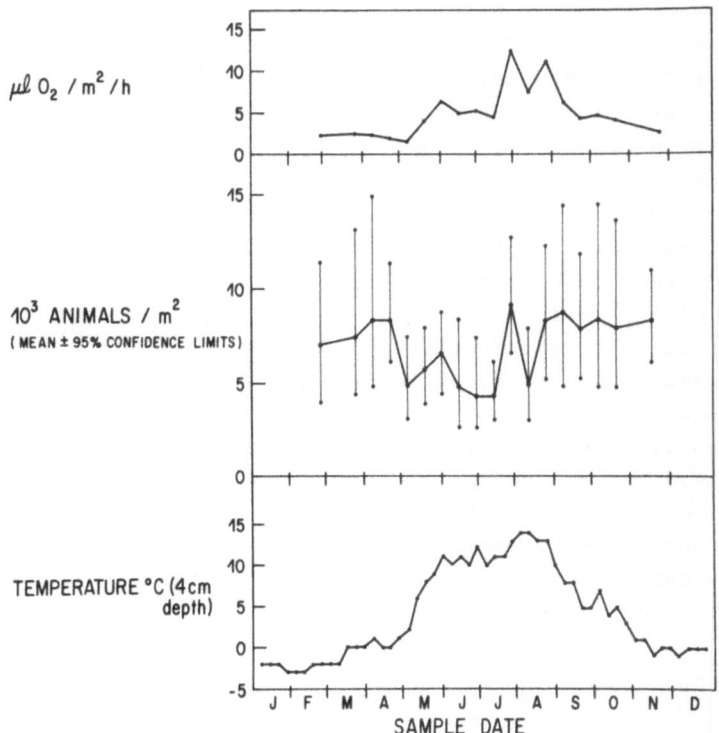

$\mu\ell\ O_2\ /\ m^2\ /\ h$

$10^3$ ANIMALS / $m^2$
( MEAN ± 95% CONFIDENCE LIMITS )

TEMPERATURE °C (4cm depth)

SAMPLE DATE

Fig.1. Population Metabolism of Adult *Ceratozetes* sp. A. (Aspen, Woodland, 1972).

material was oven dried at 105 °C to constant weight and the result expressed in g/$m^2$ of input (Table 2).

When the litter input 'was converted into calories using a conversion factor of 4.72 Kcal/g (Peterson et al., 1970), an estimate of 1454 Kcal/$m^2$/y was derived. Similarly when the oxygen consumption of *Cerato-*

T a b l e   2.    Aspen site litter input for 1972, Alberta, Canada

| Category | g/$m^2$ Oven dry weight<br>mean    95% confidence limits<br>(number of samples) |
|---|---|
| Above ground standing crops<br>   of herbs and grasses | 93.4 ± 12.6 (16) |
| Overstory leaf fall | 214.7 ± 39.2 ( 4) |
| Estimate of Autumn Input | 308.1 |

*zetes* sp. A. was converted to calories using a conversion factor of 4.7 cal/ml (Petrusewicz and Macfadyen, 1970), a utilization of 0.17 $Kcal/m^2/y$ was estimated which constitutes 0.012 percent of the input. These results are in agreement with other workers who have found that oribatid mite respiration consumes very little of the energy input into the soil. For instance Webb (1970) found that for all the life stages of *Nothrus silvestris* in a heathland, their respiration consumed only 0.2 percent of the litter input. Likewise Berthet (1964) working in a woodland estimated that the total adult oribatid mite respiration consumed 1.8 percent of the leaf fall.

In general, gas chromatographic microrespirometry can be applied to soil animals to assess their metabolic role and could also be used to determine the effect of fluctuating temperatures and the concentration of atmospheric constituents on metabolic activity. Presently, the technique is being used to assess the metabolic role of various oribatid species including immature forms. These data will be integrated into information on population parameters and the biological roles of selected species so that a more complete picture of the function of oribatid mites in decomposition can be achieved.

**Acknowledgements**

The author is deeply indebted for the aid and encouragement of Dr. D. Parkinson. The advice of Dr. I. Hodkinson and Dr. G. Pritchard concerning this manuscript is appreciated. Thanks is also given to A. Carter and D. Lousier, who helped in the collection of the litter input data, and my wife, Deborah, who assisted in the calculations. During the course of this work, the author was supported by an Izaak Walton Killam Memorial Scholarship; additional financial aid came from Dr. D. Parkinson (N.R.C., A-2257) and Dr. J. B. Cragg (N.R.C., A-3760).

R e f e r e n c e s

B e r t h e t , P. (1964): L'activité des Oribatides (*Acari: Oribatei*) d'une chênaie. Mem. Inst. Roy. Sci. Nat. Belg. 152, 1-152.
D a s h , M. C. and C r a g g , Y. B. (1972): Ecology of *Enchytraeidae (Oligochaeta)* in Canadian Rocky Mountain soils. Pedobiologia 12, 323-335.

E n g e l m a n n , M. D. (1961): The role of soil arthropods in the energetics of an old field community. Ecological Monographs 31, 221-238.

G é r a r d , G. and B e r t h e t , P. (1966): The statistical study of microdistribution of *Oribatei (Acari)*. Part II: The transformation of the data. Oikos 17, 142-149.

M a c f a d y e n , A. (1961): Control of humidity in tree funnel-type extractors for soil arthropods. In: P.W. Murphy (Ed.), Progress in soil zoology, Butterworths, London, 158-168 pp.

M i t c h e l l , M. J. (1973): An improved method for microrespirometry using gas chromatography. Soil Biol. Biochem. 5, 271-274.

N i e l s e n , E. T. and E v a n s , D. G. (1960): Duration of the pupal stage of *Aëdes taeniorhynchus* with a discussion of the velocity of development. Oikos 11, 200-222.

P e t e r s o n , E. B., C h a n , Y. H. and C r a g g , Y. B. (1970): Above ground standing crops, leaf area, and caloric value in an aspen clone near Calgary, Alberta. Can. J. Bot. 48, 1459-1469.

P e t r u s e w i c z , K. and M a c f a d y e n , A. (1970): Productivity of terrestrial animals; principles and methods. I.B.P. Handbook 13, Oxford-Edinburg, Blackwell, 190 pp.

W e b b , N. R. (1969): Temperature and respiratory metabolism in a species of soil mite. In: G.O. Evans (Ed.), Proceedings of the 2nd International Congress of Acarology, 1967, Akadémiae Kiadó, Budapest, 61-66 pp.

W e b b , N. R. (1970): Population metabolism of *Nothrus silvestris* Nicolet *(Acari)*. Oikos 21, 155-159.

W o o d , T. G. and L a w t o n , G. H. (1973): Experimental studies on the respiratory rates of mites *(Acari)* from beech-woodland leaf litter. Oecologia 12, 169-191.

D i s c u s s i o n

H. J a n e t s c h e k : Welches ist der untere Tamperaturlimit Ihrer Methode? Der von Ihnen festgestellte Anteil Ihrer Untersuchungsobjekte am "energy turnover" ist sehr gering. Kann daraus gefolgert werden, dass sie bodenbiologisch unbedeutend sind?

M. M i t c h e l l : There is no lower temperature limit for G.C. technique. It is true that the oribatid population metabolism is not important in energy flow. However, the qualitative and quantitative effect of their feeding may be important in changing the biomass and species composition of the microflora.

J. A. Wallwork : My question is not really related to the main theme of your paper, but I would like to know how you prepare immature oribatid mites for stereoscan photography without their collapsing.

M. Mitchell : The technique is that of Sleiffer in which a series of ethanol dehydrations are used after fixing in formalin. The final dehydration is in xylen from which the specimen is airdried and then coated in gold.

# Fonctions des lombriciens
# IV. Corrections et utilisations
# des distorsions causées
# par les méthodes de capture

M. B. BOUCHÉ
I. N. R. A. Faune du Sol,
Dijon

## INTRODUCTION

Pour les lombriciens, le Programme Biologique International a permis l'accumulation d'informations nouvelles caractérisées par la confrontation de techniques, parfois originales, et par la mesure de peuplements dans des milieux variés. Or, les erreurs introduites par les techniques sont considérables: les données observées peuvent sous-estimer de 400 % la biomasse d'une espèce et de seulement 20 % celle d'une autre (Bouché, 1969). Comment dès lors comparer, ou simplement inscrire dans un tableau, des résultats obtenus entre deux stations ou espèces différentes sans uniformiser la qualité des données et sans indiquer les étapes de cette uniformisation? En géodrilologie, la comparaison directe de données brutes n'a pas de sens.

Le but de cet article est de montrer qu'il est possible de corriger les données pour une approximation plus correcte et même d'utiliser les imperfections des méthodes.

J'utiliserai partiellement une écriture codifiée par ailleurs dans le présent ouvrage (Bouché, F.L.-V).

## MÉTHODES D'ESTIMATION QUANTITATIVE STATIONNELLE

Nous n'examinerons ici que les méthodes donnant des résultats utilisables pour estimer le niveau de population dans une station (méthode quantitative stationnelle: Bouché, sous presse). Ces méthodes appartiennent à deux familles: celle comportant un prélèvement de sol puis une extraction des animaux par tri (= procédés physiques) et celle faisant intervenir un agent physique ou chimique pour faire sortir les lombriciens à la surface du sol où ils sont collectés (procédés éthologiques). Par concision, je ne décrirai et critiquerai pas ces méthodes, ce qui est fait pour l'essentiel par Satchell (1969, 1971), Bouché (1969, 1972) et Bouché et Beugnot (1972). Je propose des abréciations en italique.

Les méthodes physiques débutent toutes par un prélèvement par labour à la bêche: $b$, labour-tri manuel: $bm$, labour-lavage-tamisage: $bl$, labour-élutriation (lévigation): $be$ (Satchell, 1971; Bouché et Beugnot, 1972), labour-flottaison: $bf$ (Raw, 1960; Gerard, 1967, qui conclut à une séparation lombriciens/sol parfaite avec une solution de $SO_4Mg$ de densité maximale 1,2, ce qui n'est pas vérifié par mes observations: même avec le BrK, de densité maximale 1,28, la totalité des lombriciens ne flotte pas ce qui est dû à la variabilité de l'endentère, Bouché et Kretzschmar, F.L.-II). La profondeur, en centimètres, du prélèvement doit être indiquée en indice (exemple: $bm_{30}$).

Les méthodes éthologiques utilisent des solutions aqueuses au permanganate de potassium: $pp$ ou au formol: $fo$; les autres méthodes (autres produits chimiques, électricité, vibrations) ne sont pas employées pour les études stationnelles. La concentration en $\%_0$ du produit doit être éventuellement indiquée en indice (exemple: $fo_{1,2}$).

Les méthodes combinées doivent être abréviées en respectant l'ordre d'intervention des techniques (exemples: la méthode physico-éthologique de Bouché (1969) et $fo, bl_{20}$ tandis que Zajonc (1971) utilise $bm_{30}, fo$).


## FACTEURS AFFECTANT L'EFFICACITÉ DES MÉTHODES
### Facteurs climatiques

Le cycle des saisons entraîne une migration verticale des animaux, une entrée en léthargie de certains et une modification des propriétés physiques du sol, de sorte que toutes les méthodes en sont affectées. Il y a donc lieu, dans les comparaisons entre stations, de considérer des résultats obtenus dans des conditions climatiques équivalentes. Une analyse en fonction du temps fait ressortir des variations liées aux saisons; notons, figure 1, la pluie de septembre 1968 à Borculo et l'influence de l'hiver en climat semi-continental (Cîteaux) et océanique (Borculo) Les variations constatées reflètent surtout celles de l'efficacité des méthodes (tableau 3) et non celles des populations qui doivent être établies par l'analyse démographique (Lavelle, 1971).

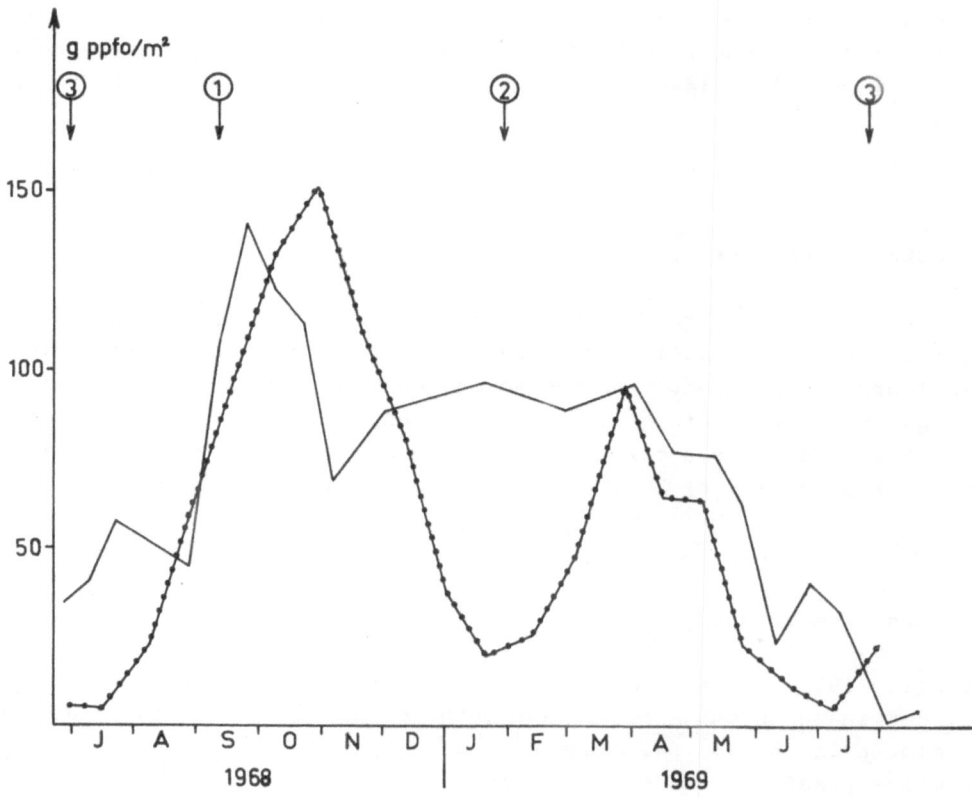

Fig.1.  Biomasses observées  par la méthode  au fcrmol à
Borculo (trait normal) at à Cîteaux (trait ponctué) pendant la période interestivale 1968-1969. 1 : pluie estivale exceptionnelle (11 % du total annuel); 2 : influence de l'hiver; 3 : léthargie estivale.

## Facteurs pédologiques et végétation

La pénétrabilité du sol  est limitante  pour les méthodes physiques; il
est  parfois impossible de prélever à cause des racines  et des roches.
Le tri manuel  est affecté  par la texture  du sol et son enracinement;
les lombriciens se reconnaissent mal  dans les sols sableux et s'agglu-
tinent dans l'argile dont ils se séparent mal; ces défauts se corrigent
par les procédés de tri élaborés (usage de dispersants, fixateurs et de

lavage). La texture influe également sur l'écoulement des solutions chimiques des méthodes *pp* et *fo*. Enfin, la nature de la couverture végétale et de la litière affectent la collecte des vers à la surface du sol.

## Facteurs économiques

Les méthodes éthologiques diminuent le travail par rapport à la méthode *bm*; elles sont donc utilisées dans les pays à main d'oeuvre coûteuse mais leurs inconvénients considérables en limitent fortement l'intérêt sauf si des corrections sont apportées (chap. IV). Les méthodes physiques nécessitent une abondante main d'oeuvre: cet inconvénient s'estompe par la mécanisation et l'automatisation du lavage-tamisage.

## Facteurs biométriques

Satchell (1971) signale que les spécimens prélevés par *bl* sont endommagés; cet inconvénient, commun aux méthodes physiques, résulte du prélèvement à la bêche. Les dénumbrements et mesures de biomasses sont néanmoins possibles (Bouché et Beugnot, 1972).

## Facteurs géodrilologiques

Les caractéristiques des lombriciens interviennent fortement dans l'efficacité des techniques. Leur taille, leur couleur et la vivacité influent sensiblement sur le tri manuel et les collectes éthologiques. Sur la base d'un large échantillonnage, Lavelle (com. pers.) obtient *Millsonia anomala* Omodeo, 1954, grosse espèce (300 à 3500 mg), avec *bl/bm* = 1 et *Chuniodrilus zielae* Omodeo, 1956 (3 à 150 mg) avec *bl/bm* = = 3,33 (dénombrements d'individus).

　　L'homochromie et la mucosité agglutinant la terre autour des vers perturbent la collecte de certains lombriciens par les procédés *bm*, *pp* et *fo*.

　　La léthargie rend impossible les captures par les méthodes éthologiques et oblige à creuser profondément pour les méthodes physiques (fig. 1).

J'ai déjà illustré (Bouché, 1969) l'influence de la lucifugie pour
$fo$ (et probablement $pp$) en montrant que $fo,bl_{20}$ donne une biomasse supérieure (167,4 g/m2)  mais un dénombrement plus faible (328,8 ind./m2)
que $bl_{60}$ (153,3 g/m2 et 394,6 ind./m2):  les grosses espèces initiale-
ment dessous - 60 cm remontent  près de la surface  mais ne sortent que
partiellement.

Le type  de galeries creusées varie;  certaines espèces  "courent"
sous la litière, d'autres ont des terriers subverticaux, tandis que les
autres vivent  dans des galeries subhorizontales débouchant peu en surface;  la pénétration des solutions $pp$ et $fo$  dépend du système souterrain.

Ces divers éléments, taille, couleur, vivacité, homochromie, mucosité, léthargie et système souterrain, se reflètent dans les catégories
écologiques des lombriciens.


## POSSIBILITÉS D'EXTRAPOLATIONS ET DE SYNTHÈSE
### Etapes de synthèse

Les étapes de synthèse relatives  aux fonctions des lombriciens ont été
décrites in Bouché (F.L.-III)  et utilisent comme données initiales des
observations morphologiques, quantitatives au(x) terrain(s),  fonctionnelles de terrain et de laboratoire.  Comme nous ne possédons jamais la
totalité des informations nécessaires et qu'il n'est pas économiquement
possible  de les établir  à chaque cas, **il est indispensable de mettre
au point  une méthodologie  critique  d'extrapolation  des informations
entre stations et populations monospécifiques différentes.**


### Catégories écologiques

Les diverses caractéristiques fonctionnelles des lombriciens,  influant
notamment  sur les méthodes de collecte,  se  reflètent  dans  les  catégories  écologiques (Bouché,  1971, 1972, 1972a,  F.L.-III).  Quelle
que soit la taxonomie phylétique, il est possible de regrouper en catégories et sous.catégories les vers de terre présentant, dans une mesure
appréciable,  les mêmes aptitudes  et propriétés en raison de leurs con-
vergences fonctionnelles: les extrapolations interspécifiques et inter-
stationnelles deviennent  sur cette base logiquement possible  (Bouché,
F.L.-III).

Tableau 1. Calcul par groupe de lots (23 indique les lots 22 23 24) de l'évolution en fonction du temps des coefficients de correction entre les 3 méthodes utilisées à Cîteaux

| Lots regroupés | % $fo/fo_{max}$ | % $fo/fo_{max}$ >40 | $\dfrac{fo}{fo,bl_{20}}$ | $\dfrac{fo}{fo,bl_{60}}$ | $\dfrac{fo,bl_{20}}{bl_{60}}$ | dates médianes |
|---|---|---|---|---|---|---|
| 23 | 4,21 | - | 14,98 | 16,73 | 1.12 | 25/06/68 |
| 26 | 3,91 | - | 15,44 | 29,09 | 1,88 | 16/07/68 |
| 29 | 15,87 | - | 5,06 | 7,93 | 1,58 | 6/08/68 |
| 32 | 40,92 | + | 2,42 | 3,27 | 1,35 | 27/08/68 |
| 35 | 66,98 | + | 1,76 | 2,02 | 1,15 | 17/09/68 |
| 38 | 88,54 | + | 1,56 | 1,71 | 1,10 | 8/10/68 |
| 41 | 100,00 | + | 1,63 | 1,53 | 0,94 | 29/10/68 |
| 44 | 74,22 | + | 1,92 | 2,19 | 1,14 | 19/11/68 |
| 47 | 54,77 | + | 2,53 | 2,64 | 1,04 | 10/12/68 |
| 50 | 25,99 | - | 3,63 | 4,72 | 1,30 | 30/12/68 |
| 53 | 13,48 | - | 6,87 | 9,61 | 1,40 | 21/01/69 |
| 56 | 17,49 | - | 3,88 | 6,31 | 1,63 | 11/02/69 |
| 59 | 33,12 | - | 3,25 | 4,26 | 1,31 | 4/03/69 |
| 62 | 64,35 | + | 2,25 | 2,21 | 0,98 | 25/03/69 |
| 65 | 40,94 | + | 2,65 | 3,74 | 1,41 | 15/04/69 |
| 68 | 43,30 | + | 2,58 | 3,56 | 1,38 | 6/05/69 |
| 71 | 15,68 | - | 4,55 | 7,21 | 1,59 | 27/05/69 |
| 74 | 8,05 | - | 7,01 | 12,70 | 1,81 | 17/06/69 |
| 77 | 3,76 | - | 18,25 | 22,89 | 1,25 | 8/07/69 |
| 80 | 0,01 | - | 2832,48 | 3994,87 | 1,41 | 29/07/69 |
| 23 à 80 $\bar{x}$ | 35,77 | +et- | 146,73 | 206,95 | 1,34 | |
| 23 à 80 $\sigma$ | 5,0 | +et- | 16,2 | 19,4 | 0,3 | |
| 32 à 47 et 62 à 68 $\bar{x}$ | 63,78 | + | 2,14 | 2,54 | 1,16 | |
| $\sigma$ | 4,4 | + | 0,6 | 0,8 | 0,4 | |

La limite $fo/fo_{max}$ % = 40 permet d'obtenir des coefficients (biomasse, peuplement) relativement réduits et homogènes. Moyenne = $\bar{x}$, écart-type = $\sigma$.

## Indices d´activités

Les méthodes éthologiques exigent un déplacement actif des animaux dépendant de la température et de l´humidité (Lakhani et Satchell, 1970),
des rythmes physiologiques et saisonniers et d´autres facteurs plus
ou moins connus. Il est présentement impossible d´analyser l´inci-
dence de chacun de ces facteurs et leur interaction, donc d´effectuer
une correction des biais techniques qui en résultent au niveau de peuplements plurispécifiques. On peut, par contre, utiliser une récolte
"éthologique", par rapport à un niveau de peuplement de référence, pour
établir un **indice d´activité**. Ainsi, pour le peuplement de Cîteaux
(tableau 1) le pourcentage des collectes $fo$ par rapport à la collecte
maximale $fo$ ($\% \, fo/fo_{max}$) reflète son activité, indépendamment de son
niveau réel.

Pour un peuplement de niveau "connu", on peut établir un **indice
d´activité relative** (rapport de la biomasse collectée par un procédé
éthologique à la biomasse totale calculée; exemple: $\dfrac{pvso,fo}{pvsc}$ soit pour
étudier les variations saisonnières (cf. $fo \, bl_{60}$ du tableau 1), soit
pour comparer des stations. Par exemple, pendant la période interestivale 1968-1969 (figure 1), les moyennes calculées à partir des inté-
grales des graphes $fo$ (et non à partir des valeurs observées, pour
tenir compte de la variabilité temporelle des prélèvements) nous don-
nent: Citeaux = 59,7 g ppfo/m2 et Borculo = 77,5 g ppfo/m2 pour des niveaux de peuplement respectivement de 13,7 g ppfo/m2 et 162 g ppfc/m2
(Bouché, F.L.-III), d´où des indices d´activité relative de 59,7/137 =
= 0,44 et 77,5/162 = 0,48 très voisins, qui semblent indiquer que les
deux peuplements ont une activité annuelle moyenne voisine, malgré des
rythmes saisonniers différents (il s´agit de premières estimations im-
précises, établies pour illustrer cet indice d´activité relative).

## Conditions de collectes et coefficients de correction

On peut voir, au tableau 1, les coefficients de correction établis par
comparaison des données obtenues sur le peuplement total de Cîteaux au
cours d´une année (rapports entre méthodes): par exemple, il faut mul-
tiplier par 16,73 la biomasse observée par $fo$ áu lot regroupé 23 pour
obtenir la biomasse observée par $bl_{60}$ (pour les lots regroupés voir
Bouché, 1969a).

Ces coefficients feront l´objet d´approximations successives
(Bouché, F.L.-V) car ils sont imparfaits:

T a b l e a u  2.   Pourcentages relatifs des catégories écologiques et indéterminables observés (biomasse) à Cîteaux en 1968-1969, période à % $fo/fo_{max}$ > 40, par 3 méthodes simultanées

| Lots regroupés | Epigés % | | | Endogés % | | | Anéciques | | | Indéterminables % | | |
|---|---|---|---|---|---|---|---|---|---|---|---|---|
| | fo | fo,bl20 | bl60 | fo | fo,bl20 | bl60 | fo | fo,bl20 | bl60 | fo | fo,bl20 | bl60 |
| 32 | 1,01 | 0,39 | 0,32 | 16,27 | 15,74 | 13,16 | 81,98 | 57,32 | 60,04 | 1,11 | 26,55 | 26,44 |
| 35 | 2,17 | 1,25 | 2,35 | 15,19 | 17,33 | 15,31 | 80,83 | 53,22 | 46,18 | 1,81 | 28,20 | 36,14 |
| 38 | 4,14 | 2,92 | 3,76 | 10,86 | 12,75 | 14,29 | 84,08 | 64,80 | 58,54 | 0,92 | 19,74 | 23,40 |
| 41 | 3,91 | 2,72 | 3,30 | 8,89 | 12,12 | 15,67 | 86,08 | 64,74 | 62,12 | 1,71 | 20,78 | 28,90 |
| 44 | 11,34 | 6,59 | 5,00 | 6,98 | 12,62 | 10,52 | 81,06 | 61,95 | 42,78 | 0,62 | 18,83 | 41,70 |
| 47 | 8,63 | 4,66 | 2,83 | 2,40 | 12,42 | 14,64 | 87,51 | 53,58 | 48,81 | 1,46 | 29,34 | 33,72 |
| 62 | 3,97 | 2,56 | 5,63 | 10,79 | 16,90 | 18,75 | 82,82 | 58,99 | 39,88 | 2,43 | 21,55 | 35,72 |
| 65 | 9,84 | 4,74 | 10,55 | 7,08 | 15,38 | 13,01 | 81,71 | 55,73 | 45,78 | 1,38 | 24,15 | 30,64 |
| 68 | 3,71 | 1,92 | 1,40 | 13,78 | 18,76 | 16,34 | 81,05 | 58,78 | 49,17 | 1,46 | 20,54 | 33,33 |

Tableau 3. Pourcentages relatifs (biomasse) des catégories écologiques observées à Cîteaux en 1968-1969, période à % $fo/fo_{max} > 40$, par 3 méthodes simultanées

| | Lots regroupés | Epigés | | | Endogés | | | Anéciques | | |
|---|---|---|---|---|---|---|---|---|---|---|
| | | fo | $fo,bl_{20}$ | $bl_{60}$ | fo | $fo,bl_{20}$ | $bl_{60}$ | fo | $fo,bl_{20}$ | $bl_{60}$ |
| Pourcentages | 32 | 1,03 | 0,53 | 0,44 | 16,45 | 21,43 | 17,89 | 82,90 | 78,04 | 81,62 |
| | 35 | 2,19 | 1,74 | 3,68 | 15,34 | 24,14 | 23,97 | 81,63 | 74,12 | 73,31 |
| | 38 | 4,17 | 3,64 | 4,91 | 10,97 | 15,89 | 18,66 | 84,86 | 80,74 | 76,42 |
| | 41 | 3,98 | 3,43 | 4,64 | 9,04 | 15,30 | 22,04 | 87,57 | 81,72 | 73,31 |
| | 44 | 11,44 | 8,12 | 8,58 | 7,04 | 15,55 | 18,04 | 81,81 | 76,32 | 73,37 |
| | 47 | 8,76 | 6,59 | 4,27 | 2,44 | 17,58 | 22,09 | 88,81 | 82,90 | 73,64 |
| | 62 | 4,07 | 3,26 | 8,76 | 11,06 | 21,54 | 29,17 | 84,88 | 75,19 | 62,04 |
| | 65 | 9,98 | 6,25 | 15,21 | 7,18 | 20,28 | 18,76 | 82,85 | 73,47 | 66,00 |
| | 68 | 3,76 | 2,42 | 2,10 | 13,98 | 23,61 | 24,51 | 82,25 | 73,97 | 73,76 |
| | $\bar{x}$ | 5,49 | 3,99 | 5,84 | 10,39 | 19,43 | 21,68 | 84,17 | 77,23 | 72,50 |
| Valeurs absolues | g ppfo/m2 | 5,25 | 7,70 | 12,85 | 9,94 | 37,60 | 47,72 | 80,62 | 149,10 | 159,58 |
| | fo $fo,bl_{20}$ | 1,46 | 1 | 0,59 | 3,78 | 1 | 0,78 | 1,84 | 1 | 0,93 |
| | fo $bl_{60}$ | 2,44 | 1,67 | 1 | 4,80 | 1,27 | 1 | 1,97 | 1,07 | 1 |

Les indéterminables ont été attribués au prorata de l'importance des catégories écologiques; $\bar{x}$ = valeur moyenne. Biomasse moyenne durant ces périodes et coefficients calculés a partir de ces données.

1°) ils s'appuient sur la comparaison avec une autre technique, elle-même source de sous-estimation; 2°) ils sont inutilisables pour les cocons, le rapport $bl/fo = \infty$; $fo$ étant nul; 3°) ils n'ont de sens que si l'on se limite à des conditions proches de l'optimum, le rapport $bl/fo$ évoluant, en dehors de celles-ci, très vite vers des valeurs élevées et même absurdes; 4°) ils ne doivent pas être établis seulement sur la journée de récolte maximale car on se trouve alors lié aux phénomènes aléatoires temporels du jour du prélèvement; 5°) la détermination de la catégorie écologique est parfois incertaine.

On constate au tableau 1 que, pendant **certaines périodes favorables de l'année**, les coefficients diminuent; la biomasse augmentant parellèlement, ils deviennent moins aléatoires et s'appliquent à des valeurs plus sûres. Ceci s'observe constamment si l'indice d'activité % $fo/fo_{max} > 40$; sous cette condition les coefficients varient peu et fournissent des moyennes établies sur des valeurs peu dispersées.

## Catégories écologiques et coefficients de correction

On observe, au tableau 2, que les vers indéterminables constituent une fraction importante pour $bl_{60}$ et $fo, bl_{20}$ (actions mécaniques); je les attribue, au tableau 3, au prorata des vers déterminés (ce procédé imprécis sera amélioré). Les pourcentages relatifs varient considérablement suivant les méthodes, notamment les épigés, peu lucifuges, ont un rendement relativement bon avec $fo$, à l'opposé des endogés apigmentés et à galeries subhorizontales.

## Extrapolations interstationnelles des coefficients de corrections

Il est rarement possible d'établir un coefficient de correction sur un échantillonnage suffisant, en fonction du temps et par diverses méthodes. Comme certaines catégories peuvent être récoltées 4,8 fois plus par une méthode que par une autre on doit utiliser les coefficients disponibles en se plaçant dans des conditions comparables.

On effectue donc les extrapolations interspécifiques, en adoptant des coefficients identiques pour des animaux à moeurs voisines (même catégorie écologique), et interstationnelles en adoptant des coefficients semblables pour les périodes ayant un indice d'activité similaire et élevé (captures suffisantes).

L'indice d'activité   suppose une étude stadiale dynamique (Bouché,
F.L.-III) qui n'est pas toujours réalisée. En adoptant des coefficients
liés à des variations mésologiques amples,  on peut corriger les données obtenues  dans  de "bonnes conditions"  sur la base  d'un jugement
empirique  (à cet égard,  l'indice $\% \ fo/fo_{max} > 40$ semble satisfaisant)
ou sur des critères climatiques. De telles démarches ont été développées
pour le P.B.I.-France (Bouché, F.L.-I et F.L.-III) et ont donné des résultats cohérents.

RÉSUMÉ


La définition  de catégories écologiques,  d'indices d'activité  et de
coefficients  de correction $\tau$,  permet  sur  une base logique,  non rigoureuse mais perfectible:  1°) d'utiliser $\tau$ entre  stations  et  taxons
(correction des mesures de populations); 2°) de comparer, dans des conditions optimales définies (indices d'activité) ou supposées correctes,
les diverses espèces capturées par une méthode éthologique (corrections
atteignant 480 %);  3°) d'établir pour chaque station un spectre écologique  (= rapport des catégories écologiques);  4°) d'établir un indice
d'activité instantané,  ou annuel moyen,  à partir des techniques éthologiques;  5°) d'effectuer  des comparaisons interstationnelles  sur la
base des critères 3 et 4,  et d'y reporter  éventuellement  des données
fonctionnelles;  6°) d'éviter de confondre  variation d'efficacité de
méthodes avec variation de populations,  et notamment dynamique des populations:  cette dernière,  qui doit être établie  sur  une base démographique, sera facilitée par $\tau$.

B i b l i o g r a p h i e

B o u c h é ,  M. B. (1969):  Comparaison critique  de méthodes  d'eva-
    luation des populations de lombricidés. Pedobiologia 9, 1/2, 26-34.
B o u c h é ,  M. B. (1972):  Contribution a l'approche  méthodologique
    de l'etude  des biocénoses. I. Vers l'analyse quantitative globale
    des prairies. Ann. zool. écol. anim. 4, 4, 529-536.
B o u c h é ,  M. B. (1972a): Lombriciens de France. Ecologie et systé-
    matique, Ed.I.N.R.A., Ann. zool. écol. anim., numéro special 72.2,
    1-671.
B o u c h é ,  M. B.:  Discussions  d'écologie.  I. Introduction.   II.

582    Bouché, M.B.

L'obtention de données ecologiques  et  l'uniformisation spatiale. (Sous presse.)

B o u c h é ,  M. B. (F.L.-I): Fonctions des lombriciens. I. Recherches françaises et resultats de la R.C.P.-40. Série des monographies de la R.C.P.-40, 4, éd. C.N.R.S., Paris. (Sous presse.)

B o u c h é ,  M. B.  K r e t z s c h m a r ,  A. (F.L.-II): Fonctions des lombriciens.  II. Recherches méthodologiques pour l'analyse du sol ingéré (étude du peuplement  de la station R.C.P.-165/P.B.I.). Rev. écol. biol. sol. (Sous presse.)

B o u c h é ,  M. B. (F.L.-III): Fonctions  des lombriciens. III. Premières  estimations  des stations francaises  du P.B.I. C.R. coll. biol. sol, Montpellier, mai 1973. (Sous presse.)

B o u c h é ,  M. B. (F.L.-V): Fonctions des lombriciens. Essais de codification des approximations successives. C.R. Ve Coll. Int. Zool. Sol, Prague, septembre 1973.

B o u c h é ,  M. B.,  B e u g n o t ,  M.  (1972):  Contribution à l'approche méthodologique  de l'étude des biocénoses. II. L'extraction des macro-éléments  du sol  par  lavage-tamisage. Ann. zool. écol. anim. 4, 4, 537-544.

G e r a r d ,  B. M. (1967): Factors affecting earthworms  in pastures. J. anim. écol. 36, 235-52.

L a k h a n i ,  K. H.,  S a t c h e l l ,  J. E. (1970):  Production by Lumbricus terrestris. J. anim. écol. 39, 2, 473-492.

L a v e l l e ,  P. (1971): Recherches  sur la démographie  d'un ver de terre d'Afrique: Millsonia anomala Omodeo  (Oligochetes, Acanthodrilidae). Bull. soc. écol. 2. 4, 302-312.

R a w ,  F. (1960): Earthworms population studies: a comparison of sampling methods. Nature, London, 187, 4733-257.

S a t c h e l l ,  J. E. (1969): Methods of sampling earthworms populations. Pedobiologia 9, 1/2, 20-25.

S a t c h e l l ,  J.E. (1971): Earthworms. In: J. P h i l l i p s o n , Methods of study in quantitative soil ecology: population, production and energy flow,  ed. Blackwell Sci. Publ., Oxford,  1o7-127.

Z a j o n c ,  I. (1971): Synusia analysis of earthworms (Lumbricidae Oligochaeta)  in  the oak-hornbeam forest  in south-west Slovakia. In: P. D u v i g n e a u d ,  Productivité des écosystèmes forestiers, Actes Coll. Bruxelles, éd. UNESCO, Paris, 443-452.

# Fonctions des lombriciens
# V. Essais de codification
# des approximations successives

M. B. BOUCHÉ
I. N. R. A. Faune du Sol,
Dijon

## INTRODUCTION

Le Programme Biologique International a permis l'accumulation d'une masse considérable de données nouvelles; malheureusement, elles sont rarement utilisables directement en raison de différences techniques et conceptuelles. Pressées par un calendrier impératif, la mise au point des techniques, leur généralisation, l'acquisition des données, l'interprétation et la synthèse, s'effectuent dans des conditions difficiles. Seules des équipes constituées avant le P.B.I. ont généralement arbitraires. Les travaux les plus élaborés (donc lentement réalisés), originaux ou isolés, risquent de ne pas être incorpores aux synthèses du P.B.I. Par ailleurs, l'objectif d'appréciation "des processus de la productivité" implique la mise en oeuvre de techniques complexes (Bouché, F.L.-III) dont chaque élément est source d'erreur; l'addition des erreurs, au cours de la cascade des calculs nécessaires, est un aspect inquiétant de nos estimations.

Il n'est nullement question de nier ici l'intérêt exceptionnel des travaux du P.B.I. (Bouché, F.L.-III), mais de reconnaître certaines difficultés afin de mieux les résoudra. Pour effectuer des comparaisons intra- et inter-stationnelles, nous devons adopter des **coefficients de corrections** pour redresser les distorsions de nos techniques, des **coefficients de transformation** pour exprimer une donnée sous différentes formes (un poids frais en poids sec ou en kcal) et, enfin, nous devons travailler par **approximations successives**. Je voudrais brièvement montrer ici les possibilités et les limites de synthèse des connaissances acquises sur les lombriciens, notamment dans le cadre du P.B.I.

ETAPES DES SYNTHÈSES

J'ai détaillé ailleurs (Bouché, F.L.-III) les liens qui réunissent actuellement les diverses étapes des recherches relatives aux roles des
lombriciens dans l'écosystème. En résumé, on effectue des mesures du
niveau et de la structure des peuplements sur lesquelles on reporte,
à partir d'études faites sur aliquotes, des données fonctionnelles de
natures variées obtenues soit directement dans le terrain d'étude du
peuplement, soit dans un autre terrain, soit en conditions artificielles. Le problème revient, si possible, à établir un lien **logique** entre
ces opérations et à les rendre **précises**.

## Liens logiques
*Données obtenues dans le terrain du peuplement*

On peut effectuer l'étude in situ de certains paramètres fonctionnels
d'un peuplement connu; la relation fonctions-lombriciens est alors directe et ne pose aucun problème si ce n'est ceux de la qualité des
échantillons, de la concordance temporo-spatiale et de l'interprétation
des résultats. Des éléments essentiels peuvent être acquis de cette
façon (structure et activité pédogénétique, influence sur la microflore,
résultantes globales, pH, etc.).

*Données obtenues dans un autre terrain que celui du peuplement*

La diversité et l'imbrication de l'activité des lombriciens (pédogenèse,
structure de l'écosystème, dynamique des cycles biogéochimiques, flux
énergétique, interrelations drilofaune-microflore) permettent d'affir-
mer que, dans chaque station du P.B.I., seuls quelques-uns de ces
aspects ont pu être abordés suivant les possibilités locales, de sorte
que nous ne possédons **nulle part** un bilan synthétique du rôle de ces
animaux. Il est dès lors essentiel de transposer les résultats acquis
d'une station à une autre. Cette interpolation peut se faire empiriquement mais aussi s'appuyer sur des raisonnements logiques qui, s'ils
ne garantissent pas l'exactitude (dépendant de critères techniques),
permettent de juger de la cohérence et de l'objectivité du système de
synthèse.

   a´) **Comparaison des nombres d´individus.** Elle a très peu de signification, le poids des vers variant entre stades de développement et
espèces: par exemple, pour le P.B.I.-France (Bouché, F.L.-III) on observe des rapports pondéraux cocon/adulte 1/600; adultes *Allolobophora
muldali/Scherotheca monspessulensis* = 1/1000.

   a´´) **Comparaison des biomasses.** Elle est la plus intéressante (les
différentes espèces ayant des propriétés tissulaires voisines) si les
expressions pondérales sont comparables (cf. chap. III) et si la classification des animaux est précise.

   b´) **Comparaison sur une base taxonomique.** On peut attribuer à une
biomasse de l´espèce $X$ de la station A les fonctions établies pour une
biomasse équivalente de $X$ de la station B; cependant, les populations
$X_A$ et $X_B$ peuvent avoir des modes de vie différents.

   b´´) **Comparaison sur une base écologique.** Outre que la comparaison
taxonomique risque d´introduire des erreurs si $X_A \neq X_B$, elle ne permet
pas de transferts fonctionnels entre espèces différentes: la classification taxonomique poursuit un but phylétique et nullement fonctionnel.
Il faut alors utiliser la classification en catégories écologiques
(Bouché, 1971, F.L.-III) qui regroupe, à partir de caractères morpho-
fonctionnels, des taxons ayant des modes de vie identiques ou voisins.
Il est dès lors possible, dans des limites appréciables, d´utiliser les
informations acquises sur une espèce pour une autre.

   En conclusion, les échanges d´informations acquises sur des stations différentes doivent se faire préférentiellement à l´aide des biomasses des catégories écologiques; cette démarche a permis d´établir des
interprétations biogéocénotiques cohérentes (Bouché, F.L.-I, F.L.-III).


*Données obtenues en conditions artificielles*

Pour mettre en évidence et quantifier certains phénomènes, on est parfois conduit à placer les animaux dans des conditions artificielles.
L´extrapolation au terrain des résultats ainsi acquis est une opération
délicate qui semble ne pouvoir s´appuyer sur aucune relation logique:
un taux respiratoire établi au Warburg s´extrapole au terrain sur la
base des courbes de température ... mais beaucoup d´autres paramètres
incontrôlés interviennent. Il nous reste parfois la possibilité d´uti-
liser des dispositifs ayant des degrés d´artificialisation décroissants
qui permettent ainsi une critique empirique de l´extrapolation.

## Précision des données

Nous venons de voir qu'il est possible, sur une base logique, d'effectuer des interpolations de données entre espèces et stations. Ceci ne
signifie pas que les valeurs et coefficients utilisés soient **exacts**.
Sur des données **observées**, on peut effectuer des **corrections**: j'ai montré, par exemple, que les distorsions considérables résultant de la
collecte au formol pouvaient être en partie redressées grâce à un coefficient de correction technique τ (Bouché, F.L.-IV); il est donc indispensable d'indiquer clairement si une donnée est **observée** et par
quelle technique (en abréviation italique, cf. F.L.-IV), si elle a été
corrigée et par quel moyen. Je propose d'accompagner les mesures des
lettres o (= observé) ou c (= corrigé) d'une part et d'indiquer le
coefficient τ et sa source d'autre part. La nature des données auxquelles s'applique τ varie, je propose d'indiquer celle-ci entre [] avant
la lettre τ; il faut ensuite signaler entre parenthèses à quelle référence bibliographique τ se rapporte. Cela conduit à: [p anéciques]
*fo* τ *fo,bl* = 2 (Bouché, F.L.-I), qui signifie: le coefficient de correction technique des anéciques pour passer des poids observés par la méthode au formol au poids corrigés par référence à la méthode bêche-la-
vage-tamisage est de 2 in Bouché (F.L.-I). Ces condensation et codification devraient faciliter nos échanges d'informations quantitatives
et nous permettre constamment d'opiner sur une valeur numérique.
     Une autre difficulté provient du mode d'expression des données. Il
est actuellement rare de ne pas voir, dans des tableaux de données pondérales comparant des écosystèmes, des ordres de grandeur différents:
des biomasses végétales en poids sec sont souvent comparées à des zoo-
masses en poids frais! Pour les lombriciens, il y a plusieurs expressions pondérales: le poids sec (ps), le poids humide vif (ph), le poids
humide après fixation au formol (pf) ou à l'alcool (pa) avec tube digestif plein (pp) ou vide (pv); il faut donc associer à une biomasse
l'unité du système métrique suivie de sa nature (par exemple: 18 g
pvs = 120 g pph signifie que 18 grammes de lombriciens tube digestif
vide en poids sec équivalent à 120 grammes de lombriciens tube digestif
plein en poids humide vif).
     La synthèse des résultats oblige à présenter les données sous une
même forme, c'est-à-dire préalablement à passer d'un mode d'expression
à un autre, ce qui implique l'adoption du coefficient π de transformation pondérale. Il faut, comme pour τ, indiquer le sens de la transformation, ce à quoi elle s'applique et sa source; exemple: [*Lumbricus
terrestris*] ps π ph = 6,37 (Lakhani et Satchell, 1970) signifie que ces
deux auteurs ont multiplié par 6,37 les poids secs de *Lumbricus terrestris* pour obtenir les poids humides vifs.

APPROXIMATIONS SUCCESSIVES

Les données de terrain doivent être généralement corrigées par $\tau$ et transformées par $\pi$ . Ces calculs portent sur des biomasses établies à des temps $t_1$, $t_2$, $t_3$ ... ou sur des productions mesurées entre les temps $t_1$, $t_2$, $t_3$ ... Biomasses et productions constituent la référence normale des extrapolations fonctionnelles de divers paramètres (tonnage de terre brassée, oxygène consommé, etc.). Nos estimations dépendent donc de $\tau$ et $\pi$ . Actuellement, il est certain que nous ne pouvons pas établir ces coefficients de façon totalement satisfaisante.

$\tau$ permet de comparer une étude assurée par une méthode de collecte à une autre étude utilisant un autre procédé et d'adapter la technique optimale (Bouché, F.L.-IV). L'évolution de nos techniques de capture et de calcul conduira à adopter des coefficients $\tau$ successifs de plus en plus élaborés pour chaque classe de poids, pour chaque espèce, pour chaque catégorie écologique et pour chaque condition de collecte.

Il en est de même de $\pi$ . Si le pourcentage d'eau chez les vers de terre semble assez constant, le contenu du tube digestif varie qualitativement et quantitativement en fonction des espèces, des variétés, des stades de développement, de l'état physiologique et du temps (Bouché et Kretzschmar, F.L.-II). Nous sommes loin de connaître les lois de ces variations: les coefficients $\pi$ actuels sont donc temporaires.

Pour les lombriciens des synthèses et des comparaisons de données intra- et inter-stationnelles sont possibles:

- elles peuvent être tentées logiquement, (Bouché F.L.-III, F.L.-IV);

- elles sont inexactes en raison des coefficients $\tau$ et $\pi$ adoptés actuellement;

- elles permettent, par **approximations successives** et à partir des **mêmes données observées,** des estimations de précision croissante.

L'adoption d'un système d'approximations successives permet d'établir une synthèse sans attendre des éléments vraiment satisfaisants et d'y integrer chaque nouvelle donnée. Un tel système est complexe car ses éléments sont eux-mêmes complexes, ce qui oblige à une codification rigoureuse indispensable pour progresser:

- Comparer dans une station des espèces de catégories écologiques différentes sur la base de biomasses collectées au formol, confronter des biomasses (et à fortiori des nombres!) observées avec des techniques différentes et dans des stations différentes, porter en tableau poids frais et sec, sont des pratiques qui devraient disparaître avec une codification qui illustrerait ces incohérences.

- Les études de populations naturelles de lombriciens étant difficiles et coûteuses ne peuvent être indéfiniment répétées. Un système d'approximations successives permettra d'utiliser l'abondante information quantitative déjà existante et de se consacrer plus activement aux interprétations fonctionnelles qu'elles autorisent.

- Les moyens modernes de calcul permettent une gestion ordonnée de nos connaissances mais exigent une méthodologie normalisée. L'application des modèles simulés d'écosystèmes notamment n'a de sens que si les informations introduites dans ces systèmes sont valables et cohérentes: l'actuel fossé existant entre systèmes simulés et données disponibles sera ainsi réduit.


## CODIFICATIONS PROPOSEES

A - E x p r e s s i o n s   q u a n t i t a t i v e s

    1° - Références générales: système métrique.

    2° - Associer aux données pondérales la lettre p et aux données numériques la lettre n avant l'indication de ses qualité.

    3° - Qualifier la donnée pondérale par l'état du tube digestif: p = plein, v = vide.

    4° - Qualifier ensuite la donnée pondérale par le mode de mesure: h = poids humide vif, f = après fixation au formol (écrire eventuellement $f_{12}$ = fixation a 12 ‰ de formaldehyde), a = - après fixation à l'alcool, s = après dessication.

    5° - Indiquer ensuite le degré de transformation: o = données effectivement observées, c = données corrigées par des coefficients $\tau$ ou/et $\pi$, e = données estimées empiriquement.

    **EXEMPLES:** ppfo = poids, tube digestif plein, pesé après fixation au formol observé.

              pvsc = poids, tube digestif vide, pesé sec, corrigé.

B - C o e f f i c i e n t s

B' - C o e f f i c i e n t s   d e   c o r r e c t i o n

    $\tau$ permet de corriger, par une multiplication, un biais provenant des methodes de collecte.

    1° - Indiquer [] la nature des données (n ou p), puis le taxon, la categorie, etc.

    2° - Indiquer la technique de collecte, puis $\tau$ , puis la technique servant de référence de correction. Les techniques (en ita-

lique) étant *pp* = permanganate de potassium; *fo* = formol;
*bm* = bêche, tri manuel; *bl* = bêche, tri après lavage-tamisage, ou des systémes combinés qui doivent être indiqués dans
l'ordre d'exécution (cf. Bouché, F.L.-IV).

3° - Eventuellement, indiquer en indice le ‰ de la solution utilisée ou la profondeur des prélèvements: $bl_{20}$ = méthode *bl*
limitée à -20 cm.

4° - Indiquer la source du coefficient.

EXEMPLE: [p *Lumbricus terrestris*] *fo* τ *fo,bl*$_{20}$ = 1,18 (Bouché, F.
L.-I).

B'' - C o e f f i c i e n t s   d e   t r a n s f o r m a t i o n
Je n'aborderai ici que le coefficient de transformation des bio-
quantités (π) qui les relie entre elles par une multiplication.

1° - Indiquer entre [] a quel groupe ce coefficient s'applique.

2° - Indiquer la bioquantité initiale, puis π, puis la bioquantité
corrigée par le coefficient, en utilisant les expressions
pondérales codifiées plus haut.

3° - Indiquer la source du coefficient.

EXEMPLE: {*Lumbricus terrestris*] pso π pfc = 6,37 (Lakhani et
Satchell, 1970).

RÉSUMÉ

La nature des divers paramètres que nécessite la synthèse écologique
des fonctions des lombriciens est analysée. Certains sont reliables logiquement d'autres seulement empiriquement. Ils sont observés à travers
des techniques qui apportent des distorsions souvent considérables mais
celles-ci peuvent être corrigées. Ces corrections ne peuvent se faire
que par approximations successives. La complexité des diverses opérations de synthèse oblige à une codification conceptuelle et à une condensation de l'écriture en symboles que l'auteur propose.


SUMMARY

The characteristics of the various parameters necessary to carry out
a comprehensive ecological study of the role of earthworms are analysed.

These parameters can be related to one another, either logically or only
empirically. They are studied by means of techniques often involving
important differences in the results which can however be compensated.
The corrections are only possible by continual approach. The different
stages of the synthesis are so intricate that a conceptual coding and
symbols are suggested by the author.

B i b l i o g r a p h i e

B o u c h é , M. B. (1971): Relations entre les structures spatiales
    et fonctionnelles des écosystèmes illustrées par le rôle pédobio-
    logiques des vers de terre. In: Pesson "La vie dans les sols", éd.
    Gauthier-Villars, Paris, 187-209.
B o u c h é , M. B. (F.L.-I): Fonctions des lombriciens. I. Recherches
    françaises et résultats de la R.C.P.-40. Série des monographies de
    la R.C.P.-40, 4, éd. C.N.R.S., Paris. (Sous presse.)
B o u c h é , M. B., K r e t z s c h m a r , A. (F.L.-II): Fonctions
    des lombriciens. II. Recherches méthodologiques pour l'analyse du
    sol ingéré (étude du peuplement de la station R.C.P.-165/P.B.I.).
    Rev. écol. biol. sol. (Sous presse.)
B o u c h é , M. B. (F.L.-III): Fonctions des lombriciens. III. Pre-
    mières estimations quantitatives des stations françaises du P.B.I.
    C.R. coll. biol. sol, Montpellier, mai 1973. (Sous presse.)
B o u c h é , M. B. (F.L.-IV): Fonctions des lombriciens. IV. Correc-
    tions et utilisations des distorsions causées par les methodes de
    capture. C.R. Ve Coll. Int. Zool. Sol, Prague, septembre 1973.
L a k h a n i , K. H., S a t c h e l l , J. E. (1970): Production by
    *Lumbricus terrestris* (L.). J. anim. écol. 39, 2, 473-492.

D i s c u s s i o n

J. W. R e y n o l d s : (pour les 2 exposés de M. B. Bouché): Your presentation discusses the use of the formalin method for extraction and
sampling. In North America, we have many earthworm families which do not
respond to this method (cf. Reynolds, 1972). Do you find this to be the
case for the non-*Lumbricidae* in your studies? Do you make the distinction between this technique as a measurement of activity as opposed to
a measurement of population?

M. B.  B o u c h é :  I have not tried this method  on any earthworm families other than *Lumbricidae*.  My papers  are only related to this family which seem to show some groups having a way of life different from
other  families  of  megadriles  (cf. Bouché, F.L.-III).  In fact, Dash
and Patra (1971),  A comparison of extraction methods  for *Megascoleci-
dae* (Olig.)  and  *Ocnerodrilidae* (Olig.)  from agricultural soils  from
Berhampur,  Onssa. Carr. sci., 41, 7, 254-255)  presented  observations
differing from our knowledge of *Lumbricidae*. Nevertheless,  it a pity
that we do not have any comparative studies of ethological  and mechanical methods  on  a large number  of  samples,  under different climatic
conditions, at different localities for families other than *Lumbricidae*.
For this reason,  we are not able  to make  a general judgement  on the
possibility  of using ethological methods  as an activity  index, since
the source of data after transformation gives reasonable estimates, etc.
For these two papers  I used information  obtained on 483 sp.m  by formalin extraction and 65 metric tons by the *bl* method (digging-washing).
This sample was taken at 8 sites during a 3 year survey.

Ph.  L e b r u n :  (pour les 2 exposés de M. B.  B o u c h é :  a) Les
espèces adaptées  a la consommation  d'essences  très coriaces comme le
chêne vert  présentent-elles des adaptations morphologiques  des pièces
buccales?

      b) Y-a-t-il corrélation entre le rythme phénologique du chêne vert
et le rythme d'activité des espèces mèditerranéennes ou bien la période
xérothermique est-elle prépondérante?

M. B.  B o u c h é :  a) Au niveau actuel  de nos connaissances, essentiellement qualitatives, les différences observées  entre formes anéciques consommatrices de feuilles tendres  ou coriaces portent essentiellement  sur la taille.  Nous passons d'animaux longs de 20 cm (feuilles
tendres) à des animaux longs de 80 cm (feuilles coriaces);  la muscula-
ture pharyngiale  très développée  chez les anéciques augmente  en proportion de la taille.  Nous essayons actuellement au laboratoire d'analyser plus scrupuleusement  et  de quantifier  ces caractères  (travaux
de M. Pons).

      b) Non,  il n'y a pas de corrélation;  dans les stations de chênes
verts, le maximum de production de feuilles mortes se situe en mai-juin
(Rapp, M., 1971 - Cycle de la matière organique  et  des éléments minéraux  dans  quelques ecosystèmes méditerranéens.  P.B.I.,  R.C.P.-40, 2,
édition C.N.R.S., Paris, 19-184);  à cette époque, les *Scherotheca* consommateurs sont inactifs (sècheresse + diapause) comme cela a été montré
par Galissian  pour une espèce différente, mais d'activité probablement
semblable  (Galissian, A., 1971 -  Diapause et régénération postérieure

chez   le  lombricide  *Eophila dollfusi*  (Tétry).  Thèse  univ. Provence,
1-243). Il faut attendre le retour des conditions favorables à l'automne
pour que les lombricides  assurent l'ingestion-enfouissage de cette litière.

Z.  M a s s o u d :  Est-ce que la methode électrique  présente  encore
un matériel "qualitativ- ou quantitativement"?

M. B. B o u c h é :  Non, sauf pour récolter les animaux pour l'élevage.

# Observations on the digestive enzymes
# of some litter-feeding animals

G. MARCUZZI AND M. TURCHETTO LAFISCA
Institute of Animal Biology,
University of Padova

The investigations have been carried out on the following species of soil and litter animals:

Isopods. - *Oniscus* sp., from S. Daniele (Veneto); *Oniscus* sp., from Gorgazze (Friuli - Venezia Giulia); *Porcellio* sp., from Cimolais (idem); *Porcellio ficulneus*, from Mashhad (Iran); they are all primary and, maybe, also secondary degraders, sometimes harmful to crops or plants in glasshouses.

Diplopods. - *Glomeris conspersa*, from Bassano (Veneto); *G. euganeorum*, from Colli Euganei (Veneto); *G. undulata*, from Bassano (Veneto) and *G. marginata* from Grange-over-Sands, England (collected by Dr.Heath); they are primary and secondary degraders.

I n s e c t a : 1. - *Coleoptera*. A) *Tenebrionidae*: larva of *Tenebrio molitor* (the common mealworm), laboratory stock; a poliphagous, almost omnivorous animal; *Tentyria grossa*, from Trapani (Sicily), a psammo-xerophil animal, almost omnivorous; *Pimelia grossa*, from Trapani, relatively psammophilous, almost omnivorous; *P. rugulosa*, from Trapani, a relatively xerophilous species, same alimentary habits; *P. grandis*, from Israel, a saprophagous insect; *Blaps sulcata*, from Israel, collected by Dr.Shkolnik, saprophagous, polyphagous; *Adesmia dilatata*, from Israel, collected by Dr.Shkolnik, polyphagous; *Centrioptera variolosa*, from Tempe, Arizona, collected by Prof.Hadley, poliphagous; and finally *Phaleria bimaculata*, from Chioggia (near Venice), halo-psammophilous, polisaprophagous, sometimes necrophagous. B) *Scarabaeidae*: larva of *Oryctomorphus bimaculatus*, from Bariloche (Southern Argentina) collected by Dr.Contreras, feeding on litter of *Notophagus*; larva of *Oryctes monoceros*, from Lamto, Ivory Coast, collected by Dr.P.Lavelle, in dead and living palms (*Borassus aethiopum* or "palm roncier"); C) *Ipidae*: *Ips typographus*, adult and larva, feeding on dead trunks of coniferous trees as *Picea excelsa*, *Pinus*, *Larix*; D) *Cerambycidae*: larva of *Rhagium* sp., feeding on recently dead trees. 2. *Diptera*: larvae of *Ti-*

*pula* sp., from Foresta Umbra (Puglie), a secondary degrader, poliphagous; larva of *Drosophila heydeni*, obtained from a laboratory stock by courtesy of Dr.Danieli; living on artificial diet (phytosaprophagous).

G a s t r o p o d s : 1. - *Prosobranchia*: *Pomatias elegans*, from Trieste, phytosaprophagous; 2. - *Pulmonata*: *Lehmannia marginata*, from Cansiglio (Veneto), phytophagous, feeding also on algae and lichens; *Limax albipes*, from Bormio; *Limax flavus*, from Bormio, feeding essentially on subterraneus parts of plants (carrots, turnips celery atc.); *L. maximus*, from Veneto, probably feeding only on living matter, both vegetal and animal; *Arion subfuscus*, from Bormio, omnivorous but particularly fungivorous; *Chilostoma (= Helicigona) cingulata* from Asiago (Veneto); *Oxychilus* sp., from Cansiglio, Veneto, mainly carnivorous but also herbovorous; *Helicigona planospira*, from Vajont (Veneto), probably phytosaprophagous; *Helix aspersa*, from Padova, phytophagous and secondarily phytosaprophagous; *H. aperta* from Lecce (Puglia) and Trapani (Sicily), feeding preferably on fresh leaves but also on dead leaves and *Basidiomycetes*; *H. pomatia*, from Padova, phytophagous; *Sphincterocheila boissieri*, from Israel, collected by Dr.Shkolnik probably a phytophage feeding on desert halophilic plants.

The following are the assayed substrates: 1. oligosaccharides: cellobiose, lactose, maltose, melezitose, melibiose, raffinose, sucrose and trehalose; 2. Polysaccharides: starch, glycogen, lichenin, chitin, holocellulose, hydrocellulose, algin, pectin, mannane (salep, i.e. tubers of Orchids, carob seeds and *Phytelephas* seed or ivory nut); 3.- esters s. lato: a) lipids (oils): olive, arachis, maize, lin,castor, sunflower, grape-seeds, rape-seed; b) esters s. str.: glycerol tributirate (= tributyrin), ethyl ester of butyric acid and Tween 80 (a mixture of poliossyethylenic esters and oleic esters of sorbitol anhydrides); c) some not well definited substances, as cork, cuticle of Agave, cuticle of *Nelumbium*, cuticle of *Hedera*, Agave wax, *Nelumbium* wax, cutine of *Agave* and cutine of *Nelumbium*; 4. - Proteins and polypeptides: gelatin, casein, albumin, sericin, salmin, elastin, fibroin, glycil-glycil alanine, glycil-glycil glycine and L leucil-glycil-glycine.

R e s u l t s : 1. - Oligosaccharases. - It is to point out the greater variety of enzymes present in insects, as compared with mollusks; information about isopods and diplopods is too scarce to reach any general conclusion. The data on mollusks indicate the absence of enzymes able to split melezitose and trehalose, whereas insects appear to be always furnished with a "trehalose". It is of interest that also saprophage macrophytophagous mites are unable to digest trehalose, which is present in large amounts in mushrooms as a reserve carbohydrate

(Luxton & Thomas, 1972, (unpublished data). As far as the paucity of
oligosaccharases in some animals is concerned, we must bear in mind
that according to Borchet some di-and trisaccharides are toxic to some
insects and, therefore, maybe, these may not possess enzymes able to
split them. For instance, lactose, mannose, rhamnose, melibiose and
raffinose are toxic to the bee; they contain galactose, which is toxic
to this animal. 2. - Polysaccharases. - Polysaccharases, followed by
oligosaccharases, are the most well-known enzymes in all terrestrial
invertebrates, partly in view of the assumed ability of these animals
(particularly phytophages and phytophages and phytosaprophages) to
digest many natural polysaccharides, particularly cellulose, partly
because students have dealt with these enzymes much more than with the
others. Certainly the problem is interesting from an ecological point
of view, since polisaccharides are widely distributed in the vegetal
kingdom and therefore, included in the diet of all soil animals, except
for some highly specialized parasites. We must recall here, besides
some recent papers on springtails and mites, earlier papers by Mansour
and Mansour-Bek (1933, 1934) and by Parkin (1940) on xylophagous or
lignicolous insect larvae, the work by Nielsen on many soil and litter
feeding animals (1962) and some others, which have employed some very
sensitive techniques, as thin-layer chromatography. Nielsen has found
in *Porcellio scaber* only an amylase, whereas Jeuniaux has found also
a chitinase. We have found chitinase (though only in traces) in a species of *Oniscus,* but, as a whole, in the species under consideration,
there was a great paucity of polisaccharases. This, however, may be
ascribed to the values of pH at which we have worked. Nielsen has found
in *Glomeris marginata* an amylase at a not specified pH, whereas we have
found in the same species at pH 7.5 a glucosidase splitting glycogen
but not starch. Parkin (1940) found in *Scolytidae* larvae enzymes splitting starch and hemicellulose A and B, and in the larva of *Rhagium
mordax* enzymes which attack - though in a small amount - starch, cellulose, and hemicellulose A and B. On the basis of our results, the larva
of *Rhagium* attacks only starch and arbutin, the adult *Ips typographus*
glycogen, and larval *Ips arbutin.* In *Tipula* larvae Nielsen found an
amylase and doubtfully a cellulase, whereas in our species we found
glycogenase and chitinase. We wish to point out that the larvae of some
lignivorous scarabeids possess a fermentation chamber which seems to be
important in the digestion of cellulose: nevertheless, some species
have a chamber which contains no cellulose-splitting bacteria, as for
instance the larva of *Oryctes momceros* that we investigated also from
the bacteriological standpoint.

Von Buddenbrock says that in gastropods amylases are widely dis-

tributed, particularly    - amylase, which in *Helix* should act after 5´.
The genus *Helix* shows a remarkable enzymatic activity (  - amylase,   -
glucosidae, chitinase, hydrocellulase  etc.)  which  has been confirmed
both  in  research work  of other authors  and by us.  Nielsen found in
gastropods enzymes splitting cellulose, pectin, xylane  and chitin: the
fact that Nielsen finds a true cellulase in animals systematically very
near to those  we have studied,  in  which  we have found no cellulase,
could be attributed to the different pH (5.5 - 6.9 Nielsen, 7.5 present
authors). From this controversial results it is possible to deduce that
the presence  of a true cellulase  in soil invertebrates  is  much discussed.

The greatest amount  of  species  we have delt  with  has not been
examined by previous authors,  so that a comparison  with other data is
practically impossible.  Incidentally  we learn from Zinkler´s research
on springtails, from Luxton´s research on mites  and from Nielsen´s observations that also  in closely related species  there are conspicuous
differences,  and the same applies  to other groups of enzymes  we have
studied (proteases, lipases).  Any generalization is therefore inadvis-
able.

From our results  we consider  that  isopods  have  very few polysaccharases; Diplopods would follow, whereas both insects and gastropods
possess a rich set of enzymes.  This possibly has an importance when we
come to consider the role  of these invertebrates  in the decomposition
of litter.  Our observations on a number of substrates, partly natural,
partly artificial,  and on a number of animals,  sometimes belonging to
closely related species (*Glomeris,  Limax,  Arion,  Helix*) suggest that
specificity of most polisaccharases  demonstrated by us is high.  It is
worthwhile to quote Dixon & Webb (1958) (cfr. Jeuniaux): "the high specificity of enzymes, i.e., the close limitation of action of each enzyme
upon a very small number of closely related substances, is one of their
characteristics. Enzyme specificity is one of most important biological
phenomens, without the ordered metabolism of living substance could not
exist, and the very life should be impossible". 3. - Lipases. - Among
all existing digestive enzymes, lipases  and  esterases are very poorly
known, especially as far as soil animals are concerned. Nothing is known
about esterases  in terrestrial isopods and myriapods.  In insects they
have been demonstrated by several authors, but no data are available on
litter feeding animals. As far as mollusks are concerned, the only data
we have are inherent *Helix pomatia*, where some enzymes have been demonstrated, which attack tributirine, methyl butirate, tween 60  and olive
oil (Wilbur & Yonge),  triolein (Ferreri, 1958)  and  tween 20 (Byers).
From our work it results that as a whole esters and tween  are splitted

much more easily  than neutral fats and waxes, what is in harmony  with previous observations by Chauvin, von Buddenbrock  and  Wilbur & Yonge. The different behaviour of animals to waxes  and neutral fats points to a great specificity  of enzymes:  this is  particularly interesting because in the past authors generally admitted  the presence of aspecific enzymes  (disregarding  their  categories)  acting upon a wide range of substances.  Also the fact that the enzymes of different animals act on a given substrate at different pH is a confirmation of specificity, and this is still more evident  when we take into consideration the effects of inhibitory  or  activant substances  on enzymatic activity (1).

(1) This part of work  is entirely due to Dr.Luisa Dalle Molle. As we have said, nothing is known about lipases of soil and litter animals: in earthworms, for example, indirect research in lipid content of leaves and faeces points  to an enzymatic activity  of the animal,  but nobody can say  whether  this activity  is due to the earthworm  or to its intestinal bacteria.  According to Martin & Juniper,  animal contribution to the attack  of wax and cutine  is probably at least  as important as that of the soil microflora. Lipids are constant components of the diet of soil animals.  They are quite abundant  in  the chemical composition of dead leaves, though not as carbohydrates, but more  as proteins, and we have demonstarted  that they can be digested  by most animals,  differently from carbohydrates,  which sometimes  are practically an inert part of diet. We think therefore that our results represent an original contribution to the knowledge  of the role of animals  in litter decomposition.

4. - Proteases. - Isopods  are  very poor  in proteases  (we found a gelatinase  and  an albuminase in *Oniscus*);  the genus *Glomeris* (Myr. Diplopods) seems to be very rich in proteases: gelatin, salmin, elastin, fibroin and albumin are digested,  practically,  with no differences in the two examined species). *Coleoptera* (both adults  and  larval)  are rather well equipped with proteases:  possibly there are more proteases than polypeptidases, but further work will be necessary to clarify this point.  Among the mollusks there is an enormous difference between *Prosobranchia (Pomatias elegans)* and  *Pulmonata:*  the former appears practically to have no proteases;  in  *Pulmonata*  both proteases  and  polypeptidases are present;  *Arionidae* and *Limacidae,* which are practically omnivorous,  seem to be richer in proteases than the other mollusk families. Here too further work should be necessary.

References

F e r r e r i , E. (1958): L'attività lipasica nell'epitelio intesti-
    nale di Helix pomatia. Boll. Soc. It. Biol. Sper. 34, 379.

L u x t o n , M. (1972): Studies on the oribatid mites of a Danish
    beech wood soil. Pedobiologia 12, 434.

M a n s o u r , K. and M a n s o u r - B e k , J. J. (1934): On the
    digestion of wood by insects. J. Exptl. Biol. 11, 243.

M a r t i n , J. T. and J u n i p e r , B. E. (1970): The cuticles
    of plants. Edinburgh.

N i e l s e n , C. O. (1962): Carbohydrases in soil and litter inverte-
    brates. Oikos, 13, 200.

P a r k i n , E. A. (1940): The digestive enzymes of some wood-boring
    beetle larvae. J. Exptl. Biol. 17, 364.

W i l b u r , K. M. and Y o n g e , C. M. (1966): Physiology of Mol-
    lusca II. New York & London.

Z i n k l e r , D. (1968): Vergleichende Untersuchungen zum Wirkungs-
    spektrum der Carbohydrasen von Collembolen (Apterygoten). Verh. D.
    Zool. Ges. Innsbruck, 640.

Z i n k l e r , D. (1970/71): Carbohydrasen streubewohnender Collem-
    bolen und Oribatiden IV. Coll. pedobiol., Dijon, IX, 329.

# The efficiency of collecting soil arthropods from the froth by flotation method at different intervals of counting

J. SINGH

Department of Entomology, Agricultural Zoology,
Banaras Hindu University, Varanasi – 5, India

INTRODUCTION

The importance of the accurate estimation of soil organisms in the eco-system has been emphasised by the International Biological Programme for productivity studies. Hence the method for a precise estimation of soil fauna receives greater attention. In an extensive study Edwards and Fletcher (1970) stated that no single method was ideal for all groups of soil invertebrates. Singh and Sharma (in prep.) studied the comparative efficacy of flotation method and Tullgren funnel and con-cluded that the former was much more efficient then the latter espe-cially in the case of arable soil with lower organic matter content. Macfadyen (1962) also showed the distinct advantages of flotation method for arthropod extraction. In quantitative studies larger sample units will give more accurate data and these can be easily processed with the flotation method. Another advantage is that the quiescent stages of arthropods can also be extracted by flotation technique as this method is not dependent on the behavioural characteristics of soil arthropods. Mukharji and Singh (1970) and again Singh and Mukharji (1971) success-fully used flotation method for extraction of soil arthropods in their studies. The present authors tried Tullgren funnels but failed to ob-tain satisfactory results from the soil at Varanasi (India) which is sandy loam and very poor in organic content (O.M. 0.5 - 1.26 %). The number of micro-arthropods obtained from Tullgren funnels was consider-ably lower and as such it had to be abandoned and the flotation method had to be resorted to.

This communication, therefore, embodies the findings from a modi-fied flotation method adopted by the authors on the basis of Ladell's flotation technique (Ladell, 1936), and also includes the efficiency of extraction of micro-arthropods by observing the froth at three dif-ferent intervals of timings viz., 1 hr, 12 hrs and 18 hrs after pro-cessing the soil in the Ladell apparatus.

SAMPLING AND EXTRACTION METHOD

The soil samples for the present investigation  were collected by using
a rectangular iron sampler  sized 10 x 7.5 x 22.5 cm  (1687 cubic cm in
volume).  The area selected for sampling  was relatively homogeneous in
nature  in a fodder crop field where a 3 Sq. m area  was demarcated and
all the sample units were collected in close proximity to each other to
make each sample unit more homogeneous. The soil samples were then processed in Ladell's apparatus by churning  with magnesium sulphate solution (sp. gr., 1.2) for 30 minutes  and frequently bubbling air through
an air pump to allow the animals to float.  Continuous churning  of the
soil and bubbling the air from the bottom produced good froth.

The modification of Ladell's technique  adopted by the authors was
that, instead of passing the froth through the flotation tank  and then
into the Buchner funnel, it was directly swept into beakers with a fine
nylon brush  (No. 6)  by pouring a gentle stream  of magnesium sulphate
solution. The froth with magnesium sulphate solution in beaker was kept
undisturbed for  some time  to allow  the roots  and soil particles to
settle.  It was then transferred to petri dishes  and inspected in different time intervals, i.e., 1 hr, 12 hrs  and  18 hrs  under a Stereo-
scopic microscope for counting.  The micro-arthropods  would float over
the froth  and  by gentle shaking  of the petri dish all living animals
would start moving over the froth,  they could be picked up with bamboo
splints and preserved for further identification in 70 per cent alcohol.


RESULTS

The data presented in the table below record the average number of soil
animals viz., *Acarina, Collembola* and  other arthropods  (pauropods,
*Symphyla, Japyx, Palpigradi* and *Pseudoscorpion*)  collected at different
time intervals. (Table 1).

It is evident  from table 1  that relatively less micro-arthropods
were collected  from  the froth examined  after 1 hr  of processing the
soil than 12 hrs. The recovery after 18 hrs was more or less similar to
that  of 12 hrs  and  statistically  there is no significant difference
between 12 hrs  and 18 hrs  and both  are highly significant over 1 hr.
The significant differences between the numbers of animals and counting
intervals are summarized in Table 1.  The results suggest that for more
efficient counting  and  collection of micro-arthropods,  the froth has
to be observed  after 12 hrs of processing the soil.  Since there is no

T a b l e   1.   Number of arthropods collected after different intervals of counting (mean values/5 samples)

| Fauna | Different time intervals | | | |
|---|---|---|---|---|
| | 1 hr | 12 hrs | 18 hrs | |
| Acarina | 45.8 | 98.8 | 99.2 | Interation |
| Collembola | 13.4 | 31.2 | 28.2 | Hours x total number |
| Other soil arthropods | 16.8 | 38.8 | 37.0 | Highly significant |
| | | | | 'F'Value = 18.30 at 4 df. |
| Total | 76.0 | 168.8* | 164.4* | S.E.±2.27 |
| | | | | C.D. at 1 % = 5.96 |

*Highly significant.

statistically significant difference  between 12 hrs and 18 hrs  it was found better to observe the froth after 12 hrs to save more time. Higher recovery recorded after  long time interval  was due to the fact  that smaller arthropods  were released when the air bubbles burst out.  This was evidenced  from the fact that more minute sized *Prostigmata (Coccotydeus* sp.) and *Onychiuridae* were obtained after 12 hrs.

CONCLUSIONS

By  modifying  Ladell's technique  the complicated procedure  was  more simplified and higher precision was obtained.  If the froth was allowed to pass through flotation tank and Buchner funnel of Ladell's apparatus, 22 litres of magnesium sulphate solution  was required to fill the flotation tank and a suction pump has to be worked for the Buchner funnel. Besides, by observing the animals  collected  over  the filter paper of the Buchner funnel, counting and picking up of animals  were found difficult  and  the number  obtained  was less.  By this simplified method a higher number  of minute  and  soft bodies micro-arthropods  could be picked up. More viable animals were collected from the froth than after filtration of the froth  through  the Buchner funnel.  By observing the froth after keeping it for 12 hrs more micro-arthropods could be counted and  thereby  a more correct estimation  of the population  of soil animals was made possible.

Acknowledgements

We are thankful  to  Dr.S.P.Mukharji,   Head,  Department of Entomology,
Faculty of Agriculture,  Banaras Hindu University  for providing us the
adequate laboratory facilities.

References

Edwards, C. A., Fletcher, K. E. (1970): Assessment
    of terrestrial invertebrate populations.  In:  J. Philipson (ed.),
    Methods of study in soil ecology 57-66, UNESCO.
Ladell, W. R. S. (1936): A new apparatus for separating insects
    and other arthropods from the soil. Ann.Appl.Biol.23 (4), 862-879.
Macfadyen, A. (1962): Soil arthropod sampling. In: J. B. Crag
    (ed.), Advances in ecological research. Academic Press, London,
    1-34.
Mukharji, S. P., Singh, J. (1970): Seasonal variations
    in the densities  of a soil arthropod population  in a rose garden
    at Varanasi (India). Pedobiologia 10 (8) 442-446.
Singh, J., Mukharji, S. P. (1971): Quantitative compo-
    sition  of soil arthropods  in  some fields  at Varanasi  (India).
    Oriental Insects 5 (4) 487-494.
Singh, J., Sharma, A. K. (in prep.): A comparative ef-
    ficiency of Ladell apparatus  and  Tullgren funnel  for extraxting
    mesofauna from sandy loam soil in India.

# Soil biology and the doctoral syndrome

J. E. SATCHELL

Institute of Terrestrial Ecology,
Merlewood Research Station,
Grange-over-Sands, Lancashire, England

(Shortened text of closing Presidential address)

The Proceedings of our colloquia represent a substantial part of the research output of a substantial proportion of the scientists engaged in soil biology. The primary function of the colloquium is to provide a forum at which we can discuss our recent research but, in addition, we have tried to give coherence to our meetings by selecting themes reflecting the current needs of soil biology and current interests in the wider fields of general ecology. The sustained growth of our member-ship indicates that this ad hoc approach serves the interests of soil biologists. However we also need from time to time to stand back from our immediate research interests and consider our basic objectives, our effectiveness in meeting them, and any changes which may be needed. Most of us are employed either by museums, academic institutes or in-stitutes concerned with various fields of applied biology so it is realistic to consider the scope and progress of soil biology in rela-tion to the functions of these institutes.

FUNCTIONS OF INSTITUTES EMPLOYING SOIL BIOLOGISTS

The function of museums is to collect and preserve objects and to en-rich our understanding of the order and relationship between them. Their function is a cultural one.

Yesterday, on our excursion to the castle of Kost, we saw a col-lection of Gothic icons depicting the Virgin and Infant Christ. What did they convey to you? Immediately, that among them were many great works of art, skillfully painted and either beautiful or curious de-pending on personal taste. Seen with a more taxonomic eye, they fell into two classes with the Christ child cradled on the Virgin's left arm or held forward on the right. The guides explained that the first represented the Madonna as Mother of God, a symbol of universal maternal care, while the second conveyed the concept of Christ presented to the world as Saviour of Mankind. To the curious and beautiful their analysis added meaning and significance.

That is the function of museums. There seems no reason other than habit why we should discriminate between artefacts and organisms. Soil biology requires a sound taxonomic basis and taxonomists provide a service to applied biologists but this is quite incidental to their main task. The function of the taxonomist and systematist is to classify objects and add to their intrinsic beauty or interest the significance of their interrelationships and this applies equally whether the objects are Gothic icons, Etruscan pottery, Collembola or Oribatid mites. The true value of taxonomic research on soil organisms can be appreciated only when this cultural role is recognised.

The function of universities and other academic institutes is threefold: to teach; to preserve and add to our collective store of knowledge; and to preserve and develop those concepts like probity of judgement and scientific rigour handed down to us from the Renaissance. Soil biology serves these functions as part of ecology rather than as a separate discipline. By developing the unifying principle of energy flow, soil biologists have contributed substantially to the formulation of a teachable framework of ecological theory. Their response to the integrative potential of systems analysis has been more conservative, partly because of the poor mathematical background of many biologists and partly because they are often interested in painting ecological miniatures to which the modeller's broad brushes are ill-suited. Nevertheless, the contribution of soil biologists in both the teaching and the development of ecological theory can be described as energetic and progressive. I shall return to the role of academic institutes in maintaining standards of scientific rigour.

The function of institutes of applied science is to identify and solve social problems. Those to which soil biology is relevant concern crop production, pollution control, biodegradation, nature conservation and some medical and veterinary problems. In the field of pest control, soil zoologists have contributed enormously to increasing world food production and are now trying to solve the insecticide pollution problems we helped to create. With some exceptions in microbiology, all applied soil biology other than pest control concerns decomposition. What can we now usefully say about the role of decomposer organisms in litter decomposition; in the maintenance of soil structure; in soil pollution control? It is surely our job as soil biologists to be able to answer such questions. I submit that we are no more able to answer them

now than we were at our first colloquium fifteen years ago.  When faced
with such questions  we can offer  a wealth of detail  about population
metabolism,  species succession,  dispersal rates,  mortality rates and
all the rest of our ecological stock-in-trade. But when asked to provide
socially relevant answers to socially relevant questions  we seem  sin-
gularly unsuccessful.


## LIMITATIONS OF TRADITIONAL TRAINING

I suggest that this weakness  in our applied research originates in the
way  we train  our research workers  and stems from what  I have called
in my title 'the doctoral syndrome'. The names given to doctoral degrees
vary widely  from  country to country  as do  the requirements  for the
degree and the form of examination,  but they share  the same basic objective of ensuring  that  the student acquires  a good technical knowledge of his subject and,  above all,  professional standards of scientific rigour.  The student  is required to write a thesis  on a subject
which in most cases  is an aspect  of the speciality  of his supervisor
and is then examined  in searching detail  by one or more senior scientists  with  specialist  knowledge  of the subject.  If  the successful
student enters the academic world  and in turn becomes a supervisor, he
is likely  to influence  his students  to choose  an aspect  of his own
speciality for their research. So the process continues with each generation penetrating deeper and deeper into narrower  and narrower fields
of enquiry.


## THE LITERATURE PROBLEM

This effect of the doctoral syndrome is reinforced by the growth of the
scientific literature which the scientist with limited reading time can
counter only  by progressively narroving his choice  of subject matter.
Price (1963)  has demonstrated  that  scientific literature  in general
grows exponentially and the following examples (Figs. 1 and 2, Table 1)
show how this applies to the soil biology literature.
    Soil Biology and Biochemistry, started in 1969 with a volume of 327
pages,  illustrates  the initial growth pattern  of a new journal.  The
numbers of pages  in the five volumes now published exhibit a close fit
to the exponential form ($r$ = .975)  with  a growth rate suggesting that

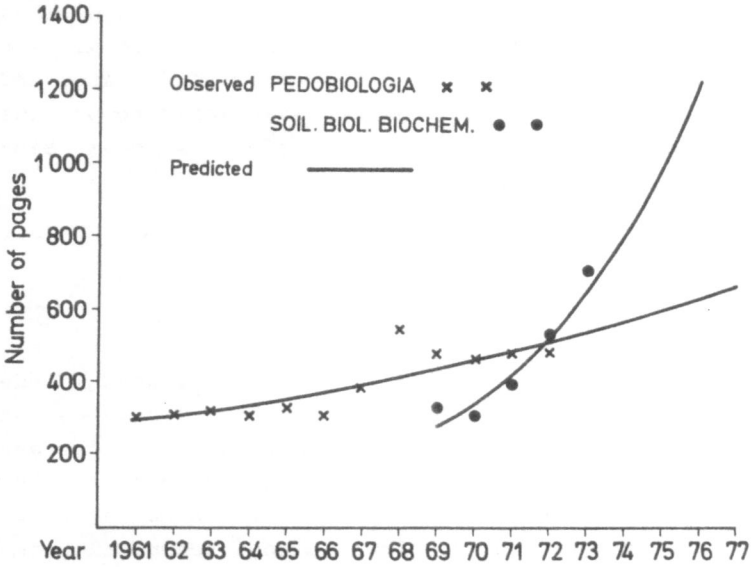

Fig.1. Growth in size of journals.

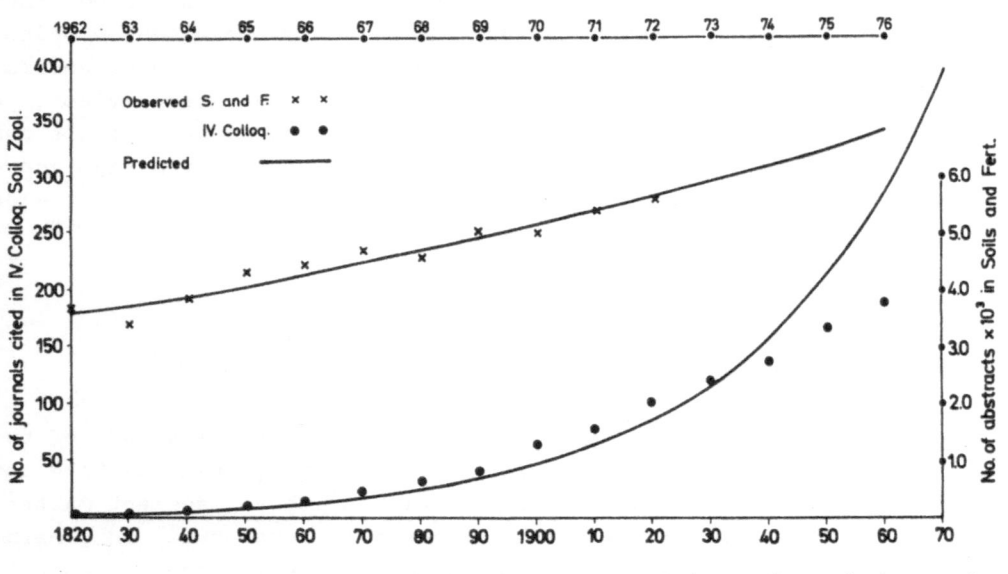

Fig.2. Growth in numbers of journals and abstracts.

T a b l e   1.   Growth rate of soil biology publications

| Publications | Linearity of transformed data* (r) | Growth constant (k/year) | Doubling time (years) |
|---|---|---|---|
| Pages in Soil. Biochem. 1969-73 | .977 | .209 | 3.3 |
| Pages in Pedobiologia 1960-72 | .842 | .051 | 13.5 |
| Journals cited in IV. Colloq. Soil Zool. Total existing each decade 1820-1969 | .997 | .081 | 23.0 |
| Abstracts in Soils and Fert. 1962-72 | .965 | .045 | 15.3 |

* to log base e.

the tenth volume will contain 1674 pages. Pedobiologia, started in 1961/62, has had a long enough run for editorial constraints to operate. Again, the numbers of pages fit the exponential form but the growth rate is slower. Nevertheless extrapolation of the growth curve would predict that the 1983 volume of Pedobiologia will be almost three times the size of the first volume.

When eventually these journals reach the maximum size decided by their editors, authors can confidently expect that new journals will be waiting to accept their manuscripts. It is difficult to analyze the growth rate of journals dealing, at least in part, with soil biology because many of those that have ceased publication are difficult to trace. However, the Proceedings of the IV Colloquium on Soil Zoology held at Dijon provide a sample of the journals cited by our members in 1969. The year in which these journals started and in some cases ended is given in the World List of Scientific Journals for 195 of the 202 journals cited. Two started in the decade 1820-1829. The number increased to the full sample of 195 in 1969, only four having expired en route. The curve of the journal-births minus journal-deaths again fits the exponential form and suggests that the Proceedings of a sim-ilar colloquium held a century hence would refer to more than six thousand journals.

The rate of literature growth in chemistry, biology and physics has been estimated by Price (1963) from the numbers of abstracts in ap-propriate abstracting journals but there is, unfortunately, no such journal devoted to soil biology. The nearest approximation in Soils and

Fertilizers which, although it includes other branches of soil science, covers soil biology with great thoroughness. The numbers of abstracts in the volumes for 1962-72 again fit an exponential curve, doubling every fifteen years. This is also the growth rate Price calculated from the numbers of abstracts in Chemical Abstracts, Biological Abstracts and Physics Abstracts and is probably a reasonable estimate of the growth of soil biology publications. Fifteen years is half the average working life of a scientist so the next generation of soil biologists will have four times as much published work to read as we have. Self-regulatory mechanisms will presumably have reduced the growth rate before then but it seems likely that soil biologists in the future will have to confine their reading to a narrower range of topics than their predecessors. When the narrowing effect of the expansion of the literature is added to the narrowing effect of the doctoral training system, soil biology seems set on a course of increasing fragmentation.

The main fields of soil biology application are agriculture, forestry, conservation and pollution control. Their problems are always complex. An example relevant to the theme of this colloquium would be how to improve the agricultural yields of a former industrial site restored to pasture. Apart from questions of residual toxicity, the problem would require a study of the primary production rates; the consumption, assimilation and defaecation rates of the herbivores; the decomposition rates of the herbivore faeces and plant litter; the exchange capacity of the substrate and the residual organic matter; the rate of leaching; the capacity of the root system to absorb fertilizers; the effects of alternative water and nutrient regimes on the primary production; the response of the stock to herbage quality; and a cost-benefit analysis.

Such a problem would be appropriate for the type of multidisciplinary team research developed during the International Biological Programme with, for example, botanists studying primary production, zoologists studying herbivores and microbiologists stydying decomposition. But the important questions in ecosystem analysis concern the interactions between the biotic and edaphic factors which together determine the productivoty and the stability of the system. A group of collaborating specialists, because of their training as specialists, cannot easily study problems which cross the boundaries of their academic disciplines nor produce answers to problems which lie outside their collective experience. Something more than a specialist training is needed for future soil biologists to contribute effectively to the multidisciplinary problems of the real world.

## AN ALTERNATIVE TRAINING FOR SOIL BIOLOGISTS

The training I propose would aim to familiarise post-graduate students with a wide field of literature, with data retrieval techniques, and with the methodology of systems analysis. It would be based on the selection by the student of a real-life biological problem such as how to grow a specific crop, conserve a species, manage a recreational area or dispose of pig dung. The student would spend the first half of his training period, say 1-2 years, preparing a model of the system, testing it and simulating different forms of management. The model would be multidisciplinary, cutting right across the conventional boundaries of academic subjects. It would be based entirely on the literature, on correspondence and on discussions with specialists. At the end of this period the student would select a variable to which the model had proved particularly sensitive or a poorly understood aspect of the model requiring the development of a sub-model. This would then be studied as a research project for the remaining 1-2 years.

Ten years ago such an approach would have been impossible. The student would have found that the information he needed either did not exist or was scattered irretrievably through the literature. We now have not only more data but review journals such as Advances in Ecology, Advances in Agronomy, and Advances in Microbiology to provide the student with up-to-date leads on where to find the information he would need. Besides the review journals we increasingly have the resources of such master papers as Kononova's (1972) schematisation of the whole process of humus formation to provide the conceptual models on which mathematical models could be based. The student would undoubtedly encounter gaps in the literature and would have to exercise judgement and use his models to test the consequences of making various assumptions.

The type of ensuing research project envisaged is illustrated by Reichle's (1972) model of earthworm feeding fitted as a sub-model into a broader system-analysis of litter decomposition. It may be argued that a student with the training described would lack experience of research in depth required for professional employment. His research experience in a narrowly defined topic would indeed be less than that of students receiving the conventional training by doctoral thesis and employers would need to give the new recruit initial help from their more experienced staff. In the long run this should however be more than balanced by the young scientist's adaptability and problem-orientated outlook. Conversely, in the academic institutes, the primary function of post-graduate biology could be clearly recognised as training and distinguished from the doctoral requirement of contributing new biological knowledge.

One  of the main virtues  of the doctoral-thesis training  is  the
emphasis placed  on scientific rigour.  It is taught by criticism  from
the student's supervisor  and fellow-students  and consists  in seeking
for faults in technique  and logic  and in statistical testing of hypotheses.  It is  well suited  to small data sets  of the kind  a student
acquires in  preparing  his thesis  but is quite unsuited  to assessing
whether the whole research project was necessary. The training proposed
would not detract  from  these  standards  of rigour  but would add the
techniques of modelling  and sensitivity analysis as aids to the development of judgement.

Soil biology is a young science and it is  no discredit that up to
now  it has sought  to establish basic information  about  the species,
numbers, biomass, behaviour  and general ecology of soil organisms. The
training that was appropriate  for this phase  when the literature  was
small may now be obsolete. I would ask you to consider whether by ceas-
ing to breed  from that sacred cow,  the doctoral thesis, we could not
give soil biology a greater relevance to the problems of our time.

R e f e r e n c e s

K o n o n o v a ,  M. M. (1972):  Current problems in the study of soil
    organic matter. Soviet Soil Sci. 4, 420-428.
P r i c e , D. J.  de  S o l l a  (1963):  Little science, big science.
    Columbia Univ. Press, New York and London.
R e i c h l e , D. E. (1972): Systems analysis as applied to ecological
    processes: a mechanism for synthesis, integration, and interpreta-
    tion  of IBP woodlands ecosystem research.  (PP. 12-26  in Systems
    Analysis in Northern Coniferous Forests.) Ed. T. R o s s w a l l ,
    Bulletin 14  of the  Ecological Research Committee  of the Swedish
    Natural Science Research Council.

# List of Participants
# Liste des participants
# Verzeichnis der Teilnehmer

d'Aguilar J., I.N.R.A., Station de Zoologie, Route de St. Cyr, 78 Versailles, France

Al-Dabbagh K.Y., University of Leicester, Department of Zoology, Adrian Building, Leicester Ley 7 RH, England

Alejnikova M.M., Institute of Biology Acad.Sci., Lobachevsky Str. 31/2, Kazan, USSR

Alvarez J., Instituti Español de Entomologia, Calle de J.Gutiérrez Abascal, 2, Madrid 6, España

Anderson J.M., Animal Ecology Research Group, Dept. of Zoology, South Parks Road, GB Oxford, England

Artemjeva I.I., Institute of Biology Acad. Sci. USSR, Lobachevsky Str. 31/2, Kazan, USSR

Athias F., I.N.R.A., Station de recherches sur la faune du sol, B.V. 1540, F-21034 Dijon-Cedex, France

Athias-Henriot C., I.N.R.A., Station de recherches sur la faune du sol, B.V. 1540, F-21034 Dijon-Cedex, France

Atlavinyte O.P., Institute of Zoology and Parasitology, Lithuanian Academy of Sciences, K.Pozhelos 54, SU-232600 Vilnius (Lith. SSR) USSR

Aucamp J.L., Institute for Zoological Research, Potchefstroom University, Potchefstroom, Rep. of South Africa

Bababekova L.A., Institute of Zoology, Academy of Sciences, Baku, USSR

Babel U., Fachgruppe 6, Universität Hohenheim, D7 Stuttgart 70, Schloss Hohenheim, DBR

Bernini F., Instituto di Zoologia dell'Universita'di Siena, Via Mattioli 4, 53100 Siena, Italy

Berthet P., Institut de Zoologie, 59 Naamsestraat, 3000 Louvain, Belgium

Bílý S., Parasitologický ústav ČSAV, Flemingovo nám. 2, 166 32 Praha 6, ČSSR

Bouché M.B., I.N.R.A., Station de recherches sur la faune du sol, B.V. 1540, F-21034 Dijon-Cedex, France

Böðvarsson H., Royal College of Forestry, Roslagsvägen, Stockholm, Sweden

Brauns A., Staatliches Naturhistorisches Museum, Pockelsstr. 10a, 3300 Braunschweig, DBR

Van de Bund Ch.F., Plantenhichte-
kundige Dienst, Geertjesweg 15,
Wageningen, Holland

Byzova J., Institute of Animal
Evolutionary Morphology and
Ecology, Lenin Avenue 33, Moscow
W-71, USSR

Coûteaux M.M., Station de recherches
cytopathologiques, I.N.R.A.
F-30380 St. Christol-les-Alès,
France

Covarrubias R., Laboratoire d'Ecologie Générale, Naamsestraat 59,
3000 Louvain, Belgium

Cykowski R.K., Zakł.Zoologii A.R.,
ul. Broniewskiego 16, 71-460
Szczecin, Poland

Dallai R., Universita di Siena,
Instituto di Zoologia, Via P.A.
Mattioli 4, Siena 53100, Italy

Dindal D.L., State University College of Forestry, Syracuse
Camous, Syracuse, New York
13210, USA

Djurkić J., Faculty of Agriculture,
Akademska 2, 21000 Novi Sad,
Yugoslavia

Domsch K., Institut für Bodenbiologie, D 33 Braunschweig, Bun-
desallee 50, DBR

Van der Drift J., Rijksinstituut
voor Natuurbeheer, Kemperbergerweg 11, Arnhem, Netherlands

Dunger W., Staatliches Museum f.
Naturkunde, Forschungsstelle
11e, Am Museum 1, Görlitz, DDR

Edwards C.A., Rothamsted Experimental Station, Herpenden,
Herts, England

Eijsackers H., Rijksinstituut voor
Natuurbeheer, Kemperbergeweg
11. Arnhem, Netherlands

Elijava J., Zoology Institute Acad.
Sci., Chavchavadze Avenue 21,
Tbilisi 30, USSR

Elijashvili T.S.,Zoology Institute
Acad. Sci., Chavchavadze Avenue
21, Tbilisi 30, USSR

Ernsting G., Zoological Department,
De Boelelaan 1087,Buitenveldert,
Amsterdam, Holland

Fletcher K.E., Rothamsted Experimental Station, Herpenden-Herts,
England

Francois J., Laboratoire de Zoologie, Faculté des Sciences
de la Vie, Boulevard Gabriel,
Dijon-21000, France

Franz H., Institut für Bodenforschung, Hochschule für Bodenkultur, Georg-Mendel-Strasse 33,
1180 Wien-XVIII, Österreich

Gadzhiev A.T., Institute of Zoology
Acad. Sci., Baku, USSR

Da Gama M., Université de Coimbra,
Institut de Zoologie, Coimbra,
Portugal

Gatilova F.G.,Institute of Biology
Acad. Sci., Kazan, USSR

Ghilarov M.S., Institute of Animal
Evolutionary Morphology and
Ecology, Lenin Avenue 33, Moscow
W-71, USSR

Girot B., Université de Bordeaux 1,
Laboratoire de Zoologie 2,
Avenue des Facultés, 33-Talence,
France

Górny M., Instytut Badawczy Leś-
nictva, ul. Kostrzewy 3, 00-973
Warszawa, Polska

Grégoire - Wibo C., Laboratoire
d'Ecologie Générale et Expérimentale, 59 Naamsestraat, 3000
Louvain, Belgium

Greenslade P.J.M., South Australian Museum, Adelaide, South Australia

Grüm L., Institute of Ecology, Polish Academy of Sciences, 05-150 Dziekanow, Poland

Halašková V., Zoologický ústav, Přírodovědecká fakulta KU, Viničná 7, Praha 2, ČSSR

Hamatová E., Ústav krajinné ekologie ČSAV, 252 43 Průhonice u Prahy, Chotobuz, ČSSR

Hauser B., Muséum d'Histoire Naturelle, Département des Arthropodes, Route de Malagnou, Ch-1211 Genève 6, Suisse

Honczarenko J., Agricultural Academy, 71-424 Szczecin, ul. Janosika 8, Poland

Huhta V., Department of Zoology, Univ. of Helsinki, Pohj. - Rautatiek. 13, SF 00100 Helsinki 10, Finland

Hůrka K., Přírodovědecká fakulta UK, Viničná 7, 128 00 Praha 2, ČSSR

Hutasse F.J., Université de Dijon, Laboratoire de Zoologie, Boulevard Gabriel, F-21000-Dijon, France

Hüther W., Sammlungen der Abteilung für Biologie, Ruhr Universität, D-463 Bochum, DBR

Chernov J.I., Institute of Animal Evolutionary Morphology and Ecology, Lenin Avenue 33, Moscow W-71, USSR

Chotko E.I., Dep.of Zoology, Academy of Sciences, BSSR, Akademicheskaja 27, Minks, USSR

Imadaté G., Biological Lab., Konodai College, Tokyo Medical and Dental Univ., 7 - Ichikawa, Chiba, Japan

Janetschek H., Institut für Zoologie der Universität Insbruck, Universitätsstrasse 4, A-6020 Insbruck, Osterreich

Joosse-van Damme J., Zoology Department V. U., Ecology Section, De Boelelaan 1087, Amsterdam - Buitenveldert, Holland

Josens, G., Laboratoire d'écologie, Université Libre de Bruxelles, Av. F. Roosewelt 50, B 1050 Bruxelles, Belgium

Kaczmarek M., PAN, Institute of Ecology, 05-150 Dziekanow Leśny, Warszawa, Poland

Kaczmarek W., PAN, Institute of Ecology, 05-150 Dziekanow Leśny, Warszawa, Poland

Kajak-Ostrihanska A., PAN, Institute of Ecology, 05-150 Dziekanow Leśny, Warszawa, Poland

Knäpper Ch., Rua Cristovao Colombo, 2075, Apto. 14, 90000 - Porto Alegre (RS), Brasilia

Korganova G.A., Institute of Animal Evolutionary Morphology and Ecology, Lenin Avenue 33, Moscow W-71, USSR

Kozlovskaja L.S., Laboratory of Biocenology, Acad.of Sci., USSR, Gubkin Str. 16 - 2, Moscow - V - 333, USSR

Krivolutsky D.A., Institute of Animal Evolutionary Morphology and Ecology, Lenin Avenue 33, Moscow - W - 71, USSR

Kubíková J., Pražské středisko památkové péče a ochrany přírody, Malé nám. 13, Praha 1, ČSSR

Kunst M., Zoologický ústav, Přírodovědecká fakulta KU, Viničná 7, 128 00  Praha 2, ČSSR

Lavelle P., Laboratoire de Zoologie de l'E.N.S., 46 rue d'Ulm, 75230 Paris, France

Lebrun P.,  Laboratoire d'Ecologie Générale  et  Experimentale, Naamsestraat 59, B-3000 Louvain, Belgium

Leman A.,  VEB Carl Zeis, Jena 69, DDR

Lofty J.F., Rothamsted Experimental Station,  Herpenden  Herts, England

Lohm U., Institute of Zoology, P.O. Boc 561, S-75122 Uppsala, Sweden

Lundvist H.,  Box 561,  S - 75122 Uppsala, Sweden

Macura J.,  Mikrobiologický  ústav ČSAV,  Budějovická 1083, 142 20 Praha 4 - Krč, ČSSR

Marcuzzi G.,  Institute of Animal Biology of the University,  Via Loredan 10, I-35100 Padova, Italy

Massoud Z., Laboratoire d'Écologie Générale  du  Muséum  National, 4 Av. du Petit Chateau, 91800 - Brunoy, France

Mejstřík V.,  Ústav krajinné ekologie ČSAV,  252 43  Průhonice u Prahy, Chotobuz, ČSSR

Mignolet R., Laboratoire d'Ecologie Générale  et  Expérimentale, Naamsestraat 59, B-3000 Louvain, Belgium

Mikulová A.,  Národní museum, Václavské nám., Praha 2, ČSSR

Mitchell M., Environmental Sciences Centre Kananaskis, The University of Calgary, CDN Seebe, Alberta, Canada

Mitra S.K.,  Zoological  Survey of India,  8,  Lindsay Street, Calcutta - 700016,  India

Moritz M.,  Zoologisches  Museum, Invalidenstr. 43,  104  Berlin, DDR

Najt J.,  Division  Entomologia, Facultad de Ciencias Naturales y Museo,  Paseo del Bosque, La Plata, Argentine

Nosek, J.,  Virologický ústav SAV, 809 00 Bratislava 9, ČSSR

Nováková E., Ústav krajinné ekologie ČSAV, Bezručova 927, 251 01 Říčany u Prahy, ČSSR

Palissa A.,  Sektion Biologie der Humboldt-Universität, Invalidenstrasse 43, 104 Berlin, DDR

Pande Y., Department of Agri. Zoology and Entomology,  University of Udaipur,  Campus - Jobner, Rajasthan, India

Parry B.W., British Museum (Natural History), Department of Zoology, Cromwell Road,  London  S.W.7. 5Bd, England

Perel T.S.,  Forest Science Lab., Acad. Sciences URSS,  SU-Uspenskoje,  Odinsovo  District., Moscow Region, URSS

Persson T.,  Institute of Zoology, Box 561, S-75122 Uppsala, Sweden

Petal J., Institute of Ecology PAN, Dziekanów Lesny  near  Warszaw, Poland

Petersen H.,  Molslaboratoriet, Femmøller,  8400  Ebeltoft, Denmark

Poboszny M.,  Institute of Animal Taxonomy, Puskin ucta 3, Budapest VIII,  Hungary

Poinsot N., Laboratoire de Biologie

Animale, Centre St. Jérome, Traverse de la Barasse, F-13013 Marseille 13, France

Pokarzhevsky A.D., Inst. of Evol. Morph. and Ecology, of Animals Acad.Sci., Lenin Av.33, Moscow-71, USSR

Pokorná J., Výzkumný ústav rostlinné výroby ČAZ, Praha 6 - Ruzyně, ČSSR

Prasse J., Sektion Pflanzenproduktion, der Martin-Luther-Universität, Halle – Wittenberg, 402-Halle/S., Weidenplan 14, DDR

Pugh G.J.F., Department of Botany, University of Nottingham, Nottingham, England

Rachmanov R.R., Inst. of Zoology, Acad. Sci., Baku, USSR

Rajski A., Agricultural Academy, 71-424 Szczecin, ul.Janosika 8, Poland

Rasulova Z.K., Inst. of Zoology, Acad. Sci., Baku, USSR

Reichle D.E., Oak Ridge National Laboratory, P.O. Box X, Oak Ridge, Tennessee 37830, USA

Reinecke J.A., Department of Zoology, University of Potchefstroom, South Africa

Reynolds J.W., University of Tennessee, 7008 Stockton Dr., USA, Knoxville, Tennessee 37919, USA

Van Rhee J.A., Rijksinstituut voor Natuurbeheer, Kemperbergerweg 11, Arnhem, Netherlands

Rusek J., Entomologický ústav ČSAV, Viničná 7, 128 00 Praha 2, ČSSR

Satchell J., Institute of Terrestrial Ecology, Merlewood Research Station, Grande-over-Sands, Lancashire, England

Sekulić R., Faculty of Agriculture, Akademska 2, 21000 Novi Sad, Yugoslavia

Selga D., Instituto Español de Entomologia, J. Gutierrez Abascal 2, Madrid 6, Spain

Sims R.W., British Museum (Natural History), Cromwell Road, London S.W.7., England

Stafford Ch.J., Rothamsted Experimental Station, Harpenden, Herts, England

Stebajeva S.K., Sibirian Branch Acad. Sci., Novosibirsk, USSR

Stomp N., 3, Rue L.Deny, Luxembourg

Striganova B.R., Institute of Animal Evolutionary Morphology and Ecology, Lenin Avenue 33, Moscow W-71, USSR

Szujecki A., Instytut ochrony lasu i drewna Akademii rolniczej, Rakowiecka 26/30, Warszawa, Poland

Thibaud J.M., 4 Avenue du Petit Chateau, Laboratoire d'Écologie, 91800 - Brunoy, France

Thiele H.U., Zoologisches Institut der Universität Köln, D5 Köln 41, Weyertal 119, DBR

Von Törne E., Rud.- Breitscheidstrasse 48, 13 Eberswalde, DDR

Utrobina N.M., Institute of Biology Acad. Sci., Galaktionova 12-8, Kazan, USSR

Vaněk J., Ústav krajinné ekologie ČSAV, Bezručova 927, 251 01 Říčany u Prahy, ČSSR

Vlijm L., Zoological Department V.U.Ecology Section de Boelelaan 1087, Amsterdam, Holland

Volz P., Ramburgstr.10, 674 Landau in der Pfalz, DBR

Wallwork J.A., Department of Zoology, Westfield College (University of London), London, N.W. 37 St., England

Wasilewska L., Institute of Ecology, Długosza 8, n.35, 01-174 Warszawa, Poland

Wauthy G., Laboratoire d'Écologie Générale et Expérimentale, Naamsestraat 59, B-3000 Louvain, Belgium

Webster J.M., Simon Frazer University, Burnaby 2 - B. Columbia, Canada

Wood T.G., Centre for Overseas Pest Research, Termite Research Unit, c/o British Museum (Nat.History), Cromwell Rd., London S.W.1, England

Zaleskaja N.T., Institute of Animal Evolutionary Morphology and Ecology, Lenin avenue 33, Moscow W-71, USSR

Zkitishivili T.D., Inst.of Zoology Acad. Sci., Tbilisi, USSR

Zajonc I., Vysoká škola polnohospodárska, Nitra, ČSSR

Ziczi A., Tiersystematisches Institut der Univ., Budapest VIII, Puskin u. 3, Hungary

# Author Index
# Index des Auteurs
# Autorenverzeichnis

# Index to Genera
# Index des Genres
# Verzeichnis der Gattungen